Springer Series on

ATOMIC, OPTICAL, AND PLASMA PHYSICS 38

Springer Series on

# ATOMIC, OPTICAL, AND PLASMA PHYSICS

The Springer Series on Atomic, Optical, and Plasma Physics covers in a comprehensive manner theory and experiment in the entire field of atoms and molecules and their interaction with electromagnetic radiation. Books in the series provide a rich source of new ideas and techniques with wide applications in fields such as chemistry, materials science, astrophysics, surface science, plasma technology, advanced optics, aeronomy, and engineering. Laser physics is a particular connecting theme that has provided much of the continuing impetus for new developments in the field. The purpose of the series is to cover the gap between standard undergraduate textbooks and the research literature with emphasis on the fundamental ideas, methods, techniques, and results in the field.

27 **Quantum Squeezing**
By P.D. Drumond and Z. Ficek

28 **Atom, Molecule, and Cluster Beams I**
Basic Theory, Production and Detection of Thermal Energy Beams
By H. Pauly

29 **Polarization, Alignment and Orientation in Atomic Collisions**
By N. Andersen and K. Bartschat

30 **Physics of Solid-State Laser Physics**
By R.C. Powell
(Published in the former Series on Atomic, Molecular, and Optical Physics)

31 **Plasma Kinetics in Atmospheric Gases**
By M. Capitelli, C.M. Ferreira, B.F. Gordiets, A.I. Osipov

32 **Atom, Molecule, and Cluster Beams II**
Cluster Beams, Fast and Slow Beams, Accessory Equipment and Applications
By H. Pauly

33 **Atom Optics**
By P. Meystre

34 **Laser Physics at Relativistic Intensities**
By A.V. Borovsky, A.L. Galkin, O.B. Shiryaev, T. Auguste

35 **Many-Particle Quantum Dynamics in Atomic and Molecular Fragmentation**
Editors: J. Ullrich and V.P. Shevelko

36 **Atom Tunneling Phenomena in Physics, Chemistry and Biology**
Editor: T. Miyazaki

37 **Charged Particle Traps**
Physics and Techniques of Charged Particle Field Confinement
By V.N. Gheorghe, F.G. Major, G. Werth

38 **Plasma Physics and Controlled Nuclear Fusion**
By K. Miyamoto

Vols. 1–26 of the former Springer Series on Atoms and Plasmas are listed at the end of the book

K. Miyamoto

# Plasma Physics and Controlled Nuclear Fusion

With 117 Figures

Springer

Professor emer. Kenro Miyamoto
Univesity of Tokyo
E-mail: miyamoto@phys.s.u-tokyo.ac.jp

Originally published in Japanese under the title "Plasma Physics and Controlled Nuclear Fusion"
by University of Tokyo Press, 2004

ISSN 1615-5653

ISBN 3-540-24217-1 Springer Berlin Heidelberg New York

Library of Congress Control Number: 2004117908



Springer is a part of Springer Science+Business Media.

springeronline.com


Printed in Germany



Typesetting and prodcution: PTP-Berlin, Protago-TEX-Production GmbH, Berlin
Cover concept by eStudio Calmar Steinen
Cover design: *design & production* GmbH, Heidelberg

Printed on acid-free paper SPIN: 11332640 57/3141/YU - 5 4 3 2 1 0

# Preface

The primary objective of these lecture notes is to present the basic theories and analytical methods of plasma physics and to provide the recent status of fusion research for graduate and advanced undergraduate students. I also hope that this text will be a useful reference for scientists and engineers working in the relevant fields.

Chapters 1–4 describe the fundamentals of plasma physics. The basic concept of the plasma and its characteristics are explained in Chaps. 1 and 2. The orbits of ions and electrons are described in several magnetic field configurations in Chap. 3, while Chap. 4 formulates the Boltzmann equation for the velocity space distribution function, which is the basic equation of plasma physics.

Chapters 5–9 describe plasmas as magnetohydrodynamic (MHD) fluids. The MHD equation of motion (Chap. 5), equilibrium (Chap. 6) and plasma transport (Chap. 7) are described by the fluid model. Chapter 8 discusses problems of MHD instabilities, i.e., whether a small perturbation will grow to disrupt the plasma or damp to a stable state. Chapter 9 describes resistive instabilities of plasmas with finite electrical resistivity.

In Chaps. 10–13, plasmas are treated by kinetic theory. The medium in which waves and perturbations propagate is generally inhomogeneous and anisotropic. It may absorb or even amplify the waves and perturbations. The cold plasma model described in Chap. 10 is applicable when the thermal velocity of plasma particles is much smaller than the phase velocity of the wave. Because of its simplicity, the dielectric tensor of cold plasma is easily derived and the properties of various waves can be discussed in the case of cold plasmas. If the refractive index becomes large and the phase velocity of the wave becomes comparable to the thermal velocity of the plasma particles, then the particles and the wave interact with each other. Chapter 11 describes Landau damping, which is the most characteristic collective phenomenon of plasmas, and also cyclotron damping. Chapter 12 discusses wave heating (wave absorption) and velocity space instabilities (amplification of perturbations) in hot plasmas, in which the thermal velocity of particles is comparable to the wave phase velocity, using the dielectric tensor of hot plasmas. Chapter 13 discusses instabilities driven by energetic particles, i.e., the fishbone instability and toroidal Alfvén eigenmodes.

In order to understand the complex nonlinear behavior of plasmas, computer simulation becomes a dominant factor in the theoretical component of plasma research, and this is briefly outlined in Chap. 14.

Chapter 15 reviews confinement research toward fusion grade plasmas. During the last decade, tokamak research has made remarkable progress. Today, realistic designs for tokamak reactors such as ITER are being actively pursued. Chapter 16 explains research work into critical features of tokamak plasmas and reactors. Non-tokamak confinement systems are also receiving great interest. The reversed field pinch and stellarators are described in Chap. 17 and inertial confinement is introduced in Chap. 18.

The reader may have the impression that there is too much mathematics in these lecture notes. However, there is a reason for this. If a graduate student tries to read and understand, for example, frequently cited short papers on the analysis of the high-$n$ ballooning mode and fishbone instability [Phys. Rev. Lett **40**, 396 (1978); ibid. **52**, 1122 (1984)] without some preparatory knowledge, he must read and understand a few tens of cited references, and references of references. I would guess that he would be obliged to work hard for a few months. Therefore, one motivation for writing this monograph is to save the student time struggling with the mathematical derivations, so that he can spend more time thinking about the physics and experimental results.

This textbook was based on lectures given at the Institute of Plasma Physics, Nagoya University, Department of Physics, University of Tokyo and discussion notes from ITER Physics Expert Group Meetings. It would give me great pleasure if the book were to help scientists make their own contributions in the field of plasma physics and fusion research.

Tokyo, November 2004 *Kenro Miyamoto*

# Contents

**Part I Plasma Physics**

**1 Nature of Plasma** ... 3
- 1.1 Introduction ... 3
- 1.2 Charge Neutrality and Landau Damping ... 5
- 1.3 Fusion Core Plasma ... 6

**2 Plasma Characteristics** ... 13
- 2.1 Velocity Space Distribution Function ... 13
- 2.2 Plasma Frequency. Debye Length ... 14
- 2.3 Cyclotron Frequency. Larmor Radius ... 15
- 2.4 Drift Velocity of Guiding Center ... 16
- 2.5 Magnetic Moment. Mirror Confinement ... 19
- 2.6 Coulomb Collision. Fast Neutral Beam Injection ... 21
- 2.7 Runaway Electron. Dreicer Field ... 27
- 2.8 Electric Resistivity. Ohmic Heating ... 28
- 2.9 Variety of Time and Space Scales in Plasmas ... 28

**3 Magnetic Configuration and Particle Orbit** ... 31
- 3.1 Maxwell Equations ... 31
- 3.2 Magnetic Surface ... 33
- 3.3 Equation of Motion of a Charged Particle ... 34
- 3.4 Particle Orbit in Axially Symmetric System ... 36
- 3.5 Drift of Guiding Center in Toroidal Field ... 38
  - 3.5.1 Guiding Center of Circulating Particles ... 39
  - 3.5.2 Guiding Center of Banana Particles ... 40
- 3.6 Orbit of Guiding Center and Magnetic Surface ... 42
- 3.7 Effect of Longitudinal Electric Field on Banana Orbit ... 44
- 3.8 Polarization Drift ... 45

**4 Velocity Space Distribution Function and Boltzmann's Equation** ... 47
- 4.1 Phase Space and Distribution Function ... 47
- 4.2 Boltzmann's Equation and Vlasov's Equation ... 48

5 **Plasma as MHD Fluid** ..... 51
5.1 Magnetohydrodynamic Equations for Two Fluids ..... 51
5.2 Magnetohydrodynamic Equations for One Fluid ..... 53
5.3 Simplified Magnetohydrodynamic Equations ..... 55
5.4 Magnetoacoustic Wave ..... 58

6 **Equilibrium** ..... 61
6.1 Pressure Equilibrium ..... 61
6.2 Equilibrium Equation for Axially Symmetric Systems ..... 63
6.3 Tokamak Equilibrium ..... 67
6.4 Upper Limit of Beta Ratio ..... 69
6.5 Pfirsch–Schlüter Current ..... 70
6.6 Virial Theorem ..... 71

7 **Plasma Transport** ..... 75
7.1 Collisional Diffusion (Classical Diffusion) ..... 77
7.1.1 Magnetohydrodynamic Treatment ..... 77
7.1.2 A Particle Model ..... 79
7.2 Neoclassical Diffusion of Electrons in a Tokamak ..... 80
7.3 Fluctuation Loss. Bohm and Gyro-Bohm Diffusion. Convective Loss ..... 83
7.4 Loss by Magnetic Fluctuation ..... 89

8 **Magnetohydrodynamic Instabilities** ..... 91
8.1 Interchange Instabilities ..... 92
8.1.1 Interchange Instability ..... 92
8.1.2 Stability Criterion for Interchange Instability. Magnetic Well ..... 95
8.2 Formulation of Magnetohydrodynamic Instabilities ..... 99
8.2.1 Linearization of Magnetohydrodynamic Equations ..... 99
8.2.2 Energy Principle ..... 102
8.3 Instabilities of a Cylindrical Plasma ..... 104
8.3.1 Instabilities of Sharp-Boundary Configuration ..... 104
8.3.2 Instabilities of Diffuse Boundary Configurations ..... 109
8.3.3 Suydam's Criterion ..... 113
8.3.4 Tokamak Configuration ..... 115
8.4 Hain–Lüst Magnetohydrodynamic Equation ..... 117
8.5 Energy Integral of Axisymmetric Toroidal System ..... 119
8.5.1 Energy Integral in Illuminating Form ..... 119
8.5.2 Energy Integral of Axisymmetric Toroidal System ..... 121
8.5.3 Energy Integral of High-$n$ Ballooning Mode ..... 126
8.6 Ballooning Instability ..... 128
8.7 Eta-i Mode Due to Density and Temperature Gradient ..... 133

**9 Resistive Instabilities** . . . . . 137
9.1 Tearing Instability . . . . . 137
9.2 Resistive Drift Instability . . . . . 142

**10 Plasma as Medium of Waves** . . . . . 147
10.1 Dispersion Equation of Waves in a Cold Plasma . . . . . 148
10.2 Properties of Waves . . . . . 152
10.2.1 Polarization and Particle Motion . . . . . 152
10.2.2 Cutoff and Resonance . . . . . 153
10.3 Waves in a Two-Component Plasma . . . . . 153
10.4 Various Waves . . . . . 158
10.4.1 Alfven Wave . . . . . 158
10.4.2 Ion Cyclotron Wave and Fast Wave . . . . . 159
10.4.3 Lower Hybrid Resonance . . . . . 161
10.4.4 Upper Hybrid Resonance . . . . . 162
10.4.5 Electron Cyclotron Wave . . . . . 162
10.5 Conditions for Electrostatic Waves . . . . . 164

**11 Landau Damping and Cyclotron Damping** . . . . . 167
11.1 Landau Damping (Amplification) . . . . . 167
11.2 Transit Time Damping . . . . . 171
11.3 Cyclotron Damping . . . . . 171
11.4 Quasi-Linear Theory of Evolution in the Distribution Function . . . . . 174

**12 Hot Plasma** . . . . . 177
12.1 Energy Flow . . . . . 178
12.2 Ray Tracing . . . . . 182
12.3 Dielectric Tensor of Hot Plasma . . . . . 183
12.4 Wave Heating in the Ion Cyclotron Frequency Range . . . . . 189
12.5 Lower Hybrid Heating . . . . . 192
12.6 Electron Cyclotron Heating . . . . . 196
12.7 Velocity Space Instabilities (Electrostatic Waves) . . . . . 199
12.7.1 Dispersion Equation of Electrostatic Wave . . . . . 199
12.7.2 Electron Beam Instability . . . . . 201
12.7.3 Various Velocity Space Instabilities . . . . . 202
12.8 Derivation of Dielectric Tensor in Hot Plasma . . . . . 202
12.8.1 Formulation of Dispersion Relation in Hot Plasma . . 202
12.8.2 Solution of Linearized Vlasov Equation . . . . . 204
12.8.3 Dielectric Tensor of Hot Plasma . . . . . 206
12.8.4 Dielectric Tensor of Bi-Maxwellian Plasma . . . . . 209
12.8.5 Dispersion Relation of Electrostatic Wave . . . . . 210

**13 Instabilities Driven by Energetic Particles** ... 215
13.1 Fishbone Instability ... 215
13.1.1 Formulation ... 215
13.1.2 MHD potential Energy ... 216
13.1.3 Kinetic Integral of Hot Component ... 218
13.1.4 Growth Rate of Fishbone Instability ... 221
13.2 Toroidal Alfvén Eigenmode ... 224
13.2.1 Toroidicity-Induced Alfvén Eigenmode ... 225
13.2.2 Instability of TAE Driven by Energetic Particles ... 229
13.2.3 Various Alfvén Modes ... 237

**14 Computer Simulation** ... 239
14.1 MHD model ... 240
14.2 Linearized Kinetic Model ... 242
14.3 Modeling Bulk Plasma and Energetic Particles ... 243
14.4 Gyrofluid/Gyro-Landau-Fluid Models ... 244
14.5 Gyrokinetic Particle Model ... 247
14.6 Full Orbit Particle Model ... 251

**Part II Controlled Nuclear Fusion**

**15 Development of Fusion Research** ... 259
15.1 From Secrecy to International Collaboration ... 260
15.2 Artsimovich Era ... 262
15.3 The Trek to Large Tokamaks Since the Oil Crisis ... 263
15.4 Alternative Approaches ... 266

**16 Tokamaks** ... 269
16.1 Tokamak Devices ... 269
16.2 Equilibrium ... 272
16.3 MHD Stability and Density Limit ... 274
16.4 Beta Limit of Elongated Plasma ... 277
16.5 Impurity Control, Scrape-Off Layer and Divertor ... 278
16.6 Confinement Scaling of L Mode ... 284
16.7 H Mode and Improved Confinement Modes ... 286
16.8 Non-Inductive Current Drive ... 293
16.8.1 Lower Hybrid Current Drive ... 293
16.8.2 Electron Cyclotron Current Drive ... 297
16.8.3 Neutral Beam Current Drive ... 300
16.8.4 Bootstrap Current ... 302
16.9 Neoclassical Tearing Mode ... 304
16.10 Tokamak Reactors ... 311

**17 RFP Stellarator** . . . . . 319
17.1 Reversed Field Pinch . . . . . 319
17.1.1 Reversed Field Pinch Configuration . . . . . 319
17.1.2 MHD Relaxation . . . . . 320
17.1.3 Confinement . . . . . 323
17.1.4 Oscillating Field Current Drive . . . . . 325
17.2 Stellarator . . . . . 325
17.2.1 Helical Field . . . . . 325
17.2.2 Stellarator Devices . . . . . 329
17.2.3 Neoclassical Diffusion in Helical Field . . . . . 331
17.2.4 Confinement of Stellarator System . . . . . 334

**18 Inertial Confinement** . . . . . 337
18.1 Pellet Gain . . . . . 337
18.2 Implosion . . . . . 342
18.3 MHD Instabilities . . . . . 345
18.4 Fast Ignition . . . . . 347

**References** . . . . . 353

**Index** . . . . . 367

# Part I

# Plasma Physics

# 1 Nature of Plasma

## 1.1 Introduction

As the temperature of a material is raised, its state changes from solid to liquid and then to gas. If the temperature is elevated further, an appreciable number of the gas atoms are ionized and a high temperature gaseous state is achieved, in which the charge numbers of ions and electrons are almost the same and charge neutrality is satisfied on a macroscopic scale.

When ions and electrons move, these charged particles interact with the Coulomb force which is a long range force and decays only as the inverse square of the distance $r$ between the charged particles. The resulting current flows due to the motion of the charged particles and Lorentz interaction takes place. Therefore many charged particles interact with each other by long range forces and various collective movements occur in the gaseous state. In typical cases, there are many kinds of instabilities and wave phenomena. The word 'plasma' is used in physics to designate this high temperature ionized gaseous state with charge neutrality and collective interaction between the charged particles and waves.

When the temperature of a gas is $T$(K), the average velocity of the thermal motion of a particle with mass $m$, that is, thermal velocity $v_{\mathrm{T}}$ is given by

$$m v_{\mathrm{T}}^2/2 = \kappa T/2 \ , \tag{1.1}$$

where $\kappa$ is the Boltzmann constant $\kappa = 1.380\,658(12) \times 10^{-23}\,\mathrm{J/K}$ and $\kappa T$ denotes the thermal energy. Therefore the unit of $\kappa T$ is the joule (J) in MKSA units. In many fields of physics, the electron volt (eV) is frequently used as the unit of energy. This is the energy required to move an electron, charge $e = 1.602\,177\,33(49) \times 10^{-19}$ coulomb, against a potential difference of 1 volt:

$$1\,\mathrm{eV} = 1.602\,177\,33(49) \times 10^{-19}\,\mathrm{J} \ .$$

The temperature corresponding to a thermal energy of 1 eV is $1.16 \times 10^4\,\mathrm{K}$ ($= e/\kappa$). The ionization energy of the hydrogen atom is 13.6 eV. Even if the thermal energy (average energy) of hydrogen gas is 1 eV, that is $T \sim 10^4\,\mathrm{K}$, there exists a small number of electrons with energy higher than 13.6 eV, which ionize the gas to a hydrogen plasma.

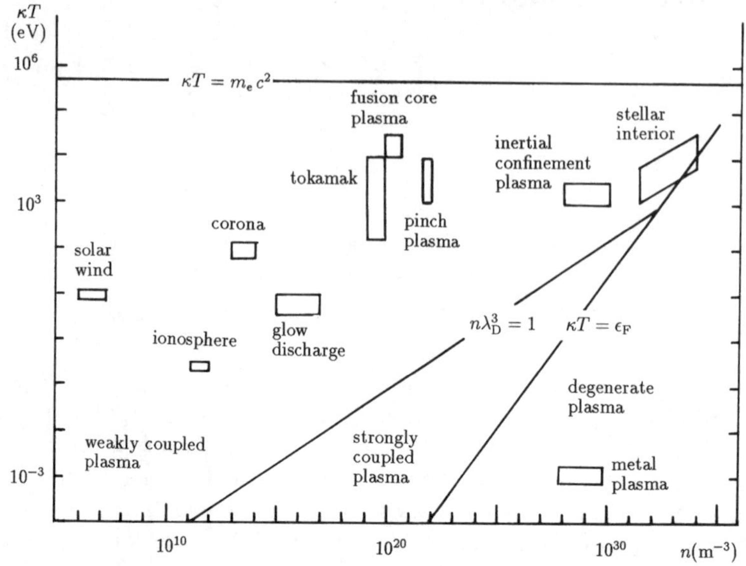


**Fig. 1.1.** Various plasma domains in the $n$–$\kappa T$ diagram

Plasmas are found in nature in various forms (see Fig. 1.1). One example is the Earth's ionosphere at altitudes of 70–500 km, with density $n \sim 10^{12}\,\mathrm{m}^{-3}$ and $\kappa T \approx 0.2\,\mathrm{eV}$. Another is the solar wind, a plasma flow originating from the sun, with $n \sim 10^6$–$10^7\,\mathrm{m}^{-3}$ and $\kappa T \approx 10\,\mathrm{eV}$. The sun's corona extending around our star has density $\sim 10^{14}\,\mathrm{m}^{-3}$ and electron temperature $\sim 100\,\mathrm{eV}$, although these values are position-dependent. The white dwarf, the final state of stellar evolution, has an electron density of $10^{35}$–$10^{36}\,\mathrm{m}^{-3}$. Various plasma domains in the diagram of electron density $n(\mathrm{m}^{-3})$ and electron temperature $\kappa T\,(\mathrm{eV})$ are shown in Fig. 1.1.

Active research in plasma physics has been motivated by the aim to create and confine hot plasmas in fusion research. In space physics and astrophysics, plasmas play important roles in studies of pulsars radiating microwaves or solar X-ray sources. Another application of plasma physics is the study of the Earth's environment in space.

Practical applications of plasma physics are MHD (magnetohydrodynamic) energy conversion for electric power generation and ion rocket engines for spacecraft. Plasma processes for the manufacture of integrated circuits have attracted much attention recently.

## 1.2 Charge Neutrality and Landau Damping

One fundamental property of plasmas is charge neutrality. Plasmas shield electric potentials applied to the plasma. When a probe is inserted into a plasma and a positive (negative) potential is applied, the probe attracts (repels) electrons and the plasma tends to shield the electric disturbance. Let us estimate the shielding length. Assume that heavy ions have uniform density ($n_\mathrm{i} = n_0$) and that there is a small perturbation in the electron density $n_\mathrm{e}$ and potential $\phi$. Since the electrons are in the Boltzmann distribution with electron temperature $T_\mathrm{e}$, the electron density $n_\mathrm{e}$ becomes

$$n_\mathrm{e} = n_0 \exp(e\phi/\kappa T_\mathrm{e}) \simeq n_0(1 + e\phi/\kappa T_\mathrm{e}) ,$$

where $\phi$ is the electrostatic potential and $e\phi/\kappa T_\mathrm{e} \ll 1$ is assumed. The equation for the electrostatic potential comes from Maxwell's equations (see Sect. 3.1),

$$\boldsymbol{E} = -\boldsymbol{\nabla}\phi , \qquad \boldsymbol{\nabla}\cdot(\epsilon_0 \boldsymbol{E}) = -\epsilon_0 \nabla^2\phi = \rho = -e(n_\mathrm{e} - n_0) = -\frac{e^2 n_0}{\kappa T_\mathrm{e}}\phi$$

and

$$\nabla^2\phi = \frac{\phi}{\lambda_\mathrm{D}^2} , \qquad \lambda_\mathrm{D} = \left(\frac{\epsilon_0 \kappa T_\mathrm{e}}{n_\mathrm{e} e^2}\right)^{1/2} = 7.45 \times 10^3 \left(\frac{1}{n_\mathrm{e}}\frac{\kappa T_\mathrm{e}}{e}\right)^{1/2} (\mathrm{m}) , \tag{1.2}$$

where $\epsilon_0$ is the dielectric constant of the vacuum and $\boldsymbol{E}$ is the electric intensity. $n_\mathrm{e}$ is in $\mathrm{m}^{-3}$ and $\kappa T_\mathrm{e}/e$ is in eV. When $n_\mathrm{e} \sim 10^{20}\,\mathrm{cm}^{-3}$, $\kappa T_\mathrm{e}/e \sim 10\,\mathrm{keV}$, then $\lambda_\mathrm{D} \sim 75\,\mu\mathrm{m}$. In the spherically symmetric case, the Laplacian $\nabla^2$ becomes

$$\nabla^2\phi = \frac{1}{r}\frac{\partial}{\partial r}\left(r\frac{\partial\phi}{\partial r}\right) ,$$

and the solution is

$$\phi = \frac{q}{4\pi\epsilon_0}\frac{\exp(-r/\lambda_\mathrm{D})}{r} .$$

It is clear from the foregoing formula that the Coulomb potential $q/4\pi\epsilon_0 r$ of a point charge is shielded out to a distance $\lambda_\mathrm{D}$. This distance $\lambda_\mathrm{D}$ is called the *Debye length*. When the plasma size is $a$ and $a \gg \lambda_\mathrm{D}$ is satisfied, the plasma is considered to be electrically neutral. If on the other hand $a < \lambda_\mathrm{D}$, individual particles are not shielded electrostatically and this state is no longer a plasma but an assembly of independent charged particles.

The number of electrons included in a sphere of radius $\lambda_\mathrm{D}$ is called the *plasma parameter* and is given by

$$n_\mathrm{e}\lambda_\mathrm{D}^3 = \left(\frac{\epsilon_0}{e}\frac{\kappa T_\mathrm{e}}{e}\right)^{3/2}\frac{1}{n_\mathrm{e}^{1/2}} . \tag{1.3}$$

When the density is increased while keeping the temperature constant, this value becomes small. If the plasma parameter is less than say $\sim 1$, the concept of Debye shielding is not applicable, since the continuity of charge density breaks down on the Debye length scale. Plasmas in the region of $n_e\lambda_D^3 > 1$ are called classical plasmas or *weakly coupled plasmas*, since the ratio of the electron thermal energy $\kappa T_e$ and the Coulomb energy between electrons $E_{\rm Coulomb} = e^2/4\pi\epsilon_0 d$, with $d \simeq n_e^{-1/3}$ the average distance between electrons with density $n_e$, is given by

$$\frac{\kappa T_e}{E_{\rm Coulomb}} = 4\pi(n_e\lambda_D^3)^{2/3}\,, \tag{1.4}$$

and $n_e\lambda_D^3 > 1$ means that the Coulomb energy is smaller than the thermal energy. The case $n_e\lambda_D^3 < 1$ corresponds to a *strongly coupled plasma* (see Fig. 1.1).

The Fermi energy of a degenerate electron gas is given by

$$\epsilon_F = \frac{h^2}{2m_e}(3\pi^2 n_e)^{2/3}\,,$$

where $h = 6.626\,075\,5(40) \times 10^{-34}$ J s is Planck's constant. When the density becomes very high, it is possible to have $\epsilon_F \geq \kappa T_e$. In this case, quantum effects dominate over thermal effects. This case is called a *degenerate electron plasma*. One example is the electron plasma in a metal. Most plasmas in magnetic confinement experiments are classical weakly coupled plasmas.

The other fundamental plasma process is collective phenomena involving the charged particles. Waves are associated with coherent motions of charged particles. When the phase velocity $v_{\rm ph}$ of a wave or perturbation is much larger than the thermal velocity $v_T$ of the charged particles, the wave propagates through the plasma media without damping or amplification. However, when the refractive index $N$ of the plasma medium becomes large and the plasma becomes hot, the phase velocity $v_{\rm ph} = c/N$ (where $c$ is the light velocity) of the wave and the thermal velocity $v_T$ become comparable ($v_{\rm ph} = c/N \sim v_T$). Then energy exchange is possible between the wave and the thermal energy of the plasma. The existence of a damping mechanism for these waves was found by L.D. Landau. The process of Landau damping involves a direct wave–particle interaction in a collisionless plasma without the need to randomize collisions. This process is the fundamental mechanism in wave heating of plasmas (wave damping) and instabilities (inverse damping of perturbations). Landau damping is described in Chaps. 11 and 12.

## 1.3 Fusion Core Plasma

Progress in plasma physics has been motivated by the desire to realize a fusion core plasma. The necessary condition for fusion core plasmas is discussed in

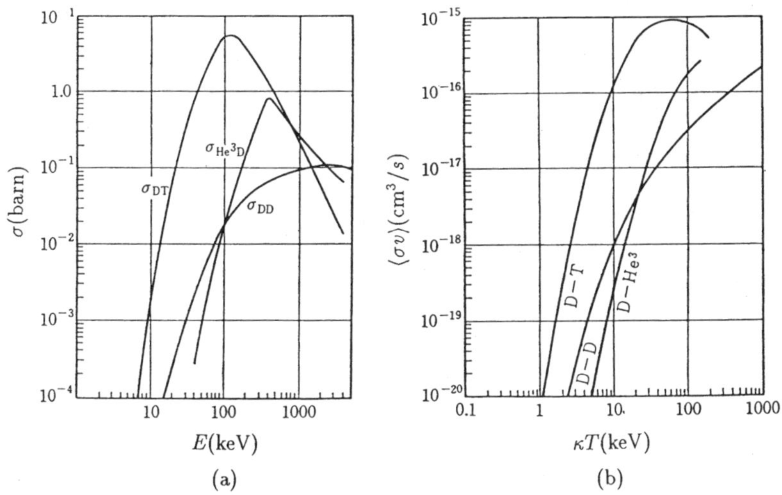


**Fig. 1.2.** (**a**) Dependence of the fusion cross-section $\sigma$ on the kinetic energy $E$ of colliding nucleons. $\sigma_{\rm DD}$ is the sum of the cross-sections of D–D reactions (1) and (2). 1 barn $= 10^{-24}\,\rm cm^2$. (**b**) Dependence of the fusion rate $\langle\sigma v\rangle$ on the ion temperature $T_{\rm i}$

this section. Nuclear fusion reactions are the fusion reactions of light nuclides to heavier ones. When the sum of the masses of nuclides after nuclear fusion is smaller than the sum before the reaction by $\Delta m$, we call this the mass defect. According to the theory of relativity, the amount of energy $(\Delta m)c^2$ ($c$ is the speed of light) is released by the nuclear fusion.

Nuclear reactions of interest for fusion reactors are as follows (D deuteron, T triton, $\rm He^3$ helium-3, Li lithium):

$$
\begin{array}{ll}
(1) & \rm D + D \longrightarrow T\ (1.01\,MeV) + p\ (3.03\,MeV)\,, \\
(2) & \rm D + D \longrightarrow He^3\ (0.82\,MeV) + n\ (2.45\,MeV)\,, \\
(3) & \rm T + D \longrightarrow He^4\ (3.52\,MeV) + n\ (14.06\,MeV)\,, \\
(4) & \rm D + He^3 \longrightarrow He^4\ (3.67\,MeV) + p\ (14.67\,MeV)\,, \\
(5) & \rm Li^6 + n \longrightarrow T + He^4 + 4.8\,MeV\,, \\
(6) & \rm Li^7 + n\ (2.5\,MeV) \longrightarrow T + He^4 + n\,,
\end{array}
$$

where p and n are the proton (hydrogen ion) and the neutron, respectively ($1\,\rm MeV = 10^6\,eV$). Since the energy released by the chemical reaction

$$\rm H_2 + \frac{1}{2}O_2 \longrightarrow H_2O$$

is 2.96 eV, the fusion energy released is about a million times as great as the chemical energy. A binding energy per nucleon is smaller in very light or very heavy nuclides and largest in nuclides with atomic mass numbers around 60. Therefore, large amounts of energy can be released when light nuclides are fused. Deuterium is abundant in nature. For example, it comprises 0.015 atom percent of the hydrogen in sea water, with a volume of about $1.35 \times 10^9\,\mathrm{km}^3$.

Although fusion energy was released in an explosive manner by the hydrogen bomb in 1951, controlled fusion is still at the research and development stage. Nuclear fusion reactions were found in the 1920s. When proton or deuteron beams collide with a light nuclide target, the beam loses its energy by ionization or elastic collisions with target nuclides, and the probability of nuclear fusion is negligible. Nuclear fusion research has been most actively pursued in the context of hot plasmas.

In fully ionized hydrogen, deuterium and tritium plasmas, the process of ionization does not occur. If the plasma is confined adiabatically in some specified region, the average energy does not decrease by elastic collision processes. Therefore, if very hot D–T plasmas or D–D plasmas are confined, the ions have large enough velocities to overcome their mutual Coulomb repulsion, so that collision and fusion take place.

Let us consider the nuclear reaction wherein D collides with T. The cross-section of T nucleons is denoted by $\sigma$. This cross-section is a function of the kinetic energy $E$ of D. The cross-section of the D–T reaction at $E = 100\,\mathrm{keV}$ is $5 \times 10^{-24}\,\mathrm{cm}^2$. The cross-sections $\sigma$ of D–T, D–D, D–$\mathrm{He}^3$ reactions versus the kinetic energy of colliding nucleons are shown in Fig. 1.2a [1.1, 1.2]. The probability of the fusion reaction per unit time in the case where a D ion with velocity $v$ collides with T ions with density of $n_\mathrm{T}$ is given by $n_\mathrm{T}\sigma v$. (We discuss the collision probability in more detail in Sect. 2.7.) When a plasma is Maxwellian with ion temperature $T_\mathrm{i}$, one must calculate the average value $\langle\sigma v\rangle$ of $\sigma v$ over the velocity space. The dependence of $\langle\sigma v\rangle$ on the ion temperature $T_\mathrm{i}$ is shown in Fig. 1.2b [1.3]. A fitting equation for $\langle\sigma v\rangle$ for the D–T reaction as a function of $\kappa T$ in units of keV is [1.4]

$$\langle\sigma v\rangle\ (\mathrm{m}^{-3}) = \frac{3.7 \times 10^{-18}}{H(\kappa T) \times (\kappa T)^{2/3}} \exp\left[-\frac{20}{(\kappa T)^{1/3}}\right], \qquad (1.5)$$

$$H(\kappa T) \equiv \frac{\kappa T}{37} + \frac{5.45}{3 + \kappa T(1 + \kappa T/37.5)^{2.8}}.$$

Figure 1.3 shows an example of an electric power plant based on a D–T fusion reactor. Fast neutrons produced in the fusion core plasma penetrate the first wall and a lithium blanket surrounding the plasma moderates the fast neutrons, converting their kinetic energy to heat. Furthermore, the lithium blanket breeds tritium due to reactions (5) and (6) above. [Triton beta-decays to $\mathrm{He}^3$ with a half-life of 12.3 yr, $\mathrm{T} \to \mathrm{He}^3 + \mathrm{e}\ (< 18.6\,\mathrm{keV})$, and tritium does not exist as a natural resource.] The lithium blanket gives up its heat to generate steam via a heat exchanger and a steam turbine generates electric power. Part of the generated electric power is used to operate the plasma heating system. As alpha particles (He ions) are charged particles, they can heat the

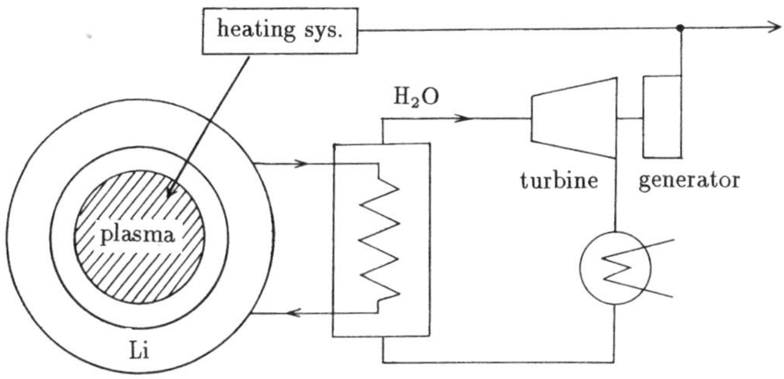


**Fig. 1.3.** Electric power plant based on a D–T fusion reactor

plasma directly by Coulomb collisions (see Sect. 2.6). The total heating power $P_{heat}$ is the sum of the $\alpha$ particle heating power $P_\alpha$ and the heating power $P_{ext}$ due to the external heating system. The total heating power needed to sustain the plasma in a steady state must be equal to the energy loss rate of the fusion core plasma. Consequently, good energy confinement (small energy loss rate) in the hot plasma is the key issue.

The thermal energy of the plasma per unit volume is $(3/2)n\kappa(T_i + T_e)$. This thermal energy is lost by thermal conduction and convective losses. The notation $P_L$ denotes these energy losses from the plasma per unit volume and unit time (power loss per unit volume). In addition to $P_L$, there is radiation loss $R$ due to electron bremsstrahlung and impurity ion radiation. The total energy confinement time $\tau_E$ is defined by

$$\tau_E \equiv \frac{(3/2)n\kappa(T_e + T_i)}{P_L + R} \simeq \frac{3n\kappa T}{P_L + R} . \tag{1.6}$$

The input heating power $P_{heat}$ required to maintain the thermal energy of the plasma is equal to $P_L + R$.

For the D–T reaction, the sum of kinetic energies $Q_\alpha = 3.52\,\text{MeV}$ of alpha particles and $Q_n = 14.06\,\text{MeV}$ of neutrons is $Q_{NF} = 17.58\,\text{MeV}$ per reaction ($Q_n : Q_\alpha = m_\alpha : m_n = 0.8 : 0.2$ due to momentum conservation). Since the densities of D ions and T ions in an equally mixed plasma are $n/2$, the number of D–T reactions per unit time and unit volume is $(n/2)(n/2)\langle\sigma v\rangle$ (refer to the discussion in Sect. 2.6), so that the fusion output power per unit volume $P_{NF}$ is given by

$$P_{NF} = (n/2)(n/2)\langle\sigma v\rangle Q_{NF} . \tag{1.7}$$

If the fusion powers due to the neutron and alpha particle are denoted by $P_n$ and $P_\alpha$ respectively, then $P_n = 0.8P_{NF}$ and $P_\alpha = 0.2P_{NF}$. Let the thermal-to-electric conversion efficiency be $\eta_{el}$ and the heating efficiency (ratio of the

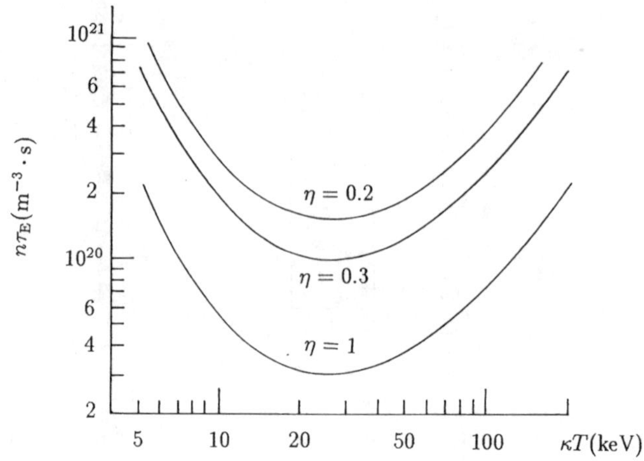


**Fig. 1.4.** Condition of D–T fusion core plasma in $n\tau_E$–$T$ diagram in the case $\eta = 0.3$, critical condition $\eta = 1$, and ignition condition $\eta = 0.2$

power deposited in the plasma to the input electric power of the heating device) by $\eta_{heat}$. When a part ($\gamma < 1$) of generated electric power is used to operate the heating system, then the available heating power to plasma is

$$(0.8\eta_{el}\gamma\eta_{heat} + 0.2)P_{NF} = \eta P_{NF} \ , \qquad \eta \equiv 0.8\gamma\eta_{el}\eta_{heat} + 0.2 \ .$$

The burning condition is

$$P_{heat} = P_L + R = \frac{3n\kappa T}{\tau_E} < \eta P_{NF} \ , \tag{1.8}$$

that is,

$$\frac{3n\kappa T}{\tau_E} < \eta \frac{Q_{NF}}{4} n^2 \langle \sigma v \rangle \ ,$$

and hence,

$$n\tau_E > \frac{12\kappa T}{\eta Q_{NF} \langle \sigma v \rangle} \ . \tag{1.9}$$

The right-hand side of (1.9) is a function of temperature $T$ alone. When $\kappa T = 10^4$ eV and $\eta \sim 0.3$ ($\gamma \sim 0.4$, $\eta_{el} \sim 0.4$, $\eta_{heat} \sim 0.8$), the necessary condition is $n\tau_E > 1.7 \times 10^{20}$ m$^{-3}$s. The burning condition of the D–T fusion plasma in the case $\eta \sim 0.3$ is shown in Fig. 1.4. In reality the plasma is hot in the core and cold at the edges. For a more accurate discussion, we must take this temperature and density profile effect into account, an analysis undertaken in Sect. 16.10.

The ratio of the fusion output power due to $\alpha$ particles to the total is $Q_\alpha/Q_{NF} = 0.2$. If the total kinetic energy (output energy) of alpha particles contributes to heating the plasma and alpha particle heating power can sustain the necessary high temperature of the plasma without heating from the outside, the plasma is in an ignited state. The condition $P_\alpha = P_{\mathrm{L}} + R$ is called the *ignition condition*, which corresponds to the case $\eta = 0.2$ in (1.8).

The condition $P_{\mathrm{heat}} = P_{\mathrm{NF}}$ is called the *break-even condition*. This corresponds to the case of $\eta = 1$ in (1.8). The ignition condition ($\eta = 0.2$) and break-even condition ($\eta = 1$) are also shown in Fig. 1.4.

# 2 Plasma Characteristics

## 2.1 Velocity Space Distribution Function

In a plasma, electrons and ions move with various velocities. The number o electrons in a unit volume is the electron density $n_\mathrm{e}$ and the number of electrons $\mathrm{d}n_\mathrm{e}(v_x)$ with the $x$ component of velocity between $v_x$ and $v_x + \mathrm{d}v_x$ is given by

$$\mathrm{d}n_\mathrm{e}(v_x) = f_\mathrm{e}(v_x)\mathrm{d}v_x \ .$$

Then $f_\mathrm{e}(v_x)$ is called the electron *velocity space distribution function.* When electrons are in a thermal equilibrium state with electron temperature $T_\mathrm{e}$, the velocity space distribution function is the Maxwell distribution:

$$f_\mathrm{e}(v_x) = n_\mathrm{e}\left(\frac{\beta}{2\pi}\right)^{1/2} \exp\left(-\frac{\beta v_x^2}{2}\right) , \qquad \beta = \frac{m_\mathrm{e}}{\kappa T_\mathrm{e}} \ .$$

From the definition, the velocity space distribution function satisfies

$$\int_{-\infty}^{\infty} f_\mathrm{e}(v_x)\mathrm{d}v_x = n_\mathrm{e} \ .$$

The Maxwell distribution function in the three-dimensional velocity space is given by

$$f_\mathrm{e}(v_x, v_y, v_z) = n_\mathrm{e}\left(\frac{m_\mathrm{e}}{2\pi\kappa T_\mathrm{e}}\right)^{3/2} \exp\left[-\frac{m_\mathrm{e}(v_x^2 + v_y^2 + v_z^2)}{2\kappa T_\mathrm{e}}\right] \ . \tag{2.1}$$

The ion distribution function is defined in the same way as for the electron. The mean of the squared velocity $v_x^2$ is given by

$$v_T^2 = \frac{1}{n}\int_{-\infty}^{\infty} v_x^2 f(v_x)\mathrm{d}v_x = \frac{\kappa T}{m} \ . \tag{2.2}$$

The pressure $p$ is

$$p = n\kappa T \ .$$

The particle flux in the $x$ direction per unit area $\Gamma_{+,x}$ is given by

$$\Gamma_{+,x} = \int_0^{\infty} v_x f(v_x)\mathrm{d}v_x = n\left(\frac{\kappa T}{2\pi m}\right)^{1/2} \ .$$

## 2.2 Plasma Frequency. Debye Length

Let us consider the case where a small perturbation occurs in a uniform plasma and the electrons in the plasma move due to the perturbation. It is assumed that the ions do not move because they have much greater mass than the electrons. Due to the displacement of electrons, electric charges appear and an electric field is induced. The electric field is given by

$$\epsilon_0 \boldsymbol{\nabla}\cdot\boldsymbol{E} = -e(n_{\mathrm{e}} - n_0) .$$

Electrons are accelerated by the electric field:

$$m_{\mathrm{e}} \frac{\mathrm{d}\boldsymbol{v}}{\mathrm{d}t} = -e\boldsymbol{E} .$$

Due to the movement of electrons, the electron density changes:

$$\frac{\partial n_{\mathrm{e}}}{\partial t} + \boldsymbol{\nabla}\cdot(n_{\mathrm{e}}\boldsymbol{v}) = 0 .$$

Writing $n_{\mathrm{e}} - n_0 = n_1$ and assuming $|n_1| \ll n_0$, we find

$$\epsilon_0 \boldsymbol{\nabla}\cdot\boldsymbol{E} = -en_1 , \qquad m_{\mathrm{e}} \frac{\partial \boldsymbol{v}}{\partial t} = -e\boldsymbol{E} , \qquad \frac{\partial n_1}{\partial t} + n_0 \boldsymbol{\nabla}\cdot\boldsymbol{v} = 0 .$$

For simplicity, the displacement is assumed to be only in the $x$ direction and sinusoidal with angular frequency $\omega$:

$$n_1(x, t) = n_1 \exp(\mathrm{i}kx - \mathrm{i}\omega t) .$$

The time derivative $\partial/\partial t$ is replaced by $-\mathrm{i}\omega$ and $\partial/\partial x$ is replaced by $\mathrm{i}k$. The electric field has only the $x$ component $E$. Then

$$\mathrm{i}k\epsilon_0 E = -en_1 , \qquad -\mathrm{i}\omega m_{\mathrm{e}} v = -eE , \qquad -\mathrm{i}\omega n_1 = -\mathrm{i}kn_0 v ,$$

so that we find

$$\omega^2 = \frac{n_0 e^2}{\epsilon_0 m_{\mathrm{e}}} . \tag{2.3}$$

This wave is called the *electron plasma wave* or *Langmuir wave* and its frequency is called the *electron plasma frequency* $\Pi_{\mathrm{e}}$:

$$\Pi_{\mathrm{e}} = \left(\frac{n_{\mathrm{e}} e^2}{\epsilon_0 m_{\mathrm{e}}}\right)^{1/2} = 5.64 \times 10^{11} \left(\frac{n_{\mathrm{e}}}{10^{20}}\right)^{1/2} \mathrm{rad/s} .$$

The following relation holds between the plasma frequency and the Debye length $\lambda_{\mathrm{D}}$:

$$\lambda_{\mathrm{D}} \Pi_{\mathrm{e}} = \left(\frac{\kappa T_{\mathrm{e}}}{m_{\mathrm{e}}}\right)^{1/2} = v_{T_{\mathrm{e}}} = 4.19 \times 10^5 \left(\frac{\kappa T_{\mathrm{e}}}{e}\right)^{1/2} m/s .$$

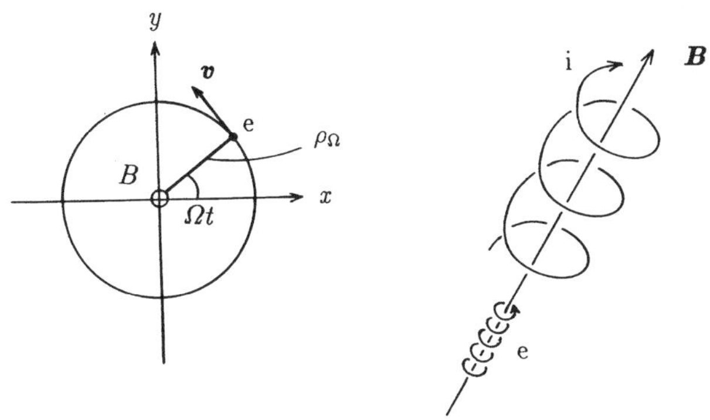


**Fig. 2.1.** Larmor motion of charged particle in magnetic field

## 2.3 Cyclotron Frequency. Larmor Radius

The equation of motion of a charged particle with mass $m$ and charge $q$ in electric and magnetic fields $\boldsymbol{E}$, $\boldsymbol{B}$ is given by

$$m\frac{\mathrm{d}\boldsymbol{v}}{\mathrm{d}t} = q(\boldsymbol{E} + \boldsymbol{v} \times \boldsymbol{B}) . \tag{2.4}$$

When the magnetic field is homogeneous and in the $z$ direction and the electric field is zero, the equation of motion becomes $\dot{\boldsymbol{v}} = (qB/m)(\boldsymbol{v} \times \boldsymbol{b})$, where $\boldsymbol{b} = \boldsymbol{B}/B$, and

$$v_x = -v_\perp \sin(\Omega t + \delta) , \qquad v_y = v_\perp \cos(\Omega t + \delta) , \qquad v_z = v_{z0} ,$$

$$\Omega = -\frac{qB}{m} . \tag{2.5}$$

The solution of these equations is a spiral motion around the magnetic line of force with angular velocity $\Omega$ (see Fig. 2.1). This motion is called *Larmor motion*. The angular frequency $\Omega$ is called *cyclotron (angular) frequency*. Denoting the radius of the orbit by $\rho_\Omega$, the centrifugal force is $mv_\perp^2/\rho_\Omega$ and the Lorentz force is $qv_\perp B$. Since the two forces must balance, we find

$$\rho_\Omega = \frac{mv_\perp}{|q|B} . \tag{2.6}$$

This radius is called the *Larmor radius*. The center of the Larmor motion is called the *guiding center*. The Larmor motion of the electron is a right-handed rotation ($\Omega_\mathrm{e} > 0$), while the Larmor motion of the ion is a left-handed rotation ($\Omega_\mathrm{i} < 0$). When $B = 1\,\mathrm{T}$, $\kappa T = 100\,\mathrm{eV}$, the values of the Larmor radius and cyclotron frequency are given in Table 2.1.

**Table 2.1.** Mass, thermal velocity, Larmor radius and cyclotron frequency of the electron and proton when $B = 1\,\mathrm{T}$, $\kappa T = 100\,\mathrm{eV}$

| | Electron | Proton |
|---|---|---|
| Mass [kg] | $9.109\,389\,7(54) \times 10^{-31}$ | $1.672\,623\,1(10) \times 10^{-27}$ |
| Thermal velocity $v_T = (\kappa T/m)^{1/2}$ | $4.2 \times 10^6\,\mathrm{m\,s^{-1}}$ | $9.8 \times 10^4\,\mathrm{m\,s^{-1}}$ |
| Larmor radius $\rho_\Omega$ | $23.8\,\mu\mathrm{m}$ | $1.02\,\mathrm{mm}$ |
| Cyclotron frequency (angular) $\Omega$ | $1.76 \times 10^{11}\,\mathrm{s^{-1}}$ | $-9.58 \times 10^7\,\mathrm{s^{-1}}$ |
| Cyclotron frequency $\Omega/2\pi$ | 28 GHz | −15.2 MHz |

## 2.4 Drift Velocity of Guiding Center

When a uniform electric field $\boldsymbol{E}$ is superposed perpendicularly to the uniform magnetic field, the equation of motion (2.4) reduces to

$$m\frac{\mathrm{d}\boldsymbol{u}}{\mathrm{d}t} = q(\boldsymbol{u} \times \boldsymbol{B}) ,$$

introducing $\boldsymbol{u}$ and $\boldsymbol{u}_\mathrm{E}$ defined by

$$\boldsymbol{v} \equiv \boldsymbol{u}_\mathrm{E} + \boldsymbol{u} , \qquad \boldsymbol{u}_\mathrm{E} \equiv \frac{\boldsymbol{E} \times \boldsymbol{b}}{B} . \tag{2.7}$$

Therefore the motion of the charged particle is a superposition of the Larmor motion and the drift motion $\boldsymbol{u}_\mathrm{E}$ of its guiding center. The direction of the guiding center drift due to $\boldsymbol{E}$ is the same for both ions and electrons (Fig. 2.2). When a gravitational field $\boldsymbol{g}$ is superposed, the force is $m\boldsymbol{g}$, which corresponds to $q\boldsymbol{E}$ in the case of an electric field. Therefore the *drift velocity of the guiding center* due to gravitation is given by

$$\boldsymbol{u}_\mathrm{g} = \frac{m}{qB}(\boldsymbol{g} \times \boldsymbol{b}) = -\frac{\boldsymbol{g} \times \boldsymbol{b}}{\Omega} . \tag{2.8}$$

Electrons and ions drift in opposite directions under gravitation and the drift velocity of the ion guiding center is much larger than the electron's (see Fig. 2.2).

When the magnetic and electric fields change slowly and gradually in time and space ($|\omega/\Omega| \ll 1$, $\rho_\Omega/R \ll 1$), the formulas for the drift velocity are valid as they are. However, because of the curvature of the magnetic field lines, a centrifugal force acts on any particle which runs along a field line with velocity $v_\parallel$. The acceleration due to the centrifugal force is

$$\boldsymbol{g}_\mathrm{curv} = \frac{v_\parallel^2}{R}\boldsymbol{n} ,$$

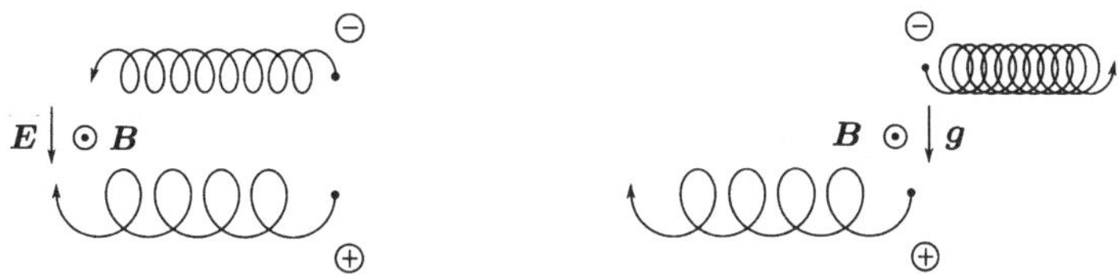


**Fig. 2.2.** Drift motion of the guiding center in electric and gravitational fields (schematic)

where $R$ is the radius of curvature of the field line and $\boldsymbol{n}$ is the unit vector running from the center of curvature to the field line (Fig. 2.3).

Furthermore, as described at the end of Sect. 2.4, the resultant effect of Larmor motion in an inhomogeneous magnetic field reduces to an acceleration

$$\boldsymbol{g}_{\nabla B} = -\frac{v_\perp^2/2}{B}\nabla B\,.$$

Therefore the drift velocity $\boldsymbol{u}_{\mathrm{G}}$ of the guiding center due to an inhomogeneous curved magnetic field is given by the *drift approximation* as

$$\boldsymbol{u}_{\mathrm{G}} = -\frac{1}{\Omega}\left(\frac{v_\parallel^2}{R}\boldsymbol{n} - \frac{v_\perp^2}{2}\frac{\nabla B}{B}\right) \times \boldsymbol{b}\,. \tag{2.9}$$

The first term is called the *curvature drift* and the second term is called *gradient B drift*. Since $\nabla \times \boldsymbol{B} = \mu_0 \boldsymbol{j}$, where $\boldsymbol{j}$ is the current density, the vector formula reduces to

$$\begin{aligned}
\frac{1}{2B}\nabla(\boldsymbol{B}\cdot\boldsymbol{B}) &= (\boldsymbol{b}\cdot\nabla)\boldsymbol{B} + \boldsymbol{b}\times(\nabla\times\boldsymbol{B}) \\
&= \frac{\partial}{\partial l}(B\boldsymbol{b}) + \boldsymbol{b}\times\mu_0\boldsymbol{j} \\
&= \frac{\partial B}{\partial l}\boldsymbol{b} + B\frac{\partial \boldsymbol{b}}{\partial l} - \mu_0\frac{\nabla p}{B} \\
&= \frac{\partial B}{\partial l}\boldsymbol{b} - B\frac{\boldsymbol{n}}{R} - \mu_0\frac{\nabla p}{B}\,,
\end{aligned}$$

where $p$ is plasma pressure and $\nabla p = \boldsymbol{j}\times\boldsymbol{B}$ holds in the equilibrium state [see (6.1) in Chap. 6]. We used the following relation (see Fig. 2.3):

$$\frac{\partial \boldsymbol{b}}{\partial l} = -\frac{\boldsymbol{n}}{R}\,,$$

where $l$ is the length along the field line. Then we have

$$\frac{\boldsymbol{n}\times\boldsymbol{b}}{R} = -\left(\frac{\nabla B}{B} + \mu_0\frac{\nabla p}{B^2}\right)\times\boldsymbol{b}\,.$$

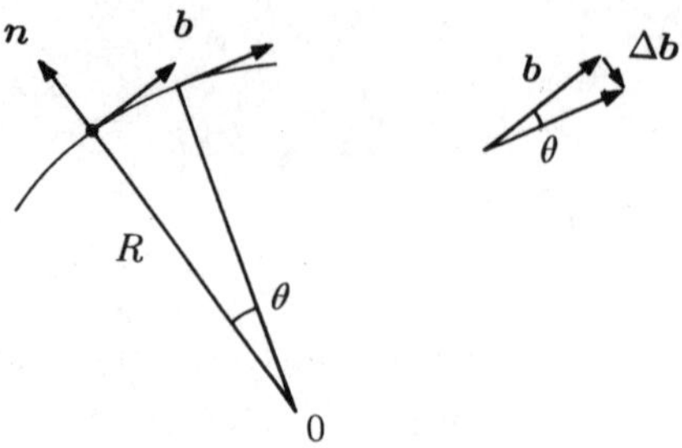


**Fig. 2.3.** Radius of curvature of the line of magnetic force

If $\nabla p$ is much smaller than $\nabla B^2/(2\mu_0)$, we find

$$\boldsymbol{u}_\mathrm{G} = -\frac{1}{\Omega}\frac{v_\parallel^2 + v_\perp^2/2}{R}(\boldsymbol{n}\times\boldsymbol{b}) .$$

The parallel motion along the magnetic field is given+by

$$m\frac{\mathrm{d}v_\parallel}{\mathrm{d}t} = qE_\parallel + mg_\parallel - \frac{mv_\perp^2/2}{B}\nabla_\parallel B .$$

Let us consider the effect on a gyrating charged particle of an inhomogeneity in the magnetic field. The $x$ component of the Lorentz force $\boldsymbol{F}_\mathrm{L} = q\boldsymbol{v}\times\boldsymbol{B}$ perpendicular to the magnetic field ($z$ direction) and the magnitude $B$ of the magnetic field near the guiding center is

$$F_{\mathrm{L}x} = qv_yB = -|q|v_\perp\cos\theta B , \qquad B = B_0 + \frac{\partial B}{\partial x}\rho_\Omega\cos\theta + \frac{\partial B}{\partial y}\rho_\Omega\sin\theta .$$

The time average of the $x$ component of the Lorentz force is given by

$$\langle F_{\mathrm{L}x}\rangle = \frac{1}{2}\frac{\partial B}{\partial x}(-|q|)v_\perp\rho_\Omega ,$$

and the $y$ component similarly, so that

$$\langle \boldsymbol{F}_\mathrm{L}\rangle_\perp = -\frac{mv_\perp^2/2}{B}\nabla_\perp B .$$

We must now estimate the time average of the $z$ component of the Lorentz force. The equation $\nabla\cdot\boldsymbol{B} = 0$ near the guiding center in Fig. 2.4 becomes $B_r/r + \partial B_r/\partial r + \partial B_z/\partial z = 0$ and we find

$$\langle F_{\mathrm{L}z}\rangle = -\langle qv_\theta B_r\rangle = |q|v_\perp\rho_\Omega\frac{\partial B_r}{\partial r} = -\frac{mv_\perp^2/2}{B}\frac{\partial B}{\partial z} ,$$

where $r \sim \rho_\Omega$ is very small and $B_r/r \approx \partial B_r/\partial r$. The required expression for $\boldsymbol{g}_{\nabla B}$ is thus derived.

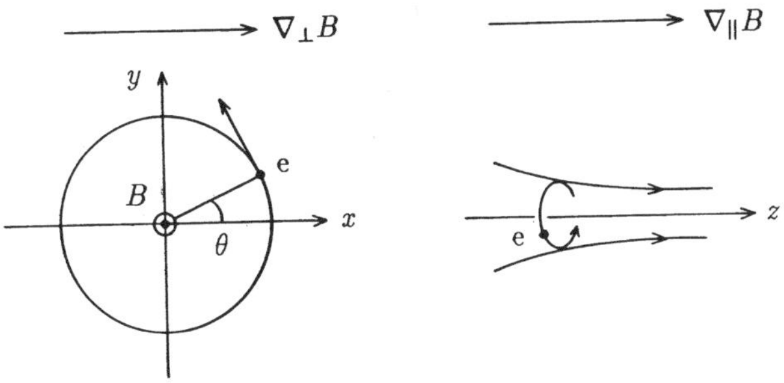


**Fig. 2.4.** Larmor motion in an inhomogeneous magnetic field

## 2.5 Magnetic Moment. Mirror Confinement

A current loop with current $I$ encircling an area $S$ has magnetic moment $\mu_{\rm m} = IS$. Since the current and area encircled by the gyrating Larmor motion are $I = q\Omega/2\pi$ and $S = \pi\rho_\Omega^2$ respectively, it has the *magnetic moment*

$$\mu_{\rm m} = \frac{q\Omega}{2\pi}\pi\rho_\Omega^2 = \frac{mv_\perp^2}{2B} \ . \tag{2.10}$$

This physical quantity is adiabatically invariant, as will be shown at the end of this section. When the magnetic field changes slowly, the magnetic moment is conserved. Therefore, if $B$ is increased, $mv_\perp^2/2 = \mu_{\rm m}B$ is also increased and the particles are heated. This kind of heating is called *adiabatic heating*.

Let us consider a mirror field as shown in Fig. 2.5, in which the magnetic field is weak at the center and strong at both ends of the mirror field. For simplicity, the electric field is assumed to be zero. Since the Lorentz force is perpendicular to the velocity, the magnetic field does not contribute to the change of kinetic energy and

$$\frac{mv_\parallel^2}{2} + \frac{mv_\perp^2}{2} = \frac{mv^2}{2} = E = \text{const.} \tag{2.11}$$

Since the magnetic moment is conserved, we find

$$v_\parallel = \pm\left(\frac{2}{m}E - v_\perp^2\right)^{1/2} = \pm\left(v^2 - \frac{2}{m}\mu_{\rm m}B\right)^{1/2} \ .$$

When the particle moves towards the open ends, the magnetic field becomes large and $v_\parallel$ becomes small or even zero. Since the force along the direction parallel to the magnetic field is $-\mu_{\rm m}\nabla_\parallel B$, both ends of the mirror field repel charged particles as a mirror reflects light. The ratio of the magnitude of the magnetic field at the open end to the central value is called the *mirror ratio*:

$$R_{\rm M} = \frac{B_{\rm M}}{B_0} \ .$$

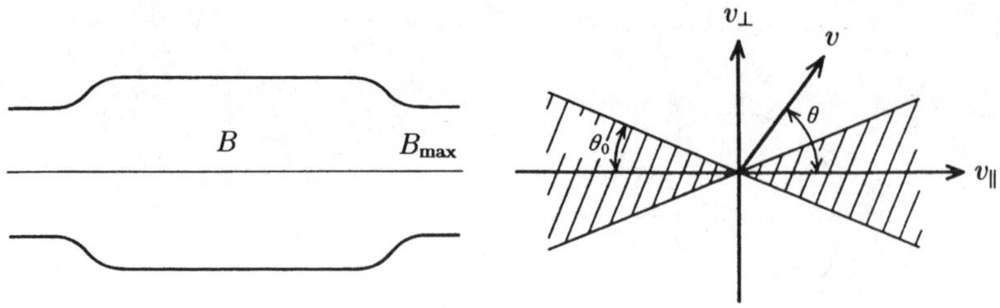


**Fig. 2.5.** Mirror field and loss cone in $v_\parallel$–$v_\perp$ space

Let us denote the parallel and perpendicular components of the velocity at the mirror center by $v_{\parallel 0}$ and $v_{\perp 0}$, respectively. The value $v_\perp^2$ at the position of maximum magnetic field $B_{\mathrm{M}}$ is given by

$$v_{\perp\mathrm{M}}^2 = \frac{B_{\mathrm{M}}}{B_0} v_{\perp 0}^2 \,.$$

This value $v_{\perp\mathrm{M}}^2$ cannot be larger than $v^2 = v_0^2$, so that the particle satisfying the following condition is reflected and trapped in the mirror field:

$$\left(\frac{v_{\perp 0}}{v_0}\right)^2 > \frac{B_0}{B_{\mathrm{M}}} = \frac{1}{R_{\mathrm{M}}} \,. \tag{2.12}$$

Particles in the region where $\sin\theta \equiv v_{\perp 0}/v_0$ satisfies

$$\sin^2\theta \le \frac{1}{R_{\mathrm{M}}}$$

are not trapped and the region is called the *loss cone* in $v_\parallel$–$v_\perp$ space (see Fig. 2.5).

Let us check the invariance of $\mu_{\mathrm{m}}$ in the presence of a slowly changing magnetic field ($|\partial B/\partial t| \ll |\Omega B|$). Taking the scalar product of $\boldsymbol{v}_\perp$ with the equation of motion gives

$$m\boldsymbol{v}_\perp \cdot \frac{\mathrm{d}\boldsymbol{v}_\perp}{\mathrm{d}t} = \frac{\mathrm{d}}{\mathrm{d}t}\left(\frac{mv_\perp^2}{2}\right) = q(\boldsymbol{v}_\perp \cdot \boldsymbol{E}_\perp) \,.$$

During one period $2\pi/|\Omega|$ of Larmor motion, the change $\Delta W_\perp$ in the kinetic energy $W_\perp = mv_\perp^2/2$ is

$$\Delta W_\perp = q\int(\boldsymbol{v}_\perp \cdot \boldsymbol{E}_\perp)\mathrm{d}t = q\oint \boldsymbol{E}_\perp \cdot \mathrm{d}\boldsymbol{s} = q\int(\nabla \times \boldsymbol{E})\cdot\boldsymbol{n}\mathrm{d}S \,,$$

where $\oint \mathrm{d}s$ is the closed line integral along the Larmor orbit and $\int \mathrm{d}S$ is the surface integral over the area encircled by the Larmor orbit. Since $\nabla \times \boldsymbol{E} = -\partial \boldsymbol{B}/\partial t$, $\Delta W_\perp$ is

$$\Delta W_\perp = -q\int \frac{\partial \boldsymbol{B}}{\partial t}\cdot \boldsymbol{n}\mathrm{d}S = |q|\pi\rho_\Omega^2 \frac{\partial B}{\partial t} \ .$$

The change $\Delta B$ in the magnetic field during one period of Larmor motion is $\Delta B = (\partial B/\partial t)(2\pi/|\Omega|)$. Hence,

$$\Delta W_\perp = \frac{m v_\perp^2}{2}\frac{\Delta B}{B} = W_\perp \frac{\Delta B}{B} \ ,$$

and

$$\mu_\mathrm{m} = \frac{W_\perp}{B} = \mathrm{const.}$$

When a system is periodic in time, the action integral $\oint p\mathrm{d}q$, in terms of the canonical variables $p, q$ is generally an adiabatic invariant. The action integral of the Larmor motion is

$$J_\perp = (-m\rho_\Omega \Omega)2\pi\rho_\Omega = -\frac{4\pi m}{q}\mu_\mathrm{m} \ .$$

$J_\perp$ is called the *transversal adiabatic invariant.*

A particle trapped in a mirror field moves back and forth along the field line, from one end to the other. The second action integral of this periodic motion, viz.,

$$J_\parallel = m\oint v_\parallel \mathrm{d}l \ , \qquad (2.13)$$

is another adiabatic invariant. $J_\parallel$ is called the *longitudinal adiabatic invariant*. As one makes the mirror length $l$ shorter, $\langle v_\parallel \rangle$ increases (for $J_\parallel = 2m\langle v_\parallel\rangle l$ is conserved), and the particles are accelerated. This phenomena is called *Fermi acceleration.*

The line of magnetic force of the mirror is convex towards the outside. The particles trapped by the mirror are subjected to curvature drift and gradient $B$ drift, so that the trapped particles move back and forth, while drifting in the $\theta$ direction. The orbit $(r, \theta)$ of the crossing point on the $z = 0$ plane of the back and forth movement is given by $J_\parallel(r, \theta, \mu_\mathrm{m}, E) = \mathrm{const.}$

## 2.6 Coulomb Collision. Fast Neutral Beam Injection

The motions of charged particles were analyzed in the previous section without considering the effects of collisions between particles. In this section, phenomena associated with *Coulomb collisions* will be discussed. Let us start from a simple model. Assume that a sphere of radius $a$ moves with velocity $v$ in a region where spheres of radius $b$ are filled with the number density $n$ (see Fig. 2.6). When the distance between the two particles becomes less than $a+b$, collision takes place. The cross-section $\sigma$ of this collision is $\sigma = \pi(a+b)^2$. Since the sphere $a$ moves through the distance $l = v\delta t$ during $\delta t$, the probability of collision with the sphere $b$ is

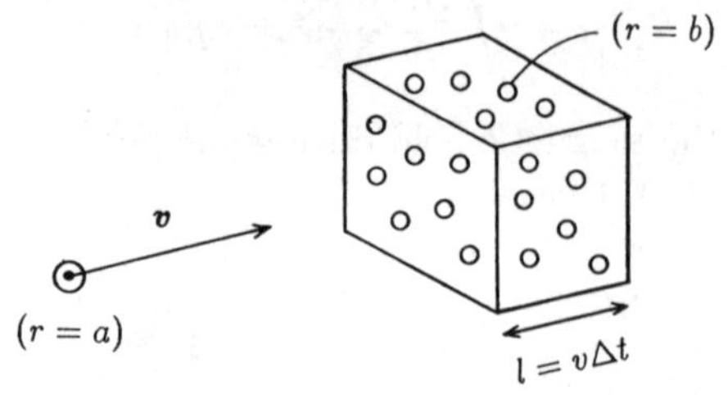


**Fig. 2.6.** Probability of collision between a sphere $a$ and spheres $b$

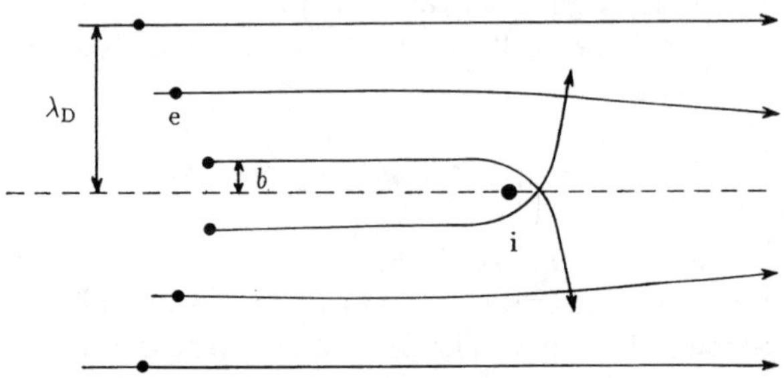


**Fig. 2.7.** Coulomb collision of electron with ion

$$nl\sigma = n\sigma v\delta t \ ,$$

since $nl$ is the number of spheres $b$ with which the sphere $a$ may collide within a unit area of incidence, and $nl\sigma$ is the total cross-section per unit area of incidence during the time $\delta t$. Therefore the collision time $\tau_{\rm coll}$, when the probability of collision becomes 1, is

$$\tau_{\rm coll} = (n\sigma v)^{-1} \ .$$

In this simple case the cross-section $\sigma$ of the collision is independent of the velocity of the incident sphere $a$. However, the cross-section depends on the incident velocity, in general.

Let us consider the strong Coulomb collision of an incident electron with ions having charge $Ze$ (see Fig. 2.7), in which the electron is strongly deflected after the collision. Such a collision can take place when the magnitude of the electrostatic potential of the electron at the closest distance $b$ is of the order of the kinetic energy of the incident electron, i.e.,

$$\frac{Ze^2}{4\pi\epsilon_0 b} = \frac{m_{\rm e}v_{\rm e}^2}{2} \ .$$

The cross-section of the strong Coulomb collision is $\sigma = \pi b^2$. The inverse of the collision time $\tau_{\rm coll}$ of the strong Coulomb collision is

$$\frac{1}{\tau_{\rm coll}} = n_{\rm i}\sigma v_{\rm e} = n_{\rm i} v_{\rm e}\pi b^2 = \frac{n_{\rm i}\pi(Ze^2)^2 v_{\rm e}}{(4\pi\epsilon_0 m_{\rm e} v_{\rm e}^2/2)^2} = \frac{Z^2 e^4 n_{\rm i}}{4\pi\epsilon_0^2 m_{\rm e}^2 v_{\rm e}^3} .$$

Since the Coulomb force is a long range interaction, a test particle is deflected by a small angle even by a distant field particle, which the test particle does not come very close to. As explained in Sect. 1.2, the Coulomb field of a field particle is not shielded inside the Debye sphere, which has radius equal to the Debye length $\lambda_{\rm D}$, and there are many field particles inside the Debye sphere in typical laboratory plasmas (weakly coupled plasmas). Accumulation of many collisions with small angle deflection results in a large effect. When the effect of the small angle deflection is taken into account, the total Coulomb cross-section increases by a factor of the *Coulomb logarithm*

$$\ln\Lambda \simeq \ln\left(\frac{2\lambda_{\rm D}}{b}\right) \simeq \int_{b/2}^{\lambda_{\rm D}} \frac{1}{r}{\rm d}r \simeq 15\text{–}20 .$$

The time derivative of the momentum $p_{\|}$ parallel to the incident direction of the electron is given by use of the *collision time* $\tau_{{\rm ei}\|}$ as follows [2.1, 2.2]:

$$\frac{{\rm d}p_{\|}}{{\rm d}t} = -\frac{p_{\|}}{\tau_{{\rm ei}\|}} , \qquad \frac{1}{\tau_{{\rm ei}\|}} = \frac{Z^2 e^4 n_{\rm i}\ln\Lambda}{4\pi\epsilon_0^2 m_{\rm e}^2 v_{\rm e}^3} , \tag{2.14}$$

where $\tau_{{\rm ei}\|}$ indicates the deceleration time of an electron by ions.

When a test particle with charge $q$, mass $m$ and velocity $v$ collides with field particles with charge $q^*$, mass $m^*$ and thermal velocity $v_T^* = (\kappa T^*/m^*)^{1/2}$, in general, the collision time of the test particle is given by [2.1, 2.2]

$$\frac{1}{\tau_{\|}} = \frac{q^2 q^{*2} n^* \ln\Lambda}{4\pi\epsilon_0^2 m m_{\rm r} v^3} = \left(\frac{qq^* n^*}{\epsilon_0 m}\right)^2 \frac{\ln\Lambda}{4\pi(m_{\rm r}/m)v^3 n^*} , \tag{2.15}$$

under the assumption that $v > v_T^*$. In this expression, $m_{\rm r}$ is the reduced mass

$$m_{\rm r} = \frac{mm^*}{m+m^*} .$$

Taking the average of $(m/2)v^2 = (3/2)\kappa T$, $1/\tau_{\|}$ becomes

$$\frac{1}{\tau_{\|}} = \frac{q^2 q^{*2} n^* \ln\Lambda}{3^{1/2} 12\pi\epsilon_0^2 (m_{\rm r}/m^{1/2})(\kappa T)^{3/2}} . \tag{2.16}$$

The inverse of the collision time, denoted by $\nu$, is called the *collision frequency*. The *mean free path* is given by $\lambda = 3^{1/2} v_T \tau$.

The collision frequency for electrons with ions is

$$\frac{1}{\tau_{{\rm ei}\|}} = \frac{Z^2 e^4 n_{\rm i}\ln\Lambda}{3^{1/2} 12\pi\epsilon_0^2 m_{\rm e}^{1/2} (\kappa T_{\rm e})^{3/2}} . \tag{2.17}$$

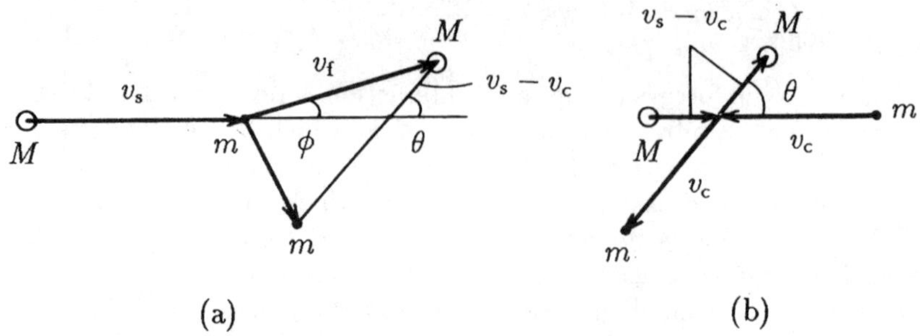


**Fig. 2.8.** Elastic collision of test particle $M$ and field particle $m$ in the laboratory system (**a**) and the center-of-mass system (**b**)

This electron-ion collision frequency is $\sim$ 1.4 times the Spitzer result [2.3] of

$$\frac{1}{\tau_{\mathrm{ei}\|\ \mathrm{Spitzer}}} = \frac{Z^2 e^4 n_{\mathrm{i}} \ln \Lambda}{9.3 \times 10 \epsilon_0^2 m_{\mathrm{e}}^{1/2} (\kappa T_{\mathrm{e}})^{3/2}} \,. \tag{2.18}$$

When an ion with charge $Z$ and mass $m_{\mathrm{i}}$ collides with the same ions, the ion–ion collision frequency is given by

$$\frac{1}{\tau_{\mathrm{ii}\|}} = \frac{Z^4 e^4 n_{\mathrm{i}} \ln \Lambda}{3^{1/2} 6\pi \epsilon_0^2 m_{\mathrm{i}}^{1/2} (\kappa T_{\mathrm{i}})^{3/2}} \,. \tag{2.19}$$

The electron–electron Coulomb collision frequency can be derived by substituting $m_{\mathrm{i}} \to m_{\mathrm{e}}$ and $Z \to 1$ into the formula for $\tau_{\mathrm{ii}\|}$, which yields

$$\frac{1}{\tau_{\mathrm{ee}\|}} = \frac{n_{\mathrm{e}} e^4 \ln \Lambda}{3^{1/2} 6\pi \epsilon_0^2 m_{\mathrm{e}}^{1/2} (\kappa T_{\mathrm{e}})^{3/2}} \,. \tag{2.20}$$

However, the case of ion-to-electron Coulomb collisions is more complicated to treat because the assumption $v_{\mathrm{i}} > v_T^*$ is no longer justified. Let us consider the case where a test particle with mass $M$ and velocity $v_{\mathrm{s}}$ collides with a field particle with the mass $m$. In the center-of-mass system, where the center of mass is at rest, the field particle $m$ moves with velocity $v_{\mathrm{c}} = -Mv_{\mathrm{s}}/(M+m)$ and the test particle $M$ moves with velocity $v_{\mathrm{s}} - v_{\mathrm{c}} = mv_{\mathrm{s}}/(M+m)$ (see Fig. 2.8).

Since the total momentum and total kinetic energy of two particles are conserved in the process of elastic collision, the speeds of the test particle and the field particle do not change and the two particles are merely deflected through an angle $\theta$ in the center-of-mass system. The velocity $v_{\mathrm{f}}$ and scattering angle $\phi$ of the test particle after the collision in the laboratory system are given by (see Fig. 2.8)

$$v_{\rm f}^2 = (v_{\rm s} - v_{\rm c})^2 + {v_{\rm c}}^2 + 2(v_{\rm s} - v_{\rm c})v_{\rm c}\cos\theta = v_{\rm s}^2 \frac{M^2 + 2Mm\cos\theta + m^2}{(M+m)^2} ,$$

$$\sin\phi = \frac{m\sin\theta}{(M^2 + 2Mm\cos\theta + m^2)^{1/2}} .$$

Denoting the momentum and kinetic energy of the test particle before and after the collision by $p_{\rm s}$, $E_{\rm s}$, and $p_{\rm f}$, $E_{\rm f}$, respectively, we find

$$\frac{\Delta E}{E_{\rm s}} \equiv \frac{E_{\rm f} - E_{\rm s}}{E_{\rm s}} = -\frac{2Mm}{(M+m)^2}(1 - \cos\theta) .$$

When the average is taken over $\theta$, we obtain the following relations in the case $m/M \ll 1$:

$$\left\langle \frac{\Delta E}{E_{\rm s}} \right\rangle \simeq -\frac{2m}{M} , \qquad \left\langle \frac{\Delta p_\parallel}{p_{\rm s}} \right\rangle \simeq -\frac{m}{M} .$$

From the foregoing discussion, the collision frequency $1/\tau_{\rm ie\parallel}$ for the situation where a heavy ion collides with light electrons is about $m_{\rm e}/m_{\rm i}$ times the value of $1/\tau_{\rm ei\parallel}$, and is given by [2.1, 2.2]

$$\frac{1}{\tau_{\rm ie\parallel}} = \frac{m_{\rm e}}{m_{\rm i}} \frac{Z^2 e^4 n_{\rm e} \ln\Lambda}{(2\pi)^{1/2} 3\pi\epsilon_0^2 m_{\rm e}^{1/2} (\kappa T_{\rm e})^{3/2}} . \tag{2.21}$$

When the parallel and perpendicular components of the momentum of a test particle are denoted by $p_\parallel$ and $p_\perp$, respectively, and the energy by $E$, we have

$$E = \frac{p_\parallel^2 + p_\perp^2}{2m} , \qquad \frac{{\rm d}p_\perp^2}{{\rm d}t} = 2m\frac{{\rm d}E}{{\rm d}t} - 2p_\parallel \frac{{\rm d}p_\parallel}{{\rm d}t} .$$

We define the *velocity diffusion time* $\tau_\perp$ in the direction perpendicular to the initial momentum and the *energy relaxation time* $\tau^\epsilon$ by

$$\frac{{\rm d}p_\perp^2}{{\rm d}t} \equiv \frac{p_\perp^2}{\tau_\perp} , \qquad \frac{{\rm d}E}{{\rm d}t} \equiv -\frac{E}{\tau^\epsilon} ,$$

respectively. $1/\tau_\perp$ and $1/\tau^\epsilon$ are given by [2.1]

$$\frac{1}{\tau_\perp} = \frac{q^2 q^{*2} n^* \ln\Lambda}{2\pi\epsilon_0^2 v (mv)^2} = \frac{q^2 q^{*2} n^* \ln\Lambda}{2\pi\epsilon_0^2 m^2 v^3} , \tag{2.22}$$

$$\frac{1}{\tau^\epsilon} = \frac{q^2 q^{*2} n^* \ln\Lambda}{4\pi\epsilon_0^2 m^* v (mv^2/2)} = \frac{q^2 q^{*2} n^* \ln\Lambda}{2\pi\epsilon_0^2 m m^* v^3} . \tag{2.23}$$

In the case of electron-to-ion collisions, we find

$$\frac{1}{\tau_{\rm ei\perp}} \simeq \frac{2}{\tau_{\rm ei\parallel}} \tag{2.24}$$

and

$$\frac{1}{\tau_{\rm ei}^{\epsilon}} \simeq \frac{m_{\rm e}}{m_{\rm i}} \frac{2}{\tau_{\rm ei\parallel}} . \tag{2.25}$$

In the case of electron-to-electron collisions and ion-to-ion collisions, we find

$$\frac{1}{\tau_{\rm ee\perp}} \simeq \frac{1}{\tau_{\rm ee\parallel}} = \frac{e^4 n_{\rm e} \ln \Lambda}{3^{1/2} 6\pi\epsilon_0^2 m_{\rm e}^{1/2} (\kappa T_{\rm e})^{3/2}} = \frac{\ln \Lambda}{32.6} \frac{\Pi_{\rm e}}{n_{\rm e}\lambda_{\rm D}^3} , \tag{2.26}$$

$$\frac{1}{\tau_{\rm ee}^{\epsilon}} \simeq \frac{1}{\tau_{\rm ee\parallel}} , \tag{2.27}$$

and

$$\frac{1}{\tau_{\rm ii\perp}} \simeq \frac{1}{\tau_{\rm ii\parallel}} = \frac{Z^4 e^4 n_{\rm i} \ln \Lambda}{3^{1/2} 6\pi\epsilon_0^2 m_{\rm i}^{1/2} (\kappa T_{\rm i})^{3/2}} , \tag{2.28}$$

$$\frac{1}{\tau_{\rm ii}{}^{\epsilon}} \simeq \frac{1}{\tau_{\rm ii\parallel}} , \tag{2.29}$$

respectively.

In the case of ion-to-electron collisions, we have the relations [2.1]

$$\frac{1}{\tau_{\rm ie\perp}} \simeq \frac{Z^2 e^4 n_{\rm e} \ln \Lambda}{(2\pi)^{3/2} \epsilon_0^2 m_{\rm e}^{1/2} E_{\rm i} (\kappa T_{\rm e})^{1/2}} \frac{m_{\rm e}}{m_{\rm i}} , \tag{2.30}$$

$$\frac{1}{\tau_{\rm ie}{}^{\epsilon}} \simeq \frac{Z^2 e^4 n_{\rm e} \ln \Lambda}{4\pi\epsilon_0^2 m_{\rm e}^{1/2} (\kappa T_{\rm e})^{3/2}} \frac{4}{3(2\pi)^{1/2}} \frac{m_{\rm e}}{m_{\rm i}} \simeq \frac{1}{\tau_{\rm ie\parallel}} \simeq \frac{m_{\rm e}}{m_{\rm i}} \frac{2.77}{\tau_{\rm ei\parallel}} , \tag{2.31}$$

where $E_{\rm i} = (3/2)\kappa T_{\rm i}$ is the kinetic energy of the ion.

High-energy neutral particle beams can be injected into plasmas across strong magnetic fields. The neutral particles are converted to high-energy ions by means of charge exchange with plasma ions or ionization. The high-energy ions (mass $m_{\rm b}$, electric charge $Z_{\rm b}e$, energy $E_{\rm b}$) running through the plasma, slow down due to Coulomb collisions with the plasma ions ($m_{\rm i}$, $Z_{\rm i}e$) and electrons ($m_{\rm e}$, $-e$) and the beam energy is thus transferred to the plasma. This method is called heating by *neutral beam injection* (NBI). The rate of change of the energy of the fast ion, that is, the heating rate of the plasma is [2.4]

$$\frac{{\rm d}E_{\rm b}}{{\rm d}t} = -\frac{E_{\rm b}}{\tau_{\rm bi}^{\epsilon}} - \frac{E_{\rm b}}{\tau_{\rm be}^{\epsilon}} , \qquad \frac{1}{\tau_{\rm bi}^{\epsilon}} = \frac{(Z_{\rm b}e)^2 (Z_{\rm i}e)^2 \ln \Lambda n_{\rm i}}{2\pi\epsilon_0^2 m_{\rm i} m_{\rm b} v_{\rm bi}^3} ,$$

and

$$\frac{{\rm d}E_{\rm b}}{{\rm d}t} = -\frac{Z_{\rm b}^2 e^4 \ln \Lambda n_{\rm e}}{4\pi\epsilon_0^2 m_{\rm e} v_{\rm bi}} \left[ \sum_{\rm i} \frac{m_{\rm e}}{m_{\rm i}} \frac{n_{\rm i} Z_{\rm i}^2}{n_{\rm e}} + \frac{4}{3\pi^{1/2}} \left( \frac{m_{\rm e} E_{\rm b}}{m_{\rm b} \kappa T_e} \right)^{3/2} \right] , \tag{2.32}$$

when the beam ion velocity $v_{\rm b}$ is much less than the plasma electron thermal velocity (say by a factor of 1/3) and much larger than the plasma ion thermal velocity (say by a factor of 2). The first term on the right-hand side is due to beam–ion collisions and the second term is due to beam–electron collisions. The critical energy $E_{\rm cr}$ of the beam ion, at which the plasma ions and electrons are heated at equal rates, is given by

$$\frac{mv_{\rm cr}^2}{2} = E_{\rm cr} = 14.8\kappa T_{\rm e} A_{\rm b} \left( \frac{1}{n_{\rm e}} \sum_{\rm i} \frac{n_{\rm i} Z_{\rm i}^2}{A_{\rm i}} \right)^{2/3} , \tag{2.33}$$

where $A_{\rm b}$, $A_{\rm i}$ are the atomic weights of the injected ion and plasma ion, respectively. When the energy of the injected ion is larger than $E_{\rm cr}$, the contribution to the electron heating is dominant. The slowing down time of the ion beam is given by

$$\tau_{\rm slowdown} = \int_{E_{\rm cr}}^{E_{\rm b}} \frac{-{\rm d}E_{\rm b}}{({\rm d}E_{\rm b}/{\rm d}t)} = \frac{\tau_{\rm be}^{\epsilon}}{1.5} \ln \left[ 1 + \left( \frac{E}{E_{\rm cr}} \right)^{3/2} \right] ,$$

$$\frac{1}{\tau_{\rm be}^{\epsilon}} = \frac{Z^2 n_{\rm e} e^4 \ln \Lambda}{(2\pi)^{1/2} 3\pi \epsilon_0^2 m_{\rm e}^{1/2} (\kappa T_{\rm e})^{3/2}} \frac{m_{\rm e}}{m_{\rm b}} , \tag{2.34}$$

where $\tau_{\rm be}^{\epsilon}$ is the energy relaxation time of the beam ion with electrons.

## 2.7 Runaway Electron. Dreicer Field

When a uniform electric field $\boldsymbol{E}$ is applied to a plasma, the motion of a test electron is

$$m_{\rm e} \frac{{\rm d}\boldsymbol{v}}{{\rm d}t} = -e\boldsymbol{E} - \frac{1}{\tau_{\rm ee}(v)} m_{\rm e} \boldsymbol{v} , \qquad \frac{1}{\tau_{\rm ee}} = n_{\rm e} \sigma v = \frac{e^4 \ln \Lambda}{2\pi \epsilon_0^2 m_{\rm e}^2 v^3} .$$

The deceleration term decreases as $v$ increases and its magnitude becomes smaller than the acceleration term $|-e\boldsymbol{E}|$ at a critical value $v_{\rm cr}$. When $v > v_{\rm cr}$, the test particle is accelerated. The deceleration term becomes smaller and the velocity starts to increase without limit. Such an electron is called a *runaway electron*. The critical velocity is given by

$$\frac{m_{\rm e} v_{\rm cr}^2}{2e} = \frac{e^2 n \ln \Lambda}{4\pi \epsilon_0^2 E} . \tag{2.35}$$

The electric field required for a given electron velocity to be $v_{\rm cr}$ is called the *Dreicer field*. Taking $\ln \Lambda = 20$, we find

$$\frac{m_{\rm e} v_{\rm cr}^2}{2e} = 5 \times 10^{-16} \frac{n}{E} ,$$

with MKS units. When $n = 10^{19}\,{\rm m}^{-3}$, $E = 1\,{\rm V/m}$, electrons with energy larger than 5 keV become runaway electrons.

## 2.8 Electric Resistivity. Ohmic Heating

When an electric field weaker than the Dreicer field is applied to a plasma, electrons are accelerated and decelerated by collisions with ions to reach an equilibrium state as follows:

$$\frac{m_{\rm e}(v_{\rm e}-v_{\rm i})}{\tau_{\rm ei}} = -eE\,.$$

The current density $j$ induced by the electric field becomes

$$j = -en_{\rm e}(v_{\rm e}-v_{\rm i}) = \frac{e^2 n_{\rm e}\tau_{\rm ei}}{m_{\rm e}}E\,.$$

The *specific electric resistivity* defined by $\eta j = E$ is [2.3]

$$\eta_{\rm Spitzer} = \frac{m_{\rm e}\nu_{\rm ei\|Spitzer}}{n_{\rm e}e^2} = \frac{(m_{\rm e})^{1/2}Ze^2\ln\Lambda}{9.3\times 10\epsilon_0^2}(\kappa T_{\rm e})^{-3/2}$$
$$= 5.2\times 10^{-5}Z\ln\Lambda\left(\frac{\kappa T_{\rm e}}{e}\right)^{-3/2}\quad(\Omega\,{\rm m})\,. \tag{2.36}$$

The specific resistivity of a plasma with $T_{\rm e} = 1\,\rm keV$ and $Z = 1$ is $\eta = 3.3\times 10^{-8}\,\Omega\,\rm m$ and is slightly larger than the specific resistivity of copper at 20°C, $1.8\times 10^{-8}\,\Omega\,\rm m$. When a current density of $j$ is induced, the power $\eta j^2$ per unit volume contributes to electron heating. This electron heating mechanism is called *ohmic heating*.

## 2.9 Variety of Time and Space Scales in Plasmas

Various kinds of plasma characteristics have been described in this chapter. Characteristic time scales are:

- period of electron plasma frequency $2\pi/\Pi_{\rm e}$,
- electron cyclotron period $2\pi/\Omega_{\rm e}$,
- ion cyclotron period $2\pi/|\Omega_{\rm i}|$,
- electron-to-ion collision time $\tau_{\rm ei}$,
- ion-to-ion collision time $\tau_{\rm ii}$,
- electron–ion thermal energy relaxation time $\tau_{\rm ei}^{\epsilon}$.

The Alfvén velocity $v_{\rm A}$, which is the propagation velocity of a magnetic perturbation, is $v_{\rm A}^2 = B^2/(2\mu_0\rho_{\rm m})$, where $\rho_{\rm m}$ is the mass density (see Chaps. 5 and 10). The Alfvén transit time $\tau_{\rm H} = L/v_{\rm A}$ is a typical magnetohydrodynamic time scale, where $L$ is the plasma size. In a medium with specific resistivity $\eta$, the electric field diffuses with a time scale of $\tau_{\rm R} = \mu_0 L^2/\eta$ (see Chap. 5). This time scale is called the resistive diffusion time.

Characteristic length scales are:

- Debye length $\lambda_{\rm D}$,
- electron Larmor radius $\rho_{\Omega\rm e}$,
- ion Larmor radius $\rho_{\Omega\rm i}$,
- electron–ion collision mean free path $\lambda_{\rm ei}$,
- plasma size $L$.

The relations between space and time scales are:

$$\lambda_{\rm D}\Pi_{\rm e} = v_{T_{\rm e}}\ , \qquad \rho_{\Omega e}\Omega_{\rm e} = v_{T_{\rm e}}\ ,$$

$$\rho_{\Omega\rm i}|\Omega_{\rm i}| = v_{T_{\rm i}}\ , \qquad \frac{\lambda_{\rm ei}}{\tau_{\rm ei}} \simeq 3^{1/2}v_{T_{\rm e}}\ ,$$

$$\frac{\lambda_{\rm ii}}{\tau_{\rm ii}} \simeq 3^{1/2}v_{T_{\rm i}}\ , \qquad \frac{L}{\tau_{\rm H}} = v_{\rm A}\ ,$$

where $v_{T_{\rm e}}$, $v_{T_{\rm i}}$ are the thermal velocities

$$v_{T_{\rm e}}^2 = \frac{\kappa T_{\rm e}}{m_{\rm e}}\ , \qquad v_{T_{\rm i}}^2 = \frac{\kappa T_{\rm i}}{m_{\rm i}}\ .$$

The drift velocity of the guiding center is

$$v_{\rm drift} \sim \frac{\kappa T}{eBL} = v_T(\rho_\Omega/L)\ .$$

Parameters of a typical D fusion grade plasma with $n_{\rm e} = 10^{20}\,{\rm m}^{-3}$, $\kappa T_{\rm e} = \kappa T_{\rm i} = 10\,{\rm keV}$, $B = 5\,{\rm T}$, $L = 1\,{\rm m}$ are as follows:

$$\frac{2\pi}{\Pi_{\rm e}} = 11.1\,{\rm ps}\ , \qquad \frac{\Pi_{\rm e}}{2\pi} = 89.8\,{\rm GHz}\ ,$$

$$\frac{2\pi}{\Omega_{\rm e}} = 7.1\,{\rm ps}\ , \qquad \frac{\Omega_{\rm e}}{2\pi} = 140\,{\rm GHz}\ ,$$

$$\frac{2\pi}{|\Omega_{\rm i}|} = 26\,{\rm ns}\ , \qquad \frac{|\Omega_{\rm i}|}{2\pi} = 38\,{\rm MHz}\ ,$$

$$\tau_{\rm H} = 0.13\,\mu{\rm s}\ , \quad \tau_{\rm R} = 1.2\times 10^3\,{\rm s}\ ,$$

$$\tau_{\perp\rm ei} = 0.12\,{\rm ms}\ , \quad \tau_{\perp\rm ii} = 7.2\,{\rm ms}\ , \quad \tau_{\rm ei}^{\epsilon} = 0.3\,{\rm s}\ ,$$

$$\rho_{\Omega\rm e} = 47.6\,\mu{\rm m}\ , \qquad \rho_{\Omega\rm i} = 2.88\,{\rm mm}\ ,$$

$$\lambda_{\rm D} = 74.5\,\mu{\rm m}\ , \quad \lambda_{\rm ei} = 8.6\,{\rm km}\ , \quad \lambda_{\rm ii} = 8.6\,{\rm km}\ .$$

The ranges of scales in time and space extend to

$$\tau_{\rm R}\Pi_{\rm e} \sim 10^{14}\ , \qquad \frac{\lambda_{\rm ei}}{\lambda_{\rm D}} \sim 1.6\times 10^8\ ,$$

and the wide range of scales suggests the variety and complexity of plasma phenomena. Equations for plasma parameters are listed in Table 2.2.

**Table 2.2.** Equations for plasma parameters (M.K.S. units). $\ln \Lambda = 20$ is assumed. $\kappa T/e$ in eV and $n_{20} \equiv n\,(\mathrm{m}^3)/10^{20}$

$$\Pi_\mathrm{e} = \left(\frac{n_\mathrm{e} e^2}{m_\mathrm{e}\epsilon_0}\right)^{1/2} = 5.64 \times 10^{11} (n_{\mathrm{e}\,20})^{1/2}$$

$$\Omega_\mathrm{e} = \frac{eB}{m_\mathrm{e}} = 1.76 \times 10^{11} B$$

$$|\Omega_\mathrm{i}| = \frac{ZeB}{m_\mathrm{i}} = 9.58 \times 10^7 (Z/A) B$$

$$\nu_{\mathrm{ei}\perp} \equiv \frac{1}{\tau_{\mathrm{ei}\perp}} = \frac{Z^2 n_\mathrm{i} e^4 \ln \Lambda}{3^{1/2} 6\pi\epsilon_0^2 m_\mathrm{e}^{1/2} (\kappa T_\mathrm{e})^{3/2}} = 8.41 \times 10^9 Z^2 \left(\frac{\kappa T_\mathrm{e}}{e}\right)^{-3/2} n_{\mathrm{i}\,20}$$

$$\nu_{\mathrm{ii}\perp} \equiv \frac{1}{\tau_{\mathrm{ii}\perp}} = \frac{Z^4 n_\mathrm{i} e^4 \ln \Lambda}{3^{1/2} 6\pi\epsilon_0^2 m_\mathrm{i}^{1/2} (\kappa T_\mathrm{i})^{3/2}} = 2.0 \times 10^8 \frac{Z^4}{A^{1/2}} \left(\frac{\kappa T_\mathrm{i}}{e}\right)^{-3/2} n_{\mathrm{i}\,20}$$

$$\nu_{\mathrm{ei}\|} \equiv 1/\tau_{\mathrm{ei}\|} = \nu_{\mathrm{ei}\perp}/2, \qquad \nu_{\mathrm{ei}\|\mathrm{spitzer}} \approx 0.7 \nu_{\mathrm{ei}\|}$$

$$\lambda_\mathrm{D} = \left(\frac{\epsilon_0 \kappa T}{n_\mathrm{e} e^2}\right)^{1/2} = 7.45 \times 10^{-7} \left(\frac{\kappa T_\mathrm{e}}{e}\right)^{1/2} n_{\mathrm{e}\,20}^{-1/2}$$

$$\rho_{\Omega_\mathrm{e}} = \frac{v_{T_\mathrm{e}}}{\Omega_\mathrm{e}} = 2.38 \times 10^{-6} \left(\frac{\kappa T_\mathrm{e}}{e}\right)^{1/2} \frac{1}{B}$$

$$\rho_{\Omega_\mathrm{i}} = \frac{v_{T_\mathrm{i}}}{\Omega_\mathrm{i}} = 1.02 \times 10^{-4} \left(\frac{A \kappa T_\mathrm{i}}{e}\right)^{1/2} \frac{1}{ZB}$$

$$\lambda_\mathrm{ei} = \left(\frac{3\kappa T_\mathrm{e}}{m_\mathrm{e}}\right)^{1/2} \tau_{\mathrm{ei}\|} = 1.73 \times 10^{-4} \left(\frac{\kappa T_\mathrm{e}}{e}\right)^2 (n_{\mathrm{e}\,20})^{-1}$$

$$v_\mathrm{A} = \left(\frac{B^2}{\mu_0 n_\mathrm{i} m_\mathrm{i}}\right)^{1/2} = 2.18 \times 10^6 \frac{B}{(A n_{\mathrm{i}\,20})^{1/2}}$$

$$v_{T_\mathrm{e}} = \left(\frac{\kappa T_\mathrm{e}}{m_\mathrm{e}}\right)^{1/2} = 4.19 \times 10^5 \left(\frac{\kappa T_\mathrm{e}}{e}\right)^{1/2}$$

$$v_{T_\mathrm{i}} = \left(\frac{\kappa T_\mathrm{i}}{m_\mathrm{i}}\right)^{1/2} = 9.79 \times 10^3 \left(\frac{\kappa T_\mathrm{i}}{Ae}\right)^{1/2}$$

$$\eta_\mathrm{Spitzer} = \frac{m_\mathrm{e} \nu_{\mathrm{ei}\|\,\mathrm{Spitzer}}}{n_\mathrm{e} e^2} = 5.2 \times 10^{-5} Z \ln \Lambda \left(\frac{\kappa T_\mathrm{e}}{e}\right)^{-3/2} \quad (\Omega\mathrm{m})$$

$$\left(\frac{v_{T_\mathrm{i}}}{v_\mathrm{A}}\right)^2 = \frac{\beta_\mathrm{i}}{2}, \qquad \left(\frac{v_\mathrm{A}}{c}\right)^2 = \left(\frac{\lambda_\mathrm{D}}{\rho_{\Omega\mathrm{e}}}\right)^2 \frac{m_\mathrm{e} n_\mathrm{e}}{m_\mathrm{i} n_\mathrm{i}}$$

$$\frac{\Pi_\mathrm{e}}{\nu_{\mathrm{ei}\perp}} = \frac{32.6}{Z \ln \Lambda} n_\mathrm{e} \lambda_\mathrm{D}^3$$

# 3 Magnetic Configuration and Particle Orbit

In this chapter, the motion of individual charged particles in more general magnetic fields is studied in detail. There are a large number of charged particles in a plasma, so movements do affect the magnetic field. But this effect is neglected here.

## 3.1 Maxwell Equations

Let $\boldsymbol{E}$, $\boldsymbol{B}$, $\boldsymbol{D}$, and $\boldsymbol{H}$ be the *electric intensity*, *magnetic induction*, *electric displacement* and *magnetic intensity*, respectively. When the *charge density* and *current density* are denoted by $\rho$ and $\boldsymbol{j}$, respectively, Maxwell's equations are

$$\nabla \times \boldsymbol{E} + \frac{\partial \boldsymbol{B}}{\partial t} = 0\,, \tag{3.1}$$

$$\nabla \times \boldsymbol{H} - \frac{\partial \boldsymbol{D}}{\partial t} = \boldsymbol{j}\,, \tag{3.2}$$

$$\nabla \cdot \boldsymbol{B} = 0\,, \tag{3.3}$$

$$\nabla \cdot \boldsymbol{D} = \rho\,. \tag{3.4}$$

$\rho$ and $\boldsymbol{j}$ satisfy the relation

$$\nabla \cdot \boldsymbol{j} + \frac{\partial \rho}{\partial t} = 0\,. \tag{3.5}$$

Equations (3.2), (3.4) and (3.5) are consistent with each other due to the Maxwell displacement current $\partial \boldsymbol{D}/\partial t$. From (3.3), the vector $\boldsymbol{B}$ can be expressed as the rotation of a vector $\boldsymbol{A}$:

$$\boldsymbol{B} = \nabla \times \boldsymbol{A}\,. \tag{3.6}$$

$\boldsymbol{A}$ is called the *vector potential*. If (3.6) is substituted into (3.1), we obtain

$$\nabla \times \left( \boldsymbol{E} + \frac{\partial \boldsymbol{A}}{\partial t} \right) = 0\,. \tag{3.7}$$

The quantity in brackets can be expressed in terms of a *scalar potential* $\phi$ and

$$\boldsymbol{E} = -\nabla\phi - \frac{\partial \boldsymbol{A}}{\partial t} . \tag{3.8}$$

Since any other set of $\phi'$ and $\boldsymbol{A}'$,

$$\boldsymbol{A}' = \boldsymbol{A} - \nabla\psi , \tag{3.9}$$

$$\phi' = \phi + \frac{\partial \psi}{\partial t} , \tag{3.10}$$

can also satisfy (3.6) and (3.8) for arbitrary $\psi$, $\phi'$ and $\boldsymbol{A}'$ are not uniquely determined.

When the medium is uniform and isotropic, $\boldsymbol{B}$ and $\boldsymbol{D}$ are expressed by

$$\boldsymbol{D} = \epsilon \boldsymbol{E} , \qquad \boldsymbol{B} = \mu \boldsymbol{H} ,$$

where $\epsilon$ and $\mu$ are called the *dielectric constant* and *permeability*, respectively. The values of $\epsilon_0$ and $\mu_0$ in vacuum are

$$\epsilon_0 = \frac{10^7}{4\pi c^2}\ \mathrm{C^2 s^2/kg\, m^3} = 8.854\ldots \times 10^{-12}\ \mathrm{F/m} ,$$

$$\mu_0 = 4\pi \times 10^{-7}\ \mathrm{kg\, m/C^2} = 1.257\ldots \times 10^{-6}\ \mathrm{H/m} ,$$

$$\frac{1}{\epsilon_0 \mu_0} = c^2 , \qquad c = 2.997\,924\,58\ (\mathrm{m/s}) \quad [\text{definition}] ,$$

where $c$ is the light speed in vacuum and C is the coulomb. Plasmas in magnetic fields are anisotropic and $\epsilon$ and $\mu$ generally have tensor form. In vacuum, (3.2) and (3.3) may be reduced to

$$\nabla \times \nabla \times \boldsymbol{A} + \frac{1}{c^2}\nabla\frac{\partial \phi}{\partial t} + \frac{1}{c^2}\frac{\partial^2 \boldsymbol{A}}{\partial t^2} = \mu_0 \boldsymbol{j} , \tag{3.11}$$

$$\nabla^2\phi + \nabla\cdot\frac{\partial \boldsymbol{A}}{\partial t} = -\frac{1}{\epsilon_0}\rho . \tag{3.12}$$

As $\phi$ and $\boldsymbol{A}$ are arbitrary up to $\psi$, as shown in (3.9) and (3.10), we impose the supplementary condition (Lorentz condition)

$$\nabla\cdot\boldsymbol{A} + \frac{1}{c}\frac{\partial \phi}{\partial t} = 0 . \tag{3.13}$$

Then (3.11) and (3.12) reduce to the wave equations

$$\nabla^2\phi - \frac{1}{c^2}\frac{\partial^2 \phi}{\partial t^2} = -\frac{1}{\varepsilon_0}\rho , \tag{3.14}$$

$$\nabla^2\boldsymbol{A} - \frac{1}{c^2}\frac{\partial^2 \boldsymbol{A}}{\partial t^2} = -\mu_0 \boldsymbol{j} . \tag{3.15}$$

In the derivation of (3.15), the vector relation

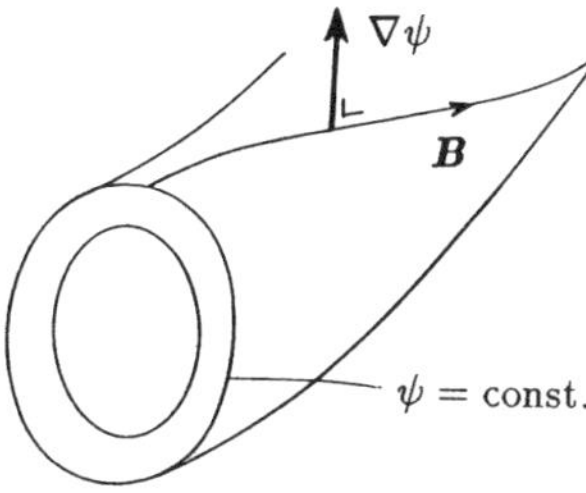


**Fig. 3.1.** Magnetic surface $\psi$ = const., normal $\nabla\psi$ and line of magnetic force

$$\nabla \times (\nabla \times \boldsymbol{a}) - \nabla(\nabla\cdot\boldsymbol{a}) = -\nabla^2 \boldsymbol{a}$$

is used, which is valid only in $(x, y, z)$ coordinates. The speed of propagation of the electromagnetic field is $1/(\mu_0\epsilon_0)^{1/2} = c$ in vacuum.

When the fields do not change in time, the field equations reduce to

$$\boldsymbol{E} = -\nabla\phi\,, \qquad \boldsymbol{B} = \nabla \times \boldsymbol{A}\,,$$

$$\nabla^2\phi = -\frac{1}{\epsilon_0}\rho\,, \qquad \nabla^2\boldsymbol{A} = -\mu\boldsymbol{j}\,, \qquad \nabla\cdot\boldsymbol{A} = 0\,, \qquad \nabla\cdot\boldsymbol{j} = 0\,.$$

## 3.2 Magnetic Surface

A *line of magnetic force* satisfies the equations

$$\frac{\mathrm{d}x}{B_x} = \frac{\mathrm{d}y}{B_y} = \frac{\mathrm{d}z}{B_z} = \frac{\mathrm{d}l}{B}\,, \tag{3.16}$$

where $l$ is the length along the line of force,

$$(\mathrm{d}l)^2 = (\mathrm{d}x)^2 + (\mathrm{d}y)^2 + (\mathrm{d}z)^2\,.$$

The *magnetic surface* $\psi(\boldsymbol{r})$ = const. is such that all magnetic lines of force lie upon on that surface. It satisfies the condition

$$\big[\nabla\psi(\boldsymbol{r})\big]\cdot\boldsymbol{B} = 0\,. \tag{3.17}$$

The vector $\nabla\psi(\boldsymbol{r})$ is normal to the magnetic surface and must be orthogonal to $\boldsymbol{B}$ (see Fig. 3.1).

In cylindrical coordinates $(r, \theta, z)$, the magnetic field $\boldsymbol{B}$ is given by

$$B_r = \frac{1}{r}\frac{\partial A_z}{\partial\theta} - \frac{\partial A_\theta}{\partial z}\,, \qquad B_\theta = \frac{\partial A_r}{\partial z} - \frac{\partial A_z}{\partial r}\,, \qquad B_z = \frac{1}{r}\frac{\partial}{\partial r}(rA_\theta) - \frac{1}{r}\frac{\partial A_r}{\partial\theta}\,. \tag{3.18}$$

For an *axisymmetric configuration* ($\partial/\partial\theta = 0$),

$$\psi(r, z) = rA_\theta(r, z) \tag{3.19}$$

satisfies the condition (3.17) of the magnetic surface:

$$B_r \frac{\partial(rA_\theta)}{\partial r} + B_\theta \cdot 0 + B_z \frac{\partial(rA_\theta)}{\partial z} = 0 \, .$$

In the case of translational symmetry ($\partial/\partial z = 0$), the magnetic surface is given by

$$\psi(r, \theta) = A_z(r, \theta) \, , \tag{3.20}$$

and in the case of helical symmetry, in which $\psi$ is a function of $r$ and $\theta - \alpha z$ alone, the magnetic surface is given by

$$\psi(r, \theta - \alpha z) = A_z(r, \theta - \alpha z) + \alpha r A_\theta(r, \theta - \alpha z) \, , \tag{3.21}$$

where $\alpha$ is the helical pitch parameter.

## 3.3 Equation of Motion of a Charged Particle

The equation of motion of a particle with mass $m$ and charge $q$ in an electromagnetic field $\boldsymbol{E}$, $\boldsymbol{B}$ is

$$m \frac{\mathrm{d}^2 \boldsymbol{r}}{\mathrm{d}t^2} = \boldsymbol{F} = q \left( \boldsymbol{E} + \frac{\mathrm{d}\boldsymbol{r}}{\mathrm{d}t} \times \boldsymbol{B} \right) \, . \tag{3.22}$$

Since the Lorentz force of the second term on the right-hand side of (3.22) is orthogonal to the velocity $\boldsymbol{v}$, the scalar product of the Lorentz force and $\boldsymbol{v}$ is zero. The kinetic energy is given by

$$\frac{mv^2}{2} - \frac{m{v_0}^2}{2} = q \int_{t=t_0}^{t} \boldsymbol{E} \cdot \boldsymbol{v} \mathrm{d}t \, .$$

When the electric field is zero, the kinetic energy of the charged particle is conserved. When generalized coordinates $q_i$ ($i = 1, 2, 3$) are used, the Lagrangian formulation must be used. The Lagrangian of a charged particle in the field with scalar and vector potentials $\phi$, $\boldsymbol{A}$ is given by

$$L(q_i, \dot{q}_i, t) = \frac{mv^2}{2} + q\boldsymbol{v} \cdot \boldsymbol{A} - q\phi \, , \tag{3.23}$$

where $\dot{q}_i$ is the time derivative of $q_i$. Lagrangians in orthogonal and cylindrical coordinates are given by

$$L(x, y, z, \dot{x}, \dot{y}, \dot{z}, t) = \frac{m}{2}(\dot{x}^2 + \dot{y}^2 + \dot{z}^2) + q(\dot{x}A_x + \dot{y}A_y + \dot{z}A_z) - q\phi \, ,$$

$$L(r,\theta,z,\dot{r},\dot{\theta},\dot{z},t) = \frac{m}{2}\left[\dot{r}^2 + (r\dot{\theta})^2 + \dot{z}^2\right] + q(\dot{r}A_r + r\dot{\theta}A_\theta + \dot{z}A_z) - q\phi\,,$$

respectively. The Lagrange equation of motion is

$$\frac{\mathrm{d}}{\mathrm{d}t}\left(\frac{\partial L}{\partial \dot{q}_i}\right) - \frac{\partial L}{\partial q_i} = 0\,. \tag{3.24}$$

Canonical transformations are more general than coordinate transformations. The Hamiltonian equation of motion is conserved under canonical transformations. In this formulation, we introduce momentum coordinates $(p_i)$, in addition to the space coordinates $(q_i)$, defined by

$$p_i \equiv \frac{\partial L}{\partial \dot{q}_i}\,, \tag{3.25}$$

and treat $p_i$ as independent variables. Then we can express $\dot{q}_i$ as a function of $(q_j, p_j, t)$ from (3.25) as follows:

$$\dot{q}_i = \dot{q}_i(q_j, p_j, t)\,. \tag{3.26}$$

The Hamiltonian $H(q_i, p_i, t)$ is given by

$$H(q_i, p_i, t) \equiv -L\big(q_i, \dot{q}_i(q_j, p_j, t), t\big) + \sum_i p_i \dot{q}_i(q_j, p_j, t)\,. \tag{3.27}$$

The $x$ component of momentum $p_x$ in orthogonal coordinates and $\theta$ component $p_\theta$ in cylindrical coordinates are given as examples:

$$p_x = m\dot{x} + qA_x\,, \qquad \dot{x} = \frac{p_x - qA_x}{m}\,,$$

$$p_\theta = mr^2\dot{\theta} + qrA_\theta\,, \qquad \dot{\theta} = \frac{p_\theta - qrA_\theta}{mr^2}\,.$$

In orthogonal coordinates, the Hamiltonian is

$$H = \frac{1}{2m}\left[(p_x - qA_x)^2 + (p_y - qA_y)^2 + (p_z - qA_z)^2\right] + q\phi(x,y,z,t)\,,$$

and in cylindrical coordinates, the Hamiltonian is

$$H = \frac{1}{2m}\left[(p_r - qA_r)^2 + \frac{(p_\theta - qrA_\theta)^2}{r^2} + (p_z - qA_z)^2\right] + q\phi(r,\theta,z,t)\,.$$

The variation of the Lagrangian $L$ is given by

$$\begin{aligned}\delta L = \sum_i \left(\frac{\partial L}{\partial q_i}\delta q_i + \frac{\partial L}{\partial \dot{q}_i}\delta \dot{q}_i\right) &= \sum_i (\dot{p}_i \delta q_i + p_i \delta \dot{q}_i) \\ &= \delta\left(\sum_i p_i \dot{q}_i\right) + \sum_i (\dot{p}_i \delta q_i - \dot{q}_i \delta p_i)\,,\end{aligned}$$

and

$$\delta\left(-L+\sum_i p_i\dot{q}_i\right)=\sum_i(\dot{q}_i\delta p_i-\dot{p}_i\delta q_i)\,,\quad \delta H(q_i,p_i,t)=\sum_i(\dot{q}_i\delta p_i-\dot{p}_i\delta q_i)\,.$$

Accordingly, the Hamiltonian equation of motion reduces to

$$\frac{\mathrm{d}q_i}{\mathrm{d}t}=\frac{\partial H}{\partial p_i}\,,\qquad \frac{\mathrm{d}p_i}{\mathrm{d}t}=-\frac{\partial H}{\partial q_i}\,. \tag{3.28}$$

In orthogonal coordinates, (3.28) is

$$\frac{\mathrm{d}x}{\mathrm{d}t}=\frac{p_x-qA_x}{m}\,,\qquad \frac{\mathrm{d}p_x}{\mathrm{d}t}=\frac{q}{m}\frac{\partial \boldsymbol{A}}{\partial x}\cdot(\boldsymbol{p}-q\boldsymbol{A})-q\frac{\partial\phi}{\partial x}\,,$$

$$\begin{aligned} m\frac{\mathrm{d}^2x}{\mathrm{d}t^2}&=\frac{\mathrm{d}p_x}{\mathrm{d}t}-q\frac{\mathrm{d}A_x}{\mathrm{d}t}\\ &=q\left\{\left(\boldsymbol{v}\cdot\frac{\partial\boldsymbol{A}}{\partial x}\right)-\frac{\partial\phi}{\partial x}-\left[\frac{\partial A_x}{\partial t}+(\boldsymbol{v}\cdot\boldsymbol{\nabla})A_x\right]\right\}\\ &=q(\boldsymbol{E}+\boldsymbol{v}\times\boldsymbol{B})_x\,, \end{aligned}$$

and it follows that (3.28) is equivalent to (3.22).

When $H$ does not depend explicitly on $t$, i.e., when $\phi$, $\boldsymbol{A}$ do not depend on $t$,

$$\frac{\mathrm{d}H(q_i,p_i)}{\mathrm{d}t}=\sum_i\left(\frac{\partial H}{\partial q_i}\frac{\mathrm{d}q_i}{\mathrm{d}t}+\frac{\partial H}{\partial p_i}\frac{\mathrm{d}p_i}{\mathrm{d}t}\right)=0\,,$$

and

$$H(q_i,p_i)=\text{const.} \tag{3.29}$$

is one integral of the Hamiltonian equations. This integral expresses the conservation of energy.

When the electromagnetic field is axially symmetric, $p_\theta$ is constant because $\partial H/\partial\theta=0$, as can be seen from (3.28), and

$$p_\theta=mr^2\dot{\theta}+qrA_\theta=\text{const.} \tag{3.30}$$

This indicates conservation of angular momentum. In the case of translational symmetry ($\partial/\partial z=0$), we have

$$p_z=m\dot{z}+qA_z=\text{const.} \tag{3.31}$$

## 3.4 Particle Orbit in Axially Symmetric System

The coordinates $(r^*,\theta^*,z^*)$ on a magnetic surface of an axially symmetric field satisfy

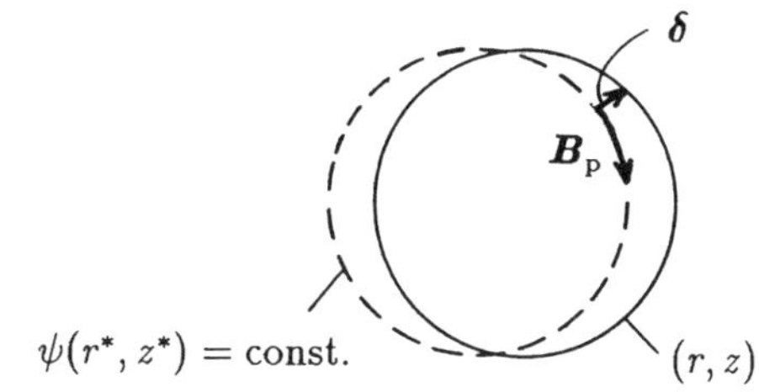


**Fig. 3.2.** Magnetic surface (*dotted line*) and particle orbit (*solid line*)

$$\psi = r^* A_\theta(r^*, z^*) = c_{\rm M} \qquad (c_{\rm M} \text{ constant}) \, .$$

On the other hand the coordinates $(r, \theta, z)$ of a particle orbit are given by the conservation of angular momentum (3.30) as follows:

$$rA_\theta(r, z) + \frac{m}{q} r^2 \dot{\theta} = \frac{p_\theta}{q} = \text{const.}$$

If $c_{\rm M}$ is chosen to be $c_{\rm M} = p_\theta/q$, the relation between the magnetic surface and the particle orbit reduces to

$$rA_\theta(r, z) - r^* A_\theta(r^*, z^*) = -\frac{m}{q} r^2 \dot{\theta} \, .$$

The distance $\boldsymbol{\delta}$ (Fig. 3.2) between the magnetic surface and the orbit is given by

$$\boldsymbol{\delta} = (r - r^*)\boldsymbol{e}_r + (z - z^*)\boldsymbol{e}_z \, , \qquad \boldsymbol{\delta} \cdot \nabla(rA_\theta) = -\frac{m}{q} r^2 \dot{\theta} \, ,$$

where $\boldsymbol{e}_r$ and $\boldsymbol{e}_z$ are unit vectors in the directions of $r$ and $z$, respectively. Since $rB_r = -\partial(rA_\theta)/\partial z$, $rB_z = \partial(rA_\theta)/\partial r$, this reduces to

$$-(z - z^*)B_r + (r - r^*)B_z = -\frac{m}{q} r \dot{\theta} \, .$$

The left-hand side of the above equation is the $\theta$ component of the vector product of $\boldsymbol{B}_{\rm p} = (B_r, B_z)$ and $\boldsymbol{\delta} = (r - r^*, z - z*)$, and

$$(\boldsymbol{B}_{\rm p} \times \boldsymbol{\delta})_\theta = -\frac{m}{q} r \dot{\theta} \, .$$

Denoting the magnitude of the poloidal component $\boldsymbol{B}_{\rm p}$ [the component in the $(rz)$ plane] of $\boldsymbol{B}$ by $B_{\rm p}$, we find the relation $-B_{\rm p}\delta = -(m/q)v_\theta$ $(v_\theta = r\dot{\theta})$ and we have

$$\delta = \frac{mv_\theta}{qB_{\rm p}} = \rho_{\Omega{\rm p}} \, .$$

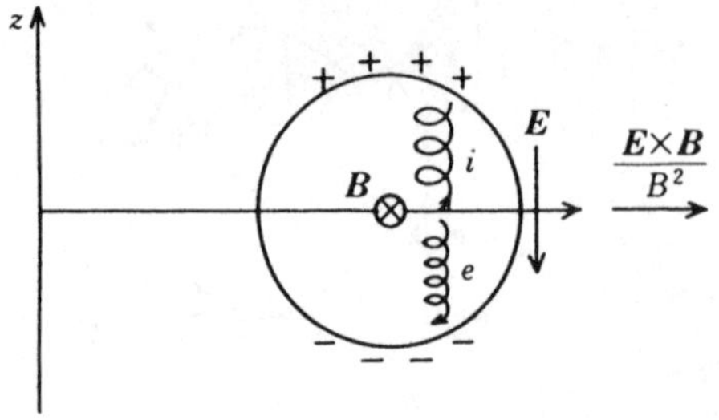


**Fig. 3.3.** Toroidal drift

This value is equal to the Larmor radius corresponding to the magnetic field $B_{\rm p}$ and the tangential velocity $v_\theta$. If $c_{\rm M}$ is chosen to be $c_{\rm M} = (p_\theta - m\langle rv_\theta\rangle)/q$, where $\langle rv_\theta\rangle$ is the average of $rv_\theta$, we find

$$\delta = \frac{m}{qB_{\rm p}}\left(v_\theta - \frac{\langle rv_\theta\rangle}{r}\right) . \tag{3.32}$$

## 3.5 Drift of Guiding Center in Toroidal Field

Let us consider the drift of the guiding center of a charged particle in a simple toroidal field ($B_r = 0$, $B_\varphi = B_0R_0/R$, $B_z = 0$), using cylindrical coordinates $(R, \varphi, z)$. The $\varphi$ component $B_\varphi$ is called the toroidal field and $B_\varphi$ decreases as $1/R$ as we move outward. The lines of magnetic force are circles around the $z$ axis, which is called the *major axis* of the torus. As described in Sect. 2.4, the drift velocity of the guiding center is given by

$$\boldsymbol{v}_{\rm G} = v_\parallel \boldsymbol{e}_\varphi + \frac{m}{qB_\varphi R}\left(v_\parallel^2 + \frac{v_\perp^2}{2}\right)\boldsymbol{e}_z ,$$

where $\boldsymbol{e}_\varphi$ is the unit vector in the $\varphi$ direction (see Fig. 3.4). Particles in this simple torus move fast in the toroidal direction and drift slowly in the $z$ direction with velocity

$$v_{\rm dr} = \frac{m}{qB_0R_0}\left(v_\parallel^2 + \frac{v_\perp^2}{2}\right) \sim \left(\frac{\rho\Omega}{R_0}\right) v . \tag{3.33}$$

This drift is called *toroidal drift*. Ions and electrons drift in opposite directions along the $z$ axis.

As a consequence of the resulting charge separation, an electric field $\boldsymbol{E}$ is induced, and both ions and electrons drift outward by $\boldsymbol{E} \times \boldsymbol{B}/B^2$ drift. Consequently, a simple toroidal field cannot confine a plasma (Fig. 3.3), unless the separated charges are cancelled or short-circuited by an appropriate

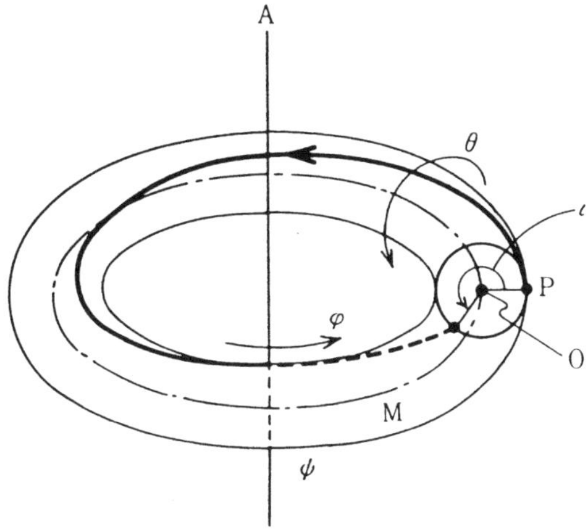


**Fig. 3.4.** Major axis A and minor axis M of a toroidal field, showing the rotational transform angle $\iota$

method. If lines of magnetic force connect the upper and lower regions as shown in Fig. 3.4, the separated charges can be short-circuited, since the charged particles can move freely along the lines of force.

If a current is induced in a toroidal plasma, a component of the magnetic field around the *magnetic axis* (also called the *minor axis*) is introduced as shown in Fig. 3.4. This component $\boldsymbol{B}_{\rm p}$ is called the *poloidal magnetic field*. The radius $R$ of the magnetic axis is called the *major radius* of the torus and the radius $a$ of the plasma cross-section is called the *minor radius*. Let $r$ be the radial coordinate in the plasma cross-section. When a line of magnetic force circles the major axis of the torus and comes back to cross the plane P, the crossing point rotates in P through an angle $\iota$ around the minor axis O. We have the relation

$$\frac{r\iota}{2\pi R} = \frac{B_{\rm p}}{B_\varphi} \,.$$

The angle $\iota$ is called the *rotational transform angle* and is given by

$$\frac{\iota}{2\pi} = \frac{R}{r}\frac{B_{\rm p}}{B_\varphi} \,. \tag{3.34}$$

$A \equiv R/a$ is called the *aspect ratio*.

### 3.5.1 Guiding Center of Circulating Particles

When a particle circulates around the torus with velocity $v_\parallel$, it takes $T \approx 2\pi R_0/v_\parallel$. Accordingly the particle rotates around the minor axis with angular velocity

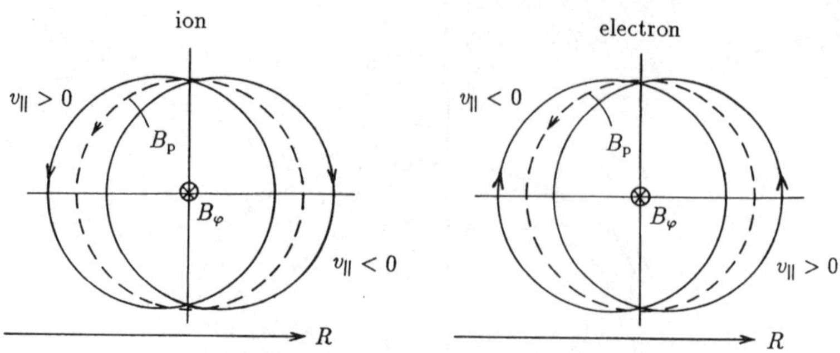


**Fig. 3.5.** Orbits (*solid lines*) of guiding center of circulating ions and electrons, showing magnetic surfaces (*dotted lines*)

$$\omega = \frac{\iota}{T} = \frac{\iota v_\parallel}{2\pi R_0} ,$$

and drifts in the $z$ direction with velocity $v_{\mathrm{dr}}$. Introducing the coordinate $x = R - R_0$, the orbit of the guiding center of the particle is given by

$$\frac{\mathrm{d}x}{\mathrm{d}t} = -\omega z , \qquad \frac{\mathrm{d}z}{\mathrm{d}t} = \omega x + v_{\mathrm{dr}} .$$

The solution is

$$\left(x + \frac{v_{\mathrm{dr}}}{\omega}\right)^2 + z^2 = r^2 .$$

If a rotational transform angle is introduced, the orbit becomes a closed circle and the center of this circle deviates from the center of the magnetic surface by the amount

$$\Delta = -\frac{v_{\mathrm{dr}}}{\omega} = -\frac{mv_\parallel}{qB_0}\frac{2\pi}{\iota}\left(1 + \frac{v_\perp^2}{2v_\parallel^2}\right) , \qquad |\Delta| \sim \rho_\Omega \left(\frac{2\pi}{\iota}\right) , \tag{3.35}$$

where $\rho_\Omega$ is the Larmor radius. As can be seen in Fig. 3.5, the sign of the deviation is $\Delta < 0$ for the case $v_\parallel > 0$, $q > 0$ (ion) since $v_{\mathrm{dr}} > 0$, $\omega > 0$ and the sign becomes $\Delta > 0$ for the case $v_\parallel < 0$ (opposite to $v_\parallel > 0$) $q > 0$ (ion).

### 3.5.2 Guiding Center of Banana Particles

When $|B_\varphi| \gg |B_{\mathrm{p}}|$, the magnitude of the toroidal field is nearly equal to $B_\varphi$ and

$$B = \frac{B_0 R_0}{R} = \frac{B_0}{1 + (r/R)\cos\theta} \simeq B_0 \left(1 - \frac{r}{R_0}\cos\theta\right) .$$

Let $l$ denote the length along the line of magnetic force and use coordinates $(r, \theta)$ for the projection of a location on the magnetic line of force onto the $(R, z)$ plane, as shown in Fig. 3.6. Since

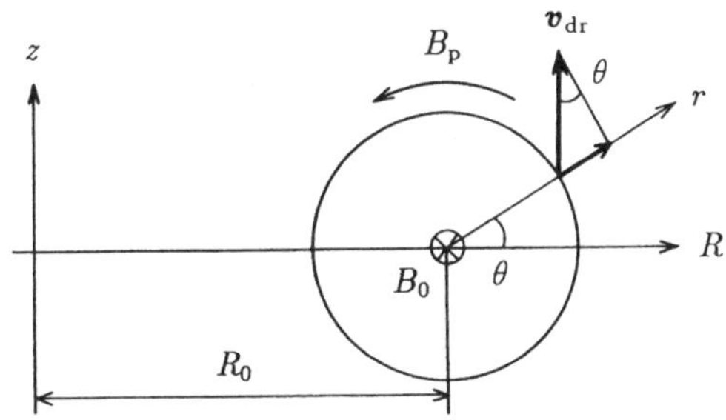


**Fig. 3.6.** $(r, \theta)$ coordinates

$$\frac{r\theta}{l} = \frac{B_\mathrm{p}}{B_0} , \qquad \theta = \frac{l}{r}\frac{B_\mathrm{p}}{B_0} = \kappa l ,$$

we find that

$$B = B_0 \left[1 - \frac{r}{R_0}\cos(\kappa l)\right] .$$

If $v_\parallel$ (component parallel to magnetic field) is much smaller than the component $v_\perp$ and satisfies the condition

$$\frac{v_\perp^2}{v^2} > 1 - \frac{r}{R} , \qquad \frac{v_\parallel^2}{v^2} < \frac{r}{R} , \tag{3.36}$$

the particle is trapped outside in the region of weak magnetic field due to the mirror effect, as described in Sect. 2.5. [The mirror ratio is $(1/R)/\big(1/(R+r)\big)$.] This particle is called a *trapped particle*. Circulating particles that are not trapped are also called untrapped particles. Since $v_\parallel^2 \ll v_\perp^2$ for the trapped particle, the $r$ component of the toroidal drift $v_\mathrm{dr}$ of the trapped particle is given by

$$\frac{\mathrm{d}r}{\mathrm{d}t} = v_\mathrm{dr}\sin\theta = \frac{m}{qB_0}\frac{v_\perp^2}{2R}\sin\theta .$$

The parallel motion of the guiding center is given by (see Sect. 2.4)

$$\frac{\mathrm{d}v_\parallel}{\mathrm{d}t} = -\frac{\mu_\mathrm{m}}{m}\frac{\partial B}{\partial l} = -\frac{\mu_\mathrm{m}}{m}\frac{r}{R}\kappa B_0 \sin\kappa l = -\frac{v_\perp^2}{2R}\frac{B_\mathrm{p}}{B_0}\sin\theta .$$

The solution is

$$\frac{\mathrm{d}}{\mathrm{d}t}\left(r + \frac{m}{qB_\mathrm{p}}v_\parallel\right) = 0 , \qquad r - r_0 = -\frac{m}{qB_\mathrm{p}}v_\parallel . \tag{3.37}$$

Here $r = r_0$ indicates the radial coordinate of the turning point by the mirror effect. Since the orbit has banana shape, the trapped particle is also called a

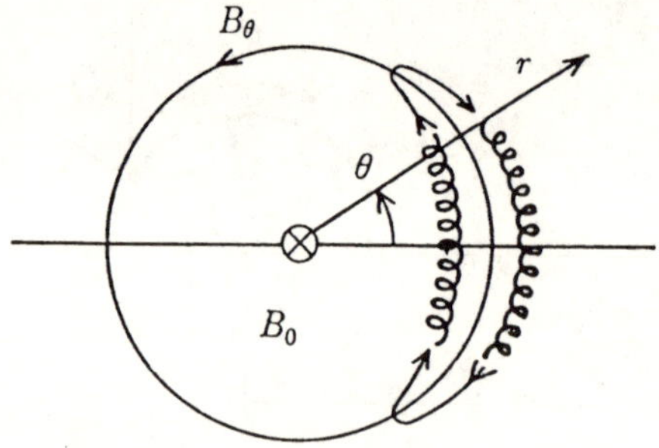


**Fig. 3.7.** Banana orbit of ion

*banana particle* (see Fig. 3.7). The width $\Delta_{\mathrm{b}}$ of the banana is given by

$$\Delta_{\mathrm{b}} = \frac{m}{qB_{\mathrm{p}}}v_\parallel \sim \frac{mv}{qB_0}\frac{v_\parallel}{v}\frac{B_0}{B_{\mathrm{p}}} \sim \frac{B_0}{B_{\mathrm{p}}}\left(\frac{r}{R}\right)^{1/2}\rho_\Omega \sim \left(\frac{R}{r}\right)^{1/2}\left(\frac{2\pi}{\iota}\right)\rho_\Omega \,. \quad (3.38)$$

## 3.6 Orbit of Guiding Center and Magnetic Surface

The velocity of the guiding center was derived in Sect. 2.4 as

$$\boldsymbol{v}_{\mathrm{G}} = v_\parallel \boldsymbol{b} + \frac{1}{B}(\boldsymbol{E}\times\boldsymbol{b}) + \frac{mv_\perp^2/2}{qB^2}(\boldsymbol{b}\times\nabla B) + \frac{mv_\parallel^2}{qB^2}\left[\boldsymbol{b}\times(\boldsymbol{b}\cdot\nabla)\boldsymbol{B}\right] , \quad (3.39)$$

$$\mu_{\mathrm{m}} = \frac{mv_\perp^2}{2B} = \mathrm{const.}$$

When the electric field $\boldsymbol{E}$ is static and expressed by $\boldsymbol{E} = -\nabla\phi$, we have conservation of energy, i.e.,

$$\frac{m}{2}(v_\parallel^2 + v_\perp^2) + q\phi = W \,.$$

Then $v_\parallel$ is expressed by

$$v_\parallel = \pm\left(\frac{2}{m}\right)^{1/2}(W - q\phi - \mu_{\mathrm{m}}B)^{1/2} \,.$$

Noting that $v_\parallel$ is a function of the coordinates, we can write

$$\begin{aligned}\nabla\times(mv_\parallel\boldsymbol{b}) &= mv_\parallel\nabla\times\boldsymbol{b} + \nabla(mv_\parallel)\times\boldsymbol{b} \\ &= mv_\parallel\nabla\times\boldsymbol{b} + \frac{1}{v_\parallel}(-q\nabla\phi - \mu_{\mathrm{m}}\nabla B)\times\boldsymbol{b}\end{aligned}$$

and

$$\frac{v_\parallel}{qB}\boldsymbol{\nabla}\times(mv_\parallel\boldsymbol{b}) = \frac{mv_\parallel^2}{qB}\boldsymbol{\nabla}\times\boldsymbol{b} + \frac{1}{B}(\boldsymbol{E}\times\boldsymbol{b}) + \frac{mv_\perp^2/2}{qB^2}(\boldsymbol{b}\times\boldsymbol{\nabla}B)\,.$$

Then eq.(3.39) for $\boldsymbol{v}_{\mathrm{G}}$ reduces to [3.1]

$$\begin{aligned}\boldsymbol{v}_{\mathrm{G}} &= v_\parallel\boldsymbol{b} + \left[\frac{v_\parallel}{qB}\boldsymbol{\nabla}\times(mv_\parallel\boldsymbol{b}) - \frac{mv_\parallel^2}{qB}\boldsymbol{\nabla}\times\boldsymbol{b}\right] + \frac{mv_\parallel^2}{qB^2}\left[\boldsymbol{b}\times(\boldsymbol{b}\cdot\boldsymbol{\nabla})\boldsymbol{B}\right] \\ &= v_\parallel\boldsymbol{b} + \frac{v_\parallel}{qB}\boldsymbol{\nabla}\times(mv_\parallel\boldsymbol{b}) - \frac{mv_\parallel^2}{qB}\left[\boldsymbol{\nabla}\times\boldsymbol{b} - \boldsymbol{b}\times(\boldsymbol{b}\cdot\boldsymbol{\nabla})\boldsymbol{b}\right]\,.\end{aligned}$$

Since we have

$$\boldsymbol{\nabla}(\boldsymbol{b}\cdot\boldsymbol{b}) = 2(\boldsymbol{b}\cdot\boldsymbol{\nabla})\boldsymbol{b} + 2\boldsymbol{b}\times(\boldsymbol{\nabla}\times\boldsymbol{b}) = 0\,,$$

because $\boldsymbol{b}\cdot\boldsymbol{b} = 1$, the square-bracketed part of the third term on the right-hand side of the equation for $\boldsymbol{v}_{\mathrm{G}}$ becomes

$$(\boldsymbol{\nabla}\times\boldsymbol{b}) - (\boldsymbol{\nabla}\times\boldsymbol{b})_\perp = (\boldsymbol{\nabla}\times\boldsymbol{b})_\parallel = \left[\boldsymbol{b}\cdot(\boldsymbol{\nabla}\times\boldsymbol{b})\right]\boldsymbol{b}\,.$$

Accordingly, within accuracy of 2nd order of Larmor radius/characteristic length of $\boldsymbol{b}$, the velocity of guiding center is reduced to

$$\boldsymbol{v}_{\mathrm{G}} = \frac{1}{1+(mv_\parallel/qB)\boldsymbol{b}\cdot\nabla\times\boldsymbol{b}}\left(v_\parallel\boldsymbol{b} + \frac{mv_\parallel}{qB}\nabla\times(v_\parallel\boldsymbol{b})\right)\,. \tag{3.40}$$

The first factor in the righthand side of (3.40) is necessary in order to conserve the phase space volume in Lagrange-Hamiltonian formulation of guiding center motion [3.2].

Since

$$\nabla\times\boldsymbol{B} = B\nabla\times\boldsymbol{b} + \nabla B\times\boldsymbol{b} = \mu_0\boldsymbol{j},$$

we have $\boldsymbol{b}\cdot\nabla\times\boldsymbol{b} = \mu_0 j_\parallel/B$. The second term of the denominator in (3.40) is usually very small compared with 1 (zero in the case of $j_\parallel = 0$). If the second term of the denominator can be neglected, eq.(3.39) for $\boldsymbol{v}_{\mathrm{G}}$ is reduced to [3.1]

$$\frac{\mathrm{d}\boldsymbol{r}_{\mathrm{G}}}{\mathrm{d}t} = \frac{v_\parallel}{B}\nabla\times\left(\boldsymbol{A} + \frac{mv_\parallel}{qB}\boldsymbol{B}\right)\,. \tag{3.41}$$

The orbit of (3.40) and (3.41) are identical when the magnetic field does not depend on time. The orbit of the guiding center is equal to the field line of the magnetic field $\boldsymbol{B}^* = \boldsymbol{\nabla}\times\boldsymbol{A}^*$ with the vector potential

$$\boldsymbol{A}^* \equiv \boldsymbol{A} + \frac{mv_\parallel}{qB}\boldsymbol{B}\,.$$

For analogous reasons to those discussed in Sect. 3.2, the *orbit surface of the drift motion* of the guiding center is given by

$$rA_\theta^*(r,z) = \text{const.} \tag{3.42}$$

## 3.7 Effect of Longitudinal Electric Field on Banana Orbit

In the tokamak configuration, a toroidal electric field is applied in order to induce the plasma current. The guiding center of a particle drifts by $\boldsymbol{E} \times \boldsymbol{B}/B^2$, but the banana center moves in a different way. The toroidal electric field can be described by

$$E_\varphi = -\frac{\partial A_\varphi}{\partial t} ,$$

in $(R, \varphi, z)$ coordinates. Since angular momentum is conserved, we can write

$$R(mR\dot{\varphi} + qA_\varphi) = \text{const.}$$

Taking the average of the foregoing equation over a Larmor period, and using the relation

$$\langle R\dot{\varphi} \rangle = \frac{B_\varphi}{B} v_\parallel ,$$

we find

$$R\left(mv_\parallel \frac{B_\varphi}{B} + qA_\varphi\right) = \text{const.} \tag{3.43}$$

For particles in banana motion ($v_\parallel \ll v_\perp$), $v_\parallel$ becomes 0 at the turning points of the banana orbit. The displacement of a turning point $(R, Z)$ per period $\Delta t$ is obtained from

$$0 = \Delta\big[RA_\varphi(R, Z)\big] = \Delta r \frac{\partial}{\partial r} RA_\varphi + \Delta t \frac{\partial}{\partial t} RA_\varphi ,$$

where $r$ is the radial coordinate of the magnetic surface. The derivatives of $RA_\varphi = \text{const.}$ with respect to $\varphi$ and $\theta$ are zero, since $RA_\varphi = \text{const.}$ is the magnetic surface. By means of the relation

$$\begin{aligned} \frac{1}{R}\frac{\partial}{\partial r}(RA_\varphi) &= \frac{1}{R}\left[\frac{\partial R}{\partial r}\frac{\partial (RA_\varphi)}{\partial R} + \frac{\partial Z}{\partial r}\frac{\partial (RA_\varphi)}{\partial Z}\right] \\ &= \cos\theta B_Z - \sin\theta B_R = \boldsymbol{B}_\text{p} \cdot \boldsymbol{e}_\theta = -B_\text{p} , \end{aligned}$$

we obtain the drift velocity

$$\frac{\Delta r}{\Delta t} = -\frac{E_\varphi}{B_\text{p}} , \tag{3.44}$$

where $\boldsymbol{e}_\theta$ is the unit vector in the $\theta$ direction (see Fig. 3.8). When the sign of $B_\text{p}$ produced by the current induced by the electric field $E_\varphi$ is taken into account ($B_\text{p} > 0$, $E_\varphi > 0$ in the case of Fig. 3.8), the sign of $\Delta r/\Delta t$ is negative and the banana center moves inward. Since $|B_\text{p}| \ll |B_\varphi| \simeq B$, the drift velocity of the banana center is $(B/B_\text{p})^2$ times as fast as the the drift velocity $E_\varphi B_\text{p}/B^2$ of the guiding center of the particle. This phenomena is called *Ware's pinch*.

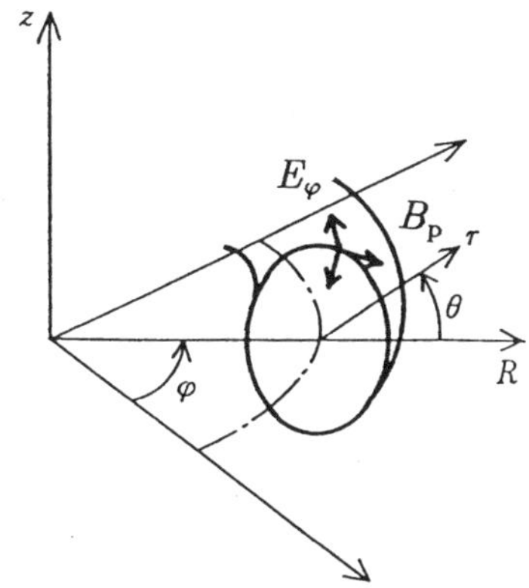


**Fig. 3.8.** Coordinate system in which the Ware pinch is analyzed

## 3.8 Polarization Drift

Let us consider the case when $\boldsymbol{E} = E_0 \exp(-\mathrm{i}\omega t)\hat{\boldsymbol{x}}$ lies in the $x$ direction and is time dependent, but $\boldsymbol{B}$ is stationary and constant in the $z$ direction. Then the equation of motion (3.22) is

$$\ddot{v}_x = \frac{q}{m}\dot{E}_x + \frac{q}{m}\dot{v}_y B = \mathrm{i}\omega\Omega\frac{E_x}{B} - \Omega^2 v_x \; ,$$

$$\ddot{v}_y = -\frac{q}{m}\dot{v}_x B = -\Omega^2\frac{E_x}{B} - \Omega^2 v_y \; .$$

When we define

$$v_{\mathrm{E}} \equiv -\frac{E_x}{B} \; , \qquad v_{\mathrm{p}} = \mathrm{i}\frac{\omega}{\Omega}\frac{E_x}{B} \; ,$$

the equation of motion reduces to

$$\ddot{v}_x = -\Omega^2(v_x - v_{\mathrm{p}}) \; , \qquad \ddot{v}_y = -\Omega^2(v_y - v_{\mathrm{E}}) \; .$$

When $\Omega^2 \gg \omega^2$, the solution is

$$v_x = v_\perp \exp(-\mathrm{i}\Omega t) + v_{\mathrm{p}} \; , \qquad v_y = v_\perp \exp(-\mathrm{i}\Omega t) + v_{\mathrm{E}} \; .$$

This solution shows that the guiding center motion consists of the usual $\boldsymbol{E} \times \boldsymbol{B}$ drift (but slowly oscillating) and the new drift along $\boldsymbol{E}$. This new term is called the *polarization drift* and is expressed by

$$\boldsymbol{v}_{\mathrm{p}} = -\frac{1}{\Omega B}\frac{\partial \boldsymbol{E}}{\partial t} \; . \tag{3.45}$$

Since $\boldsymbol{v}_{\mathrm{p}}$ is in opposite directions for ions and electrons, there is a *polarization current*

$$\boldsymbol{j}_{\mathrm{p}} = en_{\mathrm{e}}(\boldsymbol{v}_{\mathrm{pi}} - \boldsymbol{v}_{\mathrm{pe}}) = \frac{n_{\mathrm{e}}(m_{\mathrm{i}} + m_{\mathrm{e}})}{B^2}\frac{\partial \boldsymbol{E}}{\partial t} = \frac{\rho_{\mathrm{m}}}{B^2}\frac{\partial \boldsymbol{E}}{\partial t}\,, \tag{3.46}$$

where $\rho_{\mathrm{m}}$ is the mass density.

# 4 Velocity Space Distribution Function and Boltzmann's Equation

A plasma consists of many ions and electrons, but the individual behavior of each particle can hardly be observed. What can be observed instead are statistical averages. In order to describe the properties of a plasma, one must define a distribution function that indicates particle number density in the phase space whose ordinates are the particle positions and velocities. The distribution function is not necessarily stationary with respect to time. In Sect. 4.1, the equation governing the distribution function $f(q_i, p_i, t)$ is derived by means of Liouville's theorem. Boltzmann's equation for the distribution function $f(\boldsymbol{x}, \boldsymbol{v}, t)$ is formulated in Sect. 4.2. When the collision term is neglected, Boltzmann's equation is called Vlasov's equation.

## 4.1 Phase Space and Distribution Function

A particle can be specified by its coordinates $(x, y, z)$, velocity $(v_x, v_y, v_z)$, and time $t$. More generally, the particle can be specified by *canonical variables* $q_1, q_2, q_3, p_1, p_2, p_3$ and $t$ in phase space. When canonical variables are used, an infinitesimal volume in phase space $\Delta = \delta q_1 \delta q_2 \delta q_3 \delta p_1 \delta p_2 \delta p_3$ is conserved (Liouville's theorem). The motion of a particle in phase space is described by Hamilton's equations

$$\frac{\mathrm{d}q_i}{\mathrm{d}t} = \frac{\partial H(q_j, p_j, t)}{\partial p_i}\,, \qquad \frac{\mathrm{d}p_i}{\mathrm{d}t} = -\frac{\partial H(q_j, p_j, t)}{\partial q_i}\,, \qquad i, j = 1, 2, 3\,. \tag{4.1}$$

The variation over time of $\Delta$ is given by

$$\begin{aligned}\frac{\mathrm{d}\Delta}{\mathrm{d}t} &= \left[\frac{\mathrm{d}(\delta q_1)}{\mathrm{d}t}\delta p_1 + \frac{\mathrm{d}(\delta p_1)}{\mathrm{d}t}\delta q_1\right]\delta q_2 \delta p_2 \delta q_3 \delta p_3 \\ &\quad + \left[\frac{\mathrm{d}(\delta q_2)}{\mathrm{d}t}\delta p_2 + \frac{\mathrm{d}(\delta p_2)}{\mathrm{d}t}\delta q_2\right]\delta q_3 \delta p_3 \delta q_1 \delta p_1 \\ &\quad + \left[\frac{\mathrm{d}(\delta q_3)}{\mathrm{d}t}\delta p_3 + \frac{\mathrm{d}(\delta p_3)}{\mathrm{d}t}\delta q_3\right]\delta q_1 \delta p_1 \delta q_2 \delta p_2\,,\end{aligned}$$

$$\frac{\mathrm{d}}{\mathrm{d}t}\delta q_i = \delta\left(\frac{\partial H}{\partial p_i}\right) = \frac{\partial^2 H}{\partial p_i \partial q_i}\delta q_i\,, \qquad \frac{\mathrm{d}}{\mathrm{d}t}\delta p_i = -\delta\left(\frac{\partial H}{\partial q_i}\right) = -\frac{\partial^2 H}{\partial q_i \partial p_i}\delta p_i\,,$$

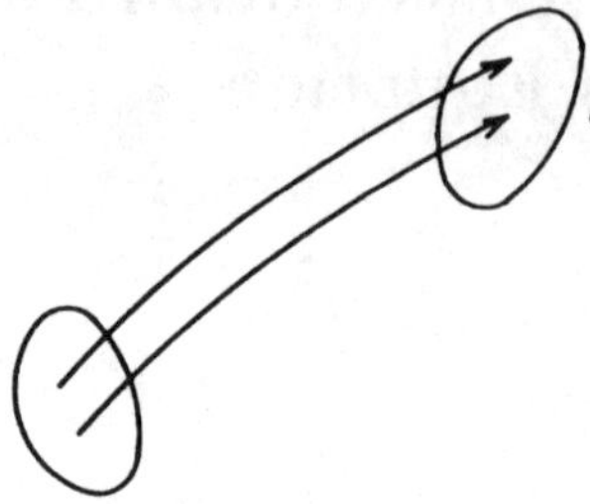


**Fig. 4.1.** Movement of particles in phase space

$$\frac{\mathrm{d}\Delta}{\mathrm{d}t} = \sum_i \left( \frac{\partial^2 H}{\partial p_i \partial q_i} - \frac{\partial^2 H}{\partial q_i \partial p_i} \right) \Delta = 0 . \tag{4.2}$$

This is a proof of Liouville's theorem (see Fig. 4.1).

Let the number of particles in a small volume of phase space be $\delta N$ given by

$$\delta N = F(q_i, p_i, t) \delta \boldsymbol{q} \delta \boldsymbol{p} , \tag{4.3}$$

where $\delta \boldsymbol{q} = \delta q_1 \delta q_2 \delta q_3$, $\delta \boldsymbol{p} = \delta p_1 \delta p_2 \delta p_3$, and $F(q_i, p_i, t)$ is the *distribution function in phase space*. If the particles move according to the equation of motion and are not scattered by collisions, the small volume in phase space is conserved. As the particle number $\delta N$ within the small phase space is conserved, the distribution function ($F = \delta N/\Delta$) is also constant, i.e.,

$$\frac{\mathrm{d}F}{\mathrm{d}t} = \frac{\partial F}{\partial t} + \sum_i \left( \frac{\partial F}{\partial q_i}\frac{\mathrm{d}q_i}{\mathrm{d}t} + \frac{\partial F}{\partial p_i}\frac{\mathrm{d}p_i}{\mathrm{d}t} \right) = \frac{\partial F}{\partial t} + \sum_i \left( \frac{\partial H}{\partial p_i}\frac{\partial F}{\partial q_i} - \frac{\partial H}{\partial q_i}\frac{\partial F}{\partial p_i} \right) = 0 . \tag{4.4}$$

In the foregoing discussion, we did not take collisions into account. If we denote the variation of $F$ due to collisions by $(\delta F/\delta t)_{\mathrm{coll}}$, (4.4) becomes

$$\frac{\partial F}{\partial t} + \sum_i \left( \frac{\partial H}{\partial p_i}\frac{\partial F}{\partial q_i} - \frac{\partial H}{\partial q_i}\frac{\partial F}{\partial p_i} \right) = \left( \frac{\delta F}{\delta t} \right)_{\mathrm{coll}} . \tag{4.5}$$

## 4.2 Boltzmann's Equation and Vlasov's Equation

Let us use the space and velocity space coordinates $x_1, x_2, x_3, v_1, v_2, v_3$ instead of canonical coordinates. The Hamiltonian is

$$H = \frac{1}{2m}(\boldsymbol{p} - q\boldsymbol{A})^2 + q\phi , \tag{4.6}$$

$$p_i = mv_i + qA_i \; , \tag{4.7}$$

$$q_i = x_i \; , \tag{4.8}$$

and

$$\frac{\mathrm{d}x_i}{\mathrm{d}t} = \frac{\partial H}{\partial p_i} = v_i \; , \tag{4.9}$$

$$\frac{\mathrm{d}p_i}{\mathrm{d}t} = -\frac{\partial H}{\partial x_i} = \sum_k \frac{p_k - qA_k}{m} q \frac{\partial A_k}{\partial x_i} - q\frac{\partial \phi}{\partial x_i} \; . \tag{4.10}$$

Consequently, (4.5) becomes

$$\frac{\partial F}{\partial t} + \sum_i v_k \frac{\partial F}{\partial x_k} + q\sum_i \left( \sum_k v_k \frac{\partial A_k}{\partial x_i} - \frac{\partial \phi}{\partial x_i} \right) \frac{\partial F}{\partial p_i} = \left( \frac{\delta F}{\delta t} \right)_{\text{coll}} \; . \tag{4.11}$$

Using (4.7) and (4.8), independent variables are transformed from $(q_i, p_i, t)$ to $(x_j, v_j, t)$ and

$$\frac{\partial v_j(x_k, p_k, t)}{\partial p_i} = \frac{1}{m}\delta_{ij} \; , \qquad \frac{\partial v_j(x_k, p_k, t)}{\partial x_i} = -\frac{q}{m}\frac{\partial A_j}{\partial x_i} \; ,$$

$$\frac{\partial v_j(x_k, p_k, t)}{\partial t} = -\frac{q}{m}\frac{\partial A_j}{\partial t} \; .$$

We denote $F(x_i, p_i, t) = F\big(x_i, p_i(x_j, v_j, t), t\big) \equiv f(x_j, v_j, t)/m^3$. Then we have $m^3 F(x_i, p_i, t) = f\big(x_j, v_j(x_i, p_i, t), t\big)$ and

$$m^3 \frac{\partial}{\partial p_i} F(x_h, p_h, t) = \frac{\partial}{\partial p_i} f\big(x_j, v_j(x_h, p_h, t), t\big) = \sum_j \frac{\partial f}{\partial v_j}\frac{\partial v_j}{\partial p_i} = \frac{\partial f}{\partial v_i}\frac{1}{m} \; ,$$

$$\begin{aligned} m^3 \frac{\partial}{\partial x_k} F(x_h, p_h, t) &= \frac{\partial}{\partial x_k} f\big(x_i, v_i(x_h, p_h, t), t\big) = \frac{\partial f}{\partial x_k} + \sum_i \frac{\partial f}{\partial v_i}\frac{\partial v_i}{\partial x_k} \\ &= \frac{\partial f}{\partial x_k} + \sum_i \frac{\partial f}{\partial v_i}\left(\frac{-q}{m}\right)\frac{\partial A_i}{\partial x_k} \; , \end{aligned}$$

$$m^3 \frac{\partial}{\partial t} F(x_h, p_h, t) = \frac{\partial}{\partial t} f\big(x_i, v_i(x_h, p_h, t), t\big) = \frac{\partial f}{\partial t} + \sum_i \frac{\partial f}{\partial v_i}\left(\frac{-q}{m}\right)\frac{\partial A_i}{\partial t} \; .$$

Accordingly, (4.11) reduces to

$$\begin{aligned} &\frac{\partial f}{\partial t} + \sum_i \frac{\partial f}{\partial v_i}\left(\frac{-q}{m}\right)\frac{\partial A_i}{\partial t} + \sum_k v_k \left[ \frac{\partial f}{\partial x_k} + \sum_i \frac{\partial f}{\partial v_i}\left(\frac{-q}{m}\right)\frac{\partial A_i}{\partial x_k} \right] \\ &\quad + \sum_i \left( \sum_k v_k \frac{\partial A_k}{\partial x_i} - \frac{\partial \phi}{\partial x_i} \right) \frac{q}{m}\frac{\partial f}{\partial v_i} = \left( \frac{\delta f}{\delta t} \right)_{\text{coll}} \; , \end{aligned}$$

$$\frac{\partial f}{\partial t} + \sum_k v_k \frac{\partial f}{\partial x_k} + \sum_i \left( -\frac{\partial A_i}{\partial t} - \sum_k v_k \frac{\partial A_i}{\partial x_k} + \sum_k v_k \frac{\partial A_k}{\partial x_i} - \frac{\partial \phi}{\partial x_i} \right) \frac{q}{m} \frac{\partial f}{\partial v_i}$$
$$= \left( \frac{\delta f}{\delta t} \right)_{\text{coll}} .$$

Since

$$\sum_k v_k \frac{\partial A_k}{\partial x_i} = \sum_k v_k \frac{\partial A_i}{\partial x_k} + \left[ \boldsymbol{v} \times (\boldsymbol{\nabla} \times \boldsymbol{A}) \right]_i = \sum_k v_k \frac{\partial A_i}{\partial x_k} + (\boldsymbol{v} \times \boldsymbol{B})_i ,$$

we have

$$\frac{\partial f}{\partial t} + \sum_i v_i \frac{\partial f}{\partial x_i} + \sum_i \frac{q}{m} (\boldsymbol{E} + \boldsymbol{v} \times \boldsymbol{B})_i \frac{\partial f}{\partial v_i} = \left( \frac{\delta f}{\delta t} \right)_{\text{coll}} . \tag{4.12}$$

This equation is called *Boltzmann's equation*. The electric charge density $\rho$ and the electric current $\boldsymbol{j}$ are expressed by

$$\rho = \sum_{\text{i,e}} q \int f \mathrm{d}v_1 \mathrm{d}v_2 \mathrm{d}v_3 , \tag{4.13}$$

$$\boldsymbol{j} = \sum_{\text{i,e}} q \int \boldsymbol{v} f \mathrm{d}v_1 \mathrm{d}v_2 \mathrm{d}v_3 . \tag{4.14}$$

Accordingly, Maxwell's equations are given by

$$\boldsymbol{\nabla} \cdot \boldsymbol{E} = \frac{1}{\epsilon_0} \sum_{\text{i,e}} q \int f \mathrm{d}\boldsymbol{v} , \tag{4.15}$$

$$\frac{1}{\mu_0} \boldsymbol{\nabla} \times \boldsymbol{B} = \epsilon_0 \frac{\partial \boldsymbol{E}}{\partial t} + \sum_{\text{i,e}} q \int \boldsymbol{v} f \mathrm{d}\boldsymbol{v} , \tag{4.16}$$

$$\boldsymbol{\nabla} \times \boldsymbol{E} = -\frac{\partial \boldsymbol{B}}{\partial t} , \tag{4.17}$$

$$\boldsymbol{\nabla} \cdot \boldsymbol{B} = 0 . \tag{4.18}$$

When the plasma is rarefied, the collision term $(\delta f / \delta t)_{\text{coll}}$ may be neglected. However, the interactions of the charged particles are still included through the internal electric and magnetic fields which are calculated from the charge and current densities by means of Maxwell's equations. The charge and current densities are expressed by the distribution functions for the electron and the ion. This equation is called the *collisionless Boltzmann equation* or *Vlasov's equation*.

When the *Fokker–Planck collision term* is adopted as the collision term in Boltzmann's equation, this equation is called the *Fokker–Planck equation* (see Sect. 16.8).

# 5 Plasma as MHD Fluid

## 5.1 Magnetohydrodynamic Equations for Two Fluids

Plasmas can be described as magnetohydrodynamic double fluids of ions and electrons with mass densities $\rho_{\rm mi}$, $\rho_{\rm me}$, charge density $\rho$, current density $\boldsymbol{j}$, flow velocities $\boldsymbol{V}_{\rm i}$, $\boldsymbol{V}_{\rm e}$, and pressures $p_{\rm i}$, $p_{\rm e}$. These physical quantities can be expressed by appropriate averages in velocity space using the velocity space distribution functions $f_i(\boldsymbol{r}, \boldsymbol{v}, t)$ of ions and electrons, which were introduced in Chap. 4. The number density of ions $n_{\rm i}$, the ion mass density $\rho_{\rm mi}$, and the ion flow velocity $\boldsymbol{V}_{\rm i}(\boldsymbol{r}, t)$ are expressed as follows:

$$n_{\rm i}(\boldsymbol{r}, t) = \int f_{\rm i}(\boldsymbol{r}, \boldsymbol{v}, t){\rm d}\boldsymbol{v} \ , \tag{5.1}$$

$$\rho_{\rm mi}(\boldsymbol{r}, t) = m_{\rm i} n_{\rm i}(\boldsymbol{r}, t) \ , \tag{5.2}$$

$$\boldsymbol{V}(\boldsymbol{r}, t) = \frac{\int \boldsymbol{v} f_{\rm i}(\boldsymbol{r}, \boldsymbol{v}, t){\rm d}\boldsymbol{v}}{\int f_{\rm i}(\boldsymbol{r}, \boldsymbol{v}, t){\rm d}\boldsymbol{v}} = \frac{1}{n_{\rm i}(\boldsymbol{r}, t)} \int \boldsymbol{v} f_{\rm i}(\boldsymbol{r}, \boldsymbol{v}, t){\rm d}\boldsymbol{v} \ . \tag{5.3}$$

We have the same expressions for electrons as for ions. Since magnetohydrodynamics treats average quantities in the velocity space, phenomena associated with the shape of the velocity space distribution function (Chap. 11) will be neglected. However, the independent variables are $\boldsymbol{r}, t$ alone, and it is possible to analyze geometrically complicated configurations.

The magnetohydrodynamic equations are:

$$\frac{\partial n_{\rm e}}{\partial t} + \nabla\cdot(n_{\rm e}\boldsymbol{V}_{\rm e}) = 0 \ , \tag{5.4}$$

$$\frac{\partial n_{\rm i}}{\partial t} + \nabla\cdot(n_{\rm i}\boldsymbol{V}_{\rm i}) = 0 \ , \tag{5.5}$$

$$n_{\rm e} m_{\rm e}\frac{{\rm d}\boldsymbol{V}_{\rm e}}{{\rm d}t} = -\nabla p_{\rm e} - e n_{\rm e}(\boldsymbol{E} + \boldsymbol{V}_{\rm e} \times \boldsymbol{B}) + \boldsymbol{R} \ , \tag{5.6}$$

$$n_{\rm i} m_{\rm i}\frac{{\rm d}\boldsymbol{V}_{\rm i}}{{\rm d}t} = -\nabla p_{\rm i} + Z e n_{\rm i}(\boldsymbol{E} + \boldsymbol{V}_{\rm i} \times \boldsymbol{B}) - \boldsymbol{R} \ . \tag{5.7}$$

Here $\boldsymbol{R}$ denotes the rate of change of momentum (density) of the electron fluid by collisions with the ion fluid. The rate of change of momentum of

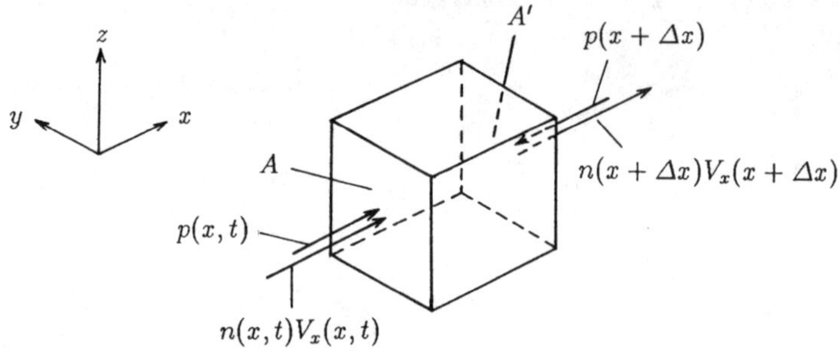


**Fig. 5.1.** Particle flux and force due to pressure

the ion fluid due to collisions with the electron fluid is $-\boldsymbol{R}$. The change in the number $n(x,y,z,t)\Delta x\Delta y\Delta z$ of particles within the region $\Delta x\Delta y\Delta z$ is the difference between the incident particle flux $n(x,y,z,t)V_x(x,y,z,t)\Delta y\Delta z$ into the surface A in Fig. 5.1 and the outgoing particle flux

$$n(x+\Delta x,y,z,t)V_x(x+\Delta x,y,z,t)\Delta y\Delta z$$

from the surface A′, that is,

$$\Big[n(x,y,z,t)V_x(x,y,z,t) - n(x+\Delta x,y,z,t)V_x(x+\Delta x,y,z,t)\Big]\Delta y\Delta z$$
$$= -\frac{\partial(nV_x)}{\partial x}\Delta x\Delta y\Delta z\,.$$

When the particle fluxes of the other surfaces are taken into account, we find the equations of continuity (5.4) and (5.5), that is,

$$\frac{\partial n}{\partial t}\Delta x\Delta y\Delta z = -\left[\frac{\partial(nV_x)}{\partial x} + \frac{\partial(nV_y)}{\partial y} + \frac{\partial(nV_z)}{\partial z}\right]\Delta x\Delta y\Delta z\,.$$

The term $-\nabla p$ in (5.6) and (5.7) is the force per unit volume of plasma due to the pressure $p$ and the second term on the right-hand side of (5.6) and (5.7) is the Coulomb force and Lorentz force per unit volume. The third term is the collision term for electron–ion collisions, as mentioned in Sect. 2.8, given by

$$\boldsymbol{R} = -n_{\mathrm{e}}m_{\mathrm{e}}(\boldsymbol{V}_{\mathrm{e}} - \boldsymbol{V}_{\mathrm{i}})\nu_{\mathrm{ei}}\,, \tag{5.8}$$

where $\nu_{\mathrm{ei}}$ is the Coulomb collision frequency of electrons with ions.

Let us consider the total time derivative on the left-hand side of the equation of motion. The flow velocity $\boldsymbol{V}$ is a function of space coordinates $\boldsymbol{r}$ and time $t$. Then the acceleration of a small volume of fluid is given by

$$\frac{\mathrm{d}\boldsymbol{V}(\boldsymbol{r},t)}{\mathrm{d}t} = \frac{\partial \boldsymbol{V}(\boldsymbol{r},t)}{\partial t} + \left(\frac{\mathrm{d}\boldsymbol{r}}{\mathrm{d}t}\cdot\nabla\right)\boldsymbol{V}(\boldsymbol{r},t) = \frac{\partial \boldsymbol{V}(\boldsymbol{r},t)}{\partial t} + \big[\boldsymbol{V}(\boldsymbol{r},t)\cdot\nabla\big]\boldsymbol{V}(\boldsymbol{r},t)\,.$$

Therefore the equations of motion (5.6) and (5.7) reduce to

$$n_\mathrm{e} m_\mathrm{e} \left[ \frac{\partial \boldsymbol{V}_\mathrm{e}}{\partial t} + (\boldsymbol{V}_\mathrm{e} \cdot \nabla) \boldsymbol{V}_\mathrm{e} \right] = -\nabla p_\mathrm{e} - e n_\mathrm{e} (\boldsymbol{E} + \boldsymbol{V}_\mathrm{e} \times \boldsymbol{B}) + \boldsymbol{R} \,, \tag{5.9}$$

$$n_\mathrm{i} m_\mathrm{i} \left[ \frac{\partial \boldsymbol{V}_\mathrm{i}}{\partial t} + (\boldsymbol{V}_\mathrm{i} \cdot \nabla) \boldsymbol{V}_\mathrm{i} \right] = -\nabla p_\mathrm{i} + Z e n_\mathrm{i} (\boldsymbol{E} + \boldsymbol{V}_\mathrm{i} \times \boldsymbol{B}) - \boldsymbol{R} \,. \tag{5.10}$$

Conservation of particle number, (5.4) and (5.5), and the equations of motion (5.9) and (5.10) can be derived from Boltzmann's equation (4.12). Integrating the Boltzmann equation over velocity space yields (5.4) and (5.5). Integrating the Boltzmann equation multiplied by $m\boldsymbol{v}$ yields (5.9) and (5.10) [5.1].

## 5.2 Magnetohydrodynamic Equations for One Fluid

Since the ion-to-electron mass ratio is $m_\mathrm{i}/m_\mathrm{e} = 1836A$, where $A$ is the atomic weight of the ion, the contribution of ions dominates the mass density of the plasma. In many cases it is more convenient to reorganize the equations of motion for two fluids into the equation of motion for one fluid and Ohm's law.

The total mass density of the plasma $\rho_\mathrm{m}$, the flow velocity of the plasma $\boldsymbol{V}$, the electric charge density $\rho$ and the current density $\boldsymbol{j}$ are defined as follows:

$$\rho_\mathrm{m} = n_\mathrm{e} m_\mathrm{e} + n_\mathrm{i} m_\mathrm{i} \,, \tag{5.11}$$

$$\boldsymbol{V} = \frac{n_\mathrm{e} m_\mathrm{e} \boldsymbol{V}_\mathrm{e} + n_\mathrm{i} m_\mathrm{i} \boldsymbol{V}_\mathrm{i}}{\rho_\mathrm{m}} \,, \tag{5.12}$$

$$\rho = -e n_\mathrm{e} + Z e n_\mathrm{i} \,, \tag{5.13}$$

$$\boldsymbol{j} = -e n_\mathrm{e} \boldsymbol{V}_\mathrm{e} + Z_\mathrm{e} n_\mathrm{i} \boldsymbol{V}_\mathrm{i} \,. \tag{5.14}$$

From (5.4) and (5.5), it follows that

$$\frac{\partial \rho_\mathrm{m}}{\partial t} + \nabla \cdot (\rho_\mathrm{m} \boldsymbol{V}) = 0 \,, \tag{5.15}$$

$$\frac{\partial \rho}{\partial t} + \nabla \cdot \boldsymbol{j} = 0 \,. \tag{5.16}$$

From (5.9) and (5.10), we find

$$\rho_\mathrm{m} \frac{\partial \boldsymbol{V}}{\partial t} + n_\mathrm{e} m_\mathrm{e} (\boldsymbol{V}_\mathrm{e} \cdot \nabla) \boldsymbol{V}_\mathrm{e} + n_\mathrm{i} m_\mathrm{i} (\boldsymbol{V}_\mathrm{i} \cdot \nabla) \boldsymbol{V}_\mathrm{i} = -\nabla (p_\mathrm{e} + p_\mathrm{i}) + \rho \boldsymbol{E} + \boldsymbol{j} \times \boldsymbol{B} \,. \tag{5.17}$$

The charge neutrality of the plasma allows us to write $n_\mathrm{e} \simeq Z n_\mathrm{i}$. Putting $\Delta n_\mathrm{e} = n_\mathrm{e} - Z n_\mathrm{i}$, we have

$$\rho_\mathrm{m} = n_\mathrm{i} m_\mathrm{i} \left( 1 + \frac{m_\mathrm{e}}{m_\mathrm{i}} Z \right) , \qquad p = p_\mathrm{i} + p_\mathrm{e} \,, \qquad \boldsymbol{V} = \boldsymbol{V}_\mathrm{i} + \frac{m_\mathrm{e} Z}{m_\mathrm{i}} (\boldsymbol{V}_\mathrm{e} - \boldsymbol{V}_\mathrm{i}) \,,$$

$$\rho = -e\Delta n_{\rm e} , \qquad \boldsymbol{j} = -en_{\rm e}(\boldsymbol{V}_{\rm e} - \boldsymbol{V}_{\rm i}) .$$

Since $m_{\rm e}/m_{\rm i} \ll 1$, the second and third terms on the left-hand side of (5.17) can be written as $(\boldsymbol{V}\cdot\Delta)\boldsymbol{V}$. Since $\boldsymbol{V}_{\rm e} = \boldsymbol{V}_{\rm i} - \boldsymbol{j}/en_{\rm e} \simeq \boldsymbol{V} - \boldsymbol{j}/en_{\rm e}$, (5.9) reduces to

$$\boldsymbol{E} + \left(\boldsymbol{V} - \frac{\boldsymbol{j}}{en_{\rm e}}\right) \times \boldsymbol{B} + \frac{1}{en_{\rm e}}\nabla p_{\rm e} - \frac{\boldsymbol{R}}{en_{\rm e}} = \frac{m_{\rm e}}{e^2 n_{\rm e}}\frac{\partial \boldsymbol{j}}{\partial t} - \frac{m_{\rm e}}{e}\frac{\partial \boldsymbol{V}}{\partial t} . \tag{5.18}$$

Using the expression for the specific resistivity $\eta$ (see Sect. 2.8), the collision term $\boldsymbol{R}$ reduces to

$$\boldsymbol{R} = n_{\rm e}\left(\frac{m_{\rm e}\nu_{\rm ei}}{n_{\rm e}e^2}\right)(-en_{\rm e})(\boldsymbol{V}_{\rm e} - \boldsymbol{V}_{\rm i}) = n_{\rm e}e\eta\boldsymbol{j} . \tag{5.19}$$

Equation (5.18) is a generalized Ohm's law. Finally, the equation of motion for the single fluid model and the generalized Ohm's law are

$$\rho_{\rm m}\left[\frac{\partial \boldsymbol{V}}{\partial t} + (\boldsymbol{V}\cdot\nabla)\boldsymbol{V}\right] = -\nabla p + \rho\boldsymbol{E} + \boldsymbol{j} \times \boldsymbol{B} , \tag{5.20}$$

$$\boldsymbol{E} + \left(\boldsymbol{V} - \frac{\boldsymbol{j}}{en_{\rm e}}\right) \times \boldsymbol{B} + \frac{1}{en_{\rm e}}\nabla p_{\rm e} - \eta\boldsymbol{j} = \frac{m_{\rm e}}{e^2 n_{\rm e}}\frac{\partial \boldsymbol{j}}{\partial t} - \frac{m_{\rm e}}{e}\frac{\partial \boldsymbol{V}}{\partial t} \simeq 0 \quad (|\omega/\Omega_{\rm e}| \ll 1) . \tag{5.21}$$

The equation of continuity and Maxwell's equations are

$$\frac{\partial \rho_{\rm m}}{\partial t} + \nabla\cdot(\rho_{\rm m}\boldsymbol{V}) = 0 , \tag{5.22}$$

$$\frac{\partial \rho}{\partial t} + \nabla\cdot\boldsymbol{j} = 0 , \tag{5.23}$$

$$\nabla \times \boldsymbol{E} = -\frac{\partial \boldsymbol{B}}{\partial t} , \tag{5.24}$$

$$\frac{1}{\mu_0}\nabla \times \boldsymbol{B} = \boldsymbol{j} + \frac{\partial \boldsymbol{D}}{\partial t} , \tag{5.25}$$

$$\nabla\cdot\boldsymbol{D} = \rho , \tag{5.26}$$

$$\nabla\cdot\boldsymbol{B} = 0 . \tag{5.27}$$

From (5.25) and (5.24), it follows that

$$\nabla \times \nabla \times \boldsymbol{E} = -\mu_0\frac{\partial \boldsymbol{j}}{\partial t} - \mu_0\epsilon_0\frac{\partial^2 \boldsymbol{E}}{\partial t^2} .$$

A typical propagation velocity for a magnetohydrodynamic wave or perturbation is the Alfvén velocity $v_{\rm A} = B/(\mu_0\rho_{\rm m})^{1/2}$ as described in Sect. 5.4. This is much smaller than the speed of light $c$ and $\omega^2/k^2 \sim v_{\rm A}^2 \ll c^2$. Since

$$\left|\nabla \times \frac{\partial \boldsymbol{B}}{\partial t}\right| = |\nabla \times \nabla \times \boldsymbol{E}| \sim k^2 |\boldsymbol{E}| \,,$$

and

$$\mu_0 \epsilon_0 \left|\frac{\partial^2 \boldsymbol{E}}{\partial t^2}\right| \sim \frac{\omega^2 |\boldsymbol{E}|}{c^2} \,,$$

the displacement current $\partial \boldsymbol{D}/\partial t$ in (5.25) is negligible. Since the ratio of the first term $(m_e/e)\partial \boldsymbol{j}/\partial t$ on the right-hand side of (5.21) to the term $(\boldsymbol{j} \times \boldsymbol{B})$ on the left-hand side is $\omega/\Omega_e$, the first term can be neglected, if $|\omega/\Omega_e| \ll 1$. The second term $(m_e/e)\partial \boldsymbol{V}/\partial t$ on the right-hand side of (5.21) is of the order of $\omega/\Omega_e$ times as large as the term $\boldsymbol{V} \times \boldsymbol{B}$ on the left-hand side. Therefore we may set the right-hand side of (5.21) almost to zero. When the term $\boldsymbol{j} \times \boldsymbol{B}$ is eliminated by means of (5.20), we find

$$\boldsymbol{E} + \boldsymbol{V} \times \boldsymbol{B} - \frac{1}{en_e} \nabla p_i - \eta \boldsymbol{j} = \frac{\Delta n_e}{n_e} \boldsymbol{E} + \frac{m_i}{e} \frac{\mathrm{d}\boldsymbol{V}}{\mathrm{d}t} \,.$$

The ratio of $(m_i/e)\mathrm{d}\boldsymbol{V}/\mathrm{d}t$ to $\boldsymbol{V} \times \boldsymbol{B}$ is around $|\omega/\Omega_i|$, and $\Delta n_e/n_e \ll 1$. When $|\omega/\Omega_i| \ll 1$, we find

$$\boldsymbol{E} + \boldsymbol{V} \times \boldsymbol{B} - \frac{1}{en_e} \nabla p_i = \eta \boldsymbol{j} \qquad (|\omega/\Omega_i| \ll 1) \,. \tag{5.28}$$

## 5.3 Simplified Magnetohydrodynamic Equations

When $|\omega/\Omega_i| \ll 1$, $|\omega/k| \ll c$, and the ion pressure term $\nabla p_i$ can be neglected in Ohm's law, the magnetohydrodynamic equations simplify as follows:

$$\boldsymbol{E} + \boldsymbol{V} \times \boldsymbol{B} = \eta \boldsymbol{j} \,, \tag{5.29}$$

$$\rho_m \left[\frac{\partial \boldsymbol{V}}{\partial t} + (\boldsymbol{V} \cdot \nabla) \boldsymbol{V}\right] = -\nabla p + \boldsymbol{j} \times \boldsymbol{B} \,, \tag{5.30}$$

$$\nabla \times \boldsymbol{B} = \mu_0 \boldsymbol{j} \,, \tag{5.31}$$

$$\nabla \times \boldsymbol{E} = -\frac{\partial \boldsymbol{B}}{\partial t} \,, \tag{5.32}$$

$$\nabla \cdot \boldsymbol{B} = 0 \,, \tag{5.33}$$

$$\frac{\partial \rho_m}{\partial t} + (\boldsymbol{V} \cdot \nabla) \rho_m + \rho_m \nabla \cdot \boldsymbol{V} = 0 \,. \tag{5.34}$$

We may add the adiabatic equation as an equation of state:

$$\frac{\mathrm{d}}{\mathrm{d}t} (p \rho_m^{-\gamma}) = 0 \,,$$

where the quantity $\gamma$ is the ratio of specific heats and $\gamma = (2+\delta)/\delta$, with $\delta$ the number of degrees of freedom, is 5/3 in the three-dimensional case $\delta = 3$. Combined with (5.34), the adiabatic equation becomes

$$\frac{\partial p}{\partial t} + (\boldsymbol{V}\cdot\boldsymbol{\nabla})p + \gamma p \boldsymbol{\nabla}\cdot\boldsymbol{V} = 0 \,. \tag{5.35}$$

Instead of this relation, we may use the simpler relation of incompressibility

$$\boldsymbol{\nabla}\cdot\boldsymbol{V} = 0 \,, \tag{5.36}$$

if $|(\mathrm{d}\rho_m/\mathrm{d}t)/\rho_m)| \ll |\boldsymbol{\nabla}\cdot\boldsymbol{V}|$.

From (5.31) and (5.32), the energy conservation law is

$$\frac{1}{\mu_0}\boldsymbol{\nabla}\cdot(\boldsymbol{E}\times\boldsymbol{B}) + \frac{\partial}{\partial t}\left(\frac{B^2}{2\mu_0}\right) + \boldsymbol{E}\cdot\boldsymbol{j} = 0 \,. \tag{5.37}$$

From (5.29), the third term on the left-hand side of (5.37) becomes

$$\boldsymbol{E}\cdot\boldsymbol{j} = \eta j^2 + (\boldsymbol{j}\times\boldsymbol{B})\cdot\boldsymbol{V} \,. \tag{5.38}$$

Using (5.30) and (5.34), the Lorentz term in (5.38) is expressed by

$$(\boldsymbol{j}\times\boldsymbol{B})\cdot\boldsymbol{V} = \frac{\partial}{\partial t}\left(\frac{\rho_{\mathrm{m}}V^2}{2}\right) + \boldsymbol{\nabla}\cdot\left(\frac{\rho_{\mathrm{m}}V^2}{2}\boldsymbol{V}\right) + \boldsymbol{V}\cdot\boldsymbol{\nabla}p \,.$$

From (5.35), it follows that

$$-\boldsymbol{\nabla}\cdot(p\boldsymbol{V}) = \frac{\partial p}{\partial t} + (\gamma-1)p\boldsymbol{\nabla}\cdot\boldsymbol{V}$$

and

$$\boldsymbol{V}\cdot\boldsymbol{\nabla}p = \frac{\partial}{\partial t}\left(\frac{p}{\gamma-1}\right) + \boldsymbol{\nabla}\cdot\left(\frac{p}{\gamma-1}+p\right)\boldsymbol{V} \,.$$

Therefore the energy conservation law (5.37) reduces to

$$\boldsymbol{\nabla}\cdot(\boldsymbol{E}\times\boldsymbol{H}) + \frac{\partial}{\partial t}\left(\frac{\rho_{\mathrm{m}}V^2}{2} + \frac{p}{\gamma-1} + \frac{B^2}{2\mu_0}\right) \tag{5.39}$$
$$+\eta j^2 + \boldsymbol{\nabla}\cdot\left(\frac{\rho_{\mathrm{m}}V^2}{2} + \frac{p}{\gamma-1} + p\right)\boldsymbol{V} = 0 \,.$$

Substituting (5.29) into (5.32) yields

$$\frac{\partial \boldsymbol{B}}{\partial t} = \boldsymbol{\nabla}\times(\boldsymbol{V}\times\boldsymbol{B}) - \eta\boldsymbol{\nabla}\times\boldsymbol{j} = \boldsymbol{\nabla}\times(\boldsymbol{V}\times\boldsymbol{B}) + \frac{\eta}{\mu_0}\Delta\boldsymbol{B} \,, \tag{5.40}$$

$$\frac{\partial \boldsymbol{B}}{\partial t} = -(\boldsymbol{V}\cdot\boldsymbol{\nabla})\boldsymbol{B} - \boldsymbol{B}(\boldsymbol{\nabla}\cdot\boldsymbol{V}) + (\boldsymbol{B}\cdot\boldsymbol{\nabla})\boldsymbol{V} + \frac{\eta}{\mu_0}\Delta\boldsymbol{B} \,. \tag{5.41}$$

Here we have used the vector formulas for $\nabla \times (\boldsymbol{V} \times \boldsymbol{B})$ and $\nabla \times (\nabla \times \boldsymbol{B})$ (refer to Table 5.1). The quantity $\eta/\mu_0 = \nu_{\mathrm{m}}$ is called the *magnetic viscosity*. Substituting (5.31) into (5.30) yields

$$\rho_{\mathrm{m}} \frac{\mathrm{d}\boldsymbol{V}}{\mathrm{d}t} = -\nabla \left( p + \frac{B^2}{2\mu_0} \right) + \frac{1}{\mu_0} (\boldsymbol{B} \cdot \nabla) \boldsymbol{B} \,. \tag{5.42}$$

The equation of motion (5.42) and the equation of magnetic diffusion (5.41) are fundamental equations of magnetohydrodynamics. Equation (5.33), the equation of continuity (5.34) and the equations of state (5.35) or (5.36) are additional equations.

The ratio $R_{\mathrm{m}}$ of the first term to the second term on the right-hand side of (5.40), defined by

$$\frac{|\nabla \times (\boldsymbol{V} \times \boldsymbol{B})|}{|\Delta \boldsymbol{B}(\eta/\mu_0)|} \approx \frac{VB/L}{(B/L^2)(\eta/\mu_0)} = \frac{\mu_0 V L}{\eta} \equiv R_{\mathrm{m}} \,, \tag{5.43}$$

is called the *magnetic Reynolds number*. The notation $L$ indicates a typical plasma size. The magnetic Reynolds number is equal to the ratio of the magnetic diffusion time $\tau_{\mathrm{R}} = \mu_0 L^2/\eta$ to the Alfvén transit time $\tau_{\mathrm{H}} = L/v_{\mathrm{A}}$ (assuming that $v \approx v_{\mathrm{A}}$), i.e., $R_{\mathrm{m}} = \tau_{\mathrm{R}}/\tau_{\mathrm{H}}$. When $R_{\mathrm{m}} \ll 1$, the magnetic field in a plasma evolves according to the diffusion equation. When $R_{\mathrm{m}} \gg 1$, it can be shown that the lines of magnetic force are frozen in the plasma. Let the magnetic flux within the surface element $\Delta S$ be $\Delta \Phi$, and take the $z$ axis in the $\boldsymbol{B}$ direction. Then $\Delta \Phi$ is

$$\Delta \Phi = \boldsymbol{B} \cdot \boldsymbol{n} \Delta S = B \Delta x \Delta y \,.$$

As the boundary of $\Delta S$ moves, the rate of change of $\Delta S$ is

$$\frac{\mathrm{d}}{\mathrm{d}t}(\Delta x) = \frac{\mathrm{d}}{\mathrm{d}t}(x + \Delta x - x) = V_x(x + \Delta x) - V_x(x) = \frac{\partial V_x}{\partial x} \Delta x \,,$$

**Table 5.1.** Vector formulas

| |
|---|
| $\boldsymbol{a} \cdot (\boldsymbol{b} \times \boldsymbol{c}) = \boldsymbol{b} \cdot (\boldsymbol{c} \times \boldsymbol{a}) = \boldsymbol{c} \cdot (\boldsymbol{a} \times \boldsymbol{b})$ |
| $\boldsymbol{a} \times (\boldsymbol{b} \times \boldsymbol{c}) = (\boldsymbol{a} \cdot \boldsymbol{c}) \boldsymbol{b} - (\boldsymbol{a} \cdot \boldsymbol{b}) \boldsymbol{c}$ |
| $(\boldsymbol{a} \times \boldsymbol{b}) \cdot (\boldsymbol{c} \times \boldsymbol{d}) = \boldsymbol{a} \cdot \boldsymbol{b} \times (\boldsymbol{c} \times \boldsymbol{d}) = \boldsymbol{a} \cdot \left[ (\boldsymbol{b} \cdot \boldsymbol{d}) \boldsymbol{c} - (\boldsymbol{b} \cdot \boldsymbol{c}) \boldsymbol{d} \right]$ |
| $= (\boldsymbol{a} \cdot \boldsymbol{c})(\boldsymbol{b} \cdot \boldsymbol{d}) - (\boldsymbol{a} \cdot \boldsymbol{d})(\boldsymbol{b} \cdot \boldsymbol{c})$ |
| $\nabla \cdot (\phi \boldsymbol{a}) = \phi \nabla \cdot \boldsymbol{a} + (\boldsymbol{a} \cdot \nabla) \phi$ |
| $\nabla \times (\phi \boldsymbol{a}) = \nabla \phi \times \boldsymbol{a} + \phi \nabla \times \boldsymbol{a}$ |
| $\nabla (\boldsymbol{a} \cdot \boldsymbol{b}) = (\boldsymbol{a} \cdot \nabla) \boldsymbol{b} + (\boldsymbol{b} \cdot \nabla) \boldsymbol{a} + \boldsymbol{a} \times (\nabla \times \boldsymbol{b}) + \boldsymbol{b} \times (\nabla \times \boldsymbol{a})$ |
| $\nabla \cdot (\boldsymbol{a} \times \boldsymbol{b}) = \boldsymbol{b} \cdot \nabla \times \boldsymbol{a} - \boldsymbol{a} \cdot \nabla \times \boldsymbol{b}$ |
| $\nabla \times (\boldsymbol{a} \times \boldsymbol{b}) = \boldsymbol{a}(\nabla \cdot \boldsymbol{b}) - \boldsymbol{b}(\nabla \cdot \boldsymbol{a}) + (\boldsymbol{b} \cdot \nabla) \boldsymbol{a} - (\boldsymbol{a} \cdot \nabla) \boldsymbol{b}$ |
| $\nabla \times \nabla \times \boldsymbol{a} = \nabla (\nabla \cdot \boldsymbol{a}) - \nabla^2 \boldsymbol{a}$ (valid only for $x, y, z$ coordinates) |

$$\frac{\mathrm{d}}{\mathrm{d}t}(\Delta S) = \left(\frac{\partial V_x}{\partial x} + \frac{\partial V_y}{\partial y}\right)\Delta x \Delta y \,.$$

The rate of change of the flux $\Delta\Phi$ is [see (5.41)]

$$\begin{aligned}\frac{\mathrm{d}}{\mathrm{d}t}(\Delta\Phi) &= \frac{\mathrm{d}B}{\mathrm{d}t}\Delta S + B\frac{\mathrm{d}}{\mathrm{d}t}(\Delta S) \\ &= \left[\frac{\mathrm{d}\boldsymbol{B}}{\mathrm{d}t} + \boldsymbol{B}(\boldsymbol{\nabla}\cdot\boldsymbol{V}) - (\boldsymbol{B}\cdot\boldsymbol{\nabla})\boldsymbol{V}\right]_z \Delta S \\ &= \frac{\eta}{\mu_0}\Delta B_z(\Delta S)\,. \end{aligned} \tag{5.44}$$

When $R_{\mathrm{m}} \to \infty$, $\eta \to 0$, the rate of change of the flux becomes zero, i.e., $\mathrm{d}(\Delta\Phi)/\mathrm{d}t \to 0$. This means that the magnetic flux is frozen in the plasma.

## 5.4 Magnetoacoustic Wave

As usual, we indicate zeroth-order quantities (in the equilibrium state) by a subscript 0 and first-order perturbation terms by a subscript 1, so that $\rho_{\mathrm{m}} = \rho_{\mathrm{m0}} + \rho_{\mathrm{m1}}$, $p = p_0 + p_1$, $\boldsymbol{V} = 0 + \boldsymbol{V}$, $\boldsymbol{B} = \boldsymbol{B}_0 + \boldsymbol{B}_1$. The case $\eta = 0$ will be considered here. We find the first-order equations as follows:

$$\frac{\partial \rho_{\mathrm{m1}}}{\partial t} + \boldsymbol{\nabla}\cdot(\rho_{\mathrm{m0}}\boldsymbol{V}) = 0\,, \tag{5.45}$$

$$\rho_{\mathrm{m0}}\frac{\partial \boldsymbol{V}}{\partial t} + \boldsymbol{\nabla}p_1 = \boldsymbol{j}_0 \times \boldsymbol{B}_1 + \boldsymbol{j}_1 \times \boldsymbol{B}_0\,, \tag{5.46}$$

$$\frac{\partial p_1}{\partial t} + (\boldsymbol{V}\cdot\boldsymbol{\nabla})p_0 + \gamma p_0 \boldsymbol{\nabla}\cdot\boldsymbol{V} = 0\,, \tag{5.47}$$

$$\frac{\partial \boldsymbol{B}_1}{\partial t} = \boldsymbol{\nabla} \times (\boldsymbol{V} \times \boldsymbol{B}_0)\,. \tag{5.48}$$

If the displacement of the plasma from the equilibrium position $\boldsymbol{r}_0$ is denoted by $\boldsymbol{\xi}(\boldsymbol{r}_0, t)$, it follows that

$$\boldsymbol{V} = \frac{\mathrm{d}\boldsymbol{\xi}}{\mathrm{d}t} \approx \frac{\partial \boldsymbol{\xi}}{\partial t}\,, \qquad \boldsymbol{\xi}(\boldsymbol{r}_0, t) = \boldsymbol{r} - \boldsymbol{r}_0\,. \tag{5.49}$$

Substituting (5.49) into (5.48), (5.45), and (5.47) yields

$$\boldsymbol{B}_1 = \boldsymbol{\nabla} \times (\boldsymbol{\xi} \times \boldsymbol{B}_0)\,, \tag{5.50}$$

$$\mu_0 \boldsymbol{j}_1 = \boldsymbol{\nabla} \times \boldsymbol{B}_1\,, \tag{5.51}$$

$$\rho_{\mathrm{m1}} = -\boldsymbol{\nabla}\cdot(\rho_{\mathrm{m0}}\boldsymbol{\xi})\,, \tag{5.52}$$

$$p_1 = -\boldsymbol{\xi}\cdot\boldsymbol{\nabla}p_0 - \gamma p_0 \boldsymbol{\nabla}\cdot\boldsymbol{\xi}\,. \tag{5.53}$$

Then (5.46) reduces to

$$\rho_{\mathrm{m0}}\frac{\partial^2\boldsymbol{\xi}}{\partial t^2} = \boldsymbol{\nabla}(\boldsymbol{\xi}\cdot\boldsymbol{\nabla}p_0 + \gamma p_0\boldsymbol{\nabla}\cdot\boldsymbol{\xi}) + \frac{1}{\mu_0}(\boldsymbol{\nabla}\times\boldsymbol{B}_0)\times\boldsymbol{B}_1 + \frac{1}{\mu_0}(\boldsymbol{\nabla}\times\boldsymbol{B}_1)\times\boldsymbol{B}_0\,. \tag{5.54}$$

Let us consider the case where $\boldsymbol{B}_0 = \text{const.}$, $p_0 = \text{const.}$ and the displacement is expressed by $\boldsymbol{\xi}(\boldsymbol{r},t) = \boldsymbol{\xi}_1 \exp \mathrm{i}(\boldsymbol{k}\cdot\boldsymbol{r} - \omega t)$. Then (5.54) reduces to

$$-\rho_{\mathrm{m0}}\omega^2\boldsymbol{\xi}_1 = -\gamma p_0(\boldsymbol{k}\cdot\boldsymbol{\xi}_1)\boldsymbol{k} - \mu_0^{-1}\Big\{\boldsymbol{k}\times\big[\boldsymbol{k}\times(\boldsymbol{\xi}_1\times\boldsymbol{B}_0)\big]\Big\}\times\boldsymbol{B}_0\,. \tag{5.55}$$

Using the vector formula $\boldsymbol{a}\times(\boldsymbol{b}\times\boldsymbol{c})$ (see Table 5.1), we can write (5.55) as

$$\begin{aligned}\big[(\boldsymbol{k}\cdot\boldsymbol{B}_0)^2 - \mu_0\omega^2\rho_{\mathrm{m0}}\big]\boldsymbol{\xi}_1 + \big[(B_0^2 + \mu_0\gamma p_0)\boldsymbol{k} - (\boldsymbol{k}\cdot\boldsymbol{B}_0)\boldsymbol{B}_0\big](\boldsymbol{k}\cdot\boldsymbol{\xi}_1)&\\ -(\boldsymbol{k}\cdot\boldsymbol{B}_0)(\boldsymbol{B}_0\cdot\boldsymbol{\xi}_1)\boldsymbol{k} = 0\,.&\end{aligned}$$

Denoting unit vectors along $\boldsymbol{k}$, $\boldsymbol{B}_0$ by $\widehat{\boldsymbol{k}} \equiv \boldsymbol{k}/k$, $\boldsymbol{b} \equiv \boldsymbol{B}_0/B_0$, respectively, and introducing

$$V \equiv \frac{\omega}{k}\,, \qquad v_{\mathrm{A}}^2 \equiv \frac{B_0^2}{\mu_0\rho_{\mathrm{m0}}}\,, \qquad \beta \equiv \frac{p_0}{B_0^2/2\mu_0}\,, \qquad \cos\theta \equiv \widehat{\boldsymbol{k}}\cdot\widehat{\boldsymbol{b}}\,,$$

we find

$$\left(\cos^2\theta - \frac{V^2}{v_{\mathrm{A}}^2}\right)\boldsymbol{\xi}_1 + \left[\left(1+\frac{\gamma\beta}{2}\right)\widehat{\boldsymbol{k}} - \cos\theta\boldsymbol{b}\right](\widehat{\boldsymbol{k}}\cdot\boldsymbol{\xi}_1) - \cos\theta(\boldsymbol{b}\cdot\boldsymbol{\xi}_1)\widehat{\boldsymbol{k}} = 0\,. \tag{5.56}$$

The scalar product of (5.56) with $\widehat{\boldsymbol{k}}$ and $\boldsymbol{b}$, and the vector product of $\widehat{\boldsymbol{k}}$ with (5.56), yield

$$\left(1+\frac{\gamma\beta}{2} - \frac{V^2}{v_{\mathrm{A}}^2}\right)(\widehat{\boldsymbol{k}}\cdot\boldsymbol{\xi}_1) - \cos\theta(\boldsymbol{b}\cdot\boldsymbol{\xi}_1) = 0\,,$$

$$\frac{\gamma\beta}{2}\cos\theta(\widehat{\boldsymbol{k}}\cdot\boldsymbol{\xi}_1) - \frac{V^2}{v_{\mathrm{A}}^2}(\boldsymbol{b}\cdot\boldsymbol{\xi}_1) = 0\,, \qquad \left(\cos^2\theta - \frac{V^2}{v_{\mathrm{A}}^2}\right)\boldsymbol{b}\cdot(\widehat{\boldsymbol{k}}\times\boldsymbol{\xi}_1) = 0\,.$$

The solutions of these equations are magnetoacoustic waves. One solution is

$$V^2 = v_{\mathrm{A}}^2\cos^2\theta\,, \qquad (\boldsymbol{\xi}_1\cdot\boldsymbol{k}) = 0\,, \qquad (\boldsymbol{\xi}_1\cdot\boldsymbol{B}_0) = 0\,. \tag{5.57}$$

Since $\boldsymbol{\xi}_1$ of this solution is orthogonal to $\boldsymbol{k}$ and $\boldsymbol{B}_0$, this is called a torsional Alfvén wave (see Sect. 10.4). The other solutions are given by

$$\left(\frac{V}{v_{\mathrm{A}}}\right)^4 - \left(1+\frac{\gamma\beta}{2}\right)\left(\frac{V}{v_{\mathrm{A}}}\right)^2 + \frac{\gamma\beta}{2}\cos^2\theta = 0\,, \qquad \boldsymbol{B}_0\cdot(\boldsymbol{k}\times\boldsymbol{\xi}_1) = 0\,. \tag{5.58}$$

If the velocity of sound is denoted by $c_{\mathrm{s}}^2 = \gamma p_0/\rho_{\mathrm{m0}}$, (5.58) becomes

$$V^4 + (v_{\mathrm{A}}^2 + c_{\mathrm{s}}^2)V^2 + v_{\mathrm{A}}^2 c_{\mathrm{s}}^2\cos^2\theta = 0$$

and

$$V_{\mathrm{f}}^2 = \frac{1}{2}(v_{\mathrm{A}}^2 + c_{\mathrm{s}}^2) + \frac{1}{2}\left[(v_{\mathrm{A}}^2 + c_{\mathrm{s}}^2)^2 - 4v_{\mathrm{A}}^2 c_{\mathrm{s}}^2 \cos^2\theta\right]^{1/2}, \tag{5.59}$$

$$V_{\mathrm{s}}^2 = \frac{1}{2}(v_{\mathrm{A}}^2 + c_{\mathrm{s}}^2) - \frac{1}{2}\left[(v_{\mathrm{A}}^2 + c_{\mathrm{s}}^2)^2 - 4v_{\mathrm{A}}^2 c_{\mathrm{s}}^2 \cos^2\theta\right]^{1/2}. \tag{5.60}$$

The solution of (5.59) is called a compressional Alfvén wave (see Sect. 10.4) and the solution of (5.60) is called a *magnetoacoustic slow wave*. The characteristic velocity

$$v_{\mathrm{A}}^2 = \frac{B^2}{\mu_0 \rho_{\mathrm{m0}}}$$

is called the *Alfvén velocity*. The plasma with zero resistivity is frozen to the magnetic field. There is tension $B^2/2\mu_0$ along the magnetic field line. As the plasma, of mass density $\rho_{\mathrm{m}}$, sticks to the field lines, the magnetoacoustic waves can be considered as waves propagating along the strings of magnetic field lines (see Sect. 10.4).

# 6 Equilibrium

In order to maintain a hot plasma, we must confine it and keep it away from the vacuum container wall. The most promising method for such confinement of a hot plasma is the use of appropriate strong magnetic fields. An equilibrium condition must be satisfied for such magnetic confinement systems.

## 6.1 Pressure Equilibrium

When a plasma is in the steady state and the fluid velocity is zero ($\boldsymbol{V} = 0$), the magnetohydrodynamic equation (5.30) yields the equilibrium equation

$$\nabla p = \boldsymbol{j} \times \boldsymbol{B} \,, \tag{6.1}$$

and

$$\nabla \times \boldsymbol{B} = \mu_0 \boldsymbol{j} \,, \tag{6.2}$$

$$\nabla \cdot \boldsymbol{B} = 0 \,, \tag{6.3}$$

$$\nabla \cdot \boldsymbol{j} = 0 \,. \tag{6.4}$$

From (6.1), we have

$$\boldsymbol{B} \cdot \nabla p = 0 \,, \tag{6.5}$$

$$\boldsymbol{j} \cdot \nabla p = 0 \,. \tag{6.6}$$

Equation (6.5) indicates that $\boldsymbol{B}$ and $\nabla p$ are orthogonal, and the surfaces of constant pressure coincide with the magnetic surfaces. Equation (6.6) shows that the current density vector $\boldsymbol{j}$ is everywhere parallel to the constant pressure surfaces. Substituting (6.2) into (6.1) yields

$$\nabla \left( p + \frac{B^2}{2\mu_0} \right) = (\boldsymbol{B} \cdot \nabla) \frac{\boldsymbol{B}}{\mu_0} = \frac{B^2}{\mu_0} \left( -\frac{1}{R} \boldsymbol{n} + \frac{\partial B / \partial l}{B} \boldsymbol{b} \right) \,. \tag{6.7}$$

The vector relations

$$\boldsymbol{B} \times (\nabla \times \boldsymbol{B}) + (\boldsymbol{B} \cdot \nabla) \boldsymbol{B} \equiv \nabla (\boldsymbol{B} \cdot \boldsymbol{B}/2) \,,$$

$$(\boldsymbol{B} \cdot \nabla) \boldsymbol{B} = B^2 \left[ (\boldsymbol{b} \cdot \nabla) \boldsymbol{b} + \boldsymbol{b} \frac{(\boldsymbol{b} \cdot \nabla) B}{B} \right] = B^2 \left( -\frac{\boldsymbol{n}}{R} + \frac{\boldsymbol{b}}{B} \frac{\partial B}{\partial l} \right)$$

are used, where $R$ is the radius of curvature of the line of magnetic force and $\boldsymbol{n}$ is the unit vector directed toward a point on the line of magnetic force from the center of curvature. $l$ is the length along the field line. We find that the right-hand side of (6.7) can be neglected when the radius of curvature is much larger than the length scale of the pressure gradient, i.e., the size of the plasma, and the variation of $\boldsymbol{B}$ along the line of magnetic force is much smaller than the variation of $\boldsymbol{B}$ in the perpendicular direction. Then (6.7) becomes

$$p + \frac{B^2}{2\mu_0} \sim \frac{B_0^2}{2\mu_0} ,$$

where $B_0$ is the the value of the magnetic field at the plasma boundary ($p = 0$). When the system is axially symmetric and $\partial/\partial z = 0$, (6.7) exactly reduces to

$$\frac{\partial}{\partial r}\left(p + \frac{B_z^2 + B_\theta^2}{2\mu_0}\right) = -\frac{B_\theta^2}{r\mu_0} . \tag{6.8}$$

Multiplying (6.8) by $r^2$ and integrating by parts, we obtain

$$\left(p + \frac{B_z^2 + B_\theta^2}{2\mu_0}\right)_{r=a} = \frac{1}{\pi a^2}\int_0^a \left(p + \frac{B_z^2}{2\mu_0}\right) 2\pi r \mathrm{d}r$$

and

$$\langle p \rangle + \frac{\langle B_z^2 \rangle}{2\mu_0} = p_a + \frac{B_z^2(a) + B_\theta^2(a)}{2\mu_0} , \tag{6.9}$$

where $\langle\ \rangle$ is the volume average and $p_a$ is the plasma pressure at the plasma boundary. As $B^2/2\mu_0$ is the pressure of the magnetic field, (6.9) is the pressure equilibrium equation. The ratio of the plasma pressure to the pressure of the external magnetic field $B_0$, viz.,

$$\beta \equiv \frac{p}{B_0^2/2\mu_0} = \frac{n(T_\mathrm{e} + T_\mathrm{i})}{B_0^2/2\mu_0} , \tag{6.10}$$

is called the *beta ratio*. For a confined plasma, $\beta$ is always smaller than 1, and is used as a figure of merit for the confining magnetic field. The ratio of the plasma pressure to the pressure of the poloidal field $B_\theta$ is called the *poloidal beta*.

When the pressure at the boundary is $p_a = 0$ and $|B_z(a) - B_z(r)| \ll |B_z(a)|$ in (6.9), the poloidal beta $\beta_\mathrm{p}$ is

$$\beta_\mathrm{p} \equiv \frac{\langle p \rangle}{B_\theta^2(a)/2\mu_0} = +\frac{B_z^2(a) - \langle B_z^2(r)\rangle}{B_\theta(a)^2} \approx 1 + \left(\frac{2B_z}{B_\theta^2}\right)_a \langle B_z(a) - B_z(r)\rangle . \tag{6.11}$$

$B_z(a)$ is the magnetic field in the direction of $z$ in the case without plasma. In the case $\beta_\mathrm{p} > 1$, the magnitude of the magnetic field $B_z(r)$ inside the plasma is smaller than that in the vacuum case $[B_z(r) < B_z(a)]$. This indicates that the plasma is *diamagnetic*. In the case $\beta_\mathrm{p} < 1$, $B_z(r)$ becomes larger than $B_z(a)$ [$B_z(r) > B_z(a)$]. When the plasma current flows along a line of magnetic force, the current produces the poloidal magnetic field, and a poloidal component of the plasma current appears and induces an additional $z$ component of the magnetic field. This is the origin of *paramagnetism* in the plasma.

## 6.2 Equilibrium Equation for Axially Symmetric Systems

Let us use cylindrical coordinates $(r, \varphi, z)$ and denote the magnetic surface by $\psi$. The magnetic surface $\psi$ in an axisymmetric system is given by [see (3.19)]

$$\psi = rA_\varphi(r, z) .$$

The $r$ and $z$ components of the magnetic field are given by

$$rB_r = -\frac{\partial\psi}{\partial z} , \qquad rB_z = \frac{\partial\psi}{\partial r} . \tag{6.12}$$

Therefore $\psi$ is also called the flux function. The relation $\boldsymbol{B}\cdot\boldsymbol{\nabla}p = 0$ follows from the equilibrium equation and is expressed by

$$-\frac{\partial\psi}{\partial z}\frac{\partial p}{\partial r} + \frac{\partial\psi}{\partial r}\frac{\partial p}{\partial z} = 0 .$$

Accordingly, $p$ is a function of $\psi$ alone, i.e.,

$$p = p(\psi) . \tag{6.13}$$

Similarly, from $\boldsymbol{j}\cdot\boldsymbol{\nabla}p = 0$ and $\boldsymbol{\nabla} \times \boldsymbol{B} = \mu_0\boldsymbol{j}$, we may write

$$-\frac{\partial p}{\partial r}\frac{\partial(rB_\varphi)}{\partial z} + \frac{\partial p}{\partial z}\frac{\partial(rB_\varphi)}{\partial r} = 0 .$$

This means that $rB_\varphi$ is a function of $\psi$ alone and

$$rB_\varphi = \frac{\mu_0 I(\psi)}{2\pi} . \tag{6.14}$$

Equation (6.14) indicates that $I(\psi)$ is the current flowing in the poloidal direction through the circular cross-section within $\psi = rA_\varphi$ (Fig. 6.1). The $r$ component of $\boldsymbol{j} \times \boldsymbol{B} = \boldsymbol{\nabla}p$ leads to the equation for $\psi$:

$$L(\psi) + \mu_0 r^2\frac{\partial p(\psi)}{\partial\psi} + \frac{\mu_0^2}{8\pi^2}\frac{\partial I^2(\psi)}{\partial\psi} = 0 , \tag{6.15}$$

where

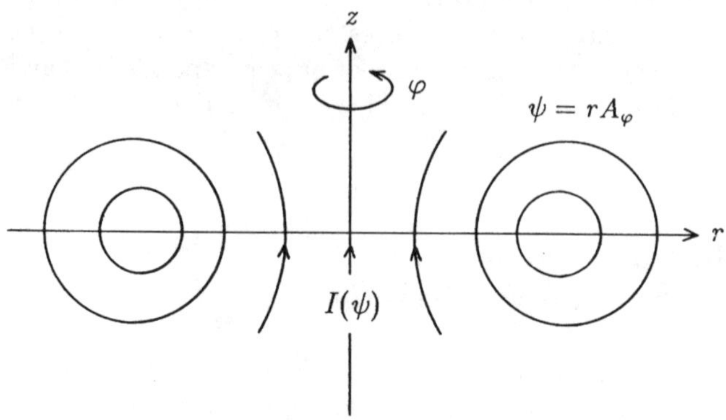


**Fig. 6.1.** Magnetic surfaces $\psi = rA_\varphi$ and $I(\psi)$

$$L(\psi) \equiv \left( r\frac{\partial}{\partial r}\frac{1}{r}\frac{\partial}{\partial r} + \frac{\partial^2}{\partial z^2} \right) \psi \, .$$

This equation is called the *Grad–Shafranov equation*. The current density is expressed in terms of the function of the magnetic surface as

$$j_r = \frac{-1}{2\pi r}\frac{\partial I(\psi)}{\partial z} \, , \qquad j_z = \frac{1}{2\pi r}\frac{\partial I(\psi)}{\partial r} \, ,$$

$$j_\varphi = \frac{-1}{\mu_0}\left( \frac{\partial}{\partial r}\frac{1}{r}\frac{\partial \psi}{\partial r} + \frac{1}{r}\frac{\partial^2 \psi}{\partial z^2} \right) = -\frac{L(\psi)}{\mu_0 r} = \frac{1}{\mu_0 r}\left[ \mu_0 r^2 p' + \frac{\mu_0^2}{8\pi^2}(I^2)' \right] \, ,$$

where the prime indicates differentiation with respect to $\psi$. Using (6.12) and (6.14), we have

$$\boldsymbol{j} = \frac{I'}{2\pi}\boldsymbol{B} + p' r \boldsymbol{e}_\varphi \, , \tag{6.16}$$

$$L(\psi) + \mu_0 r j_\varphi = 0 \, . \tag{6.17}$$

When the unit vectors in the directions of $r, \varphi, z$ are denoted by $\boldsymbol{e}_r$, $\boldsymbol{e}_\varphi$, $\boldsymbol{e}_z$, respectively, we have $\nabla\varphi = \boldsymbol{e}_\varphi / R$, $\boldsymbol{e}_r \times \boldsymbol{e}_\varphi = \boldsymbol{e}_z$, $\boldsymbol{e}_z \times \boldsymbol{e}_\varphi = -\boldsymbol{e}_r$. Therefore, from (6.12) and (6.14), $\boldsymbol{B}$ can be expressed as

$$\boldsymbol{B} = \frac{\mu_0 I(\psi)}{2\pi}\nabla\varphi + \nabla\psi \times \nabla\varphi \, . \tag{6.18}$$

$p(\psi)$ and $I^2(\psi)$ are arbitrary functions of $\psi$. When they are linear or quadratic functions of $\psi$, (6.15) becomes a linear differential equation. Let us consider a simple linear case for $\psi$. At the plasma boundary $\psi = \psi_\mathrm{b}$, we let $p_\mathrm{b} = p(\psi_\mathrm{b})$ and $I_\mathrm{b}^2 = I^2(\psi_\mathrm{b})$, i.e.,

$$p(\psi) = p_\mathrm{b} - \frac{a}{\mu_0 R^2}(\psi - \psi_\mathrm{b}) \, , \tag{6.19}$$

$$I^2(\psi) = I_b^2 - \frac{8\pi^2}{\mu_0^2} b(\psi - \psi_b) . \tag{6.20}$$

Then (6.15) and (6.17) reduce to

$$L(\psi) = a\frac{r^2}{R^2} + b = -\mu_0 r j_\varphi . \tag{6.21}$$

We set the position of the magnetic axis to $(R, 0)$. The function

$$\psi - \psi_0 = \frac{b+a}{1+\epsilon}\left[\frac{1}{2}\left(1 + c\frac{r^2 - R^2}{R^2}\right) z^2 + \frac{\epsilon}{8R^2}(r^2 - R^2)^2 \right.$$
$$\left. + \frac{(1+\epsilon)b - (1-c)(b+a)}{24(b+a)R^4}(r^2 - R^2)^3\right] \tag{6.22}$$

is then the solution of (6.21), which is correct up to the cube of $(r-R), z$ [6.1]. $\epsilon, c$ are constant and $\psi_0 = \psi(R, 0)$. When the coefficient of the third term on the right-hand side of (6.22) is 0, i.e.,

$$(1+\epsilon)b - (1-c)(b+a) = 0 \quad \longrightarrow \quad \epsilon = -(c-1)\frac{a}{b} - c , \tag{6.23}$$

equation (6.22) becomes the exact Solovev solution of Grad–Shafranov (6.21) [6.1]. When we set $c = R^2/(R^2 - R_x^2)$, $\epsilon$ becomes

$$\epsilon = -\left(\frac{a}{b} + \frac{R^2}{R_x^2}\right)\frac{R_x^2}{R^2 - R_x^2} ,$$

due to (6.23), and then (6.22) reduces to

$$\psi = \frac{b}{2}\left(1 - \frac{r^2}{R_x^2}\right) z^2 + \frac{a + (R^2/R_x^2)b}{8R^2}\left[(r^2 - R^2)^2 - (R^2 - R_x^2)^2\right] . \tag{6.24}$$

Equation (6.24) is an exact equilibrium solution in the interior region of the plasma surrounded by the conductive wall specified by $\psi(r, z) = \psi_b$. The surface $\psi(r, z) = 0$ is the separatrix surface (see Fig. 6.2 and Sect. 16.5). The separatrix points X are located at $(R_x, \pm Z_x)$, where

$$Z_x = \left[-\left(\frac{a}{b} + \frac{R^2}{R_x^2}\right)\frac{1}{2}\left(1 - \frac{R_x^2}{R^2}\right)\right]^{1/2} R_x .$$

The maximum value $R_{max}$ of $r$ within the separatrix surface is

$$R_{max} = \left(2 - \frac{R_x^2}{R^2}\right)^{1/2} R .$$

When we set the separatrix surface as the plasma boundary ($\psi_b = 0$), the aspect ratio $A$, elongation ratio $\kappa_s$, and central poloidal beta $\beta_{p0}$ are

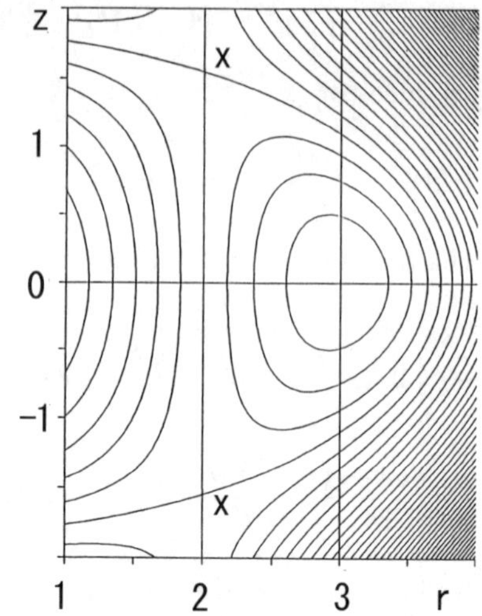


**Fig. 6.2.** The contour (magnetic surface) of the flux function $\psi$ of (6.24) in the case $a/b = 4.4$, $R = 3$, $R_{\rm x} = 2$. X indicates the separatrix points and the magnetic surface passing through the X points is the separatix surface

$$\frac{1}{A} = \frac{R_{\rm max} - R_{\rm x}}{2R} = \frac{(2 - R_{\rm x}^2/R^2)^{1/2} - R_{\rm x}/R}{2},$$

$$\kappa_{\rm s} = \frac{2Z_{\rm x}}{R_{\rm max} - R_{\rm x}} = \frac{AZ_{\rm x}}{R},$$

$$\beta_{p0} \equiv \frac{p(R,0) - p_{\rm b}}{B_z^2(R_{\rm x},0)/2\mu_0} = \frac{a}{a + (R^2/R_{\rm x}^2)b}.$$

When $A$ and $\kappa_{\rm s}$ are specified, $\beta_{p0}$ is fixed. To avoid this inadequacy, Weening [6.2] added an additional particular solution $r^2\ln(r^2/R_\alpha^2) - r^2$ to the Solovev solution (6.24), i.e.,

$$\psi = \frac{b+d}{2}\left(1 - \frac{r^2}{R_{\rm x}^2}\right)z^2 + \frac{a + (R^2/R_{\rm x}^2)(b+d)}{8R^2}\left[(r^2 - R^2)^2 - (R^2 - R_{\rm x}^2)^2\right]$$
$$-\frac{d}{4}\left[r^2\ln\frac{r^2}{R_{\rm x}^2} - (r^2 - R_{\rm x}^2)\right]. \qquad (6.25)$$

When the plasma boundary is chosen to be the separatrix $\psi(r,z) = 0$, the aspect ratio $A$, elongation ratio $\kappa_{\rm s}$ and central poloidal beta $\beta_{p0}$ are

$$\frac{Z_{\rm x}^2}{R_{\rm x}^2} = -\frac{1}{2}\left(\frac{a}{b+d} + \frac{R^2}{R_{\rm x}^2}\right)\left(1 - \frac{R_{\rm x}^2}{R^2}\right),$$

$$\frac{R_{\rm max}^2}{R^2} = \left(2 - \frac{R_{\rm x}^2}{R^2}\right) + \frac{2d[x\ln x/(x-1) - 1]}{a + (R^2/R_{\rm x}^2)(b+d)}, \qquad x \equiv \frac{R_{\rm max}^2}{R_{\rm x}^2},$$

$$\frac{1}{A} = \frac{R_{\max}/R - R_{\mathrm{x}}/R}{2} , \qquad \kappa_{\mathrm{s}} = \frac{AZ_{\mathrm{x}}}{R} ,$$

$$\beta_{p0} = \frac{a}{a + (R^2/R_{\mathrm{x}}^2)(b+d)} \left\{ 1 + \frac{2d\big[\ln(R^2/R_{\mathrm{x}}^2) - (1 - R^2/R_{\mathrm{x}}^2)\big]}{\big[a + (R^2/R_{\mathrm{x}}^2)(b+d)\big](1 - R_{\mathrm{x}}^2/R^2)} \right\} .$$

The magnetic surface $\psi$, the magnetic field $\boldsymbol{B}$, and the pressure $p$ in a translationally symmetric system $(\partial/\partial z = 0)$ are given by

$$\psi = A_z(r, \theta) ,$$

$$B_r = \frac{1}{r}\frac{\partial \psi}{\partial \theta} , \qquad B_\theta = -\frac{\partial \psi}{\partial r} , \qquad B_z = \frac{\mu_0}{2\pi} I(\psi) ,$$

$$p = p(\psi) .$$

The equilibrium equation reduces to

$$\frac{1}{r}\frac{\partial}{\partial r}\left( r \frac{\partial \psi}{\partial r} \right) + \frac{1}{r^2}\frac{\partial^2 \psi}{\partial \theta^2} + \mu_0 \frac{\partial p(\psi)}{\partial \psi} + \frac{\mu_0^2}{8\pi^2}\frac{\partial I^2(\psi)}{\partial \psi} = 0 , \tag{6.26}$$

$$\boldsymbol{j} = \frac{1}{2\pi} I' \boldsymbol{B} + p' \boldsymbol{e}_z , \qquad \Delta \psi + \mu_0 j_z = 0 . \tag{6.27}$$

It is possible to derive a similar equilibrium equation for a helically symmetric system.

## 6.3 Tokamak Equilibrium

The poloidal magnetic field produced by a plasma current $I_{\mathrm{p}}$ inside the plasma ring is stronger than that outside the plasma ring. For the tokamak equilibrium, one must therefore add a vertical field to reduce the poloidal field inside the ring and to inrease the poloidal field outside the ring, as shown in Fig. 6.3. Let us estimate the required vertical field $B_\perp$.

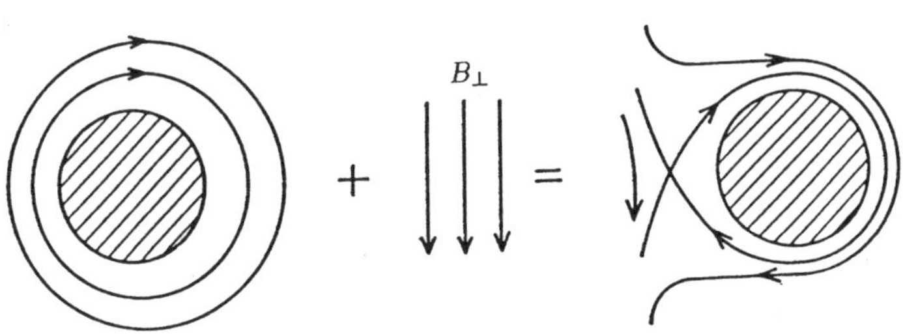


**Fig. 6.3.** Poloidal magnetic field due to the combined plasma current and vertical field

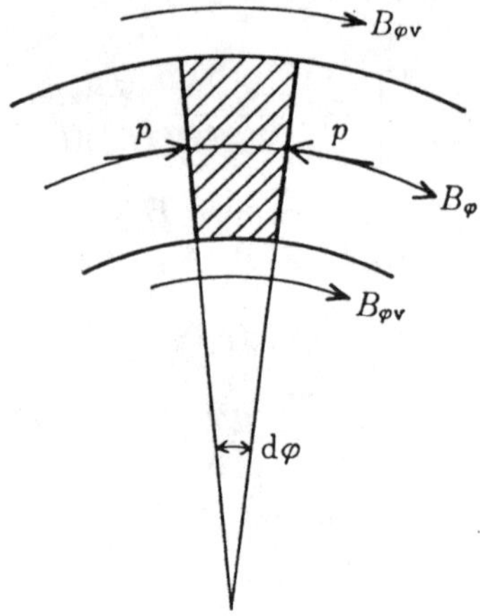


**Fig. 6.4.** Equilibrium of forces acting on a toroidal plasma

The *hoop force* by which the current ring of a plasma tends to expand is given by

$$F_\mathrm{h} = -\frac{\partial}{\partial R}\frac{L_\mathrm{p} I_\mathrm{p}^2}{2}\bigg|_{L_\mathrm{p} I_\mathrm{p}=\mathrm{const.}} = \frac{1}{2} I_\mathrm{p}^2 \frac{\partial L_\mathrm{p}}{\partial R} \, ,$$

where $L_\mathrm{p}$ is the self-inductance of the current ring,

$$L_\mathrm{p} = \mu_0 R \left( \ln \frac{8R}{a} + \frac{l_\mathrm{i}}{2} - 2 \right) . \tag{6.28}$$

In this expression, $\mu_0 R[\ln(8R/a) - 2]$ is the inductance due to the magnetic field energy outside the plasma and $\mu_0 R l_\mathrm{i}/2$ is the inductance due to the magnetic field energy inside the plasma, with

$$l_\mathrm{i} \equiv \frac{2\pi \int_0^a B_\mathrm{p}^2(\rho)\rho \mathrm{d}\rho}{\pi a^2 B_\mathrm{p}^2(a)} . \tag{6.29}$$

Accordingly, the hoop force is

$$F_\mathrm{h} = \frac{\mu_0 I_\mathrm{p}^2}{2} \left( \ln \frac{8R}{a} + \frac{l_\mathrm{i}}{2} - 1 \right) .$$

The outward force $F_\mathrm{p}$ exerted by the plasma pressure is (Fig. 6.4)

$$F_\mathrm{p} = \langle p \rangle \pi a^2 2\pi .$$

The inward (contractive) force $F_{B1}$ due to the tension of the toroidal field inside the plasma is

$$F_{B1} = -\frac{\langle B_\varphi^2 \rangle}{2\mu_0} 2\pi^2 a^2 ,$$

and the outward force $F_{B2}$ due to the pressure caused by the external magnetic field is

$$F_{B2} = \frac{B_{\varphi\mathrm{v}}^2}{2\mu_0} 2\pi^2 a^2 .$$

The force $F_\mathrm{I}$ acting on the plasma due to the vertical field $B_\perp$ is

$$F_\mathrm{I} = I_\mathrm{p} B_\perp 2\pi R .$$

Balancing these forces gives

$$\frac{\mu_0 I_\mathrm{p}^2}{2}\left(\ln \frac{8R}{a} + \frac{l_\mathrm{i}}{2} - 1\right) + 2\pi^2 a^2 \left(\langle p \rangle + \frac{B_{\varphi\mathrm{v}}^2}{2\mu_0} - \frac{\langle B_\varphi^2 \rangle}{2\mu_0}\right) + 2\pi R I_\mathrm{p} B_\perp = 0 ,$$

and the amount of $B_\perp$ required is

$$B_\perp = -\frac{\mu_0 I_\mathrm{p}}{4\pi R}\left(\ln \frac{8R}{a} + \frac{l_\mathrm{i}}{2} - 1 + \beta_\mathrm{p} - \frac{1}{2}\right) = -\frac{\mu_0 I_\mathrm{p}}{4\pi R}\left(\ln \frac{8R}{a} + \Lambda - \frac{1}{2}\right) , \tag{6.30}$$

where

$$\beta_\mathrm{p} = \frac{\langle p \rangle}{B_\mathrm{p}^2(a)/2\mu_0} , \tag{6.31}$$

$$\Lambda = \beta_\mathrm{p} + l_\mathrm{i}/2 - 1 . \tag{6.32}$$

Equation (6.9) has been used for the derivation. The equilibrium of a tokamak with circular cross-section is discussed in detail in the references [6.3, 6.4].

## 6.4 Upper Limit of Beta Ratio

In the last section, the necessary vertical field $B_\perp$ for plasma equilibrium is given by

$$B_\perp = B_a \frac{a}{2R}\left(\ln \frac{8R}{a} + \Lambda - \frac{1}{2}\right) .$$

The direction of $B_\perp$ is opposite to that of $B_\omega$ produced by the plasma current inside the torus, so that the resultant poloidal field becomes zero at some points in the inside region of the torus and a separatrix is formed. When the plasma pressure is increased and $\beta_\mathrm{p}$ becomes large, the required amount of $B_\perp$ is increased and the separatrix shifts toward the plasma. For simplicity, let us consider a sharp-boundary model in which the plasma pressure is constant inside the plasma boundary, and in which the boundary encloses a plasma current $I_\mathrm{p}$. Then the pressure-balance equation is

$$\frac{B_\omega^2}{2\mu_0} + \frac{B_{\varphi\mathrm{v}}^2}{2\mu_0} \approx p + \frac{B_{\varphi\mathrm{i}}^2}{2\mu_0} , \tag{6.33}$$

where $B_\omega$ is the poloidal field outside the plasma and $B_{\varphi\mathrm{v}}$, $B_{\varphi\mathrm{i}}$ are the $\varphi$ components of the field outside and inside the plasma boundary, respectively. $B_{\varphi\mathrm{v}}$ and $B_{\varphi\mathrm{i}}$ are proportional to $1/r$, according to (6.14). If the values of $B_{\varphi\mathrm{v}}$, $B_{\varphi\mathrm{i}}$ at $r = R$ are denoted by $B^0_{\varphi\mathrm{v}}$, $B^0_{\varphi\mathrm{i}}$, respectively, (6.33) may be written as

$$B_\omega^2 = 2\mu_0 p - \left[(B^0_{\varphi\mathrm{v}})^2 - (B^0_{\varphi\mathrm{i}})^2\right]\left(\frac{R}{r}\right)^2 .$$

The upper limit of the plasma pressure is determined by the condition that the resultant poloidal field at $r = r_{\mathrm{min}}$ inside the torus should be zero, i.e.,

$$2\mu_0 p_{\mathrm{max}} \frac{r^2_{\mathrm{min}}}{R^2} = (B^0_{\varphi\mathrm{v}})^2 - (B^0_{\varphi\mathrm{i}})^2 . \tag{6.34}$$

As $r$ is expressed by $r = R + a\cos\omega$, (6.33) reduces (using $r_{\mathrm{min}} = R - a$) to

$$B_\omega^2 = 2\mu_0 p_{\mathrm{max}}\left(1 - \frac{r^2_{\mathrm{min}}}{r^2}\right) = 8\mu_0 p_{\mathrm{max}} \frac{a}{R}\cos^2\frac{\omega}{2} .$$

Here $a/R \ll 1$ is assumed. From the relation $\oint B_\omega a \mathrm{d}\omega = \mu_0 I_{\mathrm{p}}$, the upper limit $\beta_{\mathrm{p}}^{\mathrm{c}}$ of the poloidal beta ratio is

$$\beta_{\mathrm{p}}^{\mathrm{c}} = \frac{\pi^2}{16}\frac{R}{a} \approx 0.5\frac{R}{a} . \tag{6.35}$$

Thus the upper limit of $\beta_{\mathrm{p}}^{\mathrm{c}}$ is half of the aspect ratio $R/a$ in this simple model. When the rotational transform angle $\iota$ and the safety factor $q_{\mathrm{s}} = 2\pi/\iota$ are introduced, we find that

$$\frac{B_\omega}{B_\varphi} = \frac{a}{R}\left(\frac{\iota}{2\pi}\right) = \frac{a}{Rq_{\mathrm{s}}} ,$$

so that $\beta$ is

$$\beta = \frac{p}{B^2/2\mu_0} \approx \frac{p}{B_\omega^2/2\mu_0}\left(\frac{B_\omega}{B_\varphi}\right)^2 = \left(\frac{a}{Rq_{\mathrm{s}}}\right)^2 \beta_{\mathrm{p}} . \tag{6.36}$$

Accordingly, the upper limit of the beta ratio is

$$\beta^{\mathrm{c}} = \frac{0.5}{q_{\mathrm{s}}^2}\frac{a}{R} . \tag{6.37}$$

## 6.5 Pfirsch–Schlüter Current

When the plasma pressure is isotropic, the current $\boldsymbol{j}$ in the plasma is given by (6.1) and (6.4) as

$$\boldsymbol{j}_\perp = \frac{\boldsymbol{b}}{B} \times \boldsymbol{\nabla} p \,, \tag{6.38}$$

$$\boldsymbol{\nabla} \cdot \boldsymbol{j}_\parallel = -\boldsymbol{\nabla} \cdot \boldsymbol{j}_\perp = -\boldsymbol{\nabla} \cdot \left( \frac{\boldsymbol{B}}{B^2} \times \boldsymbol{\nabla} p \right) = -\boldsymbol{\nabla} p \cdot \boldsymbol{\nabla} \times \left( \frac{\boldsymbol{B}}{B^2} \right) .$$

Then the component of the plasma current parallel to the magnetic field $j_\parallel$ is

$$\boldsymbol{\nabla} \cdot \boldsymbol{j}_\parallel = -\boldsymbol{\nabla} p \cdot \left[ \left( \boldsymbol{\nabla} \frac{1}{B^2} \times \boldsymbol{B} \right) + \frac{\mu_0 \boldsymbol{j}}{B^2} \right] = 2 \boldsymbol{\nabla} p \cdot \frac{\boldsymbol{\nabla} B \times \boldsymbol{B}}{B^3} \,, \tag{6.39}$$

$$\frac{\partial j_\parallel}{\partial s} = 2 \boldsymbol{\nabla} p \cdot \frac{(\boldsymbol{\nabla} B \times \boldsymbol{b})}{B^2} \,, \tag{6.40}$$

where $s$ is the length along a line of magnetic force. In the zeroth-order approximation, we can put $B \propto 1/R$, $p = p(r)$, and

$$\frac{\partial}{\partial s} = \frac{\partial \theta}{\partial s} \frac{\partial}{\partial \theta} = \frac{\iota}{2\pi R} \frac{\partial}{\partial \theta} \,,$$

where $\iota$ is the rotational transform angle. When $s$ increases by $2\pi R$, $\theta$ increases by $\iota$. Then (6.40) reduces to

$$\frac{\iota}{2\pi R} \frac{\partial j_\parallel}{\partial \theta} = -\frac{\partial p}{\partial r} \frac{2}{RB} \sin\theta \,,$$

that is,

$$j_\parallel = \frac{4\pi}{\iota B} \frac{\partial p}{\partial r} \cos\theta \,. \tag{6.41}$$

This current is called the *Pfirsch–Schlüter current* [6.5]. These formulas are very important, and will be used to estimate the diffusion coefficient of a toroidal plasma in Chap. 7. The Pfirsch–Schlüter current is due to the short circuiting, along magnetic field lines, of toroidal drift polarization charges. The resulting current is inversely proportional to $\iota$.

## 6.6 Virial Theorem

The equation of equilibrium $\boldsymbol{j} \times \boldsymbol{B} = (\boldsymbol{\nabla} \times \boldsymbol{B}) \times \boldsymbol{B} = \boldsymbol{\nabla} p$ can be reduced to

$$\sum_i \frac{\partial}{\partial x_i} T_{ik} - \frac{\partial p}{\partial x_k} = 0 \,, \tag{6.42}$$

where

$$T_{ik} = \frac{1}{\mu_0} \left( B_i B_k - \frac{1}{2} B^2 \delta_{ik} \right) . \tag{6.43}$$

The symbol $\delta_{ik}$ is the Kronecker delta matrix. This is called the magnetic stress tensor. From the relation (6.42), we have

$$\int_{S} \left[ \left( p + \frac{B^2}{2\mu_0} \right) \boldsymbol{n} - \frac{\boldsymbol{B}(\boldsymbol{B}\cdot\boldsymbol{n})}{\mu_0} \right] \mathrm{d}S = 0 \,, \tag{6.44}$$

where $\boldsymbol{n}$ is the outward unit normal to the closed surface surrounding a volume V. Since

$$\sum_i \frac{\partial}{\partial x_i} \left[ x_k \left( T_{ik} - p\delta_{ik} \right) \right] = (T_{kk} - p) + x_k \sum_i \frac{\partial}{\partial x_i} (T_{ik} - p\delta_{ik}) = (T_{kk} - p) \,,$$

it follows that

$$\int_{V} \left( 3p + \frac{B^2}{2\mu_0} \right) \mathrm{d}V = \int_{S} \left[ \left( p + \frac{B^2}{2\mu_0} \right) (\boldsymbol{r}\cdot\boldsymbol{n}) - \frac{(\boldsymbol{B}\cdot\boldsymbol{r})(\boldsymbol{B}\cdot\boldsymbol{n})}{\mu_0} \right] \mathrm{d}S \,. \tag{6.45}$$

This is called the virial theorem. When a plasma fills a finite region with $p = 0$ outside the region, and no solid conductor carries the current anywhere inside or outside the plasma, the magnitude of the magnetic field is of the order of $1/r^3$, so the surface integral approaches zero as the plasma surface approaches infinity ($r \to 0$). This contradicts the fact that the volume integral of (6.45) is positive definite. In other words, a plasma of finite extent cannot be in equilibrium unless there exist solid conductors to carry the current.

Let us apply the virial theorem (6.45) to a volume element of an axisymmetric plasma bounded by a closed toroidal surface $S_{\mathrm{t}}$ formed by rotating an arbitrarily shaped contour $\mathrm{l_t}$ (see Fig. 6.5a). We denote the unit normal and tangent of the contour $\mathrm{l_t}$ by $\boldsymbol{n}$ and $\boldsymbol{l}$ respectively and a surface element of the transverse cross-section by $\mathrm{d}S_\varphi$. The volume and the surface element are related by

$$\mathrm{d}V = 2\pi r \mathrm{d}S_\varphi \,.$$

The magnetic field $\boldsymbol{B}$ is expressed by

$$\boldsymbol{B} = B_\varphi \boldsymbol{e}_\varphi + \boldsymbol{B}_{\mathrm{p}} \,,$$

where $\boldsymbol{B}_{\mathrm{p}}$ is the poloidal field, $B_\varphi$ is the magnitude of the toroidal field, and $\boldsymbol{e}_\varphi$ is the unit vector in the $\varphi$ direction.

Notice the two relations

$$\int r^\alpha (\boldsymbol{r}\cdot\boldsymbol{n}) \mathrm{d}S_{\mathrm{t}} = (\alpha + 3) \int r^\alpha \mathrm{d}V \,, \tag{6.46}$$

$$\begin{aligned} \int r^\alpha (\boldsymbol{e}_r\cdot\boldsymbol{n}) \mathrm{d}S_{\mathrm{t}} &= \int \nabla\cdot(r^\alpha \boldsymbol{e}_r) \mathrm{d}V = \int \frac{1}{r}\frac{\partial}{\partial r} r^{\alpha+1} \mathrm{d}V \\ &= (\alpha + 1) \int r^{(\alpha-1)} \mathrm{d}V \\ &= 2\pi(\alpha + 1) \int r^\alpha \mathrm{d}S_\varphi \,, \end{aligned} \tag{6.47}$$

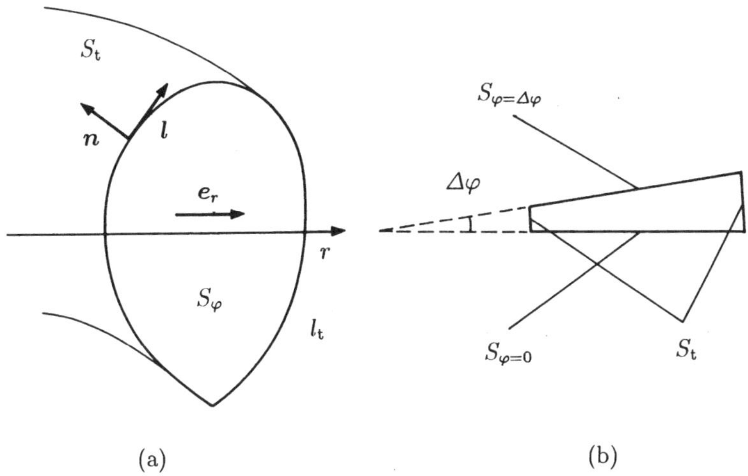


**Fig. 6.5.** Integration region of the virial theorem: (**a**) (6.45) and (**b**) (6.44)

where $\boldsymbol{e}_r$ is the unit vector in the $r$ direction. Applying (6.45) to the full torus surrounded by $S_\mathrm{t}$, we obtain

$$\int \left(3p + \frac{B_\varphi^2 + B_\mathrm{p}^2}{2\mu_0}\right) \mathrm{d}V = \int \left[\left(p + \frac{B_\varphi^2 + B_\mathrm{p}^2}{2\mu_0}\right)(\boldsymbol{n}\cdot\boldsymbol{r}) - \frac{B_n(\boldsymbol{B}\cdot\boldsymbol{r})}{\mu_0}\right] \mathrm{d}S_\mathrm{t}$$
$$= \int \left[\left(p + \frac{B_l^2 - B_n^2}{2\mu_0}\right)(\boldsymbol{n}\cdot\boldsymbol{r}) - \frac{B_n B_l}{\mu_0}(\boldsymbol{l}\cdot\boldsymbol{r})\right] \mathrm{d}S_\mathrm{t}$$
$$+ \int \frac{B_\varphi^2}{2\mu_0}(\boldsymbol{n}\cdot\boldsymbol{r})\mathrm{d}S_\mathrm{t} \,, \qquad (6.48)$$

because $\boldsymbol{B}_\mathrm{p} = B_l\boldsymbol{l} + B_n\boldsymbol{n}$ (see Fig. 6.5a). When the vacuum toroidal field (without plasma) is denoted by $B_{\varphi 0}$, this is given by $B_{\varphi 0} = \mu_0 I/(2\pi r)$, where $I$ is the total coil current generating the toroidal field. Using (6.47), (6.48) reduces to

$$\int \left(3p + \frac{B_\mathrm{p}^2 + B_\varphi^2 - B_{\varphi 0}^2}{2\mu_0}\right) 2\pi r \mathrm{d}S_\varphi \qquad (6.49)$$
$$= \int \left[\left(p + \frac{B_l^2 - B_n^2}{2\mu_0}\right)(\boldsymbol{n}\cdot\boldsymbol{r}) - \frac{B_n B_l}{\mu_0}(\boldsymbol{l}\cdot\boldsymbol{r})\right] \mathrm{d}S_\mathrm{t} \,.$$

Applying (6.44) to the sector surrounded by $\varphi = 0$, $\varphi = \Delta\varphi$ and $S_\mathrm{t}$ (see Fig. 6.5b) and taking its $r$ component gives [6.6]

$$-\Delta\varphi \int \left(p + \frac{B^2}{2\mu_0} - \frac{B_\varphi^2}{\mu_0}\right) \mathrm{d}S_\varphi$$

$$+\frac{\Delta\varphi}{2\pi} \int \left[\left(p + \frac{B^2}{2\mu_0}\right)(\boldsymbol{n}\cdot\boldsymbol{e}_r) - \frac{(\boldsymbol{B}\cdot\boldsymbol{e}_r)(\boldsymbol{B}\cdot\boldsymbol{n})}{\mu_0}\right] \mathrm{d}S_\mathrm{t} = 0 ,$$

$$2\pi \int \left(p + \frac{B_\mathrm{p}^2 - B_\varphi^2 + B_{\varphi 0}^2}{2\mu_0}\right) \mathrm{d}S_\varphi \tag{6.50}$$

$$= \int \left[\left(p + \frac{B_l^2 - B_n^2}{2\mu_0}\right)(\boldsymbol{n}\cdot\boldsymbol{e}_r) - \frac{B_l B_n}{\mu_0}(\boldsymbol{l}\cdot\boldsymbol{e}_r)\right] \mathrm{d}S_\mathrm{t} = 0 .$$

Equations (6.49) and (6.50) are used to measure the poloidal beta ratio (6.31) and the internal plasma inductance per unit length (6.29) of arbitrarily shaped axisymmetric toroidal plasmas by means of magnetic probes surrounding the plasma.

# 7 Plasma Transport

Transport in plasmas is one of the most important subjects in fusion research, with theoretical and experimental investigations being carried out concurrently. Although a general discussion of transport or confinement requires consideration of the various instabilities (which will be studied in subsequent chapters), it is also important to consider simple but fundamental diffusion for the ideal stable cases. A typical example (Sect. 7.1) is classical diffusion, in which collisions between electrons and ions are the dominant effect. Section 7.2 describes the neoclassical diffusion of toroidal plasmas confined in a tokamak, for both the rare-collisional and collisional regions. Sometimes the transport in an unstable plasma may be studied in a general way, without recourse to detailed knowledge of instabilities. In this manner, transport caused by fluctuations in a plasma is explained in Sects. 7.3 and 7.4.

The transport equation for particles is (see Sect. 5.1)

$$\frac{\partial}{\partial t}n(\boldsymbol{r},t) + \boldsymbol{\nabla}\cdot\big[n(\boldsymbol{r},t)\boldsymbol{V}(\boldsymbol{r},t)\big] = 0\,, \tag{7.1}$$

provided that the processes of ionization of neutrals and recombination of ions are negligible. In many cases, the particle flux $\boldsymbol{\Gamma} = n\boldsymbol{V}$ is given by

$$n(\boldsymbol{r},t)\boldsymbol{V}(\boldsymbol{r},t) = -D(\boldsymbol{r},t)\boldsymbol{\nabla}n(\boldsymbol{r},t)\,,$$

where $D$ is the diffusion coefficient. (Additional terms may be necessary in more general cases.)

The *diffusion coefficient* $D$ and *particle confinement time* $\tau_\mathrm{p}$ are related by the diffusion equation of the plasma density $n$ as follows:

$$\boldsymbol{\nabla}\cdot\big[D\boldsymbol{\nabla}n(\boldsymbol{r},t)\big] = \frac{\partial}{\partial t}n(\boldsymbol{r},t)\,.$$

Substituting $n(\boldsymbol{r},t) = n(\boldsymbol{r})\exp(-t/\tau_\mathrm{p})$ into the diffusion equation yields

$$\boldsymbol{\nabla}\cdot\big[D\boldsymbol{\nabla}n(\boldsymbol{r})\big] = -\frac{1}{\tau_\mathrm{p}}n(\boldsymbol{r})\,.$$

When $D$ is constant and the plasma column is a cylinder of radius $a$, the diffusion equation reduces to

$$\frac{1}{r}\frac{\partial}{\partial r}\left(r\frac{\partial n}{\partial r}\right) + \frac{1}{D\tau_{\mathrm{p}}}n = 0 \, .$$

The solution satisfying the boundary condition $n(a) = 0$ is

$$n = n_0 J_0\left(\frac{2.4r}{a}\right)\exp\left(-\frac{t}{\tau_{\mathrm{p}}}\right) ,$$

and the particle confinement time is

$$\tau_{\mathrm{p}} = \frac{a^2}{2.4^2 D} = \frac{a^2}{5.8D} , \tag{7.2}$$

where $J_0$ is the zeroth-order Bessel function. The relation (7.2) between the particle confinement time $\tau_{\mathrm{p}}$ and $D$ holds generally, with only a slight modification of the numerical factor. This formula is frequently used to obtain the diffusion coefficient from the observed values of the plasma radius and particle confinement time.

The equation of energy balance is given by [7.1]

$$\frac{\partial}{\partial t}\left(\frac{3}{2}n\kappa T\right) + \nabla\cdot\left(\frac{3}{2}\kappa T n \boldsymbol{V}\right) + \nabla\cdot\boldsymbol{q} = Q - p\nabla\cdot\boldsymbol{V} - \sum_{ij} \Pi_{ij}\frac{\partial V_i}{\partial x_j} \, . \tag{7.3}$$

The first term on the right-hand side is the heat generation due to particle collisions per unit volume per unit time, the second term is the work done by pressure, and the third term is viscous heating. The first term on the left-hand side is the time derivative of the thermal energy per unit volume, the second term is convective energy loss, and the third term is conductive energy loss. Denoting the *thermal conductivity* by $\kappa_{\mathrm{T}}$, the *thermal flux* due to heat conduction may be expressed by

$$\boldsymbol{q} = -\kappa_{\mathrm{T}}\nabla(\kappa T) \, .$$

If the *convective loss*, the second term in the left-hand side of eq.(7.3), is neglected and the terms on the right-hand side are also negligible, we find that

$$\frac{\partial}{\partial t}\left(\frac{3}{2}n\kappa T\right) - \nabla\cdot\kappa_{\mathrm{T}}\nabla(\kappa T) = 0 \, .$$

In the case $n =$ const., this equation reduces to

$$\frac{\partial}{\partial t}\left(\frac{3}{2}\kappa T\right) = \nabla\cdot\left[\frac{\kappa_{\mathrm{T}}}{n}\nabla(\kappa T)\right] \, .$$

When the thermal diffusion coefficient $\chi_{\mathrm{T}}$ is defined by

$$\chi_{\mathrm{T}} = \frac{\kappa_{\mathrm{T}}}{n} ,$$

the same equation in $\kappa T$ is obtained as (7.1). In the case $\chi_{\rm T} = $ const., the solution is

$$\kappa T = \kappa T_0 J_0 \left(\frac{2.4}{a} r\right) \exp\left(-\frac{t}{\tau_{\rm E}}\right) , \qquad \tau_{\rm E} = \frac{a^2}{5.8(2/3)\chi_{\rm T}} . \tag{7.4}$$

The term $\tau_{\rm E}$ is called the *energy confinement time.*

## 7.1 Collisional Diffusion (Classical Diffusion)

### 7.1.1 Magnetohydrodynamic Treatment

A magnetohydrodynamic treatment is applicable to diffusion phenomena when the electron-to-ion collision frequency is large and the mean free path is shorter than the *connection length* of the inside regions of good curvature and the outside region of bad curvature of the torus, i.e.,

$$\frac{v_{T_{\rm e}}}{\nu_{\rm ei}} < \frac{2\pi R}{\iota} , \qquad \nu_{\rm ei} > \nu_{\rm p} \equiv \frac{1}{R}\frac{\iota}{2\pi} v_{T_{\rm e}} = \frac{1}{R}\frac{\iota}{2\pi}\left(\frac{\kappa T_{\rm e}}{m_{\rm e}}\right)^{1/2} .$$

The MHD treatment can be applied to plasma diffusion. $v_{T_{\rm e}}$ is the electron thermal velocity and $\nu_{\rm ei}$ is the electron-to-ion collisional frequency. From Ohm's law (5.28),

$$\boldsymbol{E} + \boldsymbol{V} \times \boldsymbol{B} - \frac{1}{en}\boldsymbol{\nabla} p_{\rm i} = \eta \boldsymbol{j} ,$$

the motion of plasma across the lines of magnetic force is expressed by

$$\begin{aligned} n\boldsymbol{v}_\perp &= \frac{1}{B}\left[\left(n\boldsymbol{E} - \frac{\kappa T_{\rm i}}{e}\boldsymbol{\nabla} n\right) \times \boldsymbol{b}\right] - \frac{m_{\rm e}\nu_{\rm ei}}{e^2}\frac{\boldsymbol{\nabla} p}{B^2} \\ &= \frac{1}{B}\left[\left(n\boldsymbol{E} - \frac{\kappa T_{\rm i}}{e}\boldsymbol{\nabla} n\right) \times \boldsymbol{b}\right] - (\rho_{\Omega \rm e})^2 \nu_{\rm ei}\left(1 + \frac{T_{\rm i}}{T_{\rm e}}\right)\boldsymbol{\nabla} n , \end{aligned} \tag{7.5}$$

where $\rho_{\Omega\rm e} = v_{T_{\rm e}}/\Omega_{\rm e}$, $v_{T_{\rm e}} = (\kappa T_{\rm e}/m_{\rm e})^{1/2}$, and $\eta = m_{\rm e}\nu_{\rm ei}/e^2 n_{\rm e}$ (see Sect. 2.8).

If the first term on the right-hand side can be neglected, the particle diffusion coefficient $D$ is given by

$$D = (\rho_{\Omega\rm e})^2 \nu_{\rm ei}\left(1 + \frac{T_{\rm i}}{T_{\rm e}}\right) . \tag{7.6}$$

The *classical diffusion coefficient* $D_{\rm ei}$ is defined by

$$D_{\rm ei} \equiv (\rho_{\Omega\rm e})^2 \nu_{\rm ei} = \frac{nT_{\rm e}}{\sigma_\perp B^2} = \frac{\beta_{\rm e}\eta_\parallel}{\mu_0} , \tag{7.7}$$

where $\sigma_\perp = n_{\rm e}e^2/(m_{\rm e}\nu_{\rm ei})$, $\eta_\parallel = 1/2\sigma_\perp$.

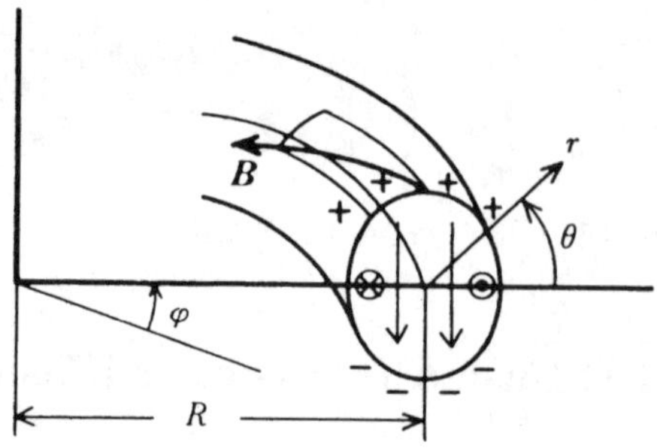


**Fig. 7.1.** Electric field in a plasma confined in a toroidal field. The symbols ⊗ and ⊙ here show the direction of the Pfirsch–Schlüter current

However, the first term on the right-hand side of (7.5) is not always negligible. In the toroidal configuration, the charge separation due to toroidal drift is not completely cancelled along the magnetic field lines due to the finite resistivity, and an electric field $\boldsymbol{E}$ arises (see Fig. 7.1). Therefore the $\boldsymbol{E} \times \boldsymbol{b}$ term in (7.5) contributes to the diffusion. Let us consider this term. From the equilibrium equation, the diamagnetic current

$$\boldsymbol{j}_\perp = \frac{\boldsymbol{b}}{B} \times \boldsymbol{\nabla} p \,, \qquad j_\perp = \left| \frac{1}{B}\frac{\partial p}{\partial r} \right| \,,$$

flows in the plasma. From $\boldsymbol{\nabla}\cdot\boldsymbol{j} = 0$, we find $\boldsymbol{\nabla}\cdot\boldsymbol{j}_\parallel = -\boldsymbol{\nabla}\cdot\boldsymbol{j}_\perp$. Using the equation $B = B_0[1 - (r/R)\cos\theta]$, $j_\parallel$ may be written as [see (6.41)]

$$j_\parallel = 2\frac{2\pi}{\iota}\frac{1}{B_0}\frac{\partial p}{\partial r}\cos\theta \,. \tag{7.8}$$

If the electrical conductivity along the lines of magnetic force is $\sigma_\parallel$, the parallel electric field is $E_\parallel = j_\parallel/\sigma_\parallel$. It is clear from Fig. 7.1 that

$$\frac{E_\theta}{E_\parallel} \approx \frac{B_0}{B_\theta} \,.$$

From $B_\theta/B_0 \approx (r/R)(\iota/2\pi)$, the $\theta$ component of the electric field is given by

$$E_\theta = \frac{B_0}{B_\theta}E_\parallel = \frac{R}{r}\frac{2\pi}{\iota}\frac{1}{\sigma_\parallel}j_\parallel = \frac{2}{\sigma_\parallel}\frac{R}{r}\left(\frac{2\pi}{\iota}\right)^2\frac{1}{B_0}\frac{\partial p}{\partial r}\cos\theta \,. \tag{7.9}$$

Accordingly, (7.5) reduces to

$$
\begin{aligned}
nV_{\mathrm{r}} &= -n\frac{E_\theta}{B} - (\rho_{\Omega\mathrm{e}})^2\nu_{\mathrm{ei}}\left(1+\frac{T_{\mathrm{i}}}{T_{\mathrm{e}}}\right)\frac{\partial n}{\partial r} \\
&= -\left[\frac{R}{r}2\left(\frac{2\pi}{\iota}\right)^2\frac{n\kappa T_{\mathrm{e}}}{\sigma_\parallel B_0^2}\cos\theta\left(1+\frac{r}{R}\cos\theta\right)+\frac{n\kappa T_{\mathrm{e}}}{\sigma_\perp B_0^2}\left(1+\frac{r}{R}\cos\theta\right)^2\right] \\
&\quad\times\left(1+\frac{T_{\mathrm{i}}}{T_{\mathrm{e}}}\right)\frac{\partial n}{\partial r}\,. \qquad (7.10)
\end{aligned}
$$

Noting that the area of a surface element depends on $\theta$, and taking the average of $nV_r$ over $\theta$, we find that

$$
\begin{aligned}
\langle nV_{\mathrm{r}}\rangle &= \frac{1}{2\pi}\int_0^{2\pi} nV_{\mathrm{r}}\left(1+\frac{r}{R}\cos\theta\right)\mathrm{d}\theta \\
&= -\frac{n\kappa T_{\mathrm{e}}}{\sigma_\perp B_0^2}\left(1+\frac{T_{\mathrm{i}}}{T_{\mathrm{e}}}\right)\left[1+\frac{2\sigma_\perp}{\sigma_\parallel}\left(\frac{2\pi}{\iota}\right)^2\right]\frac{\partial n}{\partial r}\,. \qquad (7.11)
\end{aligned}
$$

The diffusion coefficient of the toroidal plasma is $\left[1+(2\pi/\iota)^2\right]$ times as large as the diffusion coefficient of (7.6). This value is called the *Pfirsch–Schlüter factor* [7.2]. When the rotational transform angle $\iota/2\pi$ is about 0.3, the Pfirsch–Schlüter factor is about 10.

### 7.1.2 A Particle Model

The classical diffusion coefficient of electrons given by

$$D_{\mathrm{ei}} = (\rho_{\Omega\mathrm{e}})^2\nu_{\mathrm{ei}}$$

is the same as the one for electrons which move in a random walk with step length equal to the Larmor radius. Let us consider a toroidal plasma. For rotational transform angle $\iota$, the displacement $\Delta$ of the electron drift surface from the magnetic surface is (see Fig. 7.2)

$$\Delta \approx \pm\rho_{\Omega\mathrm{e}}\frac{2\pi}{\iota}\,. \qquad (7.12)$$

The $\pm$ signs depend on whether the direction of electron motion is parallel or antiparallel to the magnetic field (see Sect. 3.5). As an electron can be transferred from one drift surface to another by collision, the step length across the magnetic field is

$$\Delta = \left(\frac{2\pi}{\iota}\right)\rho_{\Omega\mathrm{e}}\,. \qquad (7.13)$$

Consequently, the diffusion coefficient is given by

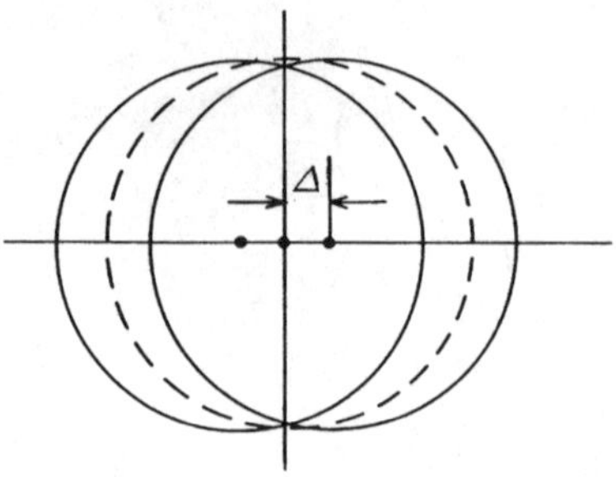


**Fig. 7.2.** Magnetic surface (*dotted line*) and drift surfaces (*solid lines*)

$$D_{\mathrm{P.S.}} = \Delta^2 \nu_{\mathrm{ei}} = \left(\frac{2\pi}{\iota}\right)^2 (\rho_{\Omega \mathrm{e}})^2 \nu_{\mathrm{ei}} \,. \tag{7.14}$$

The Pfirsch–Schlüter factor has thus been reduced, since we assume that $|2\pi/\iota| \gg 1$. The diffusion coefficient of (7.14) is called the Pfirsch–Schlüter diffusion coefficient [7.2].

## 7.2 Neoclassical Diffusion of Electrons in a Tokamak

The magnitude $B$ of the magnetic field of a tokamak is given by

$$B = \frac{RB_0}{R(1+\epsilon_{\mathrm{t}}\cos\theta)} = B_0(1-\epsilon_{\mathrm{t}}\cos\theta) \,, \tag{7.15}$$

where

$$\epsilon_{\mathrm{t}} = \frac{r}{R} \,. \tag{7.16}$$

Consequently, when the perpendicular component $v_\perp$ of an electron's velocity is much larger than the parallel component $v_\parallel$, i.e., when

$$\left(\frac{v_\perp}{v}\right)^2 > \frac{R}{R+r} \,,$$

that is,

$$\frac{v_\perp}{v_\parallel} > \frac{1}{\epsilon_{\mathrm{t}}^{1/2}} \,, \tag{7.17}$$

the electron is trapped outside of the torus, where the magnetic field is weak. Such an electron drifts in a banana orbit (see Fig. 3.7). In order to complete a circuit of the banana orbit, the effective collision time $\tau_{\mathrm{eff}} = 1/\nu_{\mathrm{eff}}$ of the trapped electron must be longer than one period $\tau_{\mathrm{b}}$ of the banana orbit, i.e.,

$$\tau_{\mathrm{b}} \approx \frac{R}{v_\parallel}\left(\frac{2\pi}{\iota}\right) = \frac{R}{v_\perp \epsilon_{\mathrm{t}}^{1/2}}\left(\frac{2\pi}{\iota}\right) \,. \tag{7.18}$$

The effective collision frequency $\nu_{\mathrm{eff}}$ of the trapped electron is the frequency at which the condition (7.17) for the trapped electron is violated by collision. As the collision frequency $\nu_{\mathrm{ei}}$ is the reciprocal of the diffusion time required to change the direction of velocity by 1 radian, the effective collision frequency $\nu_{\mathrm{eff}}$ is given by

$$\nu_{\mathrm{eff}} = \frac{1}{\epsilon_{\mathrm{t}}}\nu_{\mathrm{ei}} \,. \tag{7.19}$$

Accordingly, if $\nu_{\mathrm{eff}} < 1/\tau_{\mathrm{b}}$, i.e.,

$$\nu_{\mathrm{ei}} < \nu_{\mathrm{b}} \equiv \frac{v_{\perp}\epsilon_{\mathrm{t}}^{3/2}}{R}\left(\frac{\iota}{2\pi}\right) = \epsilon_{\mathrm{t}}^{3/2}\frac{1}{R}\left(\frac{\iota}{2\pi}\right)\left(\frac{\kappa T_{\mathrm{e}}}{m_{\mathrm{e}}}\right)^{1/2} , \tag{7.20}$$

the trapped electron can travel the entire banana orbit. When the trapped electron collides, it can shift its position by an amount equal to the banana width (see Sect. 3.5.2)

$$\Delta_{\mathrm{b}} = \frac{mv_{\parallel}}{eB_{\mathrm{p}}} \approx \frac{mv_{\perp}}{eB}\frac{v_{\parallel}}{v_{\perp}}\frac{B}{B_{\mathrm{p}}} \approx \rho_{\Omega\mathrm{e}}\epsilon_{\mathrm{t}}^{1/2}\frac{R}{r}\frac{2\pi}{\iota} = \left(\frac{2\pi}{\iota}\right)\epsilon_{\mathrm{t}}^{-1/2}\rho_{\Omega\mathrm{e}} \,. \tag{7.21}$$

As the number of trapped electrons is $\epsilon_{\mathrm{t}}^{1/2}$ times the total number of electrons, the trapped-electron contribution to diffusion is

$$\begin{aligned} D_{\mathrm{G.S.}} &= \epsilon_{\mathrm{t}}^{1/2}\Delta_{\mathrm{b}}^{2}\nu_{\mathrm{eff}} = \epsilon_{\mathrm{t}}^{1/2}\left(\frac{2\pi}{\iota}\right)^{2}\epsilon_{\mathrm{t}}^{-1}(\rho_{\Omega\mathrm{e}})^{2}\frac{1}{\epsilon_{\mathrm{t}}}\nu_{\mathrm{ei}} \\ &= \epsilon_{\mathrm{t}}^{-3/2}\left(\frac{2\pi}{\iota}\right)^{2}(\rho_{\Omega\mathrm{e}})^{2}\nu_{\mathrm{ei}} \,. \end{aligned} \tag{7.22}$$

This diffusion coefficient, introduced by Galeev and Sagdeev [7.3], is $\epsilon_{\mathrm{t}}^{-3/2} = (R/r)^{3/2}$ times as large as the diffusion coefficient for the collisional case. This derivation is semi-quantitative. A more rigorous discussion is given in [7.3].

As discussed in Sect. 7.1, the MHD treatment is applicable if the electron-to-ion collision frequency is larger than the frequency $\nu_{\mathrm{p}}$ given by

$$\nu_{\mathrm{p}} = \frac{1}{R}\frac{\iota}{2\pi}v_{T_{\mathrm{e}}} = \frac{1}{R}\left(\frac{\iota}{2\pi}\right)\left(\frac{\kappa T_{\mathrm{e}}}{m_{\mathrm{e}}}\right)^{1/2} . \tag{7.23}$$

When the electron-to-ion collision frequency is smaller than the frequency

$$\nu_{\mathrm{b}} = \epsilon_{\mathrm{t}}^{3/2}\nu_{\mathrm{p}} \,, \tag{7.24}$$

the electron can complete a banana orbit. The diffusion coefficients are

$$D_{\mathrm{P.S.}} = \left(\frac{2\pi}{\iota}\right)^{2}(\rho_{\Omega\mathrm{e}})^{2}\nu_{\mathrm{ei}} \,, \qquad \nu_{\mathrm{ei}} > \nu_{\mathrm{p}} \,, \tag{7.25}$$

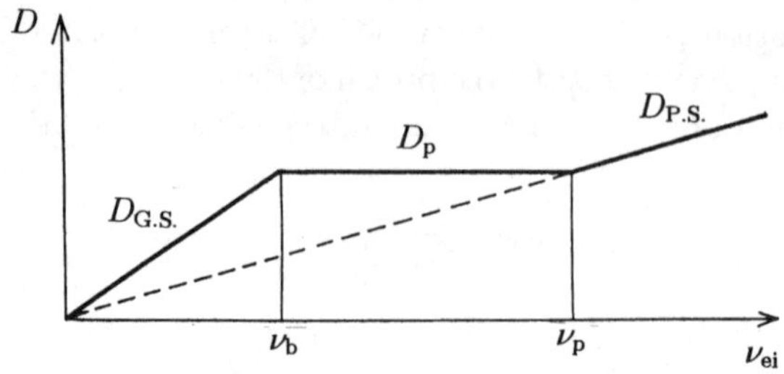


**Fig. 7.3.** Dependence of the diffusion coefficient on collision frequency in a tokamak. $\nu_{\text{p}} = (\iota/2\pi)v_{T_{\text{e}}}/R$, $\nu_{\text{b}} = \epsilon_{\text{t}}^{3/2}\nu_{\text{p}}$

$$D_{\text{G.S.}} = \epsilon_{\text{t}}^{-3/2}\left(\frac{2\pi}{\iota}\right)^2 (\rho_{\Omega\text{e}})^2\nu_{\text{ei}}\ , \qquad \nu_{\text{ei}} < \nu_{\text{b}} = \epsilon_{\text{t}}^{3/2}\nu_{\text{p}}\ . \tag{7.26}$$

If $\nu_{\text{ei}}$ is in the region $\nu_{\text{b}} < \nu_{\text{ei}} < \nu_{\text{p}}$, it is not possible to treat the diffusion phenomena of electrons in this region by means of a simple model. In this region we must resort to the drift approximation of Vlasov's equation. The result is that the diffusion coefficient is not sensitive to the collision frequency in this region and is given by [7.3, 7.4]

$$D_{\text{p}} = \left(\frac{2\pi}{\iota}\right)^2 (\rho_{\Omega\text{e}})^2\nu_{\text{p}}\ , \qquad \nu_{\text{p}} > \nu_{\text{ei}} > \nu_{\text{b}} = \epsilon_{\text{t}}^{3/2}\nu_{\text{p}}\ . \tag{7.27}$$

The dependence of the diffusion coefficient on the collision frequency is shown in Fig. 7.3. The region $\nu_{\text{ei}} > \nu_{\text{p}}$ is called the *MHD region* or *collisional region*. The region $\nu_{\text{p}} > \nu_{\text{ei}} > \nu_{\text{b}}$ is the *plateau region* or *intermediate region*; and the region $\nu_{\text{ei}} < \nu_{\text{b}}$ is called the *banana region* or *rare collisional region*. These diffusion processes are called neoclassical diffusion. There is an excellent review on neoclassical diffusion in [7.4].

The reason why the electron–electron collision frequency does not affect the electron particle diffusion coefficient is that the center-of-mass velocity is not changed by the Coulomb collision.

The neoclassical thermal diffusion coefficient $\chi_{T_{\text{e}}}$ is of the same order as the particle diffusion coefficient ($\chi_{T_{\text{e}}} \sim D_{\text{e}}$). Although an ion collision with the same ion species does not affect the ion's particle diffusion coefficient, it does contribute thermal diffusion processes, if a temperature gradient exists. Even if the ions are the same species as each other, it is possible to distinguish hot ions (with larger thermal velocity) and cold ions. Accordingly the ion's thermal diffusion coefficient in the banana region is given by $\chi_{\text{Ti}} \sim \epsilon_{\text{t}}^{-3/2}(2\pi/\iota)^2\rho_{\Omega\text{i}}^2\nu_{\text{ii}}$, and $\chi_{\text{Ti}} \sim (m_{\text{i}}/m_{\text{e}})^{1/2}D_{\text{ie}}$ ($D_{\text{ie}} \sim D_{\text{ei}}$). Therefore the ion's thermal diffusion coefficient is about $(m_{\text{i}}/m_{\text{e}})^{1/2}$ times as large as the ion's particle diffusion coefficient.

## 7.3 Fluctuation Loss. Bohm and Gyro-Bohm Diffusion. Convective Loss

In the foregoing sections we have discussed diffusion due to binary collisions and have derived the confinement times for such diffusion as an ideal case. However, a plasma will in many cases be more or less unstable, and fluctuations in the density and electric field will induce collective motions of particles and bring about anomalous losses. We will study such losses here.

Assume the plasma density $n(\boldsymbol{r},t)$ consists of the zeroth-order term $n_0(\boldsymbol{r},t)$ and first-order perturbation terms $\tilde{n}_k(\boldsymbol{r},t) = n_k \exp \mathrm{i}(\boldsymbol{k}\cdot\boldsymbol{r} - \omega_k t)$ and

$$n = n_0 + \sum_k \tilde{n}_k \,. \tag{7.28}$$

Since $n$ and $n_0$ are real, the following relations hold:

$$\tilde{n}_{-k} = (\tilde{n}_k)^* \,, \qquad n_{-k} = n_k^* \,, \qquad \omega_{-k} = -\omega_k^* \,,$$

where the asterisk denotes the complex conjugate. $\omega_k$ is generally complex, with $\omega_k = \omega_{k\mathrm{r}} + \mathrm{i}\gamma_k$ and

$$\omega_{-k\mathrm{r}} = -\omega_{k\mathrm{r}} \,, \qquad \gamma_{-k} = \gamma_k \,.$$

The plasma is forced to move by the perturbation. When the velocity is expressed by

$$\boldsymbol{V}(\boldsymbol{r},t) = \sum_k \tilde{\boldsymbol{V}}_k = \sum_k \boldsymbol{V}_k \exp \mathrm{i}(\boldsymbol{k}\cdot\boldsymbol{r} - \omega_k t) \,, \tag{7.29}$$

then $\boldsymbol{V}_{-k} = \boldsymbol{V}_k^*$ and the equation of continuity

$$\frac{\partial n}{\partial t} + \nabla\cdot(n\boldsymbol{V}) = 0$$

may be written as

$$\frac{\partial n_0}{\partial t} + \sum_k \frac{\partial \tilde{n}_k}{\partial t} + \nabla\cdot\left(\sum_k n_0 \tilde{\boldsymbol{V}}_k + \sum_{k,k'} \tilde{n}_k \tilde{\boldsymbol{V}}_{k'}\right) = 0 \,.$$

When the first- and second-order terms are separated,

$$\sum_k \frac{\partial \tilde{n}_k}{\partial t} + \nabla\cdot \sum_k n_0 \tilde{\boldsymbol{V}}_k = 0 \,, \tag{7.30}$$

$$\frac{\partial n_0}{\partial t} + \nabla\cdot\left(\sum_{k,k'} \tilde{n}_k \tilde{\boldsymbol{V}}_{k'}\right) = 0 \,. \tag{7.31}$$

Here we have assumed that the time derivative of $n_0$ is second order. The time average of the product of (7.30) and $\tilde{n}_{-k}$ becomes

$$\begin{cases} \gamma_k|n_k|^2 + \boldsymbol{\nabla} n_0 \cdot \mathrm{Re}(n_k \boldsymbol{V}_{-k}) + n_0 \boldsymbol{k} \cdot \mathrm{Im}(n_k \boldsymbol{V}_{-k}) = 0 \,, \\ \omega_{k\mathrm{r}}|n_k|^2 + \boldsymbol{\nabla} n_0 \cdot \mathrm{Im}(n_k \boldsymbol{V}_{-k}) - n_0 \boldsymbol{k} \cdot \mathrm{Re}(n_k \boldsymbol{V}_{-k}) = 0 \,. \end{cases} \tag{7.32}$$

Taking the time average of (7.31), we find that

$$\frac{\partial n_0}{\partial t} + \boldsymbol{\nabla} \cdot \left[ \sum_k \mathrm{Re}(n_k \boldsymbol{V}_{-k}) \exp(2\gamma_k t) \right] = 0 \,. \tag{7.33}$$

The diffusion equation is

$$\frac{\partial n_0}{\partial t} = \boldsymbol{\nabla} \cdot (D \boldsymbol{\nabla} n_0)$$

and the particle flux $\boldsymbol{\Gamma}$ is

$$\boldsymbol{\Gamma} = -D \boldsymbol{\nabla} n_0 = \sum_k \mathrm{Re}(n_k \boldsymbol{V}_{-k}) \exp 2\gamma_k t \,. \tag{7.34}$$

Equation (7.32) alone is not enough to determine the quantity

$$\boldsymbol{\nabla} n_0 \cdot \mathrm{Re}(n_k \boldsymbol{V}_{-k}) \exp(2\gamma_k t) \,.$$

Putting

$$\beta_k = \frac{n_0 \boldsymbol{k} \cdot \mathrm{Im}(n_k \boldsymbol{V}_{-k})}{\boldsymbol{\nabla} n_0 \cdot \left[\mathrm{Re}(n_k \boldsymbol{V}_{-k})\right]} \,,$$

equation (7.34) reduces to

$$D|\boldsymbol{\nabla} n_0|^2 = \sum_k \frac{\gamma_k |n_k|^2 \exp 2\gamma_k t}{1 + \beta_k}$$

and

$$D = \sum_k \gamma_k \frac{|\tilde{n}_k|^2}{|\boldsymbol{\nabla} n_0|^2} \frac{1}{1 + \beta_k} \,. \tag{7.35}$$

This is the anomalous diffusion coefficient due to fluctuation loss.

Let us consider the case in which the fluctuation $\tilde{\boldsymbol{E}}_k$ of the electric field is electrostatic and can be expressed by a potential $\tilde{\phi}_k$. Then the perturbed electric field is expressed by

$$\tilde{\boldsymbol{E}}_k = -\boldsymbol{\nabla} \tilde{\phi}_k = -\mathrm{i} \boldsymbol{k} \phi_k \exp \mathrm{i}(\boldsymbol{k} \cdot \boldsymbol{r} - \omega_k t) \,.$$

The electric field results in an $\tilde{\boldsymbol{E}}_k \times \boldsymbol{B}$ drift, i.e.,

$$\tilde{\boldsymbol{V}}_k = \frac{\tilde{\boldsymbol{E}}_k \times \boldsymbol{B}}{B^2} = -\mathrm{i} \frac{(\boldsymbol{k} \times \boldsymbol{b}) \tilde{\phi}_k}{B} \,, \tag{7.36}$$

where $\boldsymbol{b} = \boldsymbol{B}/B$. Equation (7.36) gives the perpendicular component of the fluctuating motion. Substituting (7.36) into (7.30) yields

$$\tilde{n}_k = \nabla n_0 \cdot \left( \frac{\boldsymbol{b} \times \boldsymbol{k}}{B} \right) \frac{\tilde{\phi}_k}{\omega_k} . \tag{7.37}$$

In general $\nabla n_0$ and $\boldsymbol{b}$ are orthogonal. Take the $z$ axis in the direction of $\boldsymbol{b}$ and the $x$ axis in the direction of $-\nabla n$, i.e., let $\nabla n = -\kappa_n n_0 \hat{\boldsymbol{x}}$, where $\kappa_n$ is the inverse of the scale of the density gradient and $\hat{\boldsymbol{x}}$ is the unit vector in the $x$ direction. Then (7.37) gives

$$\frac{\tilde{n}_k}{n_0} = \frac{\kappa_n}{B} \frac{k_y}{\omega_k} \tilde{\phi}_k = k_y \kappa_n \frac{\kappa T_e}{eB\omega_k} \frac{e\tilde{\phi}_k}{\kappa T_e} = \frac{\omega_k^*}{\omega_k} \frac{e\tilde{\phi}_k}{\kappa T_e} ,$$

where $k_y$ is the $y$ (poloidal) component of the propagation vector $\boldsymbol{k}$. The quantity

$$\omega_k^* \equiv k_y \kappa_n \frac{\kappa T_e}{eB}$$

is called the *drift frequency*. If the frequency $\omega_k$ is real (i.e., if $\gamma_k = 0$), $\tilde{n}_k$ and $\tilde{\phi}_k$ have the same phase and the fluctuation does not contribute to anomalous diffusion, as is clear from (7.35). When $\gamma_k > 0$, so that $\omega$ is complex, there is a phase difference between $\tilde{n}_k$ and $\tilde{\phi}_k$, and the fluctuation in the electric field contributes to anomalous diffusion. (When $\gamma_k < 0$, the amplitude of the fluctuation is damped and does not contribute to diffusion.) Using the real parameters $A_k$, $\alpha_k$ of $\omega_k = \omega_{kr} + i\gamma_k = \omega_k^* A_k \exp i\alpha_k$ ($A_k > 0$, $\alpha_k$ are both real), $\tilde{\boldsymbol{V}}_k$ is expressed by

$$\begin{aligned}\tilde{\boldsymbol{V}}_k &= -i(\boldsymbol{k} \times \boldsymbol{b}) \frac{\kappa T_e}{eB} \frac{\tilde{\phi}_k}{\kappa T_e} = -i(\boldsymbol{k} \times \boldsymbol{b}) \frac{\kappa T_e}{eB} \frac{\tilde{n}_k}{n_0} \frac{\omega_{kr} + \gamma_k i}{\omega_k^*} \\ &= -i(\boldsymbol{k} \times \boldsymbol{b}) \frac{\kappa T_e}{eB} \frac{\tilde{n}_k}{n_0} A_k \exp i\alpha_k ,\end{aligned}$$

$$\tilde{V}_{kx} = k_y \frac{\tilde{n}_k}{n_0} \frac{\kappa T_e}{eB} \frac{\gamma_k - \omega_{kr} i}{\omega_k^*} = k_y \frac{\tilde{n}_k}{n_0} \frac{\kappa T_e}{eB} (-iA_k \exp i\alpha_k) .$$

Then the diffusion coefficient may be obtained from (7.34) as follows:

$$\begin{aligned}D = \frac{1}{\kappa_n n_0} \mathrm{Re}(\tilde{n}_k \tilde{V}_{-kx}) &= \left( \sum_k \frac{k_y \gamma_k}{\kappa_n \omega_k^*} \left| \frac{\tilde{n}_k}{n_0} \right|^2 \right) \frac{\kappa T_e}{eB} \\ &= \left( \sum_k \frac{k_y}{\kappa_n} A_k \sin \alpha_k \left| \frac{\tilde{n}_k}{n_0} \right|^2 \right) \frac{\kappa T_e}{eB} .\end{aligned} \tag{7.38}$$

The anomalous diffusion coefficient due to fluctuation loss increases with time [from (7.35) and (7.38)] and eventually the term with the maximum growth rate $\gamma_k > 0$ becomes dominant. However, the amplitude $|\tilde{n}_k|$ will saturate due to nonlinear effects and the saturated amplitude will be of the order of

$$|\tilde{n}_k| \approx |\boldsymbol{\nabla} n_0|\Delta x \approx \frac{\kappa_n}{k_x} n_0 \,.$$

$\Delta x$ is the correlation length of the fluctuation and the inverse of the propagation constant $k_x$ in the $x$ direction. Then (7.35) yields

$$D = \frac{\gamma_k}{\kappa_n^2}\left|\frac{\tilde{n}_k}{n_0}\right|^2 \approx \frac{\gamma_k}{k_x^2} \,. \tag{7.39}$$

When the dimensionless coefficient in brackets in eq.(7.38) is assumed to be at its maximum of 1/16, we have the *Bohm diffusion coefficient*

$$D_{\mathrm{B}} = \frac{1}{16}\frac{\kappa T_{\mathrm{e}}}{eB} \,. \tag{7.40}$$

Equation (7.40) thus gives the largest possible diffusion coefficient.

When the density and potential fluctuations $\tilde{n}_k$, $\tilde{\phi}_k$ are measured, $\tilde{\boldsymbol{V}}_k$ can be calculated using (7.36), and the estimated outward particle flux $\Gamma$ using (7.34). The diffusion coefficient $D$ can then be compared with the values obtained by experiment. As the relation between $\tilde{n}_k$ and $\tilde{\phi}_k$ is given by (7.37), the phase difference will indicate whether $\omega_k$ is real (oscillatory mode) or $\gamma_k > 0$ (growing mode), so that this equation is very useful in interpreting experimental results.

We consider the example of a fluctuation driven by ion-temperature-gradient drift instability (see Sect. 8.7). The mode is described by

$$\phi(r, \theta, z) = \sum \phi_{mn}(r) \exp(-\mathrm{i}m\theta + \mathrm{i}nz/R) \,.$$

The growth rate of the fluctuation has maximum at around

$$k_\theta = -\frac{\mathrm{i}}{r}\frac{\partial}{\partial\theta} = -\frac{m}{r}$$

or [7.5]

$$|k_\theta| = \frac{m}{r} \sim \frac{\alpha_\theta}{\rho_{\mathrm{i}}} \,, \qquad \alpha_\theta = 0.7\text{–}0.8 \,.$$

Then the correlation length $\Delta_\theta$ in the $\theta$ direction is $\Delta_\theta \sim \rho_{\mathrm{i}}/\alpha_\theta$, where $\rho_{\mathrm{i}}$ is the ion Larmor radius.

The propagation constant $k_\parallel$ along the line of magnetic force near the rational surface $q(r_m) = m/n$ is

$$\begin{aligned} k_\parallel = -\mathrm{i}\boldsymbol{b}\cdot\boldsymbol{\nabla} &= \frac{B_\theta}{B}\left(\frac{-m}{r}\right) + \frac{B_{\mathrm{t}}}{B}\left(\frac{n}{R}\right) \approx \frac{1}{R}\left[n - \frac{m}{q(r)}\right] \\ &= \frac{m}{rR}\frac{rq'}{q^2}(r - r_m) = \frac{s}{Rq}k_\theta(r - r_m) \,, \end{aligned}$$

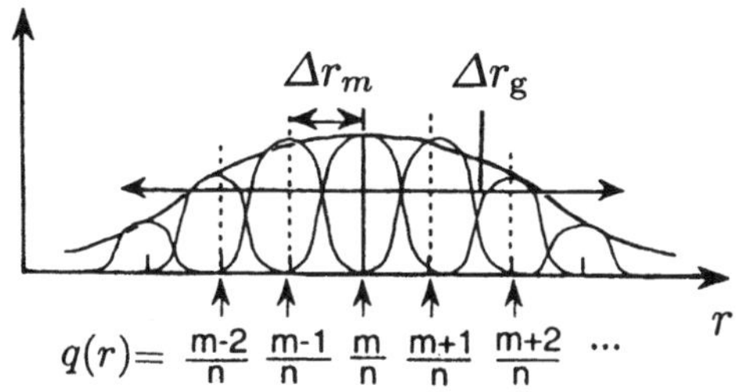


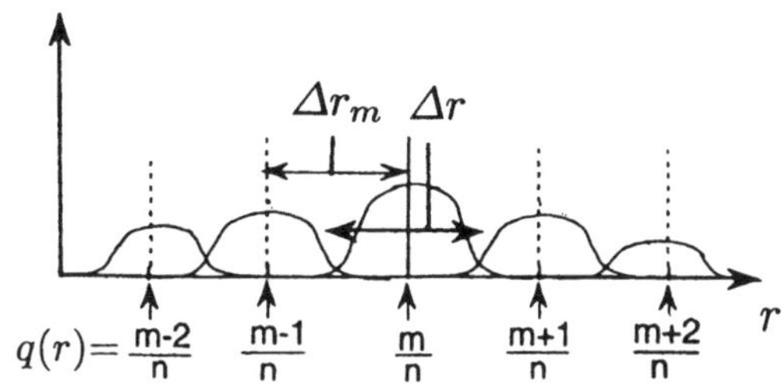


**Fig. 7.4.** *Upper*: the radial width of eigenmode $\Delta r$ is larger than the radial separation of the rational surfaces $\Delta r_m$. A semi-global eigenmode structure $\Delta r_{\rm g}$ arises due to the mode couplings. *Lower*: the radial width of eigenmode $\Delta r$ is smaller than the radial separation of the rational surfaces $\Delta r_m$. Modes with radial width $\Delta r$ are mutually independent

where $q(r) \equiv (r/R)(B_{\rm t}/B_\theta)$ is the safety factor, with $B_\theta$ and $B_{\rm t}$ the poloidal and toroidal fields, respectively, and $s$ is the shear parameter $s \equiv rq'/q$ (see Sect. 8.3.3). $|k_\parallel|$ is larger than the inverse of the connection length $qR$ of the torus and is less than the inverse of, say, the pressure gradient scale $L_{\rm p}$, that is

$$\frac{1}{qR} < |k_\parallel| < \frac{1}{L_{\rm p}} \, .$$

The radial width $\Delta r = |r - r_m|$ of the mode near the rational surface $r = r_m$ is expected to be roughly

$$\Delta r = |r - r_m| = \frac{Rq}{s}\frac{k_\parallel}{k_\theta} = \frac{\rho_{\rm i}}{s\alpha_\theta} \sim O(\rho_{\rm i}/s) \, .$$

The estimated radial width of the eigenmode of ion-temperature-gradient driven drift turbulence is given by [7.6, 7.7]

$$\Delta r = \rho_{\rm i} \left(\frac{qR}{sL_{\rm p}}\right)^{1/2} \left(\frac{\gamma_k}{\omega_{kr}}\right)^{1/2} \, .$$

The radial separation length $\Delta r_m$ between adjacent rational surfaces $r_m$ and $r_{m+1}$ is

$$q'\Delta r_m = q(r_{m+1}) - q(r_m) = \frac{m+1}{n} - \frac{m}{n} = \frac{1}{n} ,$$

$$\Delta r_m = \frac{1}{nq'} = \frac{m/n}{rq'}\frac{r}{m} \sim \frac{1}{sk_\theta} .$$

When the mode width $\Delta r$ is larger than the radial separation of the rational surface $\Delta r_m$, the different modes overlap each other and toroidal mode coupling takes place (see Fig. 7.4). The half-width $\Delta r_{\mathrm{g}}$ of the envelope of coupled modes is estimated to be [7.8, 7.9]

$$\Delta r_{\mathrm{g}} = \left(\frac{\rho_{\mathrm{i}} L_{\mathrm{p}}}{s}\right)^{1/2} .$$

The radial correlation length becomes a large multiple of $\Delta r_{\mathrm{g}}$ [$\Delta r_{\mathrm{g}}/\Delta r \sim (L_{\mathrm{p}}/\rho_{\mathrm{i}})^{1/2}$] and the radial propagation constant becomes $k_r \sim 1/\Delta r_{\mathrm{g}}$. In this case, the diffusion coefficient $D$ is

$$D = (\Delta r_{\mathrm{g}})^2 \gamma_k \sim \frac{\rho_{\mathrm{i}} L_{\mathrm{p}}}{s}\omega_k^* \sim \frac{\kappa T}{eB}\frac{\alpha_\theta}{s} ,$$

where $\omega_k^*$ is the drift frequency (Sects. 8.7 and 9.2). This coefficient is of Bohm type.

When the mode width $\Delta r$ is less than $\Delta r_m$ (weak shear case), there is no coupling between different modes and the radial correlation length is

$$\Delta r = \rho_{\mathrm{i}} \left(\frac{qR}{sL_{\mathrm{p}}}\right)^{1/2} .$$

The diffusion coefficient $D$ in this case is

$$D \sim (\Delta r)^2 \omega_k^* \sim \rho_{\mathrm{i}}^2 \left(\frac{qR}{sL_{\mathrm{p}}}\right)\left(\frac{k_\theta \kappa T}{eBL_{\mathrm{p}}}\right) \sim \frac{\kappa T}{eB}\frac{\rho_{\mathrm{i}}}{L_{\mathrm{p}}}\left(\frac{\alpha_\theta qR}{sL_{\mathrm{p}}}\right) \propto \frac{\kappa T}{eB}\frac{\rho_{\mathrm{i}}}{L_{\mathrm{p}}} . \quad (7.41)$$

This is called a gyro-Bohm-type diffusion coefficient. The transport in toroidal systems may be expected to become small in the weak shear region of the negative shear configuration near the minimum $q$ position (see Sect. 16.7).

Next, let us consider *stationary convective losses* across the magnetic flux. Even if fluctuations in the density and electric field are not observed at a fixed position, the plasma may be able to move across the magnetic field and escape continuously. When a stationary electric field exists and the equipotential surfaces do not coincide with the magnetic surfaces $\phi =$ const., the $\boldsymbol{E} \times \boldsymbol{B}$ drift is normal to the electric field $\boldsymbol{E}$, which is itself normal to the equipotential surface. Consequently, the plasma drifts along the equipotential surfaces (see Fig. 7.5) which cross the magnetic surfaces. The resulting loss is called stationary convective loss. The particle flux is given by

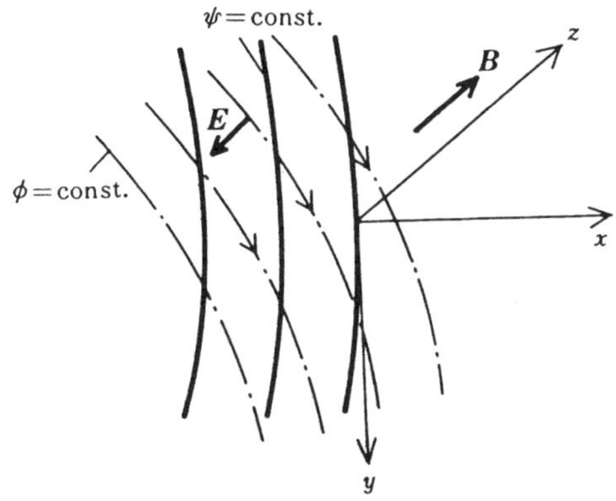


**Fig. 7.5.** Magnetic surface $\psi$ = const. and electric field equipotential $\phi$ = const. The plasma moves along the equipotential surfaces by virtue of $\boldsymbol{E} \times \boldsymbol{B}$

$$\Gamma_k = n_0 \frac{E_y}{B} \, . \tag{7.42}$$

The losses due to diffusion by binary collision are proportional to $B^{-2}$, but fluctuation or convective losses are proportional to $B^{-1}$. Even if the magnetic field is increased, the loss due to fluctuations does not decrease rapidly.

## 7.4 Loss by Magnetic Fluctuation

When the magnetic field in a plasma fluctuates, the lines of magnetic force will wander radially. Denote the radial shift of the field line by $\Delta r$ and the radial component of magnetic fluctuation $\delta \boldsymbol{B}$ by $\delta B_r$, respectively. Then we find

$$\Delta r = \int_0^L b_r \mathrm{d}l \, ,$$

where $b_r = \delta B_r / B$ and $l$ is the length along the line of magnetic force. The ensemble average of $(\Delta r)^2$ is given by

$$\langle (\Delta r)^2 \rangle = \left\langle \int_0^L b_r \, \mathrm{d}l \int_0^L b_r \, \mathrm{d}l' \right\rangle = \left\langle \int_0^L \mathrm{d}l \int_0^L \mathrm{d}l' \, b_r(l) \, b_r(l') \right\rangle$$

$$= \left\langle \int_0^L \mathrm{d}l \int_{-l}^{L-l} \mathrm{d}s \, b_r(l) \, b_r(l+s) \right\rangle \approx L \left\langle b_r^2 \right\rangle l_{\mathrm{corr}} \, ,$$

where $l_{\mathrm{corr}}$ is

$$l_{\text{corr}} = \frac{\left\langle \int_{-\infty}^{\infty} b_r(l)\, b_r(l+s)\, \mathrm{d}s \right\rangle}{\langle b_r^2 \rangle} .$$

If electrons run along the lines of magnetic force with the velocity $v_{T_e}$, the diffusion coefficient $D_e$ of the electrons becomes [7.10]

$$D_e = \frac{\langle (\Delta r)^2 \rangle}{\Delta t} = \frac{L}{\Delta t} \langle b_r^2 \rangle l_{\text{corr}} = v_{T_e} l_{\text{corr}} \left\langle \left( \frac{\delta B_r}{B} \right)^2 \right\rangle . \tag{7.43}$$

We may take $l_{\text{corr}} \sim R$ in the case of a tokamak and $l_{\text{corr}} \sim a$ in the case of the reverse field pinch (RFP, see Sect. 17.1).

# 8 Magnetohydrodynamic Instabilities

The stability of plasmas in magnetic fields is one of the primary research subjects in the area of controlled thermonuclear fusion, and both theoretical and experimental investigations have been actively pursued. If a plasma is free from all possible instabilities and if the confinement is dominated by neoclassical diffusion in the banana region, then the energy confinement time $\tau_{\rm E}$ is given by

$$\tau_{\rm E} \approx \frac{(3/2)a^2}{5.8\chi_{\rm G.S.}} \approx \frac{(3/2)}{5.8}\left(\frac{\iota}{2\pi}\right)^2 \epsilon^{3/2}\left(\frac{a}{\rho_{\Omega \rm i}}\right)^2 \frac{1}{\nu_{\rm ii}} \ ,$$

where $a$ is the plasma radius, $\rho_{\Omega \rm i}$ the ion Larmor radius, and $\nu_{\rm ii}$ the ion–ion collision frequency. For such an ideal case, a device of a reasonable size satisfies the ignition condition. (For example, with $B = 5\,\rm T$, $a = 1\,\rm m$, $T_{\rm i} = 20\,\rm keV$, $\iota/2\pi \approx 1/3$, and inverse aspect ratio $\epsilon = 0.2$, the value of $n\tau_{\rm E} \sim 3.5 \times 10^{20}\,\rm cm^{-3}s$.)

A plasma consists of many moving charged particles and has many magnetohydrodynamic degrees of freedom as well as degrees of freedom in velocity space. When a certain mode of perturbation grows, it enhances diffusion. Heating a plasma increases the kinetic energy of the charged particles but at the same time may induce fluctuations in the electric and magnetic fields, which in turn augment anomalous diffusion.

Therefore, it is very important to determine whether any particular perturbed mode is stable (damping mode) or unstable (growing mode). In the stability analysis, it is assumed that the deviation from the equilibrium state is small, so that a linearized approximation can be used. In this chapter we consider instabilities that can be described by linearized magnetohydrodynamic equations. These instabilities are called the *magnetohydrodynamic instabilities* (MHD instabilities) or *macroscopic instabilities*.

A small perturbation $\boldsymbol{F}(\boldsymbol{r}, t)$ of the first order is expanded in terms of its Fourier components,

$$\boldsymbol{F}(\boldsymbol{r}, t) = \boldsymbol{F}(\boldsymbol{r}) \exp(-{\rm i}\omega t) \ , \qquad \omega = \omega_{\rm r} + {\rm i}\omega_{\rm i} \ ,$$

and each term can be treated independently in the linearized approximation. The dispersion equation is solved for $\omega$ and the stability of the perturbation depends on the sign of the imaginary part $\omega_{\rm i}$ (unstable for $\omega_{\rm i} > 0$ and stable

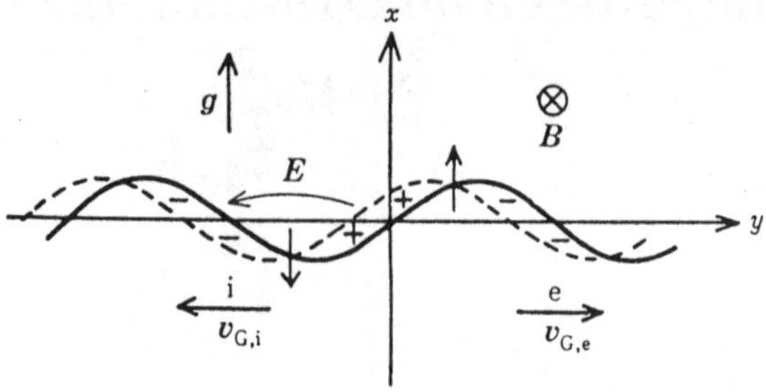


**Fig. 8.1.** Ion and electron drifts and the resulting electric field for interchange instability

for $\omega_i < 0$). When $\omega_r \neq 0$, the perturbation is oscillatory, and when $\omega_r = 0$, it grows or damps monotonically.

In the following sections, typical MHD instabilities are introduced. In Sect. 8.1, interchange instability is explained in an intuitive manner. In Sect. 8.2, the magnetohydrodynamic equations are linearized and the boundary conditions are implemented. The stability criterion is deduced from the energy principle in (8.45)–(8.48). In Sect. 8.3, a cylindrical plasma is studied as an important example, and the associated energy integrals are derived. Furthermore, important stability conditions, the Kruskal–Shafranov limit (8.66) and the Suydam criterion (8.98), are described. Tokamaks with large aspect ratios are treated approximately as cylindrical plasmas and their stabilities are examined. The MHD equation of motion in cylindrical coordinates (Sect. 8.4), the energy integral of axisymmetric toroidal systems (Sect. 8.5), ballooning instability (Sect. 8.6), and the $\eta_i$ mode due to density and temperature gradients (Sect. 8.7) are described in this chapter. It should be understood that there are many other instabilities. General reviews of MHD instabilities may be found in [8.1].

## 8.1 Interchange Instabilities

### 8.1.1 Interchange Instability

Let $x = 0$ be the boundary between plasma and vacuum and let the $z$ axis be taken in the direction of the magnetic field $\boldsymbol{B}$. The plasma region is $x < 0$ and the vacuum region is $x > 0$. It is assumed that the acceleration $\boldsymbol{g}$ is applied in the $x$ direction (see Fig. 8.1). Ions and electrons drift in opposite directions to each other, due to the acceleration, with drift velocities

$$\boldsymbol{v}_{G,i} = \frac{M}{e}\frac{\boldsymbol{g}\times\boldsymbol{B}}{B^2}, \qquad \boldsymbol{v}_{G,e} = -\frac{m}{e}\frac{\boldsymbol{g}\times\boldsymbol{B}}{B^2},$$

where $M$ and $m$ are the masses of the ion and electron, respectively. Let us assume that, due to a perturbation, the boundary of the plasma is displaced from the surface $x = 0$ by the amount

$$\delta x = a(t)\sin(k_y y) \ .$$

The charge separation due to the opposite ion and electron drifts yields an electric field. The resulting $(\boldsymbol{E} \times \boldsymbol{B})$ drift enhances the original perturbation if the acceleration $\boldsymbol{g}$ is directed outward from the plasma. We see that this is the same as saying that the magnetic flux originally inside but near the plasma boundary is displaced so that it is outside the boundary, while the flux outside moves in to fill the depression thereby left in the boundary. Owing to this geometrical picture of the process, this type of instability has come to be called *interchange instability*. As the perturbed plasma boundary has the form of flutes along the lines of magnetic force, this instability is also called a *flute instability*. Similar phenomena occur in hydrodynamics when a dense fluid is supported against gravity by a fluid with lower density. Therefore, interchange instability is also called *Rayleigh–Taylor instability*.

The drift due to the acceleration produces a surface charge on the plasma, of charge density

$$\sigma_s = \sigma(t)\cos(k_y y)\delta(x) \tag{8.1}$$

(see Fig. 8.1). The electrostatic potential $\phi$ of the induced electric field $\boldsymbol{E} = -\boldsymbol{\nabla}\phi$ is given by

$$\epsilon_\perp \frac{\partial^2 \phi}{\partial y^2} + \frac{\partial}{\partial x}\left(\epsilon_\perp \frac{\partial \phi}{\partial x}\right) = -\sigma_s \ . \tag{8.2}$$

The boundary condition is

$$\epsilon_0 \left(\frac{\partial \phi}{\partial x}\right)_{+0} - \left(\epsilon_\perp \frac{\partial \phi}{\partial x}\right)_{-0} = -\sigma_s \ , \qquad \phi_{+0} = \phi_{-0} \ .$$

Under the assumption $k_y > 0$, the solution $\phi$ is

$$\phi = \frac{\sigma(t)}{k_y(\epsilon_0 + \epsilon_\perp)} \cos(k_y y)\exp(-k_y|x|) \ . \tag{8.3}$$

The velocity of the boundary $\mathrm{d}(\delta x)/\mathrm{d}t$ is equal to $\boldsymbol{E} \times \boldsymbol{B}/B^2$ at $x = 0$, with $\boldsymbol{E}$ found from the potential (8.3). The velocity is

$$\frac{\mathrm{d}a(t)}{\mathrm{d}t}\sin(k_y y) = \frac{\sigma(t)}{(\epsilon_0 + \epsilon_\perp)B}\sin(k_y y) \ . \tag{8.4}$$

The charge flux in the $y$ direction is

$$ne|\boldsymbol{v}_{\mathrm{G,i}}| = \frac{\rho_{\mathrm{m}} g}{B} \ ,$$

where $\rho_{\mathrm{m}} = nM$. Accordingly, the rate of change of charge density is

$$\frac{\mathrm{d}\sigma(t)}{\mathrm{d}t}\cos(k_y y) = \frac{\rho_{\mathrm{m}} g}{B} a(t) \frac{\mathrm{d}}{\mathrm{d}y}\sin(k_y y) \tag{8.5}$$

and

$$\frac{\mathrm{d}^2 a}{\mathrm{d}t^2} = \frac{\rho_{\mathrm{m}} g k_y}{(\epsilon_0 + \epsilon_\perp)B^2} a \,. \tag{8.6}$$

The solution is in the form $a \propto \exp \gamma t$ and the growth rate $\gamma$ is given by

$$\gamma = \left[\frac{\rho_{\mathrm{m}}}{(\epsilon_0 + \epsilon_\perp)B^2}\right]^{1/2} (g k_y)^{1/2} \,. \tag{8.7}$$

In the low-frequency case (compared with the ion cyclotron frequency), the dielectric constant is given by

$$\epsilon_\perp = \epsilon_0 \left(1 + \frac{\rho_{\mathrm{m}}}{B^2 \epsilon_0}\right) \gg \epsilon_0 \,, \tag{8.8}$$

as will be explained in Chap. 10. Hence, the growth rate $\gamma$ is [8.2]

$$\gamma = (g k_y)^{1/2} \,. \tag{8.9}$$

When the acceleration is outward, a perturbation with the propagation vector $\boldsymbol{k}$ normal to the magnetic field $\boldsymbol{B}$ is unstable, as shown in Fig. 8.2a, i.e.,

$$(\boldsymbol{k} \cdot \boldsymbol{B}) = 0 \,. \tag{8.10}$$

However, if the acceleration is inward ($g < 0$), $\gamma$ of (8.9) is imaginary and the perturbation is oscillatory and stable.

The origin of interchange instability is charge separation due to the acceleration. When the lines of magnetic force are curved, as shown in Fig. 8.2, the charged particles are subjected to a centrifugal force. If the magnetic lines of force are convex outward (Fig. 8.2a), this centrifugal acceleration induces interchange instability. If the lines are concave outward, the plasma is stable. Accordingly, the plasma is stable when the magnitude $B$ of the magnetic field increases outward. In other words, if $B$ is a minimum at the plasma region, the plasma is stable. This is the *minimum-B condition* for stability. A more general treatment of the interchange instability is described in [8.1].

The drift motion of charged particles is expressed by (see Chap. 3)

$$\boldsymbol{v}_{\mathrm{G}} = \frac{\boldsymbol{E} \times \boldsymbol{b}}{B} + \frac{\boldsymbol{b}}{\Omega} \times \left(\boldsymbol{g} + \frac{(v_\perp^2/2) + v_\parallel^2}{R}\boldsymbol{n}\right) + v_\parallel \boldsymbol{b} \,,$$

where $\boldsymbol{n}$ is the unit normal vector from the center of curvature to a point on a line of magnetic force, and $R$ is the radius of curvature of the line of magnetic force. The equivalent acceleration is

$$\boldsymbol{g} = \frac{(v_\perp^2/2) + v_\parallel^2}{R}\boldsymbol{n} \,. \tag{8.11}$$

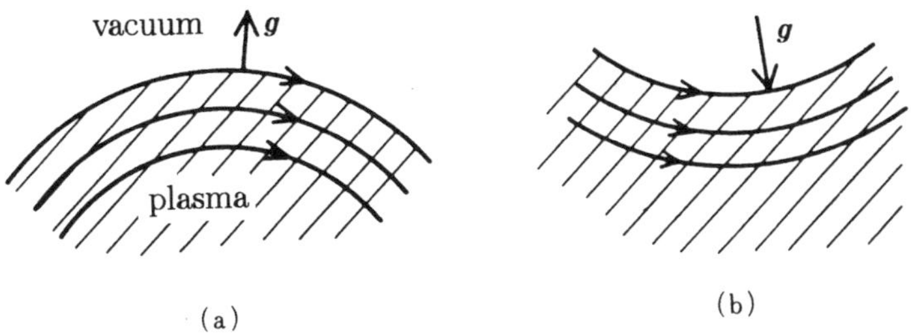


**Fig. 8.2.** Centrifugal force due to the curvature of a line of magnetic force

The growth rate becomes $\gamma \approx (a/R)^{1/2}(v_{\rm T}/a)$ in this case.

When the growth rate $\gamma \sim (gk_y)^{1/2}$ is not very large and the ion Larmor radius $\rho_\Omega^{\rm i}$ is large enough to satisfy

$$(k_y \rho_\Omega^{\rm i})^2 > \frac{\gamma}{|\Omega_{\rm i}|} ,$$

the perturbation is stabilized [8.3]. When the ion Larmor radius becomes large, the average perturbation electric field felt by the ions is different from that felt by the electrons, and the $\boldsymbol{E} \times \boldsymbol{B}/B^2$ drift velocities of the ion and the electrons are different. The charge separation thereby induced has opposite phase from the charge separation due to acceleration and stabilizes the instability.

### 8.1.2 Stability Criterion for Interchange Instability. Magnetic Well

Let us assume that a magnetic line of force has 'good' curvature at one place B and 'bad' curvature at another place A (Fig. 8.3b). Then the directions of the centrifugal force at A and B are opposite, as is the charge separation. The charge can easily be short-circuited along the magnetic lines of force, so that the problem of stability has a different aspect. Let us here consider perturbations in which the magnetic flux of region 1 is interchanged with that of region 2 and the plasma in region 2 is interchanged with the plasma in region 1 (interchange perturbations, Fig. 8.3b). It is assumed that the plasma is low-beta so that the magnetic field is nearly identical to the vacuum field. Any deviation from the vacuum magnetic field is accompanied by an increase in the magnetic energy of the disturbed field. This is a consequence of Maxwell's equations. We now show that the most dangerous perturbations are those which exchange equal magnetic fluxes.

The energy $Q_{\rm M}$ of the magnetic field inside a magnetic tube is

$$Q_{\rm M} = \int {\rm d}\boldsymbol{r} \frac{B^2}{2\mu_0} = \int {\rm d}l S \frac{B^2}{2\mu_0} , \tag{8.12}$$

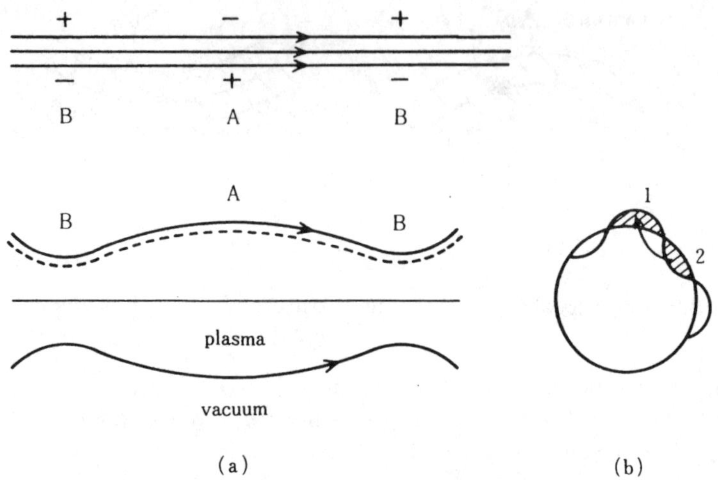


**Fig. 8.3.** Charge separation in interchange instability. (**a**) The *lower figure* shows the unstable part A and the stable part B along a magnetic line of force. The *upper figure* shows the charge separation due to the acceleration along a flute. (**b**) Cross-section of the perturbed plasma

where $l$ is length taken along a line of magnetic force and $S$ is the cross-section of the magnetic tube. As the magnetic flux $\Phi = \boldsymbol{B} \cdot \boldsymbol{S}$ is constant, the energy is

$$Q_{\mathrm{M}} = \frac{\Phi^2}{2\mu_0} \int \frac{\mathrm{d}l}{S} .$$

The change $\delta Q_{\mathrm{M}}$ in the magnetic energy due to the interchange of the fluxes in regions 1 and 2 is

$$\delta Q_{\mathrm{M}} = \frac{1}{2\mu_0} \left[ \left( \Phi_1^2 \int_2 \frac{\mathrm{d}l}{S} + \Phi_2^2 \int_1 \frac{\mathrm{d}l}{S} \right) - \left( \Phi_1^2 \int_1 \frac{\mathrm{d}l}{S} + \Phi_2^2 \int_2 \frac{\mathrm{d}l}{S} \right) \right] . \tag{8.13}$$

If the exchanged fluxes $\Phi_1$ and $\Phi_2$ are the same, the energy change $\delta Q_{\mathrm{M}}$ is zero, so that perturbations resulting in $\Phi_1 = \Phi_2$ are the most dangerous.

The kinetic energy $Q_{\mathrm{p}}$ of a plasma of volume $\mathcal{V}$ is

$$Q_{\mathrm{p}} = \frac{nT\mathcal{V}}{\gamma - 1} = \frac{p\mathcal{V}}{\gamma - 1} , \tag{8.14}$$

where $\gamma$ is the specific heat ratio. As the perturbation is adiabatic,

$$p\mathcal{V}^{\gamma} = \mathrm{const.}$$

is conserved during the interchange process. The change in the plasma energy is

$$\delta Q_{\rm p} = \frac{1}{\gamma - 1}\left(p'_2\mathcal{V}_2 - p_1\mathcal{V}_1 + p'_1\mathcal{V}_1 - p_2\mathcal{V}_2\right),$$

where $p'_2$ is the pressure after interchange from the region $\mathcal{V}_1$ to $\mathcal{V}_2$ and $p'_1$ is the pressure after interchange from the region $\mathcal{V}_2$ to $\mathcal{V}_1$. Because of adiabaticity, we have

$$p'_2 = p_1\left(\frac{\mathcal{V}_1}{\mathcal{V}_2}\right)^\gamma, \qquad p'_1 = p_2\left(\frac{\mathcal{V}_2}{\mathcal{V}_1}\right)^\gamma,$$

and $\delta Q_{\rm p}$ becomes

$$\delta Q_{\rm p} = \frac{1}{\gamma - 1}\left[p_1\left(\frac{\mathcal{V}_1}{\mathcal{V}_2}\right)^\gamma \mathcal{V}_2 - p_1\mathcal{V}_1 + p_2\left(\frac{\mathcal{V}_2}{\mathcal{V}_1}\right)^\gamma \mathcal{V}_1 - p_2\mathcal{V}_2\right]. \tag{8.15}$$

Setting

$$p_2 = p_1 + \delta p\,, \qquad \mathcal{V}_2 = \mathcal{V}_1 + \delta\mathcal{V}\,,$$

we can write $\delta Q_{\rm p}$ as

$$\delta Q_{\rm p} = \delta p \delta\mathcal{V} + \gamma p\frac{(\delta\mathcal{V})^2}{\mathcal{V}}\,. \tag{8.16}$$

Since the stability condition is $\delta Q_{\rm p} > 0$, a sufficient condition is

$$\delta p \delta\mathcal{V} > 0\,.$$

Since the volume is

$$\mathcal{V} = \int {\rm d}l S = \Phi\int\frac{{\rm d}l}{B}\,,$$

the stability condition for interchange instability is written as

$$\delta p\,\delta\int\frac{{\rm d}l}{B} > 0\,.$$

Usually, the plasma pressure $p$ decreases ($\delta p < 0$), so that the stability condition is

$$\delta\int\frac{{\rm d}l}{B} < 0\,, \tag{8.17}$$

in the outward direction [8.4].

The integral is to be taken only over the plasma region. Let the volume inside a magnetic surface $\psi$ be $V$ and the magnetic flux in the toroidal direction $\varphi$ inside the magnetic surface $\psi$ be $\Phi$. We define the *specific volume* $U$ by

$$U = \frac{{\rm d}V}{{\rm d}\Phi}\,. \tag{8.18}$$

If the unit vector of the magnetic field $\boldsymbol{B}$ is denoted by $\boldsymbol{b}$ and the normal unit vector of the infinitesimal cross-sectional area ${\rm d}S$ is denoted by $\boldsymbol{n}$ (see Fig. 8.4), then we have

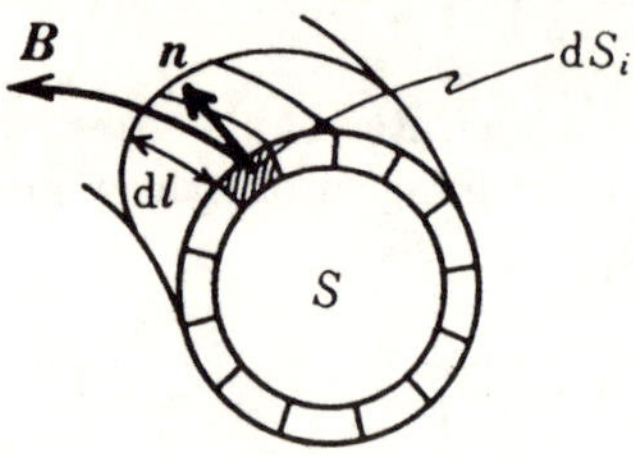


**Fig. 8.4.** Specific volume of a toroidal field

$$\mathrm{d}V = \int \sum_i (\boldsymbol{b}\cdot\boldsymbol{n})_i \mathrm{S}_i \mathrm{d}l \ , \qquad \mathrm{d}\Phi = \sum_i (\boldsymbol{b}\cdot\boldsymbol{n})_i B_i \mathrm{d}S_i \ .$$

When lines of magnetic force close upon a single circuit of the torus, the specific volume $U$ is

$$U = \frac{\oint \left[ \sum_i (\boldsymbol{b}\cdot\boldsymbol{n})_i \mathrm{d}S_i \right] \mathrm{d}l}{\sum_i (\boldsymbol{b}\cdot\boldsymbol{n})_i B_i \mathrm{d}S_i} = \frac{\sum_i (\boldsymbol{b}\cdot\boldsymbol{n})_i B_i \mathrm{d}S_i \oint \frac{\mathrm{d}l}{B_i}}{\sum_i (\boldsymbol{b}\cdot\boldsymbol{n})_i B_i \mathrm{d}S_i} \ .$$

As the integral over $l$ is carried out along a small tube of the magnetic field, $\sum_i (\boldsymbol{b}\cdot\boldsymbol{n})_i \mathrm{d}S_i B_i$ is independent of $l$ (conservation of magnetic flux). As $\oint \mathrm{d}l/B_i$ on the same magnetic surface is constant, $U$ reduces to

$$U = \oint \frac{\mathrm{d}l}{B} \ .$$

When the lines of magnetic force close after $N$ circuits, $U$ is

$$U = \frac{1}{N} \int_N \frac{\mathrm{d}l}{B} \ . \tag{8.19}$$

When the lines of magnetic force are *not* closed, $U$ is given by

$$U = \lim_{N\to\infty} \frac{1}{N} \int_N \frac{\mathrm{d}l}{B} \ .$$

Therefore, $U$ may be considered to be an average of $1/B$. When $U$ decreases outward, it means that the magnitude $B$ of the magnetic field increases outward in an average sense, so that the plasma region is the so-called *average minimum-B* region. In other words, the stability condition for interchange instability reduces to the average minimum-$B$ condition:

$$\frac{\mathrm{d}U}{\mathrm{d}\Phi} = \frac{\mathrm{d}^2 V}{\mathrm{d}\Phi^2} < 0 \,. \tag{8.20}$$

When the values of $U$ on the magnetic axis and on the outermost magnetic surface are $U_0$ and $U_a$, respectively, we define a *magnetic well depth* $-\Delta U/U$ as

$$-\frac{\Delta U}{U} = \frac{U_0 - U_a}{U_0} \,. \tag{8.21}$$

## 8.2 Formulation of Magnetohydrodynamic Instabilities

### 8.2.1 Linearization of Magnetohydrodynamic Equations

Plasma stability problems can be studied by analyzing infinitesimal perturbations from the equilibrium state. If the mass density, pressure, flow velocity, and magnetic field are denoted by $\rho_{\mathrm{m}}$, $p$, $\boldsymbol{V}$, and $\boldsymbol{B}$, respectively, the equation of motion, conservation of mass, Ohm's law, and the adiabatic relation are

$$\rho_{\mathrm{m}}\frac{\partial \boldsymbol{V}}{\partial t} = -\nabla p + \boldsymbol{j}\times\boldsymbol{B} \,, \qquad \frac{\partial \rho_{\mathrm{m}}}{\partial t} + \nabla\cdot(\rho_{\mathrm{m}}\boldsymbol{V}) = 0 \,,$$

$$\boldsymbol{E} + \boldsymbol{V}\times\boldsymbol{B} = 0 \,, \qquad \left(\frac{\partial}{\partial t} + \boldsymbol{V}\cdot\nabla\right)(p\rho_{\mathrm{m}}^{-\gamma}) = 0 \,,$$

respectively, where $\gamma$ is the specific heat ratio. Maxwell's equations are then

$$\nabla\times\boldsymbol{E} = -\frac{\partial \boldsymbol{B}}{\partial t} \,, \qquad \nabla\times\boldsymbol{B} = \mu_0\boldsymbol{j} \,, \qquad \nabla\cdot\boldsymbol{B} = 0 \,.$$

These are the magnetohydrodynamic equations of a plasma with zero specific resistivity (see Sect. 5.2). The values of $\rho_{\mathrm{m}}$, $p$, $\boldsymbol{V}$, and $\boldsymbol{B}$ in the equilibrium state are $\rho_{\mathrm{m}0}$, $p_0$, $\boldsymbol{V}_0 = 0$, and $\boldsymbol{B}_0$, respectively. The first-order small quantities are $\rho_{\mathrm{m}1}$, $p_1$, $\boldsymbol{V}_1 = \boldsymbol{V}$, and $\boldsymbol{B}_1$. The zeroth-order equations are

$$\nabla p_0 = \boldsymbol{j}_0\times\boldsymbol{B}_0 \,, \qquad \nabla\times\boldsymbol{B}_0 = \mu_0\boldsymbol{j}_0 \,, \qquad \nabla\cdot\boldsymbol{B}_0 = 0 \,. \tag{8.22}$$

The first-order linearized equations are

$$\frac{\partial \rho_{\mathrm{m}1}}{\partial t} + \nabla\cdot(\rho_{\mathrm{m}0}\boldsymbol{V}) = 0 \,, \tag{8.23}$$

$$\rho_{\mathrm{m}0}\frac{\partial \boldsymbol{V}}{\partial t} + \nabla p_1 = \boldsymbol{j}_0\times\boldsymbol{B}_1 + \boldsymbol{j}_1\times\boldsymbol{B}_0 \,, \tag{8.24}$$

$$\frac{\partial p_1}{\partial t} + (\boldsymbol{V}\cdot\nabla)p_0 + \gamma p_0\nabla\cdot\boldsymbol{V} = 0 \,, \tag{8.25}$$

$$\frac{\partial \boldsymbol{B}_1}{\partial t} = \nabla\times(\boldsymbol{V}\times\boldsymbol{B}_0) \,. \tag{8.26}$$

If the displacement of the plasma from the equilibrium position $\boldsymbol{r}_0$ is denoted by $\boldsymbol{\xi}(\boldsymbol{r}_0, t)$, it follows that

$$\boldsymbol{\xi}(\boldsymbol{r}_0, t) = \boldsymbol{r} - \boldsymbol{r}_0 , \qquad \boldsymbol{V} = \frac{\mathrm{d}\boldsymbol{\xi}}{\mathrm{d}t} \approx \frac{\partial \boldsymbol{\xi}}{\partial t} . \tag{8.27}$$

Equation (8.26) reduces to

$$\frac{\partial \boldsymbol{B}_1}{\partial t} = \boldsymbol{\nabla} \times \left( \frac{\partial \boldsymbol{\xi}}{\partial t} \times \boldsymbol{B}_0 \right)$$

and

$$\boldsymbol{B}_1 = \boldsymbol{\nabla} \times (\boldsymbol{\xi} \times \boldsymbol{B}_0) . \tag{8.28}$$

From $\mu_0 \boldsymbol{j} = \boldsymbol{\nabla} \times \boldsymbol{B}$, it follows that

$$\mu_0 \boldsymbol{j}_1 = \boldsymbol{\nabla} \times \boldsymbol{B}_1 . \tag{8.29}$$

Equations (8.23) and (8.25) yield

$$\rho_{\mathrm{m}1} = -\boldsymbol{\nabla} \cdot (\rho_{\mathrm{m}0} \boldsymbol{\xi}) , \tag{8.30}$$

$$p_1 = -\boldsymbol{\xi} \cdot \boldsymbol{\nabla} p_0 - \gamma p_0 \boldsymbol{\nabla} \cdot \boldsymbol{\xi} . \tag{8.31}$$

Substituting these equations into (8.24) gives

$$\begin{aligned} \rho_{\mathrm{m}0} \frac{\partial^2 \boldsymbol{\xi}}{\partial t^2} &= \boldsymbol{\nabla}(\boldsymbol{\xi} \cdot \boldsymbol{\nabla} p_0 + \gamma p_0 \boldsymbol{\nabla} \cdot \boldsymbol{\xi}) + \frac{1}{\mu_0}(\boldsymbol{\nabla} \times \boldsymbol{B}_0) \times \boldsymbol{B}_1 + \frac{1}{\mu_0}(\boldsymbol{\nabla} \times \boldsymbol{B}_1) \times \boldsymbol{B}_0 \\ &= -\boldsymbol{\nabla}\left( p_1 + \frac{\boldsymbol{B}_0 \cdot \boldsymbol{B}_1}{\mu_0} \right) + \frac{1}{\mu_0}\big[ (\boldsymbol{B}_0 \cdot \boldsymbol{\nabla}) \boldsymbol{B}_1 + (\boldsymbol{B}_1 \cdot \boldsymbol{\nabla}) \boldsymbol{B}_0 \big] . \end{aligned} \tag{8.32}$$

This is the linearized equation of motion in terms of $\boldsymbol{\xi}$.

Next let us consider the boundary conditions. Where the plasma contacts an ideal conductor, the tangential component of the electric field is zero, i.e., $\boldsymbol{n} \times \boldsymbol{E} = 0$. This is equivalent to $\boldsymbol{n} \times (\boldsymbol{\xi} \times \boldsymbol{B}_0) = 0$, $\boldsymbol{n}$ being taken in the outward direction. The conditions $(\boldsymbol{\xi} \cdot \boldsymbol{n}) = 0$ and $(\boldsymbol{B}_1 \cdot \boldsymbol{n}) = 0$ must also be satisfied.

When the plasma is in contact with the vacuum, the total pressure must be continuous at the boundary surface between plasma and vacuum and

$$p - p_0 + \frac{B_{\mathrm{in}}^2 - B_{0,\mathrm{in}}^2}{2\mu_0} = \frac{B_{\mathrm{ex}}^2 - B_{0,\mathrm{ex}}^2}{2\mu_0} ,$$

where $\boldsymbol{B}_{\mathrm{in}}$ and $\boldsymbol{B}_{0,\mathrm{in}}$ give the internal magnetic field of the plasma and $\boldsymbol{B}_{\mathrm{ex}}$ and $\boldsymbol{B}_{0,\mathrm{ex}}$ give the external field. When $\boldsymbol{B}_{\mathrm{in}}(\boldsymbol{r})$, $\boldsymbol{B}_{\mathrm{ex}}(\boldsymbol{r})$ and $p(\boldsymbol{r})$ are expanded in $\boldsymbol{\xi} = \boldsymbol{r} - \boldsymbol{r}_0$, with $f(\boldsymbol{r}) = f_0(\boldsymbol{r}_0) + (\boldsymbol{\xi} \cdot \boldsymbol{\nabla}) f_0(\boldsymbol{r}) + f_1$, the boundary condition reduces to

$$-\gamma p_0 \nabla \cdot \boldsymbol{\xi} + \frac{\boldsymbol{B}_{0,\text{in}} \cdot [\boldsymbol{B}_{1,\text{in}} + (\boldsymbol{\xi} \cdot \nabla)\boldsymbol{B}_{0,\text{in}}]}{\mu_0} = \frac{\boldsymbol{B}_{0,\text{ex}} \cdot [\boldsymbol{B}_{1,\text{ex}} + (\boldsymbol{\xi} \cdot \nabla)\boldsymbol{B}_{0,\text{ex}}]}{\mu_0} . \tag{8.33}$$

From Maxwell's equations, we have the following relations:

$$\boldsymbol{n}_0 \cdot (\boldsymbol{B}_{0,\text{in}} - \boldsymbol{B}_{0,\text{ex}}) = 0 , \tag{8.34}$$

$$\boldsymbol{n}_0 \times (\boldsymbol{B}_{0,\text{in}} - \boldsymbol{B}_{0,\text{ex}}) = \mu_0 \boldsymbol{K} , \tag{8.35}$$

where $\boldsymbol{K}$ is the surface current. Ohm's law yields

$$\boldsymbol{E}_{\text{in}} + \boldsymbol{V} \times \boldsymbol{B}_{0,\text{in}} = 0 , \tag{8.36}$$

in the plasma. As the electric field $\boldsymbol{E}^*$ in coordinates moving with the plasma is $\boldsymbol{E}^* = \boldsymbol{E} + \boldsymbol{V} \times \boldsymbol{B}_0$ and the tangential component of the electric field $\boldsymbol{E}^*$ is continuous across the plasma boundary. The boundary condition can be written as

$$\boldsymbol{E}_{\text{t}} + (\boldsymbol{V} \times \boldsymbol{B}_{0,\text{ex}})_{\text{t}} = 0 , \tag{8.37}$$

where the subscript t indicates the tangential component. Since the normal component of $\boldsymbol{B}$ is given by the tangential component of $\boldsymbol{E}$ by the relation $\nabla \times \boldsymbol{E} = -\partial \boldsymbol{B}/\partial t$, (8.37) reduces to

$$(\boldsymbol{n}_0 \cdot \boldsymbol{B}_{1,\text{ex}}) = \boldsymbol{n}_0 \cdot \nabla \times (\boldsymbol{\xi} \times \boldsymbol{B}_{0,\text{ex}}) . \tag{8.38}$$

The electric field $\boldsymbol{E}_{\text{ex}}$ and the magnetic field $\boldsymbol{B}_{\text{ex}}$ in the external (vacuum) region can be expressed in terms of a vector potential:

$$\boldsymbol{E}_{\text{ex}} = -\frac{\partial \boldsymbol{A}}{\partial t} , \qquad \boldsymbol{B}_{1,\text{ex}} = \nabla \times \boldsymbol{A} , \qquad \nabla \cdot \boldsymbol{A} = 0 .$$

If no current flows in the vacuum region, $\boldsymbol{A}$ satisfies

$$\nabla \times \nabla \times \boldsymbol{A} = 0 . \tag{8.39}$$

Using the vector potential, we may express (8.37) as

$$\boldsymbol{n}_0 \times \left( -\frac{\partial \boldsymbol{A}}{\partial t} + \boldsymbol{V} \times \boldsymbol{B}_{0,\text{ex}} \right) = 0 .$$

For $\boldsymbol{n}_0 \cdot \boldsymbol{B}_{0,\text{in}} = \boldsymbol{n} \cdot \boldsymbol{B}_{0,\text{ex}} = 0$, the boundary condition is

$$\boldsymbol{n}_0 \times \boldsymbol{A} = -\xi_n \boldsymbol{B}_{0,\text{ex}} . \tag{8.40}$$

The boundary condition at the wall of an ideal conductor is

$$\boldsymbol{n} \times \boldsymbol{A} = 0 . \tag{8.41}$$

The stability problem now becomes one of solving (8.32) and (8.39) under the boundary conditions (8.33) and (8.38) or (8.40) and (8.41). When a normal mode $\boldsymbol{\xi}(\boldsymbol{r}, t) = \boldsymbol{\xi}(\boldsymbol{r}) \exp(-i\omega t)$ is considered, the problem reduces to the eigenvalue problem $\rho_0 \omega^2 \boldsymbol{\xi} = -\boldsymbol{F}(\boldsymbol{\xi})$. If any eigenvalue is negative, the plasma is unstable; if all the eigenvalues are positive, the plasma is stable.

### 8.2.2 Energy Principle

The eigenvalue problem is complicated and difficult to solve in general. When we introduce a potential energy associated with the displacement $\boldsymbol{\xi}$, the stability problem can be simplified. The equation of motion has the form

$$\rho_{\mathrm{m}0}\frac{\partial^2\boldsymbol{\xi}}{\partial t^2} = \boldsymbol{F}(\boldsymbol{\xi}) = -\widehat{\boldsymbol{K}}\cdot\boldsymbol{\xi} \,, \tag{8.42}$$

where $\widehat{\boldsymbol{K}}$ is a linear operator. When this equation is integrated, the equation of energy conservation becomes:

$$\frac{1}{2}\int \rho_{\mathrm{m}0}\left(\frac{\partial\boldsymbol{\xi}}{\partial t}\right)^2 \mathrm{d}\boldsymbol{r} + \frac{1}{2}\int \boldsymbol{\xi}\cdot\widehat{\boldsymbol{K}}\boldsymbol{\xi}\,\mathrm{d}\boldsymbol{r} = \mathrm{const.}$$

The kinetic energy $T$ and the potential energy $W$ are

$$T \equiv \frac{1}{2}\int \rho_{\mathrm{m}0}\left(\frac{\partial\boldsymbol{\xi}}{\partial t}\right)^2 \mathrm{d}\boldsymbol{r}\,, \qquad W \equiv \frac{1}{2}\int \boldsymbol{\xi}\cdot\widehat{\boldsymbol{K}}\boldsymbol{\xi}\mathrm{d}\boldsymbol{r} = -\frac{1}{2}\int \boldsymbol{\xi}\cdot F(\boldsymbol{\xi})\mathrm{d}\boldsymbol{r}\,,$$

respectively. Accordingly, if $W > 0$ for all possible displacements, the system is stable. This is the stability criterion of the *energy principle* [8.5]. $W$ is called the *energy integral*.

It is possible to prove that the operator $\widehat{\boldsymbol{K}}$ is Hermitian, i.e., self-adjoint [8.6, 8.7]. A displacement $\boldsymbol{\eta}$ and a vector potential $\boldsymbol{Q}$ are introduced which satisfy the same boundary conditions as $\boldsymbol{\xi}$ and $\boldsymbol{A}$, i.e.,

$$\boldsymbol{n}_0 \times \boldsymbol{Q} = -\eta_n \boldsymbol{B}_{0,\mathrm{ex}}$$

at the plasma–vacuum boundary and

$$\boldsymbol{n}_0 \times \boldsymbol{Q} = 0$$

at the conducting wall. Substituting (8.32), the integral in the plasma region $V_{\mathrm{in}}$ is seen to be

$$\begin{aligned}\int_{V_{\mathrm{in}}} \boldsymbol{\eta}\cdot\widehat{\boldsymbol{K}}\boldsymbol{\xi}\,\mathrm{d}\boldsymbol{r} = \int_{V_{\mathrm{in}}} \Bigg\{ &\gamma p_0(\nabla\cdot\boldsymbol{\eta})(\nabla\cdot\boldsymbol{\xi}) + (\nabla\cdot\boldsymbol{\eta})(\boldsymbol{\xi}\cdot\nabla p_0) \\ &+ \frac{1}{\mu_0}\left[\nabla\times(\boldsymbol{\eta}\times\boldsymbol{B}_0)\right]\cdot\nabla\times(\boldsymbol{\xi}\times\boldsymbol{B}_0) \\ &- \frac{1}{\mu_0}\left[\boldsymbol{\eta}\times(\nabla\times\boldsymbol{B}_0)\right]\cdot\nabla\times(\boldsymbol{\xi}\times\boldsymbol{B}_0)\Bigg\}\mathrm{d}\boldsymbol{r} \\ + \int_S \boldsymbol{n}_0\cdot\boldsymbol{\eta}\left[\frac{\boldsymbol{B}_{0,\mathrm{in}}\cdot\nabla\times(\boldsymbol{\xi}\times\boldsymbol{B}_{0,\mathrm{in}})}{\mu_0} - \gamma p_0(\nabla\cdot\boldsymbol{\xi}) - (\boldsymbol{\xi}\cdot\nabla p_0)\right]\mathrm{d}S\,.\end{aligned} \tag{8.43}$$

Next let us consider the surface integral in (8.43). Due to the boundary condition $\boldsymbol{n}_0 \times \boldsymbol{Q} = -\eta_n \boldsymbol{B}_{0,\mathrm{ex}}$, we find that

$$\begin{aligned}\int_S \eta_n \boldsymbol{B}_{0,\mathrm{ex}}\cdot\boldsymbol{B}_{1,\mathrm{ex}} \mathrm{d}S &= \int_S \eta_n \boldsymbol{B}_{0,\mathrm{ex}}(\nabla \times \boldsymbol{A})\,\mathrm{d}S = -\int_S (\boldsymbol{n}_0 \times \boldsymbol{Q})\cdot(\nabla \times \boldsymbol{A})\,\mathrm{d}S \\ &= -\int_S \boldsymbol{n}_0\cdot\big[\boldsymbol{Q} \times (\nabla \times \boldsymbol{A})\big]\,\mathrm{d}S = \int_{V_\mathrm{ex}} \nabla\cdot\big[\boldsymbol{Q} \times (\nabla \times \boldsymbol{A})\big]\,\mathrm{d}\boldsymbol{r} \\ &= \int_{V_\mathrm{ex}} \big[(\nabla \times \boldsymbol{Q})\cdot(\nabla \times \boldsymbol{A}) - \boldsymbol{Q}\cdot\nabla \times (\nabla \times \boldsymbol{A})\big]\,\mathrm{d}\boldsymbol{r} \\ &= \int_{V_\mathrm{ex}} (\nabla \times \boldsymbol{Q})\cdot(\nabla \times \boldsymbol{A})\,\mathrm{d}\boldsymbol{r}\,.\end{aligned}$$

From the boundary condition (8.33), the difference between the foregoing surface integral and the surface integral in (8.43) reduces to

$$\begin{aligned}\int \eta_n &\left[\frac{\boldsymbol{B}_{0,\mathrm{in}}\cdot\boldsymbol{B}_{1,\mathrm{in}} - \boldsymbol{B}_{0,\mathrm{ex}}\cdot\boldsymbol{B}_{1,\mathrm{ex}}}{\mu_0} - \gamma p_0(\nabla\cdot\boldsymbol{\xi}) - (\boldsymbol{\xi}\cdot\nabla)p_0\right]\mathrm{d}S \\ &= \int_S \eta_n(\boldsymbol{\xi}\cdot\nabla)\left(\frac{B_{0,\mathrm{ex}}^2}{2\mu_0} - \frac{B_{0,\mathrm{in}}^2}{2\mu_0} - p_0\right)\mathrm{d}S \\ &= \int_S \eta_n \xi_n \frac{\partial}{\partial n}\left(\frac{B_{0,\mathrm{ex}}^2}{2\mu_0} - \frac{B_{0,\mathrm{in}}^2}{2\mu_0} - p_0\right)\mathrm{d}S\,,\end{aligned}$$

where the relation $\boldsymbol{n}_0 \times \nabla(p_0 + B_{0,\mathrm{in}}^2/2\mu_0 - B_{0,\mathrm{ex}}^2/2\mu_0) = 0$ has been used. The region of integration $V_\mathrm{ex}$ is the region outside the plasma. Finally, the energy integral reduces to

$$\begin{aligned}\int_{V_\mathrm{in}} \boldsymbol{\eta}\cdot\widehat{\boldsymbol{K}}\boldsymbol{\xi}\,\mathrm{d}\boldsymbol{r} = \int_{V_\mathrm{in}} &\left\{\gamma p_0(\nabla\cdot\boldsymbol{\eta})(\nabla\cdot\boldsymbol{\xi}) + \frac{1}{\mu_0}\big[\nabla \times (\boldsymbol{\eta} \times \boldsymbol{B}_0)\big]\cdot\big[\nabla \times (\boldsymbol{\xi} \times \boldsymbol{B}_0)\big]\right. \\ &\left. +(\nabla\cdot\boldsymbol{\eta})(\boldsymbol{\xi}\cdot\nabla p_0) - \frac{1}{\mu_0}\big[\boldsymbol{\eta} \times (\nabla \times \boldsymbol{B}_0)\big]\cdot\nabla \times (\boldsymbol{\xi} \times \boldsymbol{B}_0)\right\}\mathrm{d}\boldsymbol{r} \\ &+\frac{1}{\mu_0}\int_{V_\mathrm{ex}} (\nabla \times \boldsymbol{Q})\cdot(\nabla \times \boldsymbol{A})\,\mathrm{d}\boldsymbol{r} \\ &+\int_S \eta_n \xi_n \frac{\partial}{\partial n}\left(\frac{B_{0,\mathrm{ex}}^2}{2\mu_0} - \frac{B_{0,\mathrm{in}}^2}{2\mu_0} - p_0\right)\mathrm{d}S\,. \end{aligned} \tag{8.44}$$

The energy integral $W$ is divided into three parts $W_\mathrm{P}$, $W_\mathrm{S}$, and $W_\mathrm{V}$, the contributions of the plasma internal region $V_\mathrm{in}$, the boundary region $S$ and the external vacuum region $V_\mathrm{ex}$, i.e.,

$$W = \frac{1}{2}\int_{V_\mathrm{in}} \boldsymbol{\xi}\cdot\widehat{\boldsymbol{K}}\boldsymbol{\xi}\,\mathrm{d}\boldsymbol{r} = W_\mathrm{p} + W_\mathrm{S} + W_\mathrm{V}\,, \tag{8.45}$$

$$W_\mathrm{p} = \frac{1}{2}\int_{V_\mathrm{in}} \Bigg\{ \gamma p_0(\boldsymbol{\nabla}\cdot\boldsymbol{\xi})^2 + \frac{1}{\mu_0}\left[\boldsymbol{\nabla}\times(\boldsymbol{\xi}\times\boldsymbol{B}_0)\right]^2 + (\boldsymbol{\nabla}\cdot\boldsymbol{\xi})(\boldsymbol{\xi}\cdot\boldsymbol{\nabla}p_0)$$

$$-\frac{1}{\mu_0}[\boldsymbol{\xi}\times(\boldsymbol{\nabla}\times\boldsymbol{B}_0)]\cdot\boldsymbol{\nabla}\times(\boldsymbol{\xi}\times\boldsymbol{B}_0)\Bigg\}\mathrm{d}\boldsymbol{r}$$

$$= \frac{1}{2}\int_{V_\mathrm{in}} \left[\frac{\boldsymbol{B}_1^2}{\mu_0} - p_1(\boldsymbol{\nabla}\cdot\boldsymbol{\xi}) - \boldsymbol{\xi}\cdot(\boldsymbol{j}_0\times\boldsymbol{B}_1)\right]\mathrm{d}\boldsymbol{r}\ , \tag{8.46}$$

$$W_\mathrm{S} = \frac{1}{2}\int_S \xi_n^2 \frac{\partial}{\partial n}\left(\frac{B_{0,\mathrm{ex}}^2}{2\mu_0} - \frac{B_{0,\mathrm{in}}^2}{2\mu_0} - p_0\right)\mathrm{d}S\ , \tag{8.47}$$

$$W_\mathrm{V} = \frac{1}{2\mu_0}\int_{V_\mathrm{ex}} (\boldsymbol{\nabla}\times\boldsymbol{A})^2\mathrm{d}\boldsymbol{r} = \int_{V_\mathrm{ex}} \frac{\boldsymbol{B}_1^2}{2\mu_0}\mathrm{d}\boldsymbol{r}\ . \tag{8.48}$$

The stability condition is $W > 0$ for all possible $\boldsymbol{\xi}$. The frequency or growth rate of a perturbation can be obtained from the energy integral. When the perturbation varies as $\exp(-\mathrm{i}\omega t)$, the equation of motion is

$$\omega^2 \rho_{\mathrm{m}0}\boldsymbol{\xi} = \widehat{\boldsymbol{K}}\boldsymbol{\xi}\ . \tag{8.49}$$

The solution of the eigenvalue problem is the same as the solution based on the calculus of variations $\delta(\omega^2) = 0$, where

$$\omega^2 = \frac{\int \boldsymbol{\xi}\cdot\widehat{\boldsymbol{K}}\boldsymbol{\xi}\mathrm{d}\boldsymbol{r}}{\int \rho_{\mathrm{m}0}\boldsymbol{\xi}^2\mathrm{d}\boldsymbol{r}}\ . \tag{8.50}$$

As $\widehat{\boldsymbol{K}}$ is a Hermitian operator, $\omega^2$ is real. In the MHD analysis of an ideal plasma with zero resistivity, the perturbation either increases or decreases monotonically, or else the perturbed plasma oscillates with constant amplitude.

## 8.3 Instabilities of a Cylindrical Plasma

### 8.3.1 Instabilities of Sharp-Boundary Configuration

Let us consider a sharp-boundary plasma of radius $a$, with a longitudinal magnetic field $B_{0z}$ inside the boundary and a longitudinal magnetic field $B_{ez}$ and an azimuthal magnetic field $B_\theta = \mu_0 I/(2\pi r)$ outside. $B_{0z}$ and $B_{ez}$ are assumed to be constant (see Fig. 8.5). We can consider the displacement

$$\boldsymbol{\xi}(r)\exp(\mathrm{i}m\theta + \mathrm{i}kz)\ . \tag{8.51}$$

Any displacement may be expressed as a superposition of such modes. Since the term in $\boldsymbol{\nabla}\cdot\boldsymbol{\xi}$ in the energy integral is positive, the incompressible perturbation is the most dangerous. We examine only the worst mode,

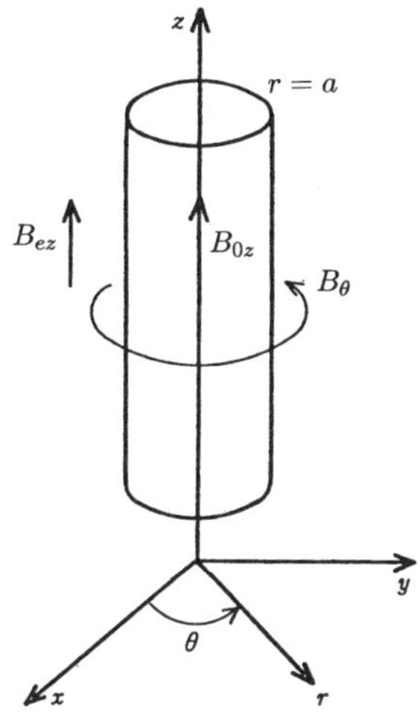


**Fig. 8.5.** Sharp-boundary plasma

$$\nabla\cdot\boldsymbol{\xi} = 0 \,. \tag{8.52}$$

The perturbation of the magnetic field $\boldsymbol{B}_1 = \nabla \times (\boldsymbol{\xi} \times \boldsymbol{B}_0)$ inside the plasma is

$$\boldsymbol{B}_1 = \mathrm{i}kB_{0z}\boldsymbol{\xi} \,. \tag{8.53}$$

The equation of motion (8.32) becomes

$$\left(-\omega^2\rho_{\mathrm{m}0} + \frac{k^2B_{0z}^2}{\mu_0}\right)\boldsymbol{\xi} = -\nabla\left(p_1 + \frac{\boldsymbol{B}_0\cdot\boldsymbol{B}_1}{\mu_0}\right) \equiv -\nabla p^* \,. \tag{8.54}$$

As $\nabla\cdot\boldsymbol{\xi} = 0$, it follows that $\Delta p^* = 0$, i.e.,

$$\left[\frac{\mathrm{d}^2}{\mathrm{d}r^2} + \frac{1}{r}\frac{\mathrm{d}}{\mathrm{d}r} - \left(k^2 + \frac{m^2}{r^2}\right)\right]p^*(r) = 0 \,. \tag{8.55}$$

The solution without singularity at $r = 0$ is given by the modified Bessel function $I_m(kr)$ of the first kind, so that $p^*(r)$ is

$$p^*(r) = p^*(a)\frac{I_m(kr)}{I_m(ka)} \,.$$

Accordingly, we find

$$\xi_r(a) = \frac{kp^*(a)/I_m(ka)}{\omega^2\rho_{\mathrm{m}0} - k^2B_0^2/\mu_0}I'_m(ka) \,. \tag{8.56}$$

As the perturbation of the vacuum magnetic field $\boldsymbol{B}_{1\mathrm{e}}$ satisfies $\nabla \times \boldsymbol{B} = 0$ and $\nabla\cdot\boldsymbol{B} = 0$, $\boldsymbol{B}_{1\mathrm{e}}$ is expressed by $\boldsymbol{B}_{1\mathrm{e}} = \nabla\psi$. The scalar magnetic potential $\psi$ satisfies $\Delta\psi = 0$ and $\psi \to 0$ as $r \to \infty$. Then

$$\psi = C\frac{K_m(kr)}{K_m(ka)}\exp(\mathrm{i}m\theta + \mathrm{i}kz) \;, \tag{8.57}$$

where $K_{\rm m}(kr)$ is the modified Bessel function of the second kind. The boundary condition (8.33) is

$$p^*(r) = p_1 + \frac{1}{\mu_0}\boldsymbol{B}_0\cdot\boldsymbol{B}_1 = \frac{1}{\mu_0}\boldsymbol{B}_{\rm e}\cdot\boldsymbol{B}_{1\rm e} + (\boldsymbol{\xi}\cdot\nabla)\left(\frac{B_{\rm e}^2}{2\mu_0} - \frac{B_0^2}{2\mu_0} - p_0\right)$$
$$= \frac{1}{\mu_0}\boldsymbol{B}_{\rm e}\cdot\boldsymbol{B}_{1\rm e} + (\boldsymbol{\xi}\cdot\nabla)\left(\frac{B_\theta^2}{2\mu_0}\right).$$

Since $B_\theta \propto 1/r$ and $\boldsymbol{B}_{1\rm e} = \nabla\psi$ with (8.57), $p^*(a)$ is given by

$$p^*(a) = \frac{i}{\mu_0}(kB_{\rm ez} + \frac{m}{a}B_\theta)C - \frac{B_\theta^2}{\mu_0 a}\xi_r(a). \tag{8.58}$$

Due to $B_{1r} = \partial\psi/\partial r$, the boundary condition (8.38) is reduced to

$$Ck\frac{K_m'(ka)}{K_m(ka)} = \mathrm{i}\left(kB_{\rm ez} + \frac{m}{a}B_\theta\right)\xi_r(a)\;. \tag{8.59}$$

From (8.56), (8.58) and (8.59), the dispersion equation is

$$\frac{\omega^2}{k^2} = \frac{B_{0z}^2}{\mu_0\rho_{\rm m0}} - \frac{\left[kB_{\rm ez} + (m/a)B_\theta\right]^2}{\mu_0\rho_{\rm m0}k^2}\frac{I_m'(ka)}{I_m(ka)}\frac{K_m(ka)}{K_m'(ka)} - \frac{B_\theta^2}{\mu_0\rho_{\rm m0}}\frac{1}{(ka)}\frac{I_m'(ka)}{I_m(ka)}\;. \tag{8.60}$$

The first and second terms represent the stabilizing effect of $B_{0z}$ and $B_{\rm ez}$, where $K_m/K_m' < 0$. If the propagation vector $\boldsymbol{k}$ is normal to the magnetic field, i.e., if

$$(\boldsymbol{k}\cdot\boldsymbol{B}_{\rm e}) = kB_{\rm ez} + \frac{m}{a}B_\theta = 0\;,$$

the second stabilizing term of (8.60) becomes zero, so that a flutelike perturbation is dangerous. The third term is the destabilizing term.

### $m = 0$ Mode with $B_{\rm ez} = 0$

Let us consider the $m = 0$ mode with $B_{\rm ez} = 0$. This azimuthally symmetric perturbation constricts the plasma like a sausage, and the mode is called the *sausage instability*. Equation (8.60) reduces to

$$\omega^2 = \frac{B_{0z}^2 k^2}{\mu_0\rho_{\rm m0}}\left[1 - \frac{B_\theta^2}{B_{0z}^2}\frac{I_0'(ka)}{(ka)I_0(ka)}\right]\;. \tag{8.61}$$

Since $I_0'(x)/xI_0(x) < 1/2$, the stability condition is

$$B_{0z}^2 > B_\theta^2/2\;.$$

### $m = 1$ Mode with $B_{ez} = 0$

The $m = 1$ mode perturbation kinks the plasma column and it is call the *kink mode*. For the $m = 1$ mode with $B_{ez} = 0$, (8.60) is

$$\omega^2 = \frac{B_{0z}^2 k^2}{\mu_0 \rho_{m0}} \left[ 1 + \frac{B_\theta^2}{B_{0z}^2} \frac{1}{(ka)} \frac{I_1'(ka)}{I_1(ka)} \frac{K_1(ka)}{K_1'(ka)} \right] . \tag{8.62}$$

For perturbations with long characteristic length, (8.62) becomes

$$\omega^2 = \frac{B_{0z}^2 k^2}{\mu_0 \rho_{m0}} \left[ 1 - \left( \frac{B_\theta}{B_{0z}} \right)^2 \ln \frac{1}{ka} \right] . \tag{8.63}$$

This dispersion equation shows that the kink mode is unstable for perturbations with long wavelength.

### Instability in the Case $|B_{ez}| > |B_\theta|$

When $|B_{ez}| \gg |B_\theta|$, the term including $|ka| \ll 1$ predominates. Expanding the modified Bessel function ($m > 0$ is assumed), we find

$$\mu_0 \rho_{m0} \omega^2 = k^2 B_{0z}^2 + \left( k B_{ez} + \frac{m}{a} B_\theta \right)^2 - \frac{m}{a^2} B_\theta^2 . \tag{8.64}$$

Then $\omega^2$ becomes minimum when $\partial \omega / \partial k = 0$, i.e.,

$$k(B_{0z}^2 + B_{ez}^2) + \frac{m}{a} B_\theta B_{ez} = 0 .$$

In this case, $\omega^2$ has the minimum value

$$\omega_{\min}^2 = \frac{B_\theta^2}{\mu_0 \rho_{m0} a^2} \left( \frac{m^2 B_{0z}^2}{B_{ez}^2 + B_{0z}^2} - m \right) = \frac{B_\theta^2}{\mu_0 \rho_{m0} a^2} m \left( m \frac{1-\beta}{2-\beta} - 1 \right) , \tag{8.65}$$

where $\beta$ is the beta ratio. Accordingly, the plasma is unstable when $0 < m < (2-\beta)/(1-\beta)$. For a low-beta plasma only the modes $m = 1$ and $m = 2$ become unstable. However, if

$$\left( \frac{B_\theta}{B_z} \right)^2 < (ka)^2 \tag{8.66}$$

is satisfied, the plasma is stable even for $m = 1$. Usually, the length of the plasma is finite so that $k$ cannot be smaller than $2\pi/L$. Accordingly, when

$$\left| \frac{B_\theta}{B_z} \right| < \frac{2\pi a}{L} ,$$

the plasma is stable. This stability condition is called the *Kruskal–Shafranov condition* [8.8, 8.9].

When a cylindrical conducting wall of radius $b$ surrounds the plasma, the scalar magnetic potential of the external magnetic field is

$$\psi = \left[ c_1 \frac{K_m(kr)}{K_m(ka)} + c_2 \frac{I_m(kr)}{I_m(ka)} \right] \exp(\mathrm{i}m\theta + \mathrm{i}kz) \,, \tag{8.67}$$

instead of (8.57). The boundary condition $B_{1\mathrm{er}} = 0$ at $r = b$ yields

$$\frac{c_1}{c_2} = -\frac{I'_m(kb)K_m(ka)}{K'_m(kb)I_m(ka)} \,.$$

The dispersion equation becomes

$$\frac{\omega^2}{k^2} = \frac{B_{0z}^2}{\mu_0\rho_{\mathrm{m}0}} - \frac{\left[kB_{\mathrm{e}z} + (m/a)B_\theta\right]^2}{\mu_0\rho_{\mathrm{m}0}k^2} \frac{I'_m(ka)}{I_m(ka)} \frac{K_m(ka)I'_m(kb) - I_m(ka)K'_m(kb)}{K'_m(ka)I'_m(kb) - I'_m(ka)K'_m(kb)}$$
$$- \frac{B_\theta^2}{\mu_0\rho_{\mathrm{m}0}} \frac{1}{ka} \frac{I'_m(ka)}{I_m(ka)} \,.$$

Expanding the modified Bessel functions under the conditions $ka \ll 1$, $kb \ll 1$, we find

$$\mu_0\rho_{\mathrm{m}0}\omega^2 = k^2 B_{0z}^2 + \frac{1 + (a/b)^{2m}}{1 - (a/b)^{2m}} \left( kB_{\mathrm{e}z} + \frac{m}{a} B_\theta \right)^2 - \frac{m}{a^2} B_\theta^2 \,.$$

The closer the wall is to the plasma boundary, the more effective is the wall stabilization.

In toroidal systems, the propagation constant is $k = n/R$, where $n$ is an integer and $R$ is the major radius of the torus. Introducing the *safety factor* $q_a$ at the plasma boundary $r = a$,

$$q_a = \frac{aB_{\mathrm{e}z}}{RB_\theta} \,, \tag{8.68}$$

$(\boldsymbol{k}\cdot\boldsymbol{B})$ may be written as

$$(\boldsymbol{k}\cdot\boldsymbol{B}) = kB_{\mathrm{e}z} + \frac{m}{a} B_\theta = \frac{nB_\theta}{a} \left( q_a + \frac{m}{n} \right) \,.$$

The Kruskal–Shafranov condition (8.66) of the $m = 1$, $n = -1$ mode can then be expressed in terms of the safety factor as

$$q_a > 1 \,. \tag{8.69}$$

This is the reason why $q_a$ is called the safety factor.

### 8.3.2 Instabilities of Diffuse Boundary Configurations

The sharp-boundary configuration treated in Sect. 8.3.1 is a special case. In most cases the plasma current decreases gradually at the boundary. Let us consider the case of a diffuse-boundary plasma whose parameters in the equilibrium state are

$$p_0(r)\,, \qquad \boldsymbol{B}_0(r) = \big(0, B_\theta(r), B_z(r)\big)\,.$$

The perturbation $\boldsymbol{\xi}$ is assumed to be

$$\boldsymbol{\xi} = \boldsymbol{\xi}(r)\exp(\mathrm{i}m\theta + \mathrm{i}kz)\,.$$

The perturbation of the magnetic field $\boldsymbol{B}_1 = \boldsymbol{\nabla} \times (\boldsymbol{\xi} \times \boldsymbol{B}_0)$ is

$$B_{1r} = \mathrm{i}(\boldsymbol{k}\cdot\boldsymbol{B}_0)\xi_r\,, \tag{8.70}$$

$$B_{1\theta} = \mathrm{i}kA - \frac{\mathrm{d}}{\mathrm{d}r}(\xi_r B_\theta)\,, \tag{8.71}$$

$$B_{1z} = -\left[\frac{\mathrm{i}mA}{r} + \frac{1}{r}\frac{\mathrm{d}}{\mathrm{d}r}(r\xi_r B_z)\right]\,, \tag{8.72}$$

where

$$(\boldsymbol{k}\cdot\boldsymbol{B}_0) = kB_z + \frac{m}{r}B_\theta\,, \tag{8.73}$$

$$A = \xi_\theta B_z - \xi_z B_\theta = (\boldsymbol{\xi} \times \boldsymbol{B}_0)_r\,. \tag{8.74}$$

Since the pressure terms

$$\gamma p_0(\boldsymbol{\nabla}\cdot\boldsymbol{\xi})^2 + (\boldsymbol{\nabla}\cdot\boldsymbol{\xi})(\boldsymbol{\xi}\cdot\boldsymbol{\nabla}p_0) = (\gamma - 1)p_0(\boldsymbol{\nabla}\cdot\boldsymbol{\xi})^2 + (\boldsymbol{\nabla}\cdot\boldsymbol{\xi})(\boldsymbol{\nabla}\cdot p_0\boldsymbol{\xi})$$

in the energy integral are nonnegative, we examine the incompressible displacement $\boldsymbol{\nabla}\cdot\boldsymbol{\xi} = 0$ again, i.e.,

$$\frac{1}{r}\frac{\mathrm{d}}{\mathrm{d}r}(r\xi_r) + \frac{\mathrm{i}m}{r}\xi_\theta + \mathrm{i}k\xi_z = 0\,. \tag{8.75}$$

From this and (8.74) for $A$, $\xi_\theta$ and $\xi_z$ are expressed in terms of $\xi_r$ and $A$ as

$$\mathrm{i}(\boldsymbol{k}\cdot\boldsymbol{B})\xi_\theta = \mathrm{i}kA - \frac{B_\theta}{r}\frac{\mathrm{d}}{\mathrm{d}r}(r\xi_r)\,, \tag{8.76}$$

$$-\mathrm{i}(\boldsymbol{k}\cdot\boldsymbol{B})\xi_z = \frac{\mathrm{i}mA}{r} + \frac{B_z}{r}\frac{\mathrm{d}}{\mathrm{d}r}(r\xi_r)\,. \tag{8.77}$$

From $\mu_0\boldsymbol{j}_0 = \boldsymbol{\nabla} \times \boldsymbol{B}_0$, it follows that

$$\mu_0 j_{0\theta} = -\frac{\mathrm{d}B_z}{\mathrm{d}r}\,, \tag{8.78}$$

$$\mu_0 j_{0z} = \frac{\mathrm{d}B_\theta}{\mathrm{d}r} + \frac{B_\theta}{r} = \frac{1}{r}\frac{\mathrm{d}}{\mathrm{d}r}(rB_\theta) \,. \tag{8.79}$$

The terms of the energy integral are given by

$$\begin{aligned} W_\mathrm{p} &= \frac{1}{4}\int_{V_\mathrm{in}} \left[\gamma p_0 |\boldsymbol{\nabla}\cdot\boldsymbol{\xi}|^2 + (\boldsymbol{\nabla}\cdot\boldsymbol{\xi}^*)(\boldsymbol{\xi}\cdot\boldsymbol{\nabla}p_0) + \frac{1}{\mu_0}|\boldsymbol{B}_1|^2 - \boldsymbol{\xi}^*\cdot(\boldsymbol{j}_0\times\boldsymbol{B}_1)\right]\mathrm{d}\boldsymbol{r} \\ &= \frac{1}{4}\int\left[-p_1(\boldsymbol{\nabla}\cdot\boldsymbol{\xi}) + \frac{1}{\mu_0}|\boldsymbol{B}_1|^2 - \boldsymbol{j}_0(\boldsymbol{B}_1\times\boldsymbol{\xi}^*)\right]\mathrm{d}\boldsymbol{r} \,, \end{aligned} \tag{8.80}$$

$$W_\mathrm{S} = \frac{1}{4}\int_S |\xi_n|^2 \frac{\partial}{\partial n}\left(\frac{B_{0,\mathrm{ex}}^2}{2\mu_0} - \frac{B_{0,\mathrm{in}}^2}{2\mu_0} - p_0\right)\mathrm{d}S \,, \tag{8.81}$$

$$W_\mathrm{V} = \frac{1}{4\mu_0}\int_{V_\mathrm{ex}} |\boldsymbol{B}_1|^2 \mathrm{d}\boldsymbol{r} \,. \tag{8.82}$$

$\xi_\theta$ and $\xi_z$ can be eliminated by means of (8.76) and (8.77), and $\mathrm{d}B_z/\mathrm{d}r$ and $\mathrm{d}B_\theta/\mathrm{d}r$ can be eliminated by means of (8.78) and (8.79) in (8.80). Then $W_\mathrm{p}$ becomes

$$\begin{aligned} W_\mathrm{p} = \frac{1}{4}\int_{V_\mathrm{in}} \Bigg\{ & \frac{(\boldsymbol{k}\cdot\boldsymbol{B})^2}{\mu_0}|\xi_r|^2 + \left(k^2 + \frac{m^2}{r^2}\right)\frac{|A|^2}{\mu_0} \\ & + \frac{1}{\mu_0}\left|B_\theta\frac{\mathrm{d}\xi_r}{\mathrm{d}r} + \xi_r\left(\mu_0 j_z - \frac{B_\theta}{r}\right)\right|^2 + \frac{1}{\mu_0}\left|\frac{\xi_r B_z}{r} + B_z\frac{\mathrm{d}\xi_r}{\mathrm{d}r}\right|^2 \\ & + \frac{2}{\mu_0}\mathrm{Re}\left\{\mathrm{i}kA^*\left[B_\theta\frac{\mathrm{d}\xi_r}{\mathrm{d}r} + \left(\mu_0 j_z - \frac{B_\theta}{r}\right)\xi_r\right] - \frac{\mathrm{i}mA^*}{r^2}\left(\xi_r B_z + rB_z\frac{\mathrm{d}\xi_r}{\mathrm{d}r}\right)\right\} \\ & + 2\mathrm{Re}\left[\xi_r^* j_{0z}\left(-B_\theta\frac{\mathrm{d}\xi_r}{\mathrm{d}r} - \frac{\xi_r\mu_0 j_z}{2} + \mathrm{i}kA\right)\right]\Bigg\}\mathrm{d}\boldsymbol{r} \,. \end{aligned}$$

The integrand of $W_\mathrm{p}$ reduces to

$$\begin{aligned} & \frac{1}{\mu_0}\left(k^2 + \frac{m^2}{r^2}\right)\left|A + \frac{\mathrm{i}kB_\theta\left(\frac{\mathrm{d}\xi_r}{\mathrm{d}r} - \frac{\xi_r}{r}\right) - \mathrm{i}m\frac{B_z}{r}\left(\frac{\mathrm{d}\xi_r}{\mathrm{d}r} + \frac{\xi_r}{r}\right)}{k^2 + (m^2/r^2)}\right|^2 \\ & \quad + \left[\frac{(\boldsymbol{k}\cdot\boldsymbol{B})^2}{\mu_0} - \frac{2j_z B_\theta}{r}\right]|\xi_r|^2 + \frac{B_z^2}{\mu_0}\left|\frac{\mathrm{d}\xi_r}{\mathrm{d}r} + \frac{\xi_r}{r}\right|^2 + \frac{B_\theta^2}{\mu_0}\left|\frac{\mathrm{d}\xi_r}{\mathrm{d}r} - \frac{\xi_r}{r}\right|^2 \\ & \quad - \frac{\left|\mathrm{i}kB_\theta\left[(\mathrm{d}\xi_r/\mathrm{d}r) - (\xi_r/r)\right] - \mathrm{i}m(B_z/r)\left[(\mathrm{d}\xi_r/\mathrm{d}r) + (\xi_r/r)\right]\right|^2}{\mu_0\left[k^2 + (m^2/r^2)\right]} \,. \end{aligned}$$

Accordingly, the integrand is a minimum when

$$A \equiv \xi_\theta B_z - \xi_z B_\theta$$
$$= -\frac{\mathrm{i}}{k^2 + (m^2/r^2)} \left[ \left( kB_\theta - \frac{m}{r} B_z \right) \frac{\mathrm{d}\xi_r}{\mathrm{d}r} - \left( kB_\theta + \frac{m}{r} B_z \right) \frac{\xi_r}{r} \right] .$$

Then $W_{\mathrm{p}}$ reduces to

$$W_{\mathrm{p}} = \frac{\pi}{2\mu_0} \int_0^a \left\{ \frac{\left| (\boldsymbol{k} \cdot \boldsymbol{B}_0)(\mathrm{d}\xi_r/\mathrm{d}r) + h(\xi_r/r) \right|^2}{k^2 + (m/r)^2} + \left[ (\boldsymbol{k} \cdot \boldsymbol{B}_0)^2 - \frac{2\mu_0 j_z B_\theta}{r} \right] |\xi_r|^2 \right\} r \mathrm{d}r , \tag{8.83}$$

where

$$h \equiv kB_z - \frac{m}{r} B_\theta .$$

Let us next determine $W_{\mathrm{S}}$. From (6.8), it follows that

$$\frac{\mathrm{d}}{\mathrm{d}r} \left( p_0 + \frac{B_z^2 + B_\theta^2}{2\mu_0} \right) = -\frac{B_\theta^2}{r\mu_0} .$$

$B_\theta^2$ is continuous across the boundary $r = a$, so that

$$\frac{\mathrm{d}}{\mathrm{d}r} \left( p_0 + \frac{B_z^2 + B_\theta^2}{2\mu_0} \right) = \frac{\mathrm{d}}{\mathrm{d}r} \left( \frac{B_{\mathrm{e}z}^2 + B_{\mathrm{e}\theta}^2}{2\mu_0} \right) .$$

Accordingly, we find

$$W_{\mathrm{S}} = 0 , \tag{8.84}$$

as is clear from (8.81).

The expression for $W_{\mathrm{V}}$ can be obtained when the quantities in (8.83) for $W_{\mathrm{p}}$ are replaced as follows:

$$\boldsymbol{j} \to 0 , \quad B_z \to B_{\mathrm{e}z} = B_{\mathrm{s}} \ (= \mathrm{const.}) , \quad B_\theta \to B_{\mathrm{e}\theta} = B_a \frac{a}{r} ,$$

$$B_{1r} = \mathrm{i}(\boldsymbol{k} \cdot \boldsymbol{B}_0)\xi_r \to B_{\mathrm{e}1r} = \mathrm{i}(\boldsymbol{k} \cdot \boldsymbol{B}_{\mathrm{e}0})\eta_r .$$

This replacement yields

$$W_{\mathrm{V}} = \frac{\pi}{2\mu_0} \int_a^b \left[ \left( kB_{\mathrm{s}} + \frac{m}{r} \frac{B_a a}{r} \right)^2 |\eta_r|^2 + \frac{1}{k^2 + (m/r)^2} \left| \left( kB_{\mathrm{s}} + \frac{m}{r} \frac{B_a a}{r} \right) \frac{\mathrm{d}\eta_r}{\mathrm{d}r} + \left( kB_{\mathrm{s}} - \frac{m}{r} \frac{B_a a}{r} \right) \frac{\eta_r}{r} \right|^2 \right] r \mathrm{d}r . \tag{8.85}$$

By partial integration, $W_{\mathrm{p}}$ is seen to be

$$W_{\mathrm{p}} = \frac{\pi}{2\mu_0}\int_0^a \left[\frac{r(\boldsymbol{k}\cdot\boldsymbol{B}_0)^2}{k^2+(m/r)^2}\left|\frac{\mathrm{d}\xi_r}{\mathrm{d}r}\right|^2 + g|\xi_r|^2\right]\mathrm{d}r \qquad (8.86)$$

$$+\frac{\pi}{2\mu_0}\frac{k^2B_{\mathrm{s}}^2-(m/a)^2B_a^2}{k^2+(m/a)^2}|\xi_r(a)|^2 ,$$

$$g = \frac{1}{r}\frac{\left[kB_z-(m/r)B_\theta\right]^2}{k^2+(m/r)^2} + r(\boldsymbol{k}\cdot\boldsymbol{B}_0)^2 - \frac{2B_\theta}{r}\frac{\mathrm{d}(rB_\theta)}{\mathrm{d}r} \qquad (8.87)$$

$$-\frac{\mathrm{d}}{\mathrm{d}r}\left[\frac{k^2B_z^2-(m/r)^2B_\theta^2}{k^2+(m/r)^2}\right] .$$

Using the notation $\zeta \equiv rB_{\mathrm{e}1r} = \mathrm{i}r(\boldsymbol{k}\cdot\boldsymbol{B}_{\mathrm{e}0})\eta_r$, we find that

$$W_{\mathrm{V}} = \frac{\pi}{2\mu_0}\int_a^b \left\{\frac{1}{r\left[k^2+(m/r)^2\right]}\left|\frac{\mathrm{d}\zeta}{\mathrm{d}r}\right|^2 + \frac{1}{r}|\zeta|^2\right\}\mathrm{d}r . \qquad (8.88)$$

The functions $\xi_r$ or $\zeta$ that will minimize $W_{\mathrm{p}}$ or $W_{\mathrm{V}}$ are the solutions of Euler's equation:

$$\frac{\mathrm{d}}{\mathrm{d}r}\left[\frac{r(\boldsymbol{k}\cdot\boldsymbol{B}_0)^2}{k^2+(m/r)^2}\frac{\mathrm{d}\xi_r}{\mathrm{d}r}\right] - g\xi_r = 0 , \qquad r \le a , \qquad (8.89)$$

$$\frac{\mathrm{d}}{\mathrm{d}r}\left\{\frac{1}{r\left[k^2+(m/r)^2\right]}\frac{\mathrm{d}\zeta}{\mathrm{d}r}\right\} - \frac{1}{r}\zeta = 0 , \qquad r > a . \qquad (8.90)$$

There are two independent solutions, which tend to $\xi_r \propto r^{m-1}, r^{-m-1}$ as $r \to 0$. As $\xi_r$ is finite at $r = 0$, the solution must satisfy the conditions

$$r \to 0 , \qquad \xi_r \propto r^{m-1} ,$$
$$r = a , \qquad \zeta(a) = \mathrm{i}a\left(kB_{\mathrm{s}} + \frac{m}{a}B_a\right)\xi_r(a) ,$$
$$r = b , \qquad \zeta(b) = 0 .$$

Using the solution of (8.90), we obtain

$$W_{\mathrm{V}} = \frac{\pi}{2\mu_0}\frac{1}{r\left[k^2+(m/r)^2\right]}\left|\frac{\mathrm{d}\zeta}{\mathrm{d}r}\zeta^*\right|_a^b . \qquad (8.91)$$

The solution of (8.90) is

$$\zeta = \mathrm{i}\frac{I_m'(kr)K_m'(kb) - K_m'(kr)I_m'(kb)}{I_m'(ka)K_m'(kb) - K_m'(ka)I_m'(kb)}r\left(kB_{\mathrm{s}} + \frac{m}{a}B_a\right)\xi_r(a) . \qquad (8.92)$$

The stability problem is now reduced to one of examining the sign of $W_{\mathrm{p}}+W_{\mathrm{V}}$. For this we use

$$\begin{cases} W_{\rm p} = \dfrac{\pi}{2\mu_0}\displaystyle\int_0^a \left( f\left|\frac{{\rm d}\xi_r}{{\rm d}r}\right|^2 + g|\xi_r|^2 \right){\rm d}r + W_a \ , \\ W_a = \dfrac{\pi}{2\mu_0}\dfrac{k^2 B_{\rm s}^2 - (m/a)^2 B_a^2}{k^2 + (m/a)^2}|\xi_r(a)|^2 \ , \\ W_{\rm V} = \dfrac{\pi}{2\mu_0}\dfrac{-1}{r\left[k^2 + (m/a)^2\right]}\left|\dfrac{{\rm d}\zeta}{{\rm d}r}\zeta^*\right|_{r=a} \ , \end{cases} \tag{8.93}$$

where

$$f = \frac{r(kB_z + (m/r)B_\theta)^2}{k^2 + (m/r)^2} \ , \tag{8.94}$$

$$g = \frac{1}{r}\frac{\left[kB_z - (m/r)B_\theta\right]^2}{k^2 + (m/r)^2} + r\left(kB_z + \frac{m}{r}B_\theta\right)^2 - \frac{2B_\theta}{r}\frac{{\rm d}(rB_\theta)}{{\rm d}r} - \frac{\rm d}{{\rm d}r}\left[\frac{k^2 B_z^2 - (m/r)^2 B_\theta^2}{k^2 + (m/r)^2}\right] \ . \tag{8.95}$$

When the equation of equilibrium $({\rm d}/{\rm d}r)(\mu_0 p + B^2/2) = -B_\theta^2/r$ is used, (8.95) of $g$ reduces to

$$g = \frac{2k^2}{k^2 + (m/r)^2}\mu_0\frac{{\rm d}p_0}{{\rm d}r} + r\left(kB_z + \frac{m}{r}B_\theta\right)^2 \frac{k^2 + (m/r)^2 - (1/r)^2}{k^2 + (m/r)^2} + \frac{(2k^2/r)\left[k^2 B_z^2 - (m/r)^2 B_\theta^2\right]}{\left[k^2 + (m/r)^2\right]^2} \ . \tag{8.96}$$

### 8.3.3 Suydam's Criterion

The function $f$ in the integrand of $W_{\rm p}$ in the last section is always $f \geq 0$, so that the term in $f$ is a stabilizing term. The first and second terms in (8.95) for $g$ are stabilizing terms, but the third and fourth terms may contribute to the instabilities. When a singular point

$$f \propto (\boldsymbol{k}\cdot\boldsymbol{B}_0)^2 = 0$$

of Euler's equation (8.89) is located at some point $r = r_0$ within the plasma region, the contribution of the stabilizing term becomes small near $r = r_0$, so that a local mode near the singular point is dangerous. In terms of the notation

$$r - r_0 = x \ , \qquad f = \alpha x^2 \ , \qquad g = \beta \ , \qquad \beta = \frac{2B_\theta^2}{B_0^2}\mu_0\frac{{\rm d}p_0}{{\rm d}r}\bigg|_{r=r_0} \ ,$$

$$\alpha = \frac{r_0}{k^2 r_0^2 + m^2}\left(kr\frac{{\rm d}B_z}{{\rm d}r} + kB_z + m\frac{{\rm d}B_\theta}{{\rm d}r}\right)^2_{r=r_0} = \frac{rB_\theta^2 B_z^2}{B^2}\left(\frac{\tilde{\mu}'}{\tilde{\mu}}\right)^2_{r=r_0} \ ,$$

with $\tilde{\mu} \equiv B_\theta / r B_z$, Euler's equation reduces to

$$\alpha \frac{\mathrm{d}}{\mathrm{d}r}\left(x^2 \frac{\mathrm{d}\xi_r}{\mathrm{d}x}\right) - \beta \xi_r = 0 \, .$$

The solution is

$$\xi_r = c_1 x^{-n_1} + c_2 x^{-n_2} \, , \tag{8.97}$$

where $n_1$ and $n_2$ are given by

$$n^2 - n - \frac{\beta}{\alpha} = 0 \, , \qquad n_i = \frac{1 \pm (1 + 4\beta/\alpha)^{1/2}}{2} \qquad (i = 1, 2) \, .$$

When $\alpha + 4\beta > 0$, $n_1$ and $n_2$ are real. The relation $n_1 + n_2 = 1$ always holds. For $n_1 < n_2$, we have the solution $x^{-n_1}$, called a *small solution*. When $n$ is complex ($n = \gamma \pm \mathrm{i}\delta$), $\xi_r$ has the form $\exp\left[(-\gamma \mp \mathrm{i}\delta) \ln x\right]$ and $\xi_r$ is oscillatory.

Let us consider a local mode $\xi_r$ which is nonzero only in the neighborhood $\varepsilon$ around $r = r_0$, and set

$$r - r_0 = \varepsilon t \, , \qquad \xi_r(r) = \xi(t) \, , \qquad \xi(1) = \xi(-1) = 0 \, .$$

Then $W_\mathrm{p}$ becomes

$$W_\mathrm{p} = \frac{\pi}{2\mu_0} \varepsilon \int_{-1}^{1} \left( \alpha t^2 \left| \frac{\mathrm{d}\xi}{\mathrm{d}t} \right|^2 + \beta |\xi|^2 \right) \mathrm{d}t + O(\varepsilon^2) \, .$$

Since Schwartz's inequality yields

$$\int_{-1}^{1} t^2 |\xi'|^2 \mathrm{d}t \int_{-1}^{1} |\xi|^2 \mathrm{d}t \geq \left| \int_{-1}^{1} t \xi' \xi^* \mathrm{d}t \right|^2 = \left( \frac{1}{2} \int_{-1}^{1} |\xi|^2 \mathrm{d}t \right)^2 \, ,$$

$W_\mathrm{p}$ is

$$W_\mathrm{p} > \frac{\pi}{2\mu_0} \frac{1}{4} (\alpha + 4\beta) \int_{-1}^{1} |\xi|^2 \mathrm{d}t \, .$$

The stability condition is $\alpha + 4\beta > 0$, i.e.,

$$\frac{r}{4} \left( \frac{\tilde{\mu}'}{\tilde{\mu}} \right)^2 + \frac{2\mu_0}{B_z^2} \frac{\mathrm{d}p_0}{\mathrm{d}r} > 0 \, . \tag{8.98}$$

$r(\tilde{\mu}'/\tilde{\mu})$ is called the *shear parameter*. Usually, the second term is negative, since, most often, $\mathrm{d}p_0/\mathrm{d}r < 0$. The first term $(\tilde{\mu}'/\tilde{\mu})^2$ represents the stabilizing effect of shear. This condition is called *Suydam's criterion* [8.10]. It is a necessary condition for stability, but is not always a sufficient condition, as Suydam's criterion is derived from consideration of local-mode behavior only. Newcomb derived the necessary and sufficient conditions for the stability of a cylindrical plasma. His twelve theorems are described in [8.11].

### 8.3.4 Tokamak Configuration

In this case the longitudinal magnetic field $B_{\mathrm{s}}$ is much larger than the poloidal magnetic field $B_\theta$. The plasma region is $r \le a$ and the vacuum region is $a \le r \le b$, with an ideal conducting wall at $r = b$. It is assumed that $ka \ll 1$, $kb \ll 1$. The function $\zeta$ in (8.91) for $W_{\mathrm{V}}$ is

$$\zeta = \mathrm{i}\frac{mB_a + kaB_{\mathrm{s}}}{1-(a/b)^{2m}}\xi_r(a)\frac{a^m}{b^m}\left(\frac{b^m}{r^m} - \frac{r^m}{b^m}\right) ,$$

from (8.92), and $W_{\mathrm{V}}$ becomes

$$W_{\mathrm{V}} = \frac{\pi}{2\mu_0}\frac{(mB_a + kaB_{\mathrm{s}})^2}{m}\xi_r^2(a)\lambda , \qquad \lambda \equiv \frac{1+(a/b)^{2m}}{1-(a/b)^{2m}} .$$

From the periodic condition for a torus, it follows that

$$\frac{2\pi n}{k} = -2\pi R \qquad (n \text{ is an integer}) ,$$

so that $(\boldsymbol{k}\cdot\boldsymbol{B})$ is given by

$$a(\boldsymbol{k}\cdot\boldsymbol{B}) = mB_a + kaB_{\mathrm{s}} = mB_a\left(1 - \frac{nq_a}{m}\right) ,$$

in terms of the safety factor. The $W_a$ term in (8.93) reduces to

$$\begin{aligned} k^2B_{\mathrm{s}}^2 - \left(\frac{m}{a}\right)^2 B_a^2 &= \left(kB_{\mathrm{s}} + \frac{m}{a}B_a\right)^2 - 2\frac{m}{a}B_a\left(kB_{\mathrm{s}} + \frac{m}{a}B_a\right) \\ &= \left(\frac{nB_a}{a}\right)^2\left[\left(1-\frac{nq_a}{m}\right)^2 - 2\left(1-\frac{nq_a}{m}\right)\right] . \end{aligned}$$

Accordingly, the energy integral becomes

$$\begin{aligned} W_{\mathrm{p}} + W_{\mathrm{V}} = {}& \frac{\pi}{2\mu_0}B_a^2\xi_r^2(a)\left[\left(1-\frac{nq_a}{m}\right)^2(1+m\lambda) - 2\left(1-\frac{nq_a}{m}\right)\right] \\ & + \frac{\pi}{2\mu_0}\int\left[f\left(\frac{\mathrm{d}\xi_r}{\mathrm{d}r}\right)^2 + g\xi_r^2\right]\mathrm{d}r . \end{aligned} \tag{8.99}$$

The first term of (8.99) is negative when

$$1 - \frac{2}{1+m\lambda} < \frac{nq_a}{m} < 1 . \tag{8.100}$$

The assumption $nq_a/m \sim 1$ corresponds to $ka \sim mB_a/B_{\mathrm{s}}$. As $B_a/B_{\mathrm{s}} \ll 1$, this is consistent with the assumption $ka \ll 1$. When $m = 1$, $(m^2-1)/m^2$ is zero in the second term of (8.96) for $g$. The magnitude of $g$ is of the order of $k^2r^2$, which is very small since $kr \ll 1$. The term in $f(\mathrm{d}\xi_r/\mathrm{d}r)^2$ can

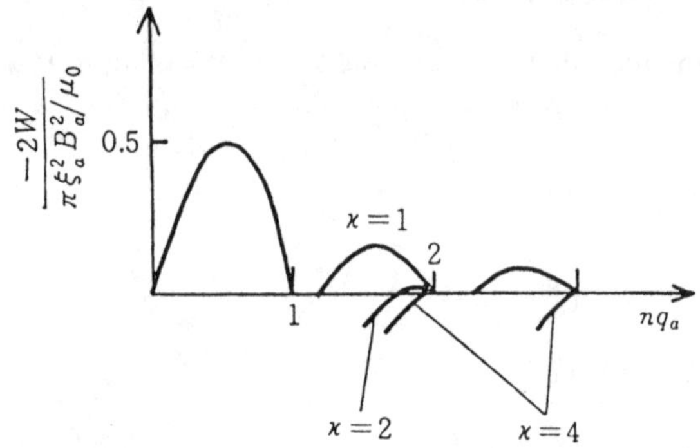


**Fig. 8.6.** Relation between the growth rate $\gamma$ and $nq_a$ for the kink instability [8.9] $-2W/(\pi\xi_a^2 B_a^2/\mu_0) = \gamma^2 a^2(\langle\rho_{m0}\rangle\mu_0/B_a^2)$

be very small if $\xi_r$ is nearly constant. Accordingly, the contribution of the integral term in (8.99) is negligible. When $m = 1$ and $a^2/b^2 < nq_a < 1$, the energy integral becomes negative ($W < 0$). The mode $m = 1$ is unstable in the region specified by (8.100), irrespective of the current distribution. The Kruskal–Shafranov condition for the mode $m = 1$ derived from the sharp-boundary configuration is also applicable to the diffuse-boundary plasma. The growth rate $\gamma^2 = -\omega^2$ is

$$\gamma^2 \simeq \frac{-W}{\int(\rho_{m0}|\boldsymbol{\xi}|^2/2)\mathrm{d}\boldsymbol{r}} = \frac{1}{\langle\rho_{m0}\rangle}\frac{B_a^2}{\mu_0 a^2}\left[2(1-nq_a) - \frac{2(1-nq_a)^2}{1-a^2/b^2}\right], \quad (8.101)$$

$$\langle\rho_{m0}\rangle = \frac{\int \rho_{m0}|\boldsymbol{\xi}|^2 2\pi r\mathrm{d}r}{\pi a^2\xi_r^2(a)} .$$

The maximum growth rate is $\gamma_{max}^2 \sim (1-a^2/b^2)B_a^2/(\mu_0\langle\rho\rangle a^2)$. When $m \neq 1$, $(m^2-1)/m^2$ in the second term of (8.96) for $g$ is large, and $g \sim 1$. Accordingly, the contribution of the integral term to $W_p$ must be checked. The region $g < 0$ is given by $\chi_1 < \chi < \chi_2$, when $\chi \equiv -krB_z/B_\theta = nq(r)$ and

$$\chi_{1,2} = m - \frac{2}{m(m^2-1)}k^2r^2 \pm \frac{2k^2r^2}{m(m^2-1)}\left[1 - \frac{m^2(m^2-1)}{2k^2r^2}\frac{\mu_0 r p_0'}{B_\theta^2}\right]^{1/2} . \quad (8.102)$$

Since $kr \ll 1$, the region $g < 0$ is narrow and close to the singular point $nq(r) = m$, and the contribution of the integral term to $W_p$ can be neglected in the case $m \neq 1$ too. Therefore, if $nq_a/m$ is in the range given by (8.100), the plasma is unstable due to the displacement $\xi_r(a)$ of the plasma boundary. When the current distribution is $j(r) = j_0\exp(-\kappa^2r^2/a^2)$ and the conducting wall is at infinity ($b = \infty$), $\gamma^2$ can be calculated from (8.101), using the solution of Euler's equation. The dependence of $\gamma^2$ on $q_a$ can then be estimated. The result is shown in Fig. 8.6 [8.9].

When the value of $nq_a/m$ is outside the region given by (8.100), the effect of the displacement of the plasma boundary is not great and the contribution of the integral term in $W_{\rm p}$ is dominant. However, the growth rate $\gamma^2$ is $k^2r^2$ times as small as that given by (8.101), as is clear from consideration of (8.102).

The characteristic of the reversed field pinch (RFP, see Sect. 17.1) is that $B_a$ and $B_{\rm s}$ are of the same order of magnitude, so that the approximation based upon $ka \ll 1$ or $B_a \ll B_{\rm s}$ can no longer be used. MHD instabilities in RFP are well analyzed in [8.12].

## 8.4 Hain–Lüst Magnetohydrodynamic Equation

When the displacement $\boldsymbol{\xi}$ is denoted by

$$\boldsymbol{\xi}(r, \theta, z, t) = \boldsymbol{\xi}(r) \exp {\rm i}(m\theta + kz - \omega t) ,$$

and the equilibrium magnetic field $\boldsymbol{B}$ is expressed by

$$\boldsymbol{B}(r) = \big(0,\ B_\theta(r),\ B_z(r)\big) ,$$

the $(r, \theta, z)$ components of the magnetohydrodynamic equation of motion are given by

$$-\mu_0\rho_{\rm m}\omega^2\xi_r = \frac{\rm d}{{\rm d}r}\left[\mu_0\gamma p(\nabla\cdot\xi) + B^2\frac{1}{r}\frac{\rm d}{{\rm d}r}(r\xi_r) + {\rm i}D(\xi_\theta B_z - \xi_z B_\theta)\right] - \left[F^2 + r\frac{\rm d}{{\rm d}r}\left(\frac{B_\theta}{r}\right)^2\right]\xi_r - 2{\rm i}k\frac{B_\theta}{r}(\xi_\theta B_z - \xi_z B_\theta) , \qquad (8.103)$$

$$-\mu_0\rho_{\rm m}\omega^2\xi_\theta = {\rm i}\frac{m}{r}\gamma\mu_0 p(\nabla\cdot\boldsymbol{\xi}) + {\rm i}DB_z\frac{1}{r}\frac{\rm d}{{\rm d}r}(r\xi_r) + 2{\rm i}k\frac{B_\theta B_z}{r}\xi_r - H^2 B_z(\xi_\theta B_z - \xi_z B_\theta) , \qquad (8.104)$$

$$-\mu_0\rho_{\rm m}\omega^2\xi_z = {\rm i}k\gamma\mu_0 p(\nabla\cdot\boldsymbol{\xi}) - {\rm i}DB_\theta\frac{1}{r}\frac{\rm d}{{\rm d}r}(r\xi_r) - 2{\rm i}k\frac{B_\theta^2}{r}\xi_r + H^2 B_\theta(\xi_\theta B_z - \xi_z B_\theta) , \qquad (8.105)$$

where

$$F = \frac{m}{r}B_\theta + kB_z = (\boldsymbol{k}\cdot\boldsymbol{B}) , \qquad D = \frac{m}{r}B_z - kB_\theta , \qquad H^2 = \left(\frac{m}{r}\right)^2 + k^2 ,$$

$$\nabla\cdot\boldsymbol{\xi} = \frac{1}{r}\frac{\mathrm{d}}{\mathrm{d}r}(r\xi_r) + \frac{\mathrm{i}m}{r}\xi_\theta + \mathrm{i}k\xi_z .$$

When $\xi_\theta$, $\xi_z$ are eliminated by (8.104), (8.105), we find

$$\frac{\mathrm{d}}{\mathrm{d}r}\left\{\frac{\mu_0\rho_\mathrm{m}\omega^2 - F^2}{\Delta}\left[\mu_0\rho_\mathrm{m}\omega^2(\gamma\mu_0 p + B^2) - \gamma\mu_0 pF^2\right]\frac{1}{r}\frac{\mathrm{d}}{\mathrm{d}r}(r\xi_r)\right\}$$
$$+\left\{\mu_0\rho_\mathrm{m}\omega^2 - F^2 - 2B_\theta\frac{\mathrm{d}}{\mathrm{d}r}\left(\frac{B_\theta}{r}\right) - \frac{4k^2}{\Delta}\frac{B_\theta^2}{r^2}(\mu_0\rho_\mathrm{m}\omega^2 B^2 - \gamma\mu_0 pF^2)\right.$$
$$\left. + r\frac{\mathrm{d}}{\mathrm{d}r}\left[\frac{2kB_\theta}{r^2\Delta}\left(\frac{m}{r}B_z - kB_\theta\right)\left[\mu_0\rho_\mathrm{m}\omega^2(\gamma\mu_0 p + B^2) - \gamma\mu_0 pF^2\right]\right]\right\}\xi_r = 0 , \tag{8.106}$$

where $\Delta$ is

$$\Delta = \mu_0^2\rho_\mathrm{m}^2\omega^4 - \mu_0\rho_\mathrm{m}\omega^2 H^2(\gamma\mu_0 p + B^2) + \gamma\mu_0 pH^2F^2 .$$

This equation was derived by Hain and Lüst [8.13]. The solution of (8.106) gives $\xi_r(r)$ in the region $0 < r < a$. The equations for the vacuum region $a < r < a_\mathrm{w}$, where $a_\mathrm{w}$ is the radius of the wall, are

$$\nabla\times\boldsymbol{B}_1 = 0 , \qquad \nabla\cdot\boldsymbol{B}_1 = 0 ,$$

so that we find

$$\boldsymbol{B}_1 = \nabla\psi , \qquad \Delta\psi = 0 ,$$

$$\psi = \left[bI_m(kr) + cK_m(kr)\right]\exp(\mathrm{i}m\theta + \mathrm{i}kz) ,$$

$$B_{1r} = \frac{\partial\psi}{\partial r} = \left[bI_m'(kr) + cK_m'(kr)\right]\exp(\mathrm{i}m\theta + \mathrm{i}kz) . \tag{8.107}$$

In the plasma region, $B_{1r}$ is given by

$$B_{1r} = \mathrm{i}(\boldsymbol{k}\cdot\boldsymbol{B})\xi_r = \mathrm{i}F\xi_r ,$$

and the boundary conditions at $r = a$ are

$$B_{1r}(a) = \mathrm{i}F\xi_r(a) , \tag{8.108}$$

$$B_{1r}'(a) = \mathrm{i}\left[F'\xi_r(a) + F\xi_r'(a)\right] , \tag{8.109}$$

and the coefficients $b$, $c$ can be fixed.

To deal with this equation as an eigenvalue problem, boundary conditions must be imposed on $\xi_r$. One is $\xi_r \propto r^{m-1}$ at $r = 0$, and the other is $B_{1r}(a_\mathrm{w}) = 0$ for the radial component of the perturbed magnetic field at the perfect conducting wall. After finding suitable $\omega^2$ to satisfy these conditions, the growth rate $\gamma^2 \equiv -\omega^2$ is obtained [8.14].

## 8.5 Energy Integral of Axisymmetric Toroidal System

### 8.5.1 Energy Integral in Illuminating Form

The energy integral (8.46) derived in Sect. 8.2, viz.,

$$W = \frac{1}{2}\int_V \left[\frac{B_1^2}{\mu_0} + \gamma p(\nabla\cdot\boldsymbol{\xi})^2 + (\nabla\cdot\boldsymbol{\xi})(\boldsymbol{\xi}\cdot\nabla p) - \boldsymbol{\xi}\cdot(\boldsymbol{j}\times\boldsymbol{B}_1)\right]\mathrm{d}\boldsymbol{r}\,, \tag{8.110}$$

can be further rearranged to the more illuminating form [8.15, 8.16]

$$W = \frac{1}{2}\int \left[\gamma p(\nabla\cdot\boldsymbol{\xi})^2 + \frac{1}{\mu_0}|\boldsymbol{B}_{1\perp}|^2 + \frac{1}{\mu_0}\left|\boldsymbol{B}_{1\parallel} - \boldsymbol{B}\frac{\mu_0(\boldsymbol{\xi}\cdot\nabla p)}{B^2}\right|^2\right.$$
$$\left. - \frac{(\boldsymbol{j}\cdot\boldsymbol{B})}{B^2}(\boldsymbol{\xi}\times\boldsymbol{B})\cdot\boldsymbol{B}_1 - 2(\boldsymbol{\xi}\cdot\nabla p)(\boldsymbol{\xi}\cdot\boldsymbol{\kappa})\right]\mathrm{d}\boldsymbol{r}\,. \tag{8.111}$$

The first term in the integrand of (8.111) is the sonic wave term. The second and third terms are Alfvén wave terms. The fourth term is the kink mode term and the last is the ballooning mode term. $\boldsymbol{\kappa}$ is the field line curvature vector. The rearrangement from (8.110) to (8.111) is described in the following. When $\boldsymbol{\xi}$ is expressed by the sum of the parallel component $\xi_\parallel \boldsymbol{b}$ and the perpendicular component $\boldsymbol{\xi}_\perp$ to the magnetic field $\boldsymbol{B} = B\boldsymbol{b}$

$$\boldsymbol{\xi} = \xi_\parallel \boldsymbol{b} + \boldsymbol{\xi}_\perp\,,$$

the last two terms of (8.110) reduce to

$$\begin{aligned}
&(\nabla\cdot\boldsymbol{\xi})(\boldsymbol{\xi}\cdot\nabla p) + (\boldsymbol{j}\times\boldsymbol{\xi})\cdot\boldsymbol{B}_1\\
&= (\boldsymbol{\xi}\cdot\nabla p)\nabla\cdot(\xi_\parallel\boldsymbol{b}) + (\boldsymbol{\xi}\cdot\nabla p)\nabla\cdot\boldsymbol{\xi}_\perp + \xi_\parallel(\boldsymbol{j}\times\boldsymbol{b})\cdot\nabla\times(\boldsymbol{\xi}\times\boldsymbol{B}) + (\boldsymbol{j}\times\boldsymbol{\xi}_\perp)\cdot\boldsymbol{B}_1\\
&= (\boldsymbol{\xi}\cdot\nabla p)(\boldsymbol{B}\cdot\nabla)\frac{\xi_\parallel}{B} + \frac{\xi_\parallel}{B}\nabla\cdot[(\boldsymbol{\xi}\times\boldsymbol{B})\times\nabla p] + (\boldsymbol{\xi}\cdot\nabla p)\nabla\cdot\boldsymbol{\xi}_\perp + (\boldsymbol{j}\times\boldsymbol{\xi}_\perp)\cdot\boldsymbol{B}_1\\
&= (\boldsymbol{\xi}\cdot\nabla p)(\boldsymbol{B}\cdot\nabla)\frac{\xi_\parallel}{B} + \frac{\xi_\parallel}{B}\nabla\cdot[(\boldsymbol{\xi}\cdot\nabla p)\boldsymbol{B}] + (\boldsymbol{\xi}\cdot\nabla p)\nabla\cdot\boldsymbol{\xi}_\perp + (\boldsymbol{j}\times\boldsymbol{\xi}_\perp)\cdot\boldsymbol{B}_1\\
&= \nabla\cdot\left[\frac{\xi_\parallel}{B}(\boldsymbol{\xi}\cdot\nabla p)\boldsymbol{B}\right] + (\boldsymbol{\xi}\cdot\nabla p)\nabla\cdot\boldsymbol{\xi}_\perp + (\boldsymbol{j}\times\boldsymbol{\xi}_\perp)\cdot\boldsymbol{B}_1\,.
\end{aligned} \tag{8.112}$$

The current density $\boldsymbol{j}$ can be expressed as the sum of components parallel and perpendicular to the magnetic field as follows:

$$\boldsymbol{j} = \sigma\boldsymbol{B} + \frac{\boldsymbol{B}\times\nabla p}{B^2}\,,$$

where

$$\sigma = \frac{\boldsymbol{j}\cdot\boldsymbol{B}}{B^2}\,.$$

The last term of (8.112) is

$$(\boldsymbol{j}\times\boldsymbol{\xi}_\perp)\cdot\boldsymbol{B}_1=\sigma(\boldsymbol{B}\times\boldsymbol{\xi}_\perp)\cdot\boldsymbol{B}_1-\frac{(\boldsymbol{\xi}_\perp\cdot\nabla p)}{B^2}\boldsymbol{B}\cdot\boldsymbol{B}_1\,,$$

and $\nabla\cdot\boldsymbol{\xi}_\perp$ in the second term of (8.112) is

$$\begin{aligned}\nabla\cdot\boldsymbol{\xi}_\perp&=\nabla\cdot\left[\frac{\boldsymbol{B}}{B^2}\times(\boldsymbol{\xi}\times\boldsymbol{B})\right]=(\boldsymbol{\xi}\times\boldsymbol{B})\cdot\nabla\times\frac{\boldsymbol{B}}{B^2}-\frac{\boldsymbol{B}}{B^2}\cdot\nabla\times(\boldsymbol{\xi}\times\boldsymbol{B})\\&=(\boldsymbol{\xi}\times\boldsymbol{B})\cdot\frac{\nabla\times\boldsymbol{B}}{B^2}-2(\boldsymbol{\xi}\times\boldsymbol{B})\cdot\frac{\nabla B}{B^3}\times\boldsymbol{B}-\frac{\boldsymbol{B}}{B^2}\nabla\times(\boldsymbol{\xi}\times\boldsymbol{B})\\&=-\frac{(\boldsymbol{\xi}\cdot\mu_0\nabla p)}{B^2}+2(\boldsymbol{\xi}\times\boldsymbol{B})\cdot\frac{\boldsymbol{B}\times\nabla B}{B^3}-\frac{\boldsymbol{B}}{B^2}\cdot\boldsymbol{B}_1\,.\end{aligned}\tag{8.113}$$

Then the energy integral (8.110) reduces to

$$\begin{aligned}W&=\frac{1}{2}\int_V\left[\gamma p(\nabla\cdot\boldsymbol{\xi})+\frac{\boldsymbol{B}_1^2}{\mu_0}-\frac{\mu_0(\boldsymbol{\xi}\cdot\nabla p)^2}{B^2}-(\boldsymbol{\xi}\cdot\nabla p)\frac{\boldsymbol{B}\cdot\boldsymbol{B}_1}{B^2}\right.\\&\left.-(\boldsymbol{\xi}_\perp\cdot\nabla p)\frac{\boldsymbol{B}\cdot\boldsymbol{B}_1}{B^2}+\sigma(\boldsymbol{B}\times\boldsymbol{\xi}_\perp)\cdot\boldsymbol{B}_1+2(\boldsymbol{\xi}\cdot\nabla p)(\boldsymbol{\xi}\times\boldsymbol{B})\cdot\frac{\boldsymbol{B}\times\nabla B}{B^3}\right]\mathrm{d}\boldsymbol{r}\\&=\frac{1}{2}\int_V\left\{\gamma p(\nabla\cdot\boldsymbol{\xi})+\frac{1}{\mu_0}\left|\boldsymbol{B}_1-\frac{\mu_0(\boldsymbol{\xi}\cdot\nabla p)}{B^2}\boldsymbol{B}\right|^2-\frac{(\boldsymbol{j}\cdot\boldsymbol{B})}{B^2}(\boldsymbol{\xi}_\perp\times\boldsymbol{B})\cdot\boldsymbol{B}_1\right.\\&\left.-2(\boldsymbol{\xi}\cdot\nabla p)\left[\frac{\mu_0(\boldsymbol{\xi}\cdot\nabla p)}{B^2}-(\boldsymbol{\xi}\times\boldsymbol{B})\cdot\frac{\boldsymbol{B}\times\nabla B}{B^3}\right]\right\}\mathrm{d}\boldsymbol{r}\,.\end{aligned}$$

By introducing a vector $\boldsymbol{\kappa}$

$$\boldsymbol{\kappa}\equiv\frac{1}{2B^4}\left[\boldsymbol{B}\times\nabla(B^2+2\mu_0 p)\right]\times\boldsymbol{B}=\frac{\mu_0\nabla p}{B^2}+\frac{(\boldsymbol{B}\times\nabla B)\times\boldsymbol{B}}{B^3}\,,\tag{8.114}$$

the last ballooning term can be expressed as

$$-2(\boldsymbol{\xi}\cdot\nabla p)(\boldsymbol{\xi}\cdot\boldsymbol{\kappa})\,,$$

since

$$(\boldsymbol{\xi}\cdot\boldsymbol{\kappa})=\frac{\mu_0(\boldsymbol{\xi}\cdot\nabla p)}{B^2}+\frac{\boldsymbol{\xi}\cdot(\boldsymbol{B}\times\nabla B)\times\boldsymbol{B}}{B^3}=\frac{\mu_0(\boldsymbol{\xi}\cdot\nabla p)}{B^2}-\frac{(\boldsymbol{\xi}\times\boldsymbol{B})\cdot(\boldsymbol{B}\times\nabla B)}{B^3}\,.\tag{8.115}$$

Equations (8.113) and (8.114) reduce to

$$\nabla\cdot\boldsymbol{\xi}_\perp+2(\boldsymbol{\xi}_\perp\cdot\boldsymbol{\kappa})=\frac{\mu_0(\boldsymbol{\xi}_\perp\cdot\nabla p)}{B^2}-\frac{\boldsymbol{B}\cdot\boldsymbol{B}_1}{B^2}\,.\tag{8.116}$$

From (6.7) of Chap. 6, we have

$$\nabla(2\mu_0 p+B^2)=2(\boldsymbol{B}\cdot\nabla)\boldsymbol{B}\,.$$

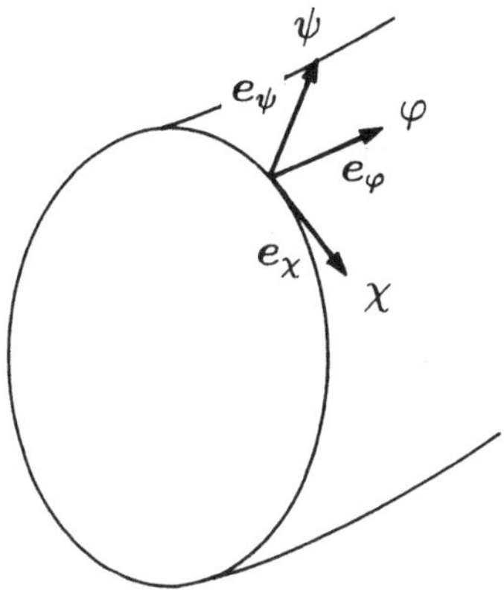


**Fig. 8.7.** Orthogonal coordinate system $(\psi, \chi, \varphi)$. $\boldsymbol{e}_\psi$, $\boldsymbol{e}_\chi$, $\boldsymbol{e}_\varphi$ are unit vectors in the $\psi$, $\chi$, and $\varphi$ directions, respectively

Then it becomes clear from (8.114) that $\boldsymbol{\kappa}$ is equal to the curvature vector as follows:

$$\begin{aligned}\boldsymbol{\kappa} &= \frac{1}{B}\left[\boldsymbol{b} \times (\boldsymbol{b}\cdot\boldsymbol{\nabla})(B\boldsymbol{b})\right] \times \boldsymbol{b} = \left\{\boldsymbol{b} \times \left[(\boldsymbol{b}\cdot\boldsymbol{\nabla})\boldsymbol{b} + \boldsymbol{b}\frac{1}{B}(\boldsymbol{b}\cdot\boldsymbol{\nabla})B\right]\right\} \times \boldsymbol{b} \\ &= \left[(\boldsymbol{b}\cdot\boldsymbol{\nabla})\boldsymbol{b}\right]_\perp = -\frac{\boldsymbol{n}}{R} ,\end{aligned}$$

where $R$ is the radius of curvature and $\boldsymbol{n}$ is the unit vector from the center of curvature to the point on the line of magnetic force (see Fig. 2.4 of Chap. 2).

### 8.5.2 Energy Integral of Axisymmetric Toroidal System

In any axisymmetric toroidal system, the energy integral may be reduced to a more convenient form. The axisymmetric magnetic field is expressed as

$$\boldsymbol{B} = \frac{\hat{I}(\psi)}{R}\boldsymbol{e}_\varphi + B_\chi \boldsymbol{e}_\chi , \qquad \hat{I}(\psi) \equiv \frac{\mu_0 I(\psi)}{2\pi} , \tag{8.117}$$

where $\varphi$ is the angle around the axis of the torus and $\psi$ is the flux function defined by

$$\psi = -RA_\varphi . \tag{8.118}$$

$R$ is the distance from the axis of symmetry and $A_\varphi$ is the $\varphi$ component of the vector potential of the magnetic field. $B_\chi$ is the poloidal component of the magnetic field. $\boldsymbol{e}_\varphi$ and $\boldsymbol{e}_\chi$ are the unit vectors in the directions of the toroidal and poloidal angles, respectively (see Fig. 8.7). The $R$ and $Z$ components of the magnetic field are given by

$$RB_R = \frac{\partial \psi}{\partial Z}\,, \qquad RB_Z = -\frac{\partial \psi}{\partial R}\,.$$

Since $\boldsymbol{\nabla}\varphi = (1/R)\boldsymbol{e}_\varphi$, the poloidal component of the magnetic field is expressed by $\boldsymbol{B}_\mathrm{p} = -\boldsymbol{\nabla}\psi \times \boldsymbol{\nabla}\varphi$. Hence,

$$\boldsymbol{B} = -\boldsymbol{\nabla}\psi \times \boldsymbol{\nabla}\varphi + \hat{I}(\psi)\boldsymbol{\nabla}\varphi\,. \tag{8.119}$$

We can introduce an orthogonal coordinate system $(\psi, \chi, \varphi)$, where $\psi =$ const. are the magnetic surfaces and $\chi, \varphi$ are the poloidal and toroidal angles, respectively. The metric for these coordinates is

$$\mathrm{d}s^2 = \left(\frac{\mathrm{d}\psi}{RB_\chi}\right)^2 + (JB_\chi \mathrm{d}\chi)^2 + (R\mathrm{d}\varphi)^2\,, \tag{8.120}$$

where the volume element is $\mathrm{d}V = J(\psi)\mathrm{d}\psi\mathrm{d}\chi\mathrm{d}\varphi$. A field line is defined by $\psi =$ const. and

$$\frac{R\mathrm{d}\varphi}{JB_\chi \mathrm{d}\chi} = \frac{B_\varphi}{B_\chi} = \frac{\hat{I}(\psi)}{RB_\chi}\,,$$

that is,

$$\frac{\mathrm{d}\varphi}{\mathrm{d}\chi} = \frac{J(\psi)\hat{I}(\psi)}{R^2} \equiv \hat{q}(\psi, \chi)\,.$$

Then the toroidal safety factor is given by

$$q(\psi) = \frac{1}{2\pi}\oint \frac{J(\psi)\hat{I}(\psi)}{R^2}\mathrm{d}\chi\,.$$

The energy integral of an axisymmetric toroidal system is given by [8.17]

$$\begin{aligned} W &= \frac{1}{2}\int_V \left[\frac{|\boldsymbol{B}_1|^2}{\mu_0} + \gamma p|(\boldsymbol{\nabla}\cdot\boldsymbol{\xi})|^2 + (\boldsymbol{\nabla}\cdot\boldsymbol{\xi}^*)(\boldsymbol{\xi}\cdot\boldsymbol{\nabla}p) - \boldsymbol{\xi}^*\cdot(\boldsymbol{j}\times\boldsymbol{B}_1)\right]\mathrm{d}\boldsymbol{r} \\ &= \int_V \left[\frac{1}{2\mu_0}\frac{B^2 k_\parallel^2}{B_\chi^2 R^2}|X|^2 + \frac{1}{2\mu_0}\frac{R^2}{J^2}\left|\frac{\partial U}{\partial \chi} - I\left(\frac{JX}{R^2}\right)'\right|^2\right. \\ &\quad + \frac{B_\chi^2}{2\mu_0}\left|\mathrm{i}nU + X' - \frac{\mu_0 j_\varphi}{RB_\chi^2}X\right|^2 + \frac{1}{2}\gamma p\left|\frac{1}{J}(JX)' + \mathrm{i}Bk_\parallel Y + \mathrm{i}nU\right|^2 \\ &\quad \left. - KXX^*\right] J\mathrm{d}\psi\mathrm{d}\chi\mathrm{d}\varphi\,. \end{aligned} \tag{8.121}$$

The derivation of (8.121) is described in the following. The notation in (8.121) will be explained as we go along.

In a general orthogonal coordinate system $(u^1, u^2, u^3)$ with metric given by

$$ds^2 = h_1^2(du^1)^2 + h_2^2(du^2)^2 + h_3^2(du^3)^2 , \qquad g^{1/2} = h_1 h_2 h_3 ,$$

the gradient operator on a scalar $\phi$ is

$$\nabla\phi = \sum \frac{1}{h_j}\frac{\partial\phi}{\partial u^j}\boldsymbol{e}_j ,$$

and the divergence and rotation operators on a vector

$$\boldsymbol{F} = F_1\boldsymbol{e}_1 + F_2\boldsymbol{e}_2 + F_3\boldsymbol{e}_3 ,$$

where $\boldsymbol{e}_j$ are unit vectors, are expressed by

$$\nabla\cdot\boldsymbol{F} = \frac{1}{g^{1/2}}\left[\frac{\partial}{\partial u^1}(h_2h_3F_1) + \frac{\partial}{\partial u^2}(h_3h_1F_1) + \frac{\partial}{\partial u^3}(h_1h_2F_3)\right] ,$$

$$\begin{aligned}\nabla\times\boldsymbol{F} = &\frac{1}{h_2h_3}\left[\frac{\partial}{\partial u^2}(h_3F_3) - \frac{\partial}{\partial u^3}(h_2F_2)\right]\boldsymbol{e}_1 \\ &+\frac{1}{h_3h_1}\left[\frac{\partial}{\partial u^3}(h_1F_1) - \frac{\partial}{\partial u^1}(h_3F_3)\right]\boldsymbol{e}_2 \\ &+\frac{1}{h_1h_2}\left[\frac{\partial}{\partial u^1}(h_2F_2) - \frac{\partial}{\partial u^2}(h_1F_1)\right]\boldsymbol{e}_3 .\end{aligned}$$

In the coordinate system $(\psi, \chi, \varphi)$, $(\boldsymbol{\xi}\cdot\nabla p)$ reduces to

$$(\boldsymbol{\xi}\cdot\nabla p) = \xi_\psi RB_\chi\frac{\partial p}{\partial\psi} = \xi_\psi RB_\chi p' .$$

The prime on $p$ means differentiation with respect to $\psi$. From (6.16) in Chap. 6, we have

$$-j_\varphi = Rp' + \frac{\hat{I}\hat{I}'}{\mu_0 R} , \qquad (8.122)$$

that is,

$$p' = -\frac{j_\varphi}{R} - \frac{\hat{I}\hat{I}'}{\mu_0 R^2} .$$

Note that $\psi$ defined by (8.118) is $-RA_\varphi$, while $\psi$ in (6.12) of Chap. 6 is $RA_\varphi$. $\nabla\cdot\boldsymbol{\xi}$ is expressed as

$$\nabla\cdot\xi = \frac{1}{J}\left[\frac{\partial}{\partial\psi}(JB_\chi R\xi_\psi) + \frac{\partial}{\partial\chi}\left(\frac{\xi_\chi}{B_\chi}\right) + \frac{\partial}{\partial\varphi}\left(\frac{J\xi_\varphi}{R}\right)\right] .$$

It is convenient to introduce

$$X \equiv RB_\chi\xi_\psi , \qquad Y \equiv \frac{\xi_\chi}{B_\chi} , \qquad U \equiv \frac{1}{RB_\chi}(B_\chi\xi_\varphi - B_\varphi\xi_\chi) = \frac{\xi_\varphi}{R} - \hat{I}R^2B_\chi\xi_\chi ,$$

whence

$$\xi_\psi = \frac{X}{RB_\chi}\,, \qquad \xi_\chi = B_\chi Y\,, \qquad \xi_\varphi = RU + \frac{\hat{I}}{R}Y\,,$$

and

$$\boldsymbol{\xi}\cdot\nabla p = Xp' = X\left(-\frac{j_\varphi}{R} - \frac{\hat{I}\hat{I}'}{\mu_0 R^2}\right)\,. \tag{8.123}$$

We analyze an individual Fourier mode $\boldsymbol{\xi} = \boldsymbol{\xi}(\psi,\chi)\exp(\mathrm{i}n\varphi)$. Then

$$(B\mathrm{i}k_\parallel)Y \equiv \left(B_\chi \frac{1}{JB_\chi}\frac{\partial}{\partial\chi} + B_\varphi \frac{1}{R}\frac{\partial}{\partial\varphi}\right)Y = \left(\frac{1}{J}\frac{\partial}{\partial\chi} + \frac{\hat{I}}{R^2}\mathrm{i}n\right)Y\,,$$

$$\frac{1}{J}\frac{\partial}{\partial\chi}Y = (B\mathrm{i}k_\parallel)Y - \mathrm{i}n\frac{\hat{I}}{R^2}Y\,.$$

Since

$$\nabla\cdot\boldsymbol{\xi} = \frac{1}{J}(JX)' + \mathrm{i}Bk_\parallel Y + \mathrm{i}nU\,, \tag{8.124}$$

it follows that

$$(\boldsymbol{\xi}\cdot\nabla p)(\nabla\cdot\boldsymbol{\xi}^*) = X\left(-\frac{j_\varphi}{R} - \frac{\hat{I}\hat{I}'}{\mu_0 R^2}\right)\left(\frac{1}{J}(JX^*)' - \mathrm{i}Bk_\parallel Y^* - \mathrm{i}nU^*\right)\,. \tag{8.125}$$

We now derive the expression for $\boldsymbol{B}_1 = \nabla\times(\boldsymbol{\xi}\times\boldsymbol{B})$:

$$(\boldsymbol{\xi}\times\boldsymbol{B})_\psi = \xi_\chi B_\varphi - \xi_\varphi B_\chi\,, \qquad (\boldsymbol{\xi}\times\boldsymbol{B})_\chi = -\xi_\psi B_\varphi\,, \qquad (\boldsymbol{\xi}\times\boldsymbol{B})_\varphi = \xi_\psi B_\chi\,,$$

$$B_{1\psi} = \frac{1}{JB_\chi R}\left[\frac{\partial X}{\partial\chi} + \frac{\partial}{\partial\varphi}\left(\frac{JB_\varphi}{R}X\right)\right] = \frac{1}{B_\chi R}\mathrm{i}Bk_\parallel X\,,$$

$$B_{1\chi} = -B_\chi\left(\mathrm{i}nU + \frac{\partial X}{\partial\psi}\right)\,,$$

$$B_{1\varphi} = \frac{R}{J}\left[-\frac{\partial}{\partial\psi}\left(\frac{J\hat{I}}{R^2}X\right) + \frac{\partial U}{\partial\chi}\right]\,.$$

The components of the current density are

$$\mu_0 j_\psi = 0\,, \qquad \mu_0 j_\chi = -B_\chi\frac{\partial}{\partial\psi}(RB_\varphi) = -B_\chi\hat{I}'\,, \qquad \mu_0 j_\varphi = \frac{R}{J}\frac{\partial}{\partial\psi}(JB_\chi^2)\,, \tag{8.126}$$

and

$$(\boldsymbol{B}_1\times\boldsymbol{\xi}^*)_\chi = \frac{R}{J}\frac{\partial U}{\partial\chi}\xi_\psi^* - \frac{R}{J}\left(\frac{IJ}{R^2}X\right)'\xi_\psi^* - \frac{\mathrm{i}Bk_\parallel}{B_\chi R}X\xi_\varphi^*\,,$$

$$(\boldsymbol{B}_1\times\boldsymbol{\xi}^*)_\varphi = \frac{\mathrm{i}Bk_\parallel}{B_\chi R}X\xi_\chi^* + (\mathrm{i}nU + X')B_\chi\xi_\psi^*\,.$$

Then,

$$
\begin{aligned}
\boldsymbol{\xi}^*\cdot(\boldsymbol{j}\times\boldsymbol{B}_1) &= \boldsymbol{j}\cdot(\boldsymbol{B}_1\times\boldsymbol{\xi}^*)\\
&= -B_\chi\frac{\hat{I}'}{\mu_0}\left[\frac{R}{J}\frac{\partial U}{\partial\chi}\xi_\psi^* - \frac{R}{J}\left(\frac{IJ}{R^2}X\right)'\xi_\psi^* - \frac{\mathrm{i}Bk_\parallel}{B_\chi R}X\xi_\varphi^*\right]\\
&\quad + j_\varphi\left[\frac{\mathrm{i}Bk_\parallel}{B_\chi R}X\xi_\chi^* + (\mathrm{i}nU + X')B_\chi\xi_\psi^*\right]\\
&= \frac{\mathrm{i}Bk_\parallel}{R}X\left[Y^*j_\varphi + \frac{\hat{I}'}{\mu_0}\left(RU^* + \frac{\hat{I}}{R}Y^*\right)\right] + \frac{\hat{I}'}{\mu_0 J}\left(\frac{IJ}{R^2}X\right)'X^*\\
&\quad - \frac{\hat{I}'}{\mu_0 J}\frac{\partial U}{\partial\chi}X^* + \mathrm{i}nUX^*\frac{1}{R}j_\varphi + X'X^*\frac{1}{R}j_\varphi\,,
\end{aligned}
$$

$$
\begin{aligned}
&(\boldsymbol{\nabla}\cdot\boldsymbol{\xi}^*)(\boldsymbol{\xi}\cdot\boldsymbol{\nabla}p) - (\boldsymbol{\xi}^*\times\boldsymbol{j})\cdot\boldsymbol{B}_1 \qquad (8.127)\\
&= \left(-\frac{j_\varphi}{R} - \frac{\hat{I}\hat{I}'}{\mu_0 R^2}\right)(XX^{*\prime} + X'X^*) + \frac{j_\varphi}{R}(\mathrm{i}nXU^* - \mathrm{i}nX^*U)\\
&\quad + \mathrm{i}n\frac{\hat{I}}{R^2}XU^*\frac{\hat{I}'}{\mu_0} + \mathrm{i}Bk_\parallel XU^*\frac{\hat{I}'}{\mu_0} + \frac{1}{J}\frac{\partial U}{\partial\chi}X^*\frac{\hat{I}'}{\mu_0}\\
&\quad + XX^*\left(-\frac{J'}{J}\frac{j_\varphi}{R} - \frac{\hat{I}'^2}{\mu_0 R^2} + \frac{\hat{I}\hat{I}'}{\mu_0 R^2}2\frac{R'}{R} - 2\frac{\hat{I}\hat{I}'}{\mu_0 R^2}\frac{J'}{J}\right)\,,
\end{aligned}
$$

and

$$
\begin{aligned}
\frac{|B_{1\varphi}|^2}{\mu_0} &= \frac{R^2}{\mu_0 J^2}\left|\frac{\partial U}{\partial\chi} - I\left(\frac{JX}{R^2}\right)' - \frac{JX}{R^2}\hat{I}'\right|^2\\
&= \frac{R^2}{\mu_0 J^2}\left|\frac{\partial U}{\partial\chi} - I\left(\frac{JX}{R^2}\right)'\right|^2 - \frac{\hat{I}'}{\mu_0 J}\left(\frac{\partial U}{\partial\chi}X^* + \frac{\partial U^*}{\partial\chi}X\right)\\
&\quad + \frac{\hat{I}\hat{I}'}{\mu_0 R^2}(X'X^* + X^{*\prime}X) + 2\frac{\hat{I}\hat{I}'}{\mu_0 J}\left(\frac{J'}{R^2} - \frac{2R'}{R^3}J\right)XX^* + \frac{\hat{I}'^2}{\mu_0 R^2}XX^*\,,
\end{aligned}
$$

$$
\begin{aligned}
\frac{|B_{1\chi}|^2}{\mu_0} &= \frac{B\chi^2}{\mu_0}|\mathrm{i}nU + X'|^2\\
&= \frac{B\chi^2}{\mu_0}\left|\mathrm{i}nU + X' - \frac{\mu_0 j_\varphi}{RB_\chi^2}X\right|^2 + (\mathrm{i}nUX^* - \mathrm{i}nU^*X)\frac{j_\varphi}{R}\\
&\quad + (X'X^* + X^{*\prime})\frac{j_\varphi}{R} - \frac{\mu_0 j_\varphi^2}{R^2B_\chi^2}XX^*\,.
\end{aligned}
$$

Finally, the energy integral of the axisymmetric toroidal system becomes

$$W = \frac{1}{2}\int_V \left[\frac{|\boldsymbol{B}_1|^2}{\mu_0} + \gamma p|(\boldsymbol{\nabla}\cdot\boldsymbol{\xi})|^2 + (\boldsymbol{\nabla}\cdot\boldsymbol{\xi}^*)(\boldsymbol{\xi}\cdot\boldsymbol{\nabla}p) - \boldsymbol{\xi}^*\cdot(\boldsymbol{j}\times\boldsymbol{B}_1)\right]\mathrm{d}\boldsymbol{r}$$

$$= \int_V \left[\frac{1}{2\mu_0}\frac{B^2 k_\parallel^2}{B_\chi^2 R^2}|X|^2 + \frac{1}{2\mu_0}\frac{R^2}{J^2}\left|\frac{\partial U}{\partial \chi} - I\left(\frac{JX}{R^2}\right)'\right|^2\right.$$

$$+\frac{B_\chi^2}{2\mu_0}\left|\mathrm{i}nU + X' - \frac{\mu_0 j_\varphi}{RB_\chi^2}X\right|^2$$

$$\left.+\frac{1}{2}\gamma p\left|\frac{1}{J}(JX)' + \mathrm{i}Bk_\parallel Y + \mathrm{i}nU\right|^2 - KXX^*\right]J\mathrm{d}\psi\mathrm{d}\chi\mathrm{d}\varphi\,, \quad (8.128)$$

where

$$K \equiv \frac{\hat{I}\hat{I}'}{\mu_0 R^2}\frac{R'}{R} + \frac{j_\varphi}{2R}\left(\frac{J'}{J} + \frac{\mu_0 j_\varphi}{RB_\chi^2}\right) = \frac{\hat{I}\hat{I}'}{\mu_0 R^2}\frac{R'}{R} + \frac{j_\varphi}{R}\left(\frac{J'}{J} + \frac{B_\chi'}{B_\chi}\right).$$

Here we have used (8.126), i.e.,

$$\mu_0 j_\varphi = \frac{R}{J}(J'B_\chi^2 + J2B_\chi B_\chi')\,.$$

### 8.5.3 Energy Integral of High-$n$ Ballooning Mode

The energy integral (8.128) is used for stability analysis of high-$n$ modes and the ballooning mode [8.17, 8.18].

The first step in minimizing $\delta W$ is to select $Y$ so that the second positive term in (8.128) vanishes ($\boldsymbol{\nabla}\cdot\boldsymbol{\xi} = 0$). The second step is to minimize with respect to $U$. The minimizing $U$ is given by

$$inU + \frac{\partial X}{\partial \psi} + \left(\frac{\mu_0 p'}{B^2} + \frac{\hat{q}'}{\hat{q}}\frac{I^2}{R^2 B^2}\right)X + \frac{I^2}{\hat{q}R^2B^2}JBk_\parallel\left(\frac{1}{n}\frac{\partial X}{\partial \psi}\right) = 0$$

When we treat ballooning modes, perturbations with large toroidal mode number $n$ and $|m - \hat{q}n| \ll n$ are important (see Sect. 8.6). After a long mathematical calculation, the energy integral with $O(1/n)$ accuracy is derived as [8.17]

$$W = \frac{\pi}{\mu_0}\int \mathrm{d}\psi\mathrm{d}\chi\left\{\frac{JB^2}{R^2B_\chi^2}\left|k_\parallel X\right|^2 + \frac{R^2B_\chi^2}{JB^2}\left|\frac{1}{n}\frac{\partial}{\partial\psi}(JBk_\parallel X)\right|^2\right.$$

$$-\frac{2J\mu_0 p'}{B^2}\left[|X|^2\frac{\partial}{\partial\psi}\left(\mu_0 p + \frac{B^2}{2}\right) - \frac{\mathrm{i}\hat{I}}{JB^2}\frac{\partial}{\partial\chi}\left(\frac{B^2}{2}\right)\frac{X^*}{n}\frac{\partial X}{\partial\psi}\right]$$

$$\left.+\frac{X^*}{n}JBk_\parallel(X\sigma') - \frac{1}{n}(P^*JBk_\parallel Q + \mathrm{c.c.})\right\}, \quad (8.129)$$

where c.c. means the conjugate complex and

$$P = X\sigma - \frac{B_\chi^2}{\hat{q}B^2}\frac{I}{n}\frac{\partial}{\partial\psi}(JBk_\parallel X) , \qquad Q = \frac{X\mu_0 p'}{B^2} + \frac{\hat{I}^2}{\hat{q}^2R^2B^2}\frac{1}{n}\frac{\partial}{\partial\psi}(JBk_\parallel X) ,$$

$$\sigma = \frac{\hat{I}\mu_0 p'}{B^2} + \hat{I}' , \qquad \mathrm{i}Jk_\parallel B = \frac{\partial}{\partial\chi} + \mathrm{i}n\hat{q} , \qquad \hat{q}(\psi,\chi) = \frac{\hat{I}J}{R^2} .$$

$\delta W$ must be minimized with respect to all periodic functions $X$ subject to an appropriate normalization

$$\pi\int J\mathrm{d}\psi\mathrm{d}\chi\rho_\mathrm{m}\left[\frac{|X|^2}{R^2B_\chi^2} + \left(\frac{RB_\chi}{B}\right)^2\left|\frac{1}{n}\frac{\partial X}{\partial\psi}\right|^2\right] = \mathrm{const.} , \qquad (8.130)$$

where $\rho_\mathrm{m}$ is the mass density and (8.130) corresponds to the total kinetic energy of transverse motion to the leading order in $1/n$. The Euler equation for the minimizing function $X(\psi,\chi)$ can be deduced form (8.129) and (8.130). As $X(\psi,\chi)$ is periodic in $\chi$, it can be expanded in a Fourier series:

$$X(\psi,\chi) = \sum_m X_m(\psi)\exp(im\chi) .$$

A continuous function $X_s(\psi)$ of $s$, which is equal to $X_m(\psi)$ for $s$ equal to an integer $m$, can be constructed and expressed as a Fourier integral:

$$X_s(\psi) = \int_{-\infty}^{\infty}\hat{X}(\psi,y)\exp(\mathrm{i}sy)\mathrm{d}y/(2\pi) , \quad \hat{X}(\psi,y) = \int_{-\infty}^{\infty}X_s(\psi)\exp(-\mathrm{i}sy)\mathrm{d}s.$$

$\hat{X}(\psi,y)$ is called the ballooning representation of $X(\psi,\chi)$. Then $X(\psi,\chi)$ reduces to

$$X(\psi,\chi) = \sum_m \exp(-im\chi)\int_{-\infty}^{\infty}\hat{X}(\psi,y)\exp(imy)\mathrm{d}y/(2\pi) . \qquad (8.131)$$

Using the delta function $\delta(x)$,

$$\frac{1}{2\pi}\sum_m \exp\left[-\mathrm{i}m(\chi-y)\right] = \sum_N \delta(y-\chi+2\pi N) ,$$

and the relation between $X(\psi,\chi)$ and $\hat{X}(\psi,y)$ is

$$X(\psi,\chi) = \sum_N \hat{X}(\psi,\chi-2\pi N) . \qquad (8.132)$$

$X(\psi,\chi)$ is expressed as an infinite sum of quasi-modes.

The Euler equation for $X(\psi,\chi)$ is converted into an identical equation for $\hat{X}(\psi,y)$, but with $\hat{X}$ in the infinite domain of $y$ and free of periodicity requirements. Let us consider $\hat{X}(\psi,y)$ with the form

$$\hat{X}(\psi, y) = F(\psi, y) \exp\left(-\mathrm{i}n \int_{y_0}^{y} \hat{q} \mathrm{d}y\right) , \tag{8.133}$$

in which the amplitude $F(\psi, y)$ is a more slowly varying function as $n \to \infty$. Then

$$\mathrm{i}Jk_{\parallel} B\hat{X}(\psi, y) = \left(\frac{\partial}{\partial y} + \mathrm{i}n\hat{q}\right) \hat{X}(\psi, y) = \frac{\partial F(\psi, y)}{\partial y} \exp\left(-\mathrm{i}n \int_{y_0}^{y} \hat{q} \mathrm{d}y\right) .$$

The leading term in the Euler equation for $\hat{X}(\psi, y)$ reduces to [8.17]

$$\begin{aligned}
&\frac{1}{J}\frac{\partial}{\partial y}\left\{\frac{1}{JR^2B_\chi^2}\left[1+\left(\frac{R^2B_\chi^2}{B}\int_{y_0}^{y}\hat{q}'\mathrm{d}y\right)^2\right]\frac{\partial F_0}{\partial y}\right\} \\
&\quad +\left[\frac{2\mu_0 p'}{B^2}\frac{\partial}{\partial \psi}\left(\mu_0 p+\frac{B^2}{2}\right)-\frac{\hat{I}\mu_0 p'}{B^4}\left(\int_{y_0}^{y}\hat{q}'\mathrm{d}y\right)\frac{1}{J}\frac{\partial B^2}{\partial y}\right]F_0 \\
&\quad +\frac{\omega^2(\psi, y_0)}{R^2B_\chi^2}\left[1+\left(\frac{R^2B_\chi^2}{B}\int_{y_0}^{y}\hat{q}'\mathrm{d}y\right)^2\right]F_0 = 0 .
\end{aligned} \tag{8.134}$$

The stability of the ballooning mode can be analyzed using this Euler equation (see Sect. 8.6) [8.18].

## 8.6 Ballooning Instability

In interchange instability, the parallel component $k_{\parallel} = (\boldsymbol{k} \cdot \boldsymbol{B})/B$ of the propagation vector is zero and an average minimum-$B$ condition may stabilize such an instability. Suydam's condition and the local-mode stability condition of the toroidal system are involved in perturbations with $k_{\parallel} = 0$. In this section, we study perturbations where $k_{\parallel} \neq 0$ but $|k_{\parallel}/k_{\perp}| \ll 1$. Although the interchange instability is stabilized by an average minimum-$B$ configuration, the perturbation with $k_{\parallel} \neq 0$ may be able to grow locally in the bad region of the average minimum-$B$ field. This type of instability is called the *ballooning mode* (see Fig. 8.8).

There is a beta limit for the stability of the ballooning mode. It will be shown that average minimum-$B$ and shear stabilize the ballooning mode. Therefore measures of the magnetic well and shear are important parameters in magnetic confinement configurations [8.19].

The energy integral $\delta W$ is given by

$$\begin{aligned}
\delta W = \frac{1}{2\mu_0}\int\Big\{&\left[\nabla\times(\boldsymbol{\xi}\times\boldsymbol{B}_0)\right]^2-\left[\boldsymbol{\xi}\times(\nabla\times\boldsymbol{B}_0)\right]\cdot\nabla\times(\boldsymbol{\xi}\times\boldsymbol{B}_0) \\
&+\gamma\mu_0 p_0(\nabla\cdot\boldsymbol{\xi})^2+\mu_0(\nabla\cdot\boldsymbol{\xi})(\boldsymbol{\xi}\cdot\nabla p_0)\Big\}\mathrm{d}\boldsymbol{r} .
\end{aligned}$$

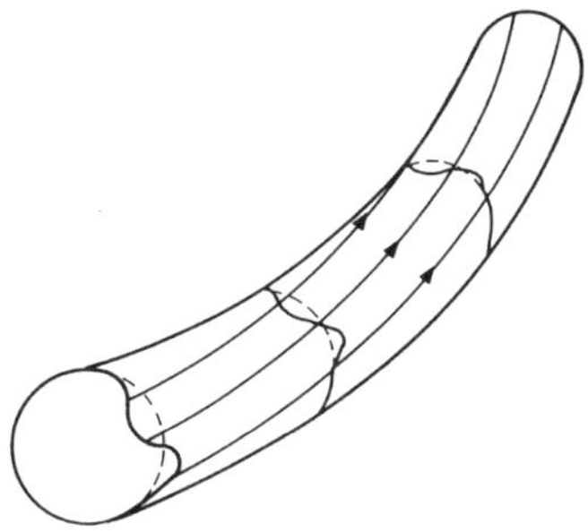

**Fig. 8.8.** Ballooning mode

Let us consider the case where $\boldsymbol{\xi}$ can be expressed as

$$\boldsymbol{\xi} = \frac{\boldsymbol{B}_0 \times \nabla\phi}{B_0^2} , \tag{8.135}$$

where $\phi$ is considered to be the time integral of the scalar electrostatic potential of the perturbed electric field. Because

$$\boldsymbol{\xi} \times \boldsymbol{B}_0 = \nabla_\perp \phi ,$$

the energy integral reduces to

$$\delta W = \frac{1}{2\mu_0} \int \left\{ (\nabla \times \nabla_\perp\phi)^2 - \left[ \frac{(\boldsymbol{B}_0 \times \nabla_\perp\phi) \times \mu_0 \boldsymbol{j}_0}{B_0^2} \right] \nabla \times \nabla_\perp \phi \right.$$
$$\left. + \gamma\mu_0 p_0 (\nabla\cdot\boldsymbol{\xi})^2 + \mu_0 (\nabla\cdot\boldsymbol{\xi})(\boldsymbol{\xi}\cdot\nabla p_0) \right\} \mathrm{d}\boldsymbol{r} .$$

$\nabla\cdot\boldsymbol{\xi}$ is given by

$$\nabla\cdot\boldsymbol{\xi} = \nabla\cdot\left(\frac{\boldsymbol{B}_0 \times \nabla\phi}{B_0^2}\right) = \nabla\phi\cdot\nabla \times \left(\frac{\boldsymbol{B}_0}{B_0^2}\right)$$
$$= \nabla\phi\cdot \left[ \left(\nabla \frac{1}{B^2}\right) \times \boldsymbol{B} + \frac{1}{B^2} \nabla \times \boldsymbol{B} \right] .$$

The second term in square brackets is negligible compared with the first term in the low beta case. Using $\nabla p_0 = \boldsymbol{j}_0 \times \boldsymbol{B}_0$, $\delta W$ is expressed as

$$\delta W = \frac{1}{2\mu_0} \int \left\{ (\nabla \times \nabla_\perp\phi)^2 + \frac{\mu_0 \nabla p_0\cdot(\nabla_\perp\phi \times \boldsymbol{B}_0)}{B_0^2} \left( \frac{\boldsymbol{B}_0\cdot\nabla \times \nabla_\perp\phi}{B_0^2} \right) \right.$$
$$- \frac{\mu_0(\boldsymbol{j}_0\cdot\boldsymbol{B}_0)}{B_0^2} \nabla_\perp\phi\cdot\nabla \times \nabla_\perp\phi + \gamma\mu_0 p_0 \left[ \nabla\left(\frac{1}{B_0^2}\right)\cdot(\boldsymbol{B}_0 \times \nabla_\perp\phi) \right]^2$$
$$\left. + \frac{\mu_0 \nabla p_0\cdot(\boldsymbol{B}_0 \times \nabla_\perp\phi)}{B_0^2} \left[ \nabla\left(\frac{1}{B_0^2}\right)\cdot(\boldsymbol{B}_0 \times \nabla_\perp\phi) \right] \right\} \mathrm{d}\boldsymbol{r} .$$

Let us use the $z$ coordinate as a length along a field line, $r$ as radial coordinate of magnetic surfaces, and $\theta$ as poloidal angle in the direction perpendicular to field lines. The $r, \theta, z$ components of $\nabla p_0$, $\boldsymbol{B}$, and $\nabla \phi$ are given approximately by

$$\nabla p_0 = (p_0', 0, 0) , \qquad \boldsymbol{B} = \left(0,\, B_\theta(r),\, B_0\left[1 - rR_c^{-1}(z)\right]\right) ,$$

$$\nabla \phi = \left(\frac{\partial \phi}{\partial r}, \frac{\partial \phi}{r\partial \theta}, \frac{\partial \phi}{\partial z}\right) , \qquad \phi(r, \theta, z) = \phi(r, z)\mathrm{Re}(\exp im\theta) .$$

$R_c(z)$ is the radius of curvature of the line of magnetic force:

$$\frac{1}{R_c(z)} = \frac{1}{R_0}\left(-w + \cos 2\pi \frac{z}{L}\right) .$$

When $R_c(z) < 0$, the curvature is said to be good. If the configuration is average minimum-$B$, $w$ and $R_0$ must be $1 > w > 0$ and $R_0 > 0$. Since $B_\theta/B_0$, $r/R_0$, $r/L$ are all small quantities, we find

$$\nabla_\perp \phi = \nabla \phi - \nabla_\parallel \phi \approx \mathrm{Re}\left(\frac{\partial \phi}{\partial r}, \frac{im}{r}\phi, 0\right) ,$$

$$\nabla \times (\nabla_\perp \phi) \approx \mathrm{Re}\left(\frac{-im}{r}\frac{\partial \phi}{\partial z}, \frac{\partial^2 \phi}{\partial z \partial r}, 0\right) ,$$

$$\boldsymbol{B}_0 \times \nabla_\perp \phi \approx \mathrm{Re}\left(\frac{-im}{r}B_0\phi, B_0\frac{\partial \phi}{\partial r}, 0\right) ,$$

and $\delta W$ reduces to

$$\delta W = \frac{1}{2\mu_0}\int \frac{m^2}{r^2}\left\{\left[\frac{\partial \phi(r,z)}{\partial z}\right]^2 - \frac{\beta}{r_p R_c(z)}\left[\phi(r,z)\right]^2\right\} 2\pi r \mathrm{d}r \mathrm{d}z ,$$

where $-p_0/p_0' = r_\mathrm{p}$ and $\beta = p_0/(B_0^2/2\mu_0)$. The second term contributes to stability in the region $R_c(z) < 0$ and contributes to instability in the region of $R_c(z) > 0$. Euler's equation is given by

$$\frac{\mathrm{d}^2\phi}{\mathrm{d}z^2} + \frac{\beta}{r_\mathrm{p}R_c(z)}\phi = 0 . \tag{8.136}$$

$R_c$ is nearly equal to $B/|\nabla B|$. Equation (8.136) is a Mathieu differential equation, whose eigenvalue is

$$w = F(\beta L^2/2\pi^2 r_\mathrm{p} R_0) .$$

Since

$$F(x) = x/4 , \qquad x \ll 1 , \qquad F(x) = 1 - x^{-1/2} , \qquad x \gg 1 ,$$

we find the approximate relation

$$\beta_{\rm c} \sim \frac{4w}{(1+3w)(1-w)^2} \frac{2\pi^2 r_{\rm p} R_0}{L^2} \ .$$

Since $w$ is of the order of $r_{\rm p}/2R_0$ and the connection length is

$$L \approx 2\pi R_0 (2\pi/\iota) \ ,$$

where $\iota$ is the rotational transform angle, the critical beta ratio $\beta_{\rm c}$ is

$$\beta_{\rm c} \sim \left(\frac{\iota}{2\pi}\right)^2 \left(\frac{r_{\rm p}}{R}\right) \ . \tag{8.137}$$

If $\beta$ is smaller than the critical beta ratio $\beta_{\rm c}$, then $\delta W > 0$, and the plasma is stable. The stability condition for the ballooning mode in the shearless case is given by [8.19]

$$\beta < \beta_{\rm c} \ .$$

In the configuration with magnetic shear, a more rigorous treatment is necessary. For ballooning modes with large toroidal mode number $n \gg 1$ and $m - nq \sim 0$ (see Sect. 8.5), the stable region in the shear parameter $S$ and the measure $\alpha$ of the pressure gradient of the ballooning mode is shown in Fig. 8.9. The shear parameter $S$ is defined by

$$S = \frac{r}{q} \frac{{\rm d}q}{{\rm d}r} \ ,$$

where $q$ is the safety factor ($q \equiv 2\pi/\iota$, $\iota$ the rotational transform angle) and the measure $\alpha$ of the pressure gradient is defined by

$$\alpha = -\frac{q^2 R}{B^2/2\mu_0} \frac{{\rm d}p}{{\rm d}r} \ .$$

The straight-line approximation to the maximum pressure gradient in the range of large positive shear ($S > 0.8$) is $\alpha \sim 0.6S$, as shown in Fig. 8.9. Since

$$\beta = \frac{1}{B_0^2/2\mu_0} \frac{1}{\pi a^2} \int_0^a p 2\pi r {\rm d}r = -\frac{1}{B_0^2/2\mu_0} \frac{1}{a^2} \int_0^a \frac{{\rm d}p}{{\rm d}r} r^2 {\rm d}r \ ,$$

the maximum ballooning-stable beta is

$$\beta = 0.6 \frac{a}{R} \left( \frac{1}{a^3} \int_0^a \frac{1}{q^3} \frac{{\rm d}q}{{\rm d}r} r^3 {\rm d}r \right) \ .$$

Under an optimum $q$ profile, the maximum beta is given by [8.20]

$$\beta_{\max} \sim 0.28 \frac{a}{R q_a} \qquad (q_a > 2) \ , \tag{8.138}$$

where $q_a$ is the safety factor at the plasma boundary. In the derivation of (8.138), it is assumed that $q_a > 2$, $q_0 = 1$.

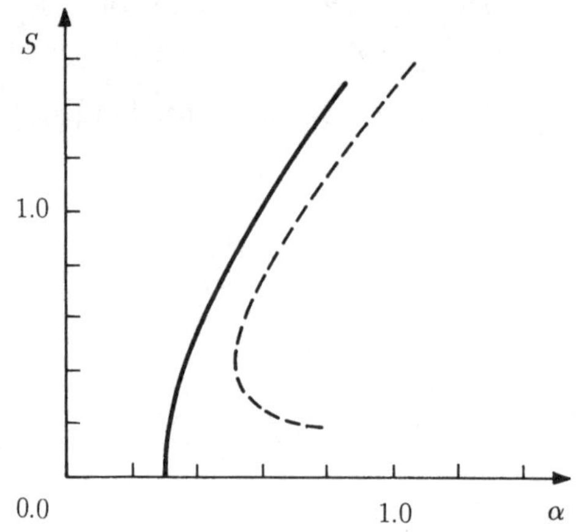


**Fig. 8.9.** Maximum stable pressure gradient $\alpha$ as a function of the shear parameter $S$ of ballooning mode. The *dotted line* is the stability boundary obtained by imposing a more restricted boundary condition on the perturbation [8.18]

It should be noted that the ballooning mode is stable in the negative shear region of $S$, as shown in Fig. 8.9. When the shear parameter $S$ is negative [$q(r)$ decreases outwardly], the outer lines of magnetic force rotate around the magnetic axis more quickly than the inner ones. When the pressure increases, the tokamak plasma tends to expand in a direction of major radius (the Shafranov shift, see the note at the end of this section). This must be counterbalanced by strengthening the poloidal field on the outside of the tokamak plasma. In the region of high pressure gradient, the necessary poloidal field increases outwardly, so on outer magnetic surfaces the magnetic field lines rotate around the magnetic axis faster than they do on inner ones and the shear parameter becomes more negative [8.21].

In reality, the shear parameter in a tokamak is positive for typical operations. However, the fact that the ballooning mode is stable in the negative shear parameter region is very important for developing a tokamak configuration that is stable against ballooning modes. Since

$$\frac{r}{Rq} = \frac{B_\theta}{B_0} = \frac{1}{B_0}\frac{\mu_0}{2\pi r}\int_0^r j(r)2\pi r \mathrm{d}r \ ,$$

the profile $q(r)$ of the safety factor is

$$\frac{1}{q(r)} = \frac{R}{2B_0}\left(\frac{\mu_0}{\pi r^2}\int_0^r j2\pi r \mathrm{d}r\right) \equiv \frac{\mu_0 R}{2B_0}\langle j(r)\rangle_r \ .$$

Therefore, a negative shear configuration can be realized by a hollow current profile. The MHD stability of a tokamak with hollow current profiles is analyzed in detail in [8.22].

**Shafranov Shift**

In the case of a tokamak with large aspect and circular plasma cross-section, the Shafranov shift $\Delta(r)$ of the center of the magnetic surface with radius $r$ from the center of the plasma boundary with radius $a$ is [8.23]

$$\frac{\mathrm{d}\Delta}{\mathrm{d}r} = \frac{1}{RrB_\theta^2}\left(\beta_\mathrm{p} B_{\theta a}^2 \int_0^r r^2 \frac{\mathrm{d}}{\mathrm{d}r}\frac{p}{\langle p\rangle}\mathrm{d}r - \int_0^r rB_\theta^2 \mathrm{d}r\right) , \tag{8.139}$$

where $B_\theta$ is the magnitude of the poloidal field at $r$ and $B_{\theta a}$ is the magnitude of the poloidal field at $r = a$. $\beta_\mathrm{p}$ is the poloidal beta and $\langle p\rangle$ is the volume average of the pressure $\langle p\rangle = \int_0^a p 2r\mathrm{d}r/a^2$. In the case of a parabolic pressure profile, the pressure term on the right-hand side of (8.139) becomes $-(a/R)\beta_\mathrm{p}(B_{\theta a}/B_\theta)^2(r/a)^3$ and in the case of a flat current profile, the second term on the right-hand side of (8.139) becomes $-(a/4R)(r/a)$. When the pressure profile is parabolic and the current profile is flat, the Shafranov shift is $\Delta/a = (a/2R)(\beta_\mathrm{p} + 1/4)\left[1 - (r/a)^2\right]$.

## 8.7 Eta-i Mode Due to Density and Temperature Gradient

Let us consider a plasma with density gradient $\mathrm{d}n_0/\mathrm{d}r$ and temperature gradient $\mathrm{d}T_{\mathrm{e}0}/\mathrm{d}r$, $\mathrm{d}T_{\mathrm{i}0}/\mathrm{d}r$ in a magnetic field in the $z$ direction. Assume that the ion density becomes $n_\mathrm{i} = n_{\mathrm{i}0} + \tilde{n}_\mathrm{i}$ by disturbance. The equation of continuity

$$\frac{\partial n_\mathrm{i}}{\partial t} + \boldsymbol{v}_\mathrm{i}\cdot\nabla n_\mathrm{i} + n_\mathrm{i}\nabla\cdot\boldsymbol{v}_\mathrm{i} = 0$$

reduces by linearization to

$$-\mathrm{i}\omega\tilde{n}_\mathrm{i} + \tilde{v}_\mathrm{r}\frac{\partial n_0}{\partial r} + n_0 i k_\parallel \tilde{v}_\parallel = 0 . \tag{8.140}$$

It is assumed that the perturbation terms change as $\exp\mathrm{i}(k_\theta r\theta + k_\parallel z - \omega t)$ and $k_\theta, k_\parallel$ are the $\theta$ and $z$ components of the propagation vector. When the perturbed electrostatic potential is denoted by $\tilde{\phi}$, the $\boldsymbol{E}\times\boldsymbol{B}$ drift velocity is $\tilde{v}_\mathrm{r} = E_\theta/B = \mathrm{i}k_\theta\tilde{\phi}/B$. Since the electron density follows a Boltzmann distribution, we find

$$\frac{\tilde{n}_\mathrm{e}}{n_0} = \frac{e\tilde{\phi}}{kT_\mathrm{e}} . \tag{8.141}$$

The component of the equation of motion parallel to the magnetic field,

$$n_\mathrm{i} m_\mathrm{i}\frac{\mathrm{d}v_\parallel}{\mathrm{d}t} = -\nabla_\parallel p_\mathrm{i} - en\nabla_\parallel\phi ,$$

reduces by linearization to

$$-\mathrm{i}\omega n_{\mathrm{i}} m_{\mathrm{i}} \tilde{v}_{\parallel} = -\mathrm{i}k_{\parallel}(\tilde{p}_{\mathrm{i}} + e n_0 \tilde{\phi}) \,. \tag{8.142}$$

Similarly the adiabatic equation

$$\frac{\partial}{\partial t}(p_{\mathrm{i}} n_{\mathrm{i}}^{-5/3}) + \boldsymbol{v}\cdot\boldsymbol{\nabla}(p_{\mathrm{i}} n_{\mathrm{i}}^{-5/3}) = 0$$

reduces to

$$-\mathrm{i}\omega\left(\frac{\tilde{p}_{\mathrm{i}}}{p_{\mathrm{i}}} - \frac{5}{3}\frac{\tilde{n}_{\mathrm{i}}}{n_{\mathrm{i}}}\right) - \frac{\mathrm{i}k_\theta \tilde{\phi}}{B}\left(\frac{\mathrm{d}T_{\mathrm{i}0}/\mathrm{d}r}{T_{\mathrm{i}0}} - \frac{2}{3}\frac{\mathrm{d}n_0/\mathrm{d}r}{n_0}\right) = 0 \,. \tag{8.143}$$

Let us define the *electron drift frequencies* $\omega_{ne}^*$, $\omega_{Te}^*$, and the *ion drift frequencies* $\omega_{ni}^*$, $\omega_{Ti}^*$ by

$$\omega_{ne}^* \equiv -\frac{k_\theta(\kappa T_{\mathrm{e}})}{eBn_{\mathrm{e}}}\frac{\mathrm{d}n_{\mathrm{e}}}{\mathrm{d}r} \,, \qquad \omega_{ni}^* \equiv \frac{k_\theta(\kappa T_{\mathrm{i}})}{eBn_{\mathrm{i}}}\frac{\mathrm{d}n_{\mathrm{i}}}{\mathrm{d}r} \,,$$

$$\omega_{Te}^* \equiv -\frac{k_\theta}{eB}\frac{\mathrm{d}(\kappa T_{\mathrm{e}})}{\mathrm{d}r} \,, \qquad \omega_{Ti}^* \equiv \frac{k_\theta}{eB}\frac{\mathrm{d}(\kappa T_{\mathrm{i}})}{\mathrm{d}r} \,.$$

The ratio of the temperature gradient to the density gradient of electrons and ions is given by

$$\eta_{\mathrm{e}} \equiv \frac{\mathrm{d}T_{\mathrm{e}}/\mathrm{d}r}{T_{\mathrm{e}}}\frac{n_{\mathrm{e}}}{\mathrm{d}n_{\mathrm{e}}/\mathrm{d}r} = \frac{\mathrm{d}\ln T_{\mathrm{e}}}{\mathrm{d}\ln n_{\mathrm{e}}} \,, \qquad \eta_{\mathrm{i}} \equiv \frac{\mathrm{d}T_{\mathrm{i}}/\mathrm{d}r}{T_{\mathrm{i}}}\frac{n_{\mathrm{i}}}{\mathrm{d}n_{\mathrm{i}}/\mathrm{d}r} = \frac{\mathrm{d}\ln T_{\mathrm{i}}}{\mathrm{d}\ln n_{\mathrm{i}}} \,,$$

respectively. The following relations hold among these values:

$$\omega_{ni}^* = -\frac{T_{\mathrm{i}}}{T_{\mathrm{e}}}\omega_{ne}^* \,, \qquad \omega_{Te}^* = \eta_{\mathrm{e}}\omega_{ne}^* \,, \qquad \omega_{Ti}^* = \eta_{\mathrm{i}}\omega_{ni}^* \,.$$

Then (8.140), (8.141), (8.142), and (8.143) reduce to

$$\frac{\tilde{n}_{\mathrm{i}}}{n_0} = \frac{\tilde{v}_{\parallel}}{\omega/k_{\parallel}} + \frac{\omega_{ne}^*}{\omega}\frac{e\tilde{\phi}}{\kappa T_{\mathrm{e}}} \,, \qquad \frac{\tilde{n}_{\mathrm{e}}}{n_0} = \frac{e\tilde{\phi}}{\kappa T_{\mathrm{e}}} \,,$$

$$\frac{\tilde{v}_{\parallel}}{\omega/k_{\parallel}} = \frac{1}{m_{\mathrm{i}}(\omega/k_{\parallel})^2}\left(e\tilde{\phi} + \frac{\tilde{p}_{\mathrm{i}}}{n_0}\right) \,, \qquad \frac{\tilde{p}_{\mathrm{i}}}{p_{\mathrm{i}0}} - \frac{5}{3}\frac{\tilde{n}}{n_0} = \frac{\omega_{ne}^*}{\omega}\left(\eta_{\mathrm{i}} - \frac{2}{3}\right)\frac{e\tilde{\phi}}{\kappa T_{\mathrm{e}}} \,.$$

The charge neutrality condition $\tilde{n}_{\mathrm{i}}/n_0 = \tilde{n}_{\mathrm{e}}/n_0$ yields the dispersion equation:

$$1 - \frac{\omega_{ne}^*}{\omega} - \left(\frac{v_{\mathrm{Ti}}}{\omega/k_{\parallel}}\right)^2\left[\frac{T_{\mathrm{e}}}{T_{\mathrm{i}}} + \frac{5}{3} + \frac{\omega_{ne}^*}{\omega}\left(\eta_{\mathrm{i}} - \frac{2}{3}\right)\right] = 0 \,,$$

where $v_{\mathrm{Ti}}^2 = \kappa T_{\mathrm{i}}/m_{\mathrm{i}}$. The solution in the case $\omega \ll \omega_{n_{\mathrm{e}}}^*$ is [8.24]

$$\omega^2 = -k_{\parallel}^2 v_{\mathrm{Ti}}^2\left(\eta_{\mathrm{i}} - \frac{2}{3}\right) \,. \tag{8.144}$$

The dispersion equation shows that this type of perturbation is unstable when $\eta_{\rm i} > 2/3$. This mode is called the $\eta_{\rm i}$ *mode* or ion temperature gradient (ITG) mode.

When the propagation velocity $|\omega/k_\parallel|$ becomes of the order of the ion thermal velocity $v_{T_{\rm i}}$, the interaction (Landau damping) between ions and wave (perturbation) becomes important, as will be described in Chap. 11, and the MHD treatment must be modified. When the value of $\eta_{\rm i}$ is not large, a kinetic treatment is needed and the threshold of $\eta_{\rm i}$ becomes $\eta_{\rm i,cr} \sim 1.5$.

# 9 Resistive Instabilities

In the last chapter, we discussed instabilities of plasmas with zero resistivity. In such cases, the conducting plasma is frozen to the line of magnetic force. However, the resistivity of a plasma is not generally zero and the plasma may therefore deviate from the magnetic line of force. Modes which are stable in the ideal case may in some instances become unstable if a finite resistivity is introduced.

Ohm's law is

$$\eta \boldsymbol{j} = \boldsymbol{E} + \boldsymbol{V} \times \boldsymbol{B} \,. \tag{9.1}$$

For simplicity we assume here that $\boldsymbol{E}$ is zero. The current density is $\boldsymbol{j} = \boldsymbol{V} \times \boldsymbol{B}/\eta$ and the $\boldsymbol{j} \times \boldsymbol{B}$ force is

$$\boldsymbol{F}_{\mathrm{s}} = \boldsymbol{j} \times \boldsymbol{B} = \frac{\boldsymbol{B}(\boldsymbol{V}\cdot\boldsymbol{B}) - \boldsymbol{V}B^2}{\eta} \,. \tag{9.2}$$

When $\eta$ tends to zero, this force becomes infinite and prevents the plasma from deviating from the line of magnetic force. When the magnitude $B$ of the magnetic field is small, this force does not become large, even if $\eta$ is small, and the plasma can deviate from the line of magnetic force. When we consider a perturbation with propagation vector $\boldsymbol{k}$, only the parallel (to $\boldsymbol{k}$) component of the zeroth-order magnetic field $\boldsymbol{B}$ affects the perturbation, as will be shown later. Even if shear exists, we can choose a propagation vector $\boldsymbol{k}$ perpendicular to the magnetic field $\boldsymbol{B}$:

$$(\boldsymbol{k}\cdot\boldsymbol{B}) = 0 \,. \tag{9.3}$$

Accordingly, if there is any force $\boldsymbol{F}_{\mathrm{dr}}$ driving the perturbation, this driving force may easily exceed the force $\boldsymbol{F}_{\mathrm{s}}$, which is very small for a perturbation where $(\boldsymbol{k}\cdot\boldsymbol{B}) = 0$, and the plasma becomes unstable. This type of instability is called *resistive instability*.

## 9.1 Tearing Instability

Let us consider a slab model in which the zeroth-order magnetic field $\boldsymbol{B}_0$ depends only on $x$, and $\boldsymbol{B}$ is given as

$$\boldsymbol{B}_0 = B_{0y}(x)\boldsymbol{e}_y + B_{0z}(x)\boldsymbol{e}_z \,. \tag{9.4}$$

From Ohm's law (9.1), we find

$$\frac{\partial \boldsymbol{B}}{\partial t} = -\nabla \times \boldsymbol{E} = \nabla \times \left[(\boldsymbol{V} \times \boldsymbol{B}) - \eta \boldsymbol{j}\right] = \nabla \times (\boldsymbol{V} \times \boldsymbol{B}) + \frac{\eta}{\mu_0}\Delta \boldsymbol{B} \,, \tag{9.5}$$

where $\eta$ is assumed to be constant. It is assumed that the plasma is incompressible. Since the growth rate of the resistive instability is small compared with the MHD characteristic rate (inverse of the Alfvén wave transit time) and the movement is slower than the sound velocity, the assumption of incompressibility is justified and it follows that

$$\nabla \cdot \boldsymbol{V} = 0 \,. \tag{9.6}$$

The magnetic field $\boldsymbol{B}$ always satisfies

$$\nabla \cdot \boldsymbol{B} = 0 \,. \tag{9.7}$$

The equation of motion is

$$\begin{aligned} \rho_{\mathrm{m}} \frac{\mathrm{d}\boldsymbol{V}}{\mathrm{d}t} &= \frac{1}{\mu_0}(\nabla \times \boldsymbol{B}) \times \boldsymbol{B} - \nabla p \\ &= \frac{1}{\mu_0}\left[(\boldsymbol{B}_0 \cdot \nabla)\boldsymbol{B}_1 + (\boldsymbol{B}_1 \cdot \nabla)\boldsymbol{B}_0 - \frac{\nabla B^2}{2}\right] - \nabla p \,. \end{aligned} \tag{9.8}$$

Let us consider the perturbation expressed by

$$f_1(\boldsymbol{r}, t) = f_1(x) \exp\left[\mathrm{i}(k_y y + k_z z) + \gamma t\right] .$$

Then (9.5) reduces to

$$\gamma B_{1x} = \mathrm{i}(\boldsymbol{k} \cdot \boldsymbol{B}) V_x + \eta \mu_0 \left(\partial^2 \partial x^2 - k^2\right) B_{1x} \,, \tag{9.9}$$

where $k^2 = k_y^2 + k_z^2$. The first term on the right-hand side of (9.8) becomes $(\boldsymbol{B}_0 \cdot \nabla)\boldsymbol{B}_1 = \mathrm{i}(\boldsymbol{k} \cdot \boldsymbol{B}_0)\boldsymbol{B}_1$. The rotation of (9.8) is

$$\mu_0 \rho_{\mathrm{m}} \gamma \nabla \times \boldsymbol{V} = \nabla \times \left[\mathrm{i}(\boldsymbol{k} \cdot \boldsymbol{B}_0)\boldsymbol{B}_1 + \left(B_{1x}\frac{\partial}{\partial x}\right)\boldsymbol{B}_0\right] . \tag{9.10}$$

Equations (9.6) and (9.7) reduce to

$$\frac{\partial B_{1x}}{\partial x} + \mathrm{i}k_y B_{1y} + \mathrm{i}k_z B_{1z} = 0 \,, \tag{9.11}$$

$$\frac{\partial V_x}{\partial x} + \mathrm{i}k_y V_y + \mathrm{i}k_z V_z = 0 \,. \tag{9.12}$$

Multiply $k_y$ and the $z$ component of (9.10) and multiply $k_z$ and the $y$ component and take the difference. Using the relations (9.11) and (9.12), we find [9.1]

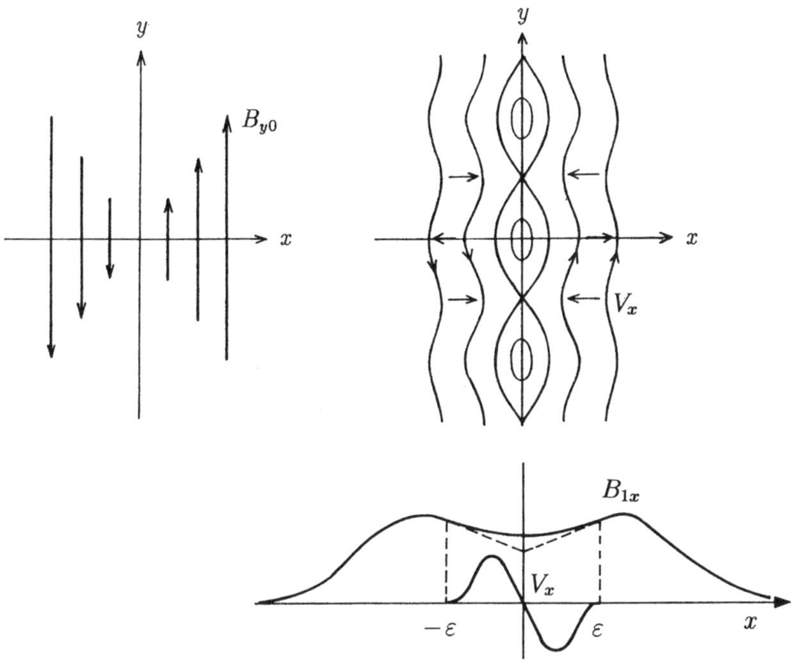


**Fig. 9.1.** Zeroth-order magnetic configuration and magnetic islands due to tearing instability. Profiles of $B_{1x}$ and $V_x$ are also shown

$$\mu_0\rho_m\gamma\left(\frac{\partial^2}{\partial x^2}-k^2\right)V_x = \mathrm{i}(\boldsymbol{k}\cdot\boldsymbol{B}_0)\left(\frac{\partial^2}{\partial x^2}-k^2\right)B_{1x} - \mathrm{i}(\boldsymbol{k}\cdot\boldsymbol{B}_0)''B_{1x} \ , \quad (9.13)$$

where the prime denotes differentiation with respect to $x$. Ohm's law and the equation of motion reduce to (9.9) and (9.13). It should be noted that the zeroth-order magnetic field $\boldsymbol{B}_0$ appears only in the form $(\boldsymbol{k}\cdot\boldsymbol{B}_0)$. When we introduce a function

$$F(x) \equiv (\boldsymbol{k}\cdot\boldsymbol{B}_0) \ , \quad (9.14)$$

the location of $F(x)=0$ is the position where resistive instabilities are likely to occur. We choose this position to be $x=0$ (see Fig. 9.1). $F(x)$ is equal to $(\boldsymbol{k}\cdot\boldsymbol{B}_0) \simeq (\boldsymbol{k}\cdot\boldsymbol{B}_0)'x$ near $x=0$. As is clear from (9.9) and (9.13), $B_{1x}$ is an even function and $V_x$ is an odd function near $x=0$.

The term $|\Delta B_{1x}| \sim |\mu_0 k_y j_{1z}|$ can be large only in the region $|x|<\varepsilon$. Since the growth rate of a resistive instability is much smaller than the MHD growth rate, the left-hand side of the equation of motion (9.13) can be neglected in the region $|x|>\varepsilon$ and we have

$$\frac{\mathrm{d}^2B_{1x}}{\mathrm{d}x^2} - k^2B_{1x} = \frac{F''}{F}B_{1x} \ , \qquad |x|>\varepsilon \ . \quad (9.15)$$

The solution in the region $x>0$ is

$$B_{1x} = \mathrm{e}^{-kx}\left[\int_{-\infty}^{x} \mathrm{e}^{2k\xi}\,\mathrm{d}\xi \int_{\infty}^{\xi}(F''/F)B_{1x}\mathrm{e}^{-k\eta}\,\mathrm{d}\eta + A\right],$$

and the solution in the region $x < 0$ is

$$B_{1x} = \mathrm{e}^{kx}\left[\int_{\infty}^{x} \mathrm{e}^{-2k\xi}\,\mathrm{d}\xi \int_{\infty}^{\xi}(F''/F)B_{1x}\mathrm{e}^{k\eta}\,\mathrm{d}\eta + B\right].$$

Let us define $\Delta'$ as the difference between $B'_{1x}(+\varepsilon)$ at $x = +\varepsilon$ and $B'_{1x}(-\varepsilon)$ at $x = -\varepsilon$, so that

$$\Delta' = \frac{B'_{1x}(+\varepsilon) - B'_{1x}(-\varepsilon)}{B_{1x}(0)}. \tag{9.16}$$

Then the value of $\Delta'$ obtained from the solutions in the region $|x| > \varepsilon$ is given by

$$\Delta' = -2k - \frac{1}{B_{1x}(0)}\left(\int_{-\infty}^{-\varepsilon} + \int_{\varepsilon}^{\infty}\right)\exp(-k|x|)(F''/F)B_{1x}\,\mathrm{d}x. \tag{9.17}$$

For a trial function

$$F(x) = F_{\mathrm{s}}x/L_{\mathrm{s}} \quad (|x| < L_{\mathrm{s}}), \qquad F(x) = F_{\mathrm{s}}x/|x| \quad (x > |L_{\mathrm{s}}|),$$

we can solve (9.15) and $\Delta'$ reduces to

$$\Delta' = \left(\frac{2\alpha}{L_{\mathrm{s}}}\right)\frac{\mathrm{e}^{-2\alpha} + (1-2\alpha)}{\mathrm{e}^{-2\alpha} - (1-2\alpha)} \approx \frac{2}{L_{\mathrm{s}}}\left(\frac{1}{\alpha} - \alpha\right).$$

Here $\alpha \equiv kL_{\mathrm{s}}$ has been used and $L_{\mathrm{s}}$ is the *shear length* defined by $L_{\mathrm{s}} = (F/F')_{x=0}$. For more general cases of $F(x)$, $B_{1x}(x)$ has a logarithmic singularity at $x = 0$, since $F''/F \propto 1/x$ generally. A method for avoiding difficulties arising from the corresponding logarithmic singularity is described in [9.2].

Equations (9.9) and (9.13) in the region $|x| < \varepsilon$ reduce to

$$\frac{\partial^2 B_{1x}}{\partial x^2} - \left(k^2 + \frac{\gamma\mu_0}{\eta}\right)B_{1x} = -\mathrm{i}\frac{\mu_0}{\eta}F'xV_x, \tag{9.18}$$

$$\frac{\partial^2 V_x}{\partial x^2} - \left[k^2 + \frac{(F')^2}{\rho_{\mathrm{m}}\eta\gamma}x^2\right]V_x = \mathrm{i}\left(F'x\frac{1}{\rho_{\mathrm{m}}\eta} - \frac{F''}{\mu_0\rho_{\mathrm{m}}\gamma}\right)B_{1x}. \tag{9.19}$$

The value of $\Delta'$ obtained from the solution in the region $|x| < \varepsilon$ is given from (9.18) as

$$\begin{aligned}\Delta' \times B_{1x}(0) &= \frac{\partial B_{1x}(+\varepsilon)}{\partial x} - \frac{\partial B_{1x}(-\varepsilon)}{\partial x} \\ &= \frac{\mu_0}{\eta}\int_{-\varepsilon}^{\varepsilon}\left[\left(\gamma + \frac{\eta}{\mu_0}k^2\right)B_{1x} - \mathrm{i}F'xV_x\right]\mathrm{d}x. \end{aligned} \tag{9.20}$$

The value for $\Delta'$ in (9.20) must be equal to the value for $\Delta'$ in (9.17). This requirement gives the eigenvalue $\gamma$, and the growth rate of this resistive instability can be obtained [9.1]. However, we shall try to deduce the growth rate in a qualitative manner in this section. In the region $|x| < \varepsilon$, it is possible to write

$$\frac{\partial^2 B_{1x}}{\partial x^2} \sim \frac{\Delta' B_{1x}}{\varepsilon} .$$

It is assumed that the three terms in (9.9), namely the induced electric field term (left-hand side), the $\boldsymbol{V} \times \boldsymbol{B}$ term (first term on the right-hand side) and Ohm's term (second term) are of the same order, i.e.,

$$\gamma B_{1x} \sim \frac{\eta}{\mu_0} \frac{\Delta' B_{1x}}{\varepsilon} , \tag{9.21}$$

$$\gamma B_{1x} \sim \mathrm{i} F' \varepsilon V_x . \tag{9.22}$$

Then (9.21) yields

$$\gamma \sim \frac{\eta}{\mu_0} \frac{\Delta'}{\varepsilon} . \tag{9.23}$$

Accordingly,

$$\Delta' > 0 \tag{9.24}$$

is the instability condition. In order to get the value of $\gamma$, $\varepsilon$ must be evaluated. Equation (9.13) reduces to

$$\mu_0 \rho_{\mathrm{m}} \gamma \left( \frac{-V_x}{\varepsilon^2} \right) \sim \mathrm{i} F' \varepsilon \frac{\Delta' B_{1x}}{\varepsilon} . \tag{9.25}$$

If the terms $V_x$, $B_{1x}$, and $\gamma$ are eliminated by (9.21), (9.22) and (9.25), we find

$$\varepsilon^5 \sim \left( \frac{\eta}{\mu_0 a^2} \right)^2 (\Delta' a) \frac{\rho_{\mathrm{m}} \mu_0}{(F' a)^2} a^5 ,$$

$$\frac{\varepsilon}{a} \sim \left[ \left( \frac{\tau_{\mathrm{A}}}{\tau_{\mathrm{R}}} \right)^2 (\Delta' a) \left( \frac{B_0}{F' a^2} \right)^2 \right]^{1/5} \sim S^{-2/5} (\Delta' a)^{1/5} \left[ \frac{B_0}{(\boldsymbol{k} \cdot \boldsymbol{B}_0)' a^2} \right]^{2/5} , \tag{9.26}$$

where the physical quantities

$$\tau_{\mathrm{R}} = \frac{\mu_0 a^2}{\eta} , \qquad \tau_{\mathrm{A}} = \frac{a}{B_0 / (\mu_0 \rho_{\mathrm{m}})^{1/2}} ,$$

are the resistive diffusion time and Alfvén transit time, respectively. The dimensionless factor

$$S = \tau_{\mathrm{R}} / \tau_{\mathrm{A}}$$

is the *magnetic Reynolds number* and $a$ is a typical plasma size. Accordingly, the growth rate $\gamma$ is given by

$$\gamma = \frac{\eta}{\mu_0 a^2}\frac{a}{\varepsilon}(\Delta' a) = \frac{(\Delta' a)^{4/5}}{\tau_{\rm R}^{3/5}\tau_{\rm A}^{2/5}}\left[\frac{(\boldsymbol{k}\cdot\boldsymbol{B}_0)' a^2}{B_0}\right]^{2/5}$$

$$= \frac{(\Delta' a)^{4/5}}{S^{3/5}}\left[\frac{(\boldsymbol{k}\cdot\boldsymbol{B}_0)' a^2}{B_0}\right]^{2/5}\frac{1}{\tau_{\rm A}}\ . \tag{9.27}$$

Since this mode very likely breaks up the plasma into a set of magnetic islands, as shown in Fig. 9.1, this mode is called the *tearing instability* [9.1].

The foregoing discussion has been based on the slab model. Let us consider this mode in a toroidal plasma. The poloidal and the toroidal components of the propagation vector $\boldsymbol{k}$ are $m/r$ and $-n/R$, respectively. Accordingly, we have the correspondences $k_y \leftrightarrow m/r$ and $k_z \leftrightarrow -n/R$, and

$$(\boldsymbol{k}\cdot\boldsymbol{B}_0) = \frac{m}{r}B_\theta - \frac{n}{R}B_{\rm z} = \frac{n}{r}B_\theta\left(\frac{m}{n} - q\right)\ , \qquad q \equiv \frac{r}{R}\frac{B_z}{B_\theta}\ .$$

Therefore weak positions for tearing instability are given by $(\boldsymbol{k}\cdot\boldsymbol{B}_0) = 0$ and these are rational surfaces satisfying $q(r_{\rm s}) = m/n$. The shear is given by

$$(\boldsymbol{k}\cdot\boldsymbol{B}_0)' = \frac{-n}{r}B_\theta\frac{{\rm d}q}{{\rm d}r}\ , \qquad \frac{(\boldsymbol{k}\cdot\boldsymbol{B}_0)' r_{\rm s}^2}{B_0} = -n\left(\frac{r_{\rm s}}{R}\right)\frac{q' r_{\rm s}}{q}\ .$$

The tearing mode is closely related to the internal disruption in a tokamak and plays an important role that we shall describe in Sect. 16.3.

It has been assumed that the specific resistivity $\eta$ and the mass density $\rho_{\rm m}$ are uniform and there is no gravitation (acceleration) $\boldsymbol{g} = 0$. If $\eta$ depends on $x$, the resistive term in (9.5) becomes $\nabla \times (\eta\nabla \times \boldsymbol{B})/\mu_0$. When there is a temperature gradient ($\eta' \neq 0$), the rippling mode with short wavelength ($kL_{\rm s} \gg 1$) may appear on the lower-resistivity side (high-temperature side) of the $x = 0$ position. When there is gravitation, the term $\rho\boldsymbol{g}$ is added to the equation of motion (9.8). If the direction of $\boldsymbol{g}$ is opposite to $\nabla\rho_{\rm m}$ ($\boldsymbol{g}$ is toward the low-density side), the gravitational interchange mode may appear [9.1].

## 9.2 Resistive Drift Instability

A finite density and temperature gradient always exists at a plasma boundary. Configurations including a gradient may be unstable under certain conditions. Let us consider a slab model. The direction of the uniform magnetic field is taken in the $z$ direction and $\boldsymbol{B}_0 = (0, 0, B_0)$. The $x$ axis is taken in the direction of the density gradient with the positive direction outward from the plasma. The pressure is $p_0(x)$ (see Fig. 9.2). The zeroth-order plasma current is $\boldsymbol{j}_0 = (0, p_0'/B_0, 0)$ and we assume that the flow velocity and the electric field are zero $\boldsymbol{V}_0 = 0$, $\boldsymbol{E}_0 = 0$ to zeroth order. The flow velocity due to classical diffusion is neglected here. Electron inertia is also neglected.

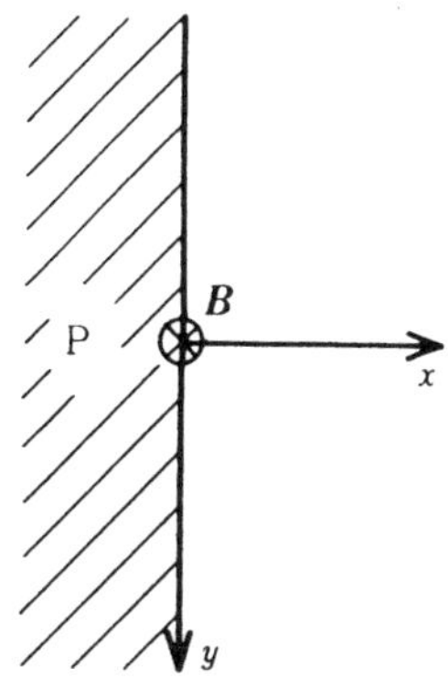


**Fig. 9.2.** Slab model of resistive drift wave

The usual relations in this configuration are

$$Mn\frac{\partial \boldsymbol{V}}{\partial t} = \boldsymbol{j} \times \boldsymbol{B} - \nabla p \,, \tag{9.28}$$

$$\boldsymbol{E} + \boldsymbol{V} \times \boldsymbol{B} = \eta \boldsymbol{j} + \frac{1}{en}(\boldsymbol{j} \times \boldsymbol{B} - \nabla p_{\rm e}) \,, \tag{9.29}$$

$$\frac{\partial n}{\partial t} + \nabla\cdot(n\boldsymbol{V}) = 0 \,, \tag{9.30}$$

$$\nabla\cdot\boldsymbol{j} = 0 \,, \tag{9.31}$$

where $M$ is the ion mass. In this configuration, electrostatic perturbations are considered here. The first-order electric field $\boldsymbol{E}_1$ is expressed by the electrostatic potential $\boldsymbol{E}_1 = -\nabla\phi_1$ and the first-order magnetic field perturbation is zero $\boldsymbol{B}_1 = 0$ $(\partial B/\partial t = \nabla \times \boldsymbol{E})$. The characteristics of the electrostatic perturbation will be explained in detail in Chap. 10. For simplicity, the ion temperature is assumed to be zero $T_{\rm i} = 0$. Let us consider the mode

$$n_1 = n_1(x) \exp {\rm i}(ky + k_{\parallel} z - \omega t) \,, \qquad \phi_1 = \phi_1(x) \exp {\rm i}(ky + k_{\parallel} z - \omega t) \,.$$

Equations (9.28) and (9.29) reduce to

$$-{\rm i}\omega M n_0 \boldsymbol{V}_1 = \boldsymbol{j}_1 \times \boldsymbol{B}_0 - \kappa T_{\rm e} \nabla n_1 \,, \tag{9.32}$$

$$\boldsymbol{j}_1 \times \boldsymbol{B}_0 - \kappa T_{\rm e} \nabla n_1 = e n_0(-\nabla\phi_1 + \boldsymbol{V}_1 \times \boldsymbol{B}_0 - \eta \boldsymbol{j}_1) \,. \tag{9.33}$$

Equations (7.32) and (7.33) yield

$${\rm i}\omega\left(\frac{M}{e}\right)\boldsymbol{V}_1 = \nabla\phi_1 - \boldsymbol{V}_1 \times \boldsymbol{B}_0 + \eta \boldsymbol{j}_1 \,. \tag{9.34}$$

When $\eta$ is small $(\nu_{\rm ei} \ll \Omega_{\rm e})$, the contribution of $\eta\boldsymbol{j}$ can be neglected in (9.34), i.e., we may write

$$V_x = -ik\frac{\phi_1}{B_0}, \qquad V_y = \left(\frac{\omega}{\Omega_\mathrm{i}}\right)\frac{k\phi_1}{B_0}, \qquad V_z = \left(-\frac{\Omega_\mathrm{i}}{\omega}\right)\frac{k_\parallel \phi_1}{B_0},$$

where $\Omega_\mathrm{i}$ is as usual the ion cyclotron frequency ($\Omega_\mathrm{i} = -eB/M$). The wave frequency $\omega$ is assumed to be low $(\omega/\Omega_\mathrm{i})^2 \ll 1$. The $x, y$ components of (9.32) and the $z$ component of (9.33) are

$$j_x = -ik\frac{\kappa T_\mathrm{e} n_1}{B_0}, \qquad j_y = kn_0\left(\frac{\omega}{\Omega_\mathrm{i}}\right)\frac{e\phi_1}{B_0}, \qquad j_z = \frac{ik_\parallel}{e\eta}\left(\kappa T_\mathrm{e}\frac{n_1}{n_0} - e\phi_1\right).$$

Since (9.31) is $j_x' + \mathrm{i}kj_y + \mathrm{i}k_\parallel j_z = 0$, and (9.30) is

$$-\mathrm{i}\omega n_1 + n_0' V_x + n_0 \mathrm{i}kV_y + n_0 \mathrm{i}k_\parallel V_z = 0,$$

we find

$$\frac{n_1}{n_0} - \left(1 + i\left(\frac{k}{k_\parallel}\right)^2 \frac{\omega}{\Omega_\mathrm{i}}\frac{en_0\eta}{B_0}\right)\frac{e\phi_1}{\kappa T_\mathrm{e}} = 0, \tag{9.35}$$

$$\frac{n_1}{n_0} - \left(\left(\frac{k_\parallel}{\omega}\right)^2\frac{\kappa T_\mathrm{e}}{M} + k^2\frac{\kappa T_\mathrm{e}}{eB_0\Omega_\mathrm{i}} + \frac{k(-n_0'/n_0)\kappa T_\mathrm{e}}{eB_0}\frac{1}{\omega}\right)\frac{e\phi_1}{\kappa T_\mathrm{e}} = 0. \tag{9.36}$$

The dispersion equation is given by the determinant of the coefficients of (9.2) and (9.2):

$$1 + i\left(\frac{k}{k_\parallel}\right)^2\frac{\omega}{\Omega_\mathrm{i}}\frac{\nu_\mathrm{ei}}{\Omega_\mathrm{e}} - \left(\frac{k_\parallel}{\omega}\right)^2 c_\mathrm{s}^2 + (k\rho_\Omega)^2 - \frac{\omega_\mathrm{e}^*}{\omega} = 0 \tag{9.37}$$

where $\eta = m_\mathrm{e}\nu_\mathrm{ei}/ne^2$, $(n_0 e\eta)/B_0 = \nu_\mathrm{ei}/\Omega_\mathrm{e}$. $c_\mathrm{s}^2 \equiv \kappa T_\mathrm{e}/M$, $\rho_\Omega \equiv c_\mathrm{s}/|\Omega_\mathrm{i}|$ and $\omega_\mathrm{e}^* \equiv k(-n_0'/n_0)(\kappa T_\mathrm{e}/eB_0)$. The *drift velocities* $\boldsymbol{v}_\mathrm{di}$ and $\boldsymbol{v}_\mathrm{de}$ of ions and electrons due to the density gradient $\boldsymbol{\nabla} n_0$ are given by

$$\boldsymbol{v}_\mathrm{di} = \frac{-(\kappa T_\mathrm{i}\boldsymbol{\nabla} n_0/n_0)\times\boldsymbol{b}}{eB_0} = \frac{-\kappa T_\mathrm{i}}{eB_0}\left(\frac{-n_0'}{n_0}\right)\boldsymbol{e}_y,$$

$$\boldsymbol{v}_\mathrm{de} = \frac{(\kappa T_\mathrm{e}\boldsymbol{\nabla} n_0/n_0)\times\boldsymbol{b}}{eB_0} = \frac{\kappa T_\mathrm{e}}{eB_0}\left(\frac{-n_0'}{n_0}\right)\boldsymbol{e}_y.$$

The *drift frequencies* of ions and electrons are defined by $\omega_\mathrm{i}^* \equiv kv_\mathrm{di}$ and $\omega_\mathrm{e}^* \equiv kv_\mathrm{de}$, respectively. As $n_0'/n_0 < 0$, $\omega_\mathrm{e}^* > 0$ and $\omega_\mathrm{i}^* = -(T_\mathrm{i}/T_\mathrm{e})\omega_\mathrm{e}^* < 0$. The dispersion equation reduces to

$$\left(\frac{\omega}{\omega_\mathrm{e}^*}\right)^2 - i\left(1 + (k\rho_\Omega)^2 - \frac{k_\parallel^2 c_\mathrm{s}^2}{\omega^2}\right)\frac{\Omega_\mathrm{e}\Omega_\mathrm{i}}{\nu_\mathrm{ei}\omega_\mathrm{e}^*}\left(\frac{k_\parallel}{k}\right)^2\left(\frac{\omega}{\omega_\mathrm{e}^*}\right) + i\frac{\Omega_\mathrm{e}\Omega_\mathrm{i}}{\nu_\mathrm{ei}\omega_\mathrm{e}^*}\left(\frac{k_\parallel}{k}\right)^2 = 0. \tag{9.38}$$

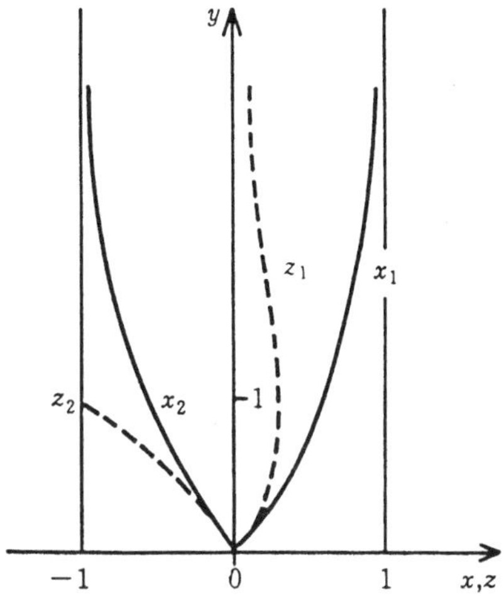


**Fig. 9.3.** Dependence of $\omega/\omega_e^* = x + iz$ on $y \propto k_\parallel/k$ for resistive drift instability

$\rho_\Omega$ is the ion Larmor radius when the ions are assumed to have the electron temperature $T_e$. Setting $\omega/\omega_e^* = x + iz$ and $-(\Omega_e\Omega_i/\nu_{ei}\omega_e^*)(k_\parallel/k)^2 = y^2$, and assuming $(k\rho_\Omega)^2 - (k_\parallel c_s/\omega)^2 \ll 1$, the dispersion equation is then

$$(x + iz)^2 + iy^2(x + iz) - iy^2 = 0 \,. \tag{9.39}$$

The dependence of the two solutions $x_1(y)$, $z_1(y)$ and $x_2(y)$, $z_2(y)$ on $y \propto (k_\parallel/k)$ is shown in Fig. 9.3. As $z_2(y) < 0$, the mode corresponding to $x_2(y), z_2(y)$ is stable. This wave propagates in the direction of the ion drift. The solution $x_1, z_1 > 0$ propagates in the direction of the electron drift and it is unstable. If the value of $(k_\parallel/k)$ is adjusted to be $y \simeq 1.3$, the $z_1$ value becomes maximum at $z_1 \approx 0.25$ and the growth rate is $\mathrm{Im}\,\omega \approx 0.25\omega_e^*$. If $\eta$ is small, the wavelength of the most unstable wave becomes long and the number of collisions required to interrupt the electron motion along the magnetic line of force is maintained. If the lower limit of $k_\parallel$ is fixed by an appropriate method, the growth rate is

$$\mathrm{Im}(\omega/\omega_e^*) \approx y^{-2} = \frac{\nu_{ei}\omega_e^*}{\Omega_e|\Omega_i|}\left(\frac{k}{k_\parallel}\right)^2 ,$$

which is proportional to $\eta \propto \nu_{ei}$. This instability is called *resistive drift instability* or *dissipative drift instability* [9.3, 9.4].

If the ion's inertia term can be neglected ($M \to 0$, $|\Omega_i| \to \infty$ in eq.(9.37)), the dispersion equation becomes $\omega^2 - \omega\omega_e^* - (k_\parallel c_s)^2 = 0$. The instability is therefore also called *collisional drift instability*.

The instability does not appear in the collisionless case in the framework of MHD theory. However, the instability may occur even in the collisionless case when it is analyzed by the kinetic theory. This instability is called *collisionless drift instability* (see Sects. 12.7 and 12.8).

# 10 Plasma as Medium of Waves

A plasma is an ensemble of an enormous number of moving ions and electrons interacting with each other. In order to describe the behavior of such an ensemble, the distribution function was introduced in Chap. 4, and Boltzmann's and Vlasov's equations were derived with respect to the distribution function. A plasma viewed as an ensemble of a large number of particles has a large number of degrees of freedom; thus the mathematical description of plasma behavior is feasible only for simplified analytical models.

In Chap. 5, statistical averages in velocity space, such as mass density, flow velocity, pressure, etc., were introduced and the magnetohydrodynamic equations for these averages were derived. We have thus obtained a mathematical description of the magnetohydrodynamic fluid model, and we have studied the equilibrium conditions, stability problems, etc., for this model in Chaps. 6–9. Since the fluid model considers only average quantities in velocity space, it is not capable of describing instabilities or damping phenomena in which the profile of the distribution function plays a significant role. The phenomena which can be handled by means of the fluid model are of low frequency (less than the ion or electron cyclotron frequency); high-frequency phenomena are not describable via this model.

In this chapter, we will focus on a model which allows us to study wave phenomena while retaining the essential features of plasma dynamics, and at the same time maintaining relative simplicity in its mathematical form. Such a model is given by a homogeneous plasma of ions and electrons at 0 K in a uniform magnetic field. In the unperturbed state, both the ions and electrons of this plasma are motionless. Any small deviation from the unperturbed state induces an electric field and a time-dependent component of the magnetic field, and movements of ions and electrons are thereby excited. The movements of the charged particles induce electric and magnetic fields which are themselves consistent with the previously induced small perturbations. This is called the kinetic model of a *cold plasma*. We will use it in this chapter to derive the dispersion relation which characterizes wave phenomena in a cold plasma.

Although this model assumes uniformity of the magnetic field and density, and also zero temperature, the cold plasma model is applicable for a non-uniform, warm plasma, if the typical length of variation of the magnetic field and the density is much larger than the wavelength, and if the phase velocity of the wave is much larger than the thermal velocity of the particles.

One may treat the plasma as a medium of electromagnetic wave propagation with a dielectric tensor $\boldsymbol{K}$. This dielectric tensor $\boldsymbol{K}$ is a function of the magnetic field and the density which may change with the position. Accordingly, plasmas are in general non-uniform, anisotropic and dispersive media.

When the temperature of plasma is finite and the thermal velocity of the particles is comparable to the phase velocity of the propagating wave, the interaction of the particles and the wave becomes important. A typical interaction is Landau damping, which is explained in Chap. 11. The general mathematical analysis of the hot-plasma wave will be discussed in Chap. 12. The plasma wave is described in more detail in [10.1–10.4].

## 10.1 Dispersion Equation of Waves in a Cold Plasma

In an unperturbed cold plasma, the particle density $n$ and the magnetic field $\boldsymbol{B}_0$ are both homogeneous in space and constant in time. The ions and electrons are motionless.

Now assume that the first-order perturbation term $\exp \mathrm{i}(\boldsymbol{k}\cdot\boldsymbol{r} - \omega t)$ is applied. The ions and electrons are forced to move by the perturbed electric field $\boldsymbol{E}$ and the induced magnetic field $\boldsymbol{B}_1$. Let us denote the velocity by $\boldsymbol{v}_k$, where the suffix $k$ indicates the species of particle (electrons, or ions of various kinds). The current $\boldsymbol{j}$ due to the particle motion is given by

$$\boldsymbol{j} = \sum_k n_k q_k \boldsymbol{v}_k \,. \tag{10.1}$$

$n_k$ and $q_k$ are the density and charge of the $k$th species, respectively. The electric displacement $\boldsymbol{D}$ is

$$\boldsymbol{D} = \epsilon_0 \boldsymbol{E} + \boldsymbol{P} \,, \tag{10.2}$$

$$\boldsymbol{j} = \frac{\partial \boldsymbol{P}}{\partial t} = -\mathrm{i}\omega \boldsymbol{P} \,, \tag{10.3}$$

where $\boldsymbol{E}$ is the electric intensity, $\boldsymbol{P}$ is the electric polarization, and $\epsilon_0$ is the dielectric constant of the vacuum. Consequently, $\boldsymbol{D}$ is expressed by

$$\boldsymbol{D} = \epsilon_0 \boldsymbol{E} + \frac{\mathrm{i}}{\omega}\boldsymbol{j} \equiv \epsilon_0 \boldsymbol{K}\cdot\boldsymbol{E} \,. \tag{10.4}$$

$\boldsymbol{K}$ is called the *dielectric tensor*. The equation of motion of a single particle of the $k$th kind is

$$m_k \frac{\mathrm{d}\boldsymbol{v}_k}{\mathrm{d}t} = q_k(\boldsymbol{E} + \boldsymbol{v}_k \times \boldsymbol{B}) \,. \tag{10.5}$$

Here $\boldsymbol{B}$ consists of $\boldsymbol{B} = \boldsymbol{B}_0 + \boldsymbol{B}_1$, where $\boldsymbol{v}_k$, $\boldsymbol{E}$, $\boldsymbol{B}_1$ are the first-order quantities. The linearized equation in these quantities is

$$-\mathrm{i}\omega m_k \boldsymbol{v}_k = q_k(\boldsymbol{E} + \boldsymbol{v}_k \times \boldsymbol{B}_0) \; . \tag{10.6}$$

When the $z$ axis is taken along the direction of $\boldsymbol{B}_0$, the solution is given by

$$\begin{cases} v_{k,x} = \dfrac{-\mathrm{i}E_x}{B_0} \dfrac{\Omega_k \omega}{\omega^2 - \Omega_k^2} - \dfrac{E_y}{B_0} \dfrac{\Omega_k^2}{\omega^2 - \Omega_k^2} \; , \\ v_{k,y} = \dfrac{E_x}{B_0} \dfrac{\Omega_k^2}{\omega^2 - \Omega_k^2} - \dfrac{\mathrm{i}E_y}{B_0} \dfrac{\Omega_k \omega}{\omega^2 - \Omega_k^2} \; , \\ v_{k,z} = \dfrac{-\mathrm{i}E_z}{B_0} \dfrac{\Omega_k}{\omega} \; , \end{cases} \tag{10.7}$$

where $\Omega_k$ is the cyclotron frequency of the charged particle of the $k$ th kind:

$$\Omega_k = -\frac{q_k B_0}{m_k} \; , \tag{10.8}$$

with $\Omega_{\mathrm{e}} > 0$ for electrons and $\Omega_{\mathrm{i}} < 0$ for ions. The components of $\boldsymbol{v}_k$ are the linear functions of $\boldsymbol{E}$ given by (10.7). Moreover, $\boldsymbol{j}$ of (10.1) and the electric displacement $\boldsymbol{D}$ of (10.4) are also linear functions of $\boldsymbol{E}$, so that the *dielectric tensor* is given by

$$\boldsymbol{K} \cdot \boldsymbol{E} = \begin{pmatrix} K_\perp & -\mathrm{i}K_\times & 0 \\ \mathrm{i}K_\times & K_\perp & 0 \\ 0 & 0 & K_\parallel \end{pmatrix} \begin{pmatrix} E_x \\ E_y \\ E_z \end{pmatrix} \; , \tag{10.9}$$

where

$$K_\perp \equiv 1 - \sum_k \frac{\Pi_k^2}{\omega^2 - \Omega_k^2} \; , \tag{10.10}$$

$$K_\times \equiv -\sum_k \frac{\Pi_k^2}{\omega^2 - \Omega_k^2} \frac{\Omega_k}{\omega} \; , \tag{10.11}$$

$$K_\parallel \equiv 1 - \sum_k \frac{\Pi_k^2}{\omega^2} \; , \tag{10.12}$$

$$\Pi_k^2 \equiv \frac{n_k q_k^2}{\epsilon_0 m_k} \; . \tag{10.13}$$

According to the Stix notation, the following quantities are introduced:

$$\begin{cases} R \equiv 1 - \displaystyle\sum_k \frac{\Pi_k^2}{\omega^2} \frac{\omega}{\omega - \Omega_k} = K_\perp + K_\times \; , \\ L \equiv 1 - \displaystyle\sum_k \frac{\Pi_k^2}{\omega^2} \frac{\omega}{\omega + \Omega_k} = K_\perp - K_\times \; . \end{cases} \tag{10.14}$$

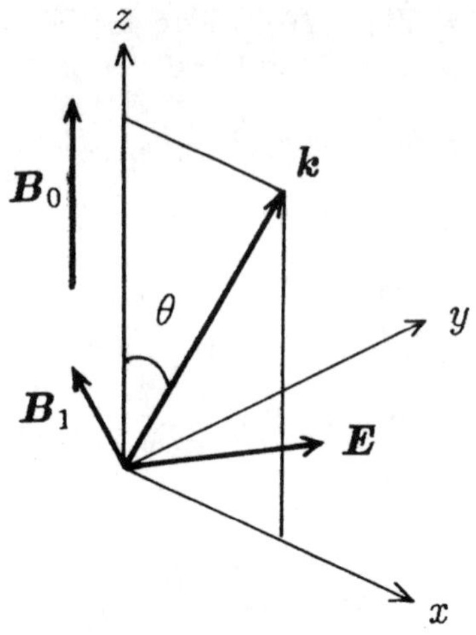


**Fig. 10.1.** Propagation vector $\boldsymbol{k}$ and $x$, $y$, $z$ coordinates

From Maxwell's equations

$$\nabla \times \boldsymbol{E} = -\frac{\partial \boldsymbol{B}}{\partial t} , \tag{10.15}$$

$$\nabla \times \boldsymbol{H} = \boldsymbol{j} + \epsilon_0 \frac{\partial \boldsymbol{E}}{\partial t} = \frac{\partial \boldsymbol{D}}{\partial t} , \tag{10.16}$$

it follows that

$$\boldsymbol{k} \times \boldsymbol{E} = \omega \boldsymbol{B}_1 , \qquad \boldsymbol{k} \times \boldsymbol{H}_1 = -\omega \epsilon_0 \boldsymbol{K} \cdot \boldsymbol{E} ,$$

and

$$\boldsymbol{k} \times (\boldsymbol{k} \times \boldsymbol{E}) + \frac{\omega^2}{c^2} \boldsymbol{K} \cdot \boldsymbol{E} = 0 . \tag{10.17}$$

Let us define a dimensionless vector

$$\boldsymbol{N} \equiv \frac{\boldsymbol{k} c}{\omega} ,$$

where $c$ is light velocity in vacuum. The absolute value $N = |\boldsymbol{N}|$ is the ratio of the light velocity to the phase velocity of the wave, i.e., $N$ is the refractive index. Using $\boldsymbol{N}$, we may write (10.17) as

$$\boldsymbol{N} \times (\boldsymbol{N} \times \boldsymbol{E}) + \boldsymbol{K} \cdot \boldsymbol{E} = 0 . \tag{10.18}$$

If the angle between $\boldsymbol{N}$ and $\boldsymbol{B}_0$ is denoted by $\theta$ (Fig. 10.1) and the $x$ axis is chosen so that $\boldsymbol{N}$ lies in the $z, x$ plane, then (10.18) may be expressed as

$$\begin{pmatrix} K_\perp - N^2\cos^2\theta & -\mathrm{i}K_\times & N^2\sin\theta\cos\theta \\ \mathrm{i}K_\times & K_\perp - N^2 & 0 \\ N^2\sin\theta\cos\theta & 0 & K_\parallel - N^2\sin^2\theta \end{pmatrix} \begin{pmatrix} E_x \\ E_y \\ E_z \end{pmatrix} = 0 \,. \tag{10.19}$$

For a nontrivial solution to exist, the determinant of the matrix must be zero, that is

$$AN^4 - BN^2 + C = 0 \,, \tag{10.20}$$

$$A = K_\perp \sin^2\theta + K_\parallel \cos^2\theta \,, \tag{10.21}$$

$$B = (K_\perp^2 - K_\times^2)\sin^2\theta + K_\parallel K_\perp (1 + \cos^2\theta) \,, \tag{10.22}$$

$$C = K_\parallel (K_\perp^2 - K_\times^2) = K_\parallel RL \,. \tag{10.23}$$

Equation (10.20) determines the relationship between the propagation vector $\boldsymbol{k}$ and the frequency $\omega$, and it is called the *dispersion equation*. The solution of (10.20) is

$$\begin{aligned} N^2 &= \frac{B \pm (B^2 - 4AC)^{1/2}}{2A} \\ &= \Bigg\{ (K_\perp^2 - K_\times^2)\sin^2\theta + K_\parallel K_\perp (1+\cos^2\theta) \\ &\quad \pm \Big[ (K_\perp^2 - K_\times^2 - K_\parallel K_\perp)^2 \sin^4\theta + 4K_\parallel^2 K_\times^2 \cos^2\theta \Big]^{1/2} \Bigg\} \\ &\quad \times \Big[ 2(K_\perp \sin^2\theta + K_\parallel \cos^2\theta) \Big]^{-1} \,. \end{aligned} \tag{10.24}$$

When the wave propagates along the line of magnetic force ($\theta = 0$), the dispersion equation (10.20) is

$$K_\parallel \left[ N^4 - 2K_\perp N^2 + (K_\perp^2 - K_\times^2) \right] = 0 \,, \tag{10.25}$$

and the solutions are

$$K_\parallel = 0 \,, \qquad N^2 = K_\perp + K_\times = R \,, \qquad N^2 = K_\perp - K_\times = L \,. \tag{10.26}$$

For the wave propagating in the direction perpendicular to the magnetic field ($\theta = \pi/2$), the dispersion equation and the solutions are given by

$$K_\perp N^4 - (K_\perp^2 - K_\times^2 + K_\parallel K_\perp) N^2 + K_\parallel (K_\perp^2 - K_\times^2) = 0 \,, \tag{10.27}$$

$$N^2 = \frac{K_\perp^2 - K_\times^2}{K_\perp} = \frac{RL}{K_\perp} \,, \qquad N^2 = K_\parallel \,. \tag{10.28}$$

## 10.2 Properties of Waves

### 10.2.1 Polarization and Particle Motion

The dispersion relation for waves in a cold plasma was derived in the previous section. We consider here the electric field of the waves and the resulting particle motion. The $y$ component of (10.19) is

$$\mathrm{i}K_\times E_x + (K_\perp - N^2)E_y = 0\,, \qquad \frac{\mathrm{i}E_x}{E_y} = \frac{N^2 - K_\perp}{K_\times}\,. \tag{10.29}$$

The relation between the components of the particle velocity is

$$\begin{aligned}\frac{\mathrm{i}v_{k,x}}{v_{k,y}} &= \frac{\mathrm{i}\left(\dfrac{-\mathrm{i}E_x}{E_y}\dfrac{\omega}{\omega^2 - \Omega_k^2} - \dfrac{\Omega_k}{\omega^2 - \Omega_k^2}\right)}{\dfrac{E_x}{E_y}\dfrac{\Omega_k}{\omega^2 - \Omega_k^2} - \mathrm{i}\dfrac{\omega}{\omega^2 - \Omega_k^2}} \\ &= \frac{(\omega + \Omega_k)(N^2 - L) + (\omega - \Omega_k)(N^2 - R)}{(\omega + \Omega_k)(N^2 - L) - (\omega - \Omega_k)(N^2 - R)}\,. \end{aligned} \tag{10.30}$$

The wave satisfying $N^2 = R$ at $\theta = 0$ has $\mathrm{i}E_x/E_y = 1$ and the electric field is right-circularly polarized. In other words, the electric field rotates in the direction of the electron Larmor motion. The motion of ions and electrons is also right-circular motion. In the wave satisfying $N^2 = L$ as $\theta \to 0$, the relation $\mathrm{i}E_x/E_y = -1$ holds and the electric field is left-circularly polarized. The motion of ions and electrons is also left-circular motion. The waves with $N^2 = R$ and $N^2 = L$ as $\theta \to 0$ are called the *R wave* and the *L wave*, respectively. The solution of the dispersion equation (10.25) at $\theta = 0$ is

$$N^2 = \frac{1}{2}\left(R + L \pm \frac{|K_\parallel|}{K_\parallel}|R - L|\right)\,, \tag{10.31}$$

so that R and L waves are exchanged when $K_\parallel$ changes sign. When $K_\times = R - L$ changes sign, R and L waves are also exchanged.

When $\theta = \pi/2$, the electric field of the wave satisfying $N^2 = K_\parallel$ is $E_x = E_y = 0$, $E_z \neq 0$. For the wave safisfying $N^2 = RL/K_\perp$, the electric field satisfies the relations

$$\mathrm{i}\frac{E_x}{E_y} = -\frac{R - L}{R + L} = -\frac{K_\times}{K_\perp}\,, \qquad E_z = 0\,.$$

The waves with $N^2 = K_\parallel$ and $N^2 = RL/K_\perp$ as $\theta \to \pi/2$ are called the ordinary wave (O) and the extraordinary wave (X), respectively. It should be pointed out that the electric field of the extraordinary wave at $\theta = \pi/2$ is perpendicular to the magnetic field ($E_z = 0$) and the electric field of the ordinary wave at $\theta = \pi/2$ is parallel to the magnetic field ($E_x = E_y = 0$). The dispersion relation (10.24) at $\theta = \pi/2$ is

$$N^2 = \frac{1}{2K_\perp}(K_\perp^2 - K_\times^2 + K_\parallel K_\perp + |K_\perp^2 - K_\times^2 - K_\parallel K_\perp|)$$
$$= \frac{1}{2K_\perp}(RL + K_\parallel K_\perp \pm |RL - K_\parallel K_\perp|) , \qquad (10.32)$$

so that the ordinary wave and the extraordinary wave are exchanged when $RL - K_\parallel K_\perp = 0$.

Besides the classification into R and L waves, and O and X waves, there is another classification, namely, that of *fast wave* and *slow wave*, following the difference in the phase velocity. Since the term inside the square root of the equation $N^2 = \left[B \pm (B^2 - 4AC)^{1/2}\right]/2A$ is always positive, as is clear from (10.24), the fast wave and slow wave do not exchange between $\theta = 0$ and $\theta = \pi/2$.

### 10.2.2 Cutoff and Resonance

The refractive index (10.24) may become infinity or zero. When $N^2 = 0$, the wave is said to be a *cutoff wave*. At cutoff, the phase velocity

$$v_{\mathrm{ph}} = \frac{\omega}{k} = \frac{c}{N} \qquad (10.33)$$

becomes infinite. It is clear from (10.20) and (10.23) that cutoff occurs when

$$K_\parallel = 0 , \qquad R = 0 , \qquad L = 0 . \qquad (10.34)$$

When $N^2 = \infty$, the wave is said to be at *resonance*. Here the phase velocity becomes zero. The wave will be absorbed by the plasma at resonance (see Chap. 11). The resonance condition is

$$\tan^2\theta = -\frac{K_\parallel}{K_\perp} . \qquad (10.35)$$

When $\theta = 0$, the resonance condition is $K_\perp = (R + L)/2 \to \pm\infty$. The condition $R \to \pm\infty$ is satisfied at $\omega = \Omega_{\mathrm{e}}$, where $\Omega_{\mathrm{e}}$ is the electron cyclotron frequency. This is called *electron cyclotron resonance*. The condition $L \to \pm\infty$ holds when $\omega = |\Omega_{\mathrm{i}}|$, and this is called *ion cyclotron resonance*.

When $\theta = \pi/2$, $K_\perp = 0$ is the resonance condition. This is called *hybrid resonance*. When waves approach a cutoff region, the wave path is curved according to Snell's refraction law and the waves are reflected (Fig. 10.2a). When waves approach a resonance region, they propagate perpendicularly toward the resonance region. The phase velocities tend to zero and the wave energy will be absorbed (Fig. 10.2b).

## 10.3 Waves in a Two-Component Plasma

Let us consider a plasma which consists of electrons and one kind of ion. Charge neutrality is expressed by

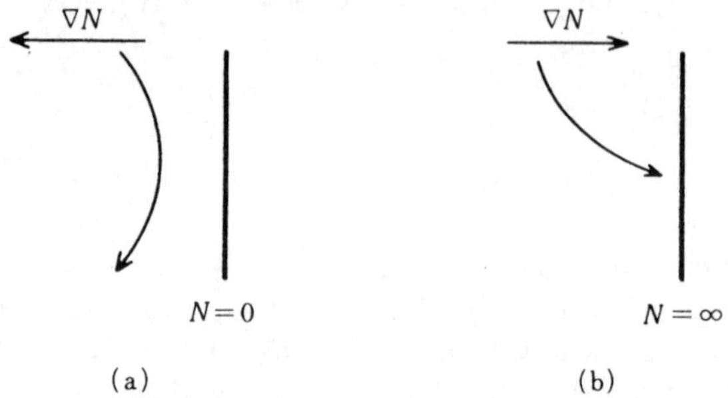


**Fig. 10.2.** Wave propagation (**a**) near cutoff region and (**b**) near a resonance region

$$n_i Z_i = n_e \,. \tag{10.36}$$

A dimensionless parameter is introduced for convenience:

$$\delta = \frac{\mu_0 (n_i m_i + n_e m_e) c^2}{B_0^2} \,. \tag{10.37}$$

The quantity defined by (10.13), which was also introduced in Sect. 2.2,

$$\Pi_e^2 = n_e e^2 / (\epsilon_0 m_e) \tag{10.38}$$

is called the electron plasma frequency. We then have the relations

$$\frac{\Pi_e^2}{\Pi_i^2} = \frac{m_i}{m_e} \gg 1 \,, \qquad \frac{\Pi_i^2 + \Pi_e^2}{|\Omega_i|\Omega_e} = \delta \approx \frac{\Pi_i^2}{\Omega_i^2} \,. \tag{10.39}$$

$K_\perp$, $K_\times$, $K_\parallel$, and $R$, $L$ are given by

$$\begin{cases} K_\perp = 1 - \dfrac{\Pi_i^2}{\omega^2 - \Omega_i^2} - \dfrac{\Pi_e^2}{\omega^2 - \Omega_e^2} \,, \\ K_\times = -\dfrac{\Pi_i^2}{\omega^2 - \Omega_i^2} \dfrac{\Omega_i}{\omega} - \dfrac{\Pi_e^2}{\omega^2 - \Omega_e^2} \dfrac{\Omega_e}{\omega} \,, \\ K_\parallel = 1 - \dfrac{\Pi_e^2 + \Pi_i^2}{\omega^2} \simeq 1 - \dfrac{\Pi_e^2}{\omega^2} \,, \end{cases} \tag{10.40}$$

$$R = 1 - \frac{\Pi_e^2 + \Pi_i^2}{(\omega - \Omega_i)(\omega - \Omega_e)} \simeq \frac{\omega^2 - (\Omega_i + \Omega_e)\omega + \Omega_i \Omega_e - \Pi_e^2}{(\omega - \Omega_i)(\omega - \Omega_e)} \,, \tag{10.41}$$

$$L = 1 - \frac{\Pi_e^2 + \Pi_i^2}{(\omega + \Omega_i)(\omega + \Omega_e)} \simeq \frac{\omega^2 + (\Omega_i + \Omega_e)\omega + \Omega_i \Omega_e - \Pi_e^2}{(\omega + \Omega_i)(\omega + \Omega_e)} \,. \tag{10.42}$$

The dispersion relations for the waves propagating parallel to $\boldsymbol{B}_0$ ($\theta = 0$) are found by setting $K_\parallel = 0$, $N^2 = R$, and $N^2 = L$. Then

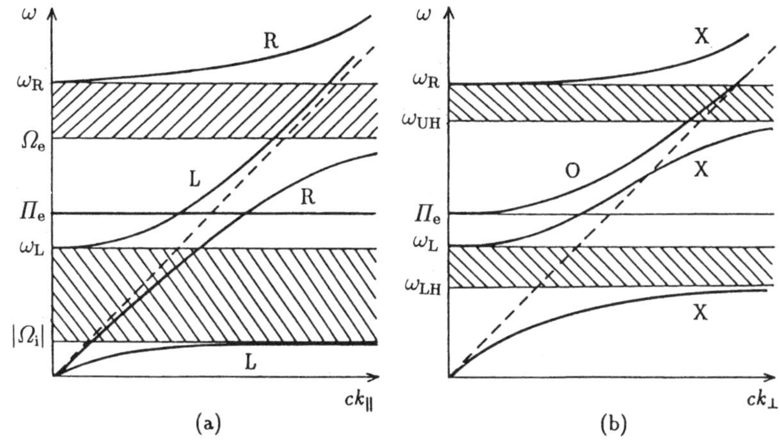


**Fig. 10.3.** (**a**) Dispersion relations ($\omega$–$ck_\parallel$) for R and L waves propagating parallel to the magnetic field ($\theta = 0$). (**b**) Dispersion relations ($\omega$–$ck_\parallel$) for O and X waves propagating perpendicularly to the magnetic field ($\theta = \pi/2$)

$$\omega^2 = \Pi_e^2 \,, \tag{10.43}$$

$$\frac{\omega^2}{c^2 k_\parallel^2} = \frac{1}{R} = \frac{(\omega - \Omega_i)(\omega - \Omega_e)}{\omega^2 - \omega\Omega_e + \Omega_e\Omega_i - \Pi_e^2} = \frac{(\omega + |\Omega_i|)(\omega - \Omega_e)}{(\omega - \omega_R)(\omega + \omega_L)} \,, \tag{10.44}$$

where $\omega_R$, $\omega_L$ are given by

$$\omega_R = \frac{\Omega_e}{2} + \left[\left(\frac{\Omega_e}{2}\right)^2 + \Pi_e^2 + |\Omega_e\Omega_i|\right]^{1/2} > 0 \,, \tag{10.45}$$

$$\omega_L = -\frac{\Omega_e}{2} + \left[\left(\frac{\Omega_e}{2}\right)^2 + \Pi_e^2 + |\Omega_e\Omega_i|\right]^{1/2} > 0 \,, \tag{10.46}$$

$$\frac{\omega^2}{c^2 k_\parallel^2} = \frac{1}{L} = \frac{(\omega + \Omega_i)(\omega + \Omega_e)}{\omega^2 + \omega\Omega_e + \Omega_e\Omega_i - \Pi_e^2} = \frac{(\omega - |\Omega_i|)(\omega + \Omega_e)}{(\omega - \omega_L)(\omega + \omega_R)} \,. \tag{10.47}$$

Note that $\Omega_e > 0$, $\Omega_i < 0$ and $\omega_R > \Omega_e$. Plots of the dispersion relations $\omega$–$ck_\parallel$ in the case of $\Omega_e > \Pi_e$ are shown in Fig. 10.3a. The dispersion relations for the waves propagating perpendicularly to $\boldsymbol{B}_0$ are found by setting $N^2 = K_\parallel$ (ordinary wave) and $N^2 = (K_\perp^2 - K_\times^2)/K_\perp$ (extraordinary wave). Then

$$\frac{\omega^2}{c^2 k_\perp^2} = \frac{1}{K_\parallel} = \left(1 - \frac{\Pi_e^2}{\omega^2}\right)^{-1} = 1 + \frac{\Pi_e^2}{c^2 k_\perp^2} \,, \tag{10.48}$$

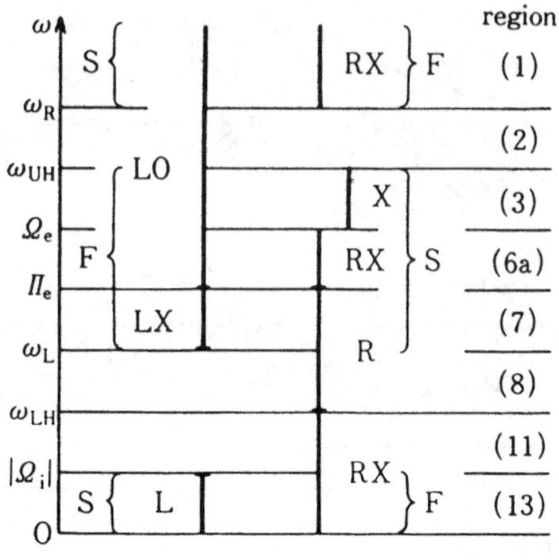


**Fig. 10.4.** The $\omega$ regions of R and L waves at $\theta = 0$, O and X waves at $\theta = \pi/2$, and F and S waves, in the case ($\omega_L < \Pi_e < \Omega_e$). *Numbers on the right* identify regions shown in the CMA diagram (Fig. 10.5)

$$\frac{\omega^2}{c^2k_\perp^2} = \frac{K_\perp}{K_\perp^2 - K_\times^2} = \frac{K_\perp}{RL}$$

$$= \frac{2(\omega^2 - \Omega_i^2)(\omega^2 - \Omega_e^2) - \Pi_e^2\left[(\omega + \Omega_i)(\omega + \Omega_e) + (\omega - \Omega_i)(\omega - \Omega_e)\right]}{2(\omega^2 - \omega_L^2)(\omega^2 - \omega_R^2)}$$

$$= \frac{\omega^4 - (\Omega_i^2 + \Omega_e^2 + \Pi_e^2)\omega^2 + \Omega_i^2\Omega_e^2 - \Pi_e^2\Omega_i\Omega_e}{(\omega^2 - \omega_L^2)(\omega^2 - \omega_R^2)} . \quad (10.49)$$

Equation (10.48) is the dispersion equation of an electron plasma wave (Langmuir wave).

Let us define $\omega_{UH}$ and $\omega_{LH}$ by

$$\omega_{UH}^2 \equiv \Omega_e^2 + \Pi_e^2 , \quad (10.50)$$

$$\frac{1}{\omega_{LH}^2} \equiv \frac{1}{\Omega_i^2 + \Pi_i^2} + \frac{1}{|\Omega_i|\Omega_e} . \quad (10.51)$$

$\omega_{UH}$ is called the *upper hybrid resonant frequency* and $\omega_{LH}$ is called the *lower hybrid resonant frequency*. Using these, we may write (10.49) as

$$\frac{\omega^2}{c^2k_\perp^2} = \frac{(\omega^2 - \omega_{LH}^2)(\omega^2 - \omega_{UH}^2)}{(\omega^2 - \omega_L^2)(\omega^2 - \omega_R^2)} . \quad (10.52)$$

We have $\omega_R > \omega_{UH} > \Pi_e$, $\Omega_e$ and $\omega_{LH}^2 < \Omega_e|\Omega_i|$, $\Omega_i^2 + \Pi^2$. Plots of the dispersion relation $\omega$–$ck_\perp$ in the case $\Omega_e > \Pi_e$ are shown in Fig. 10.3b. The gradient $\omega/ck$ in the $\omega$–$ck_\perp$ diagram is the ratio of the phase velocity $v_{ph}$ to

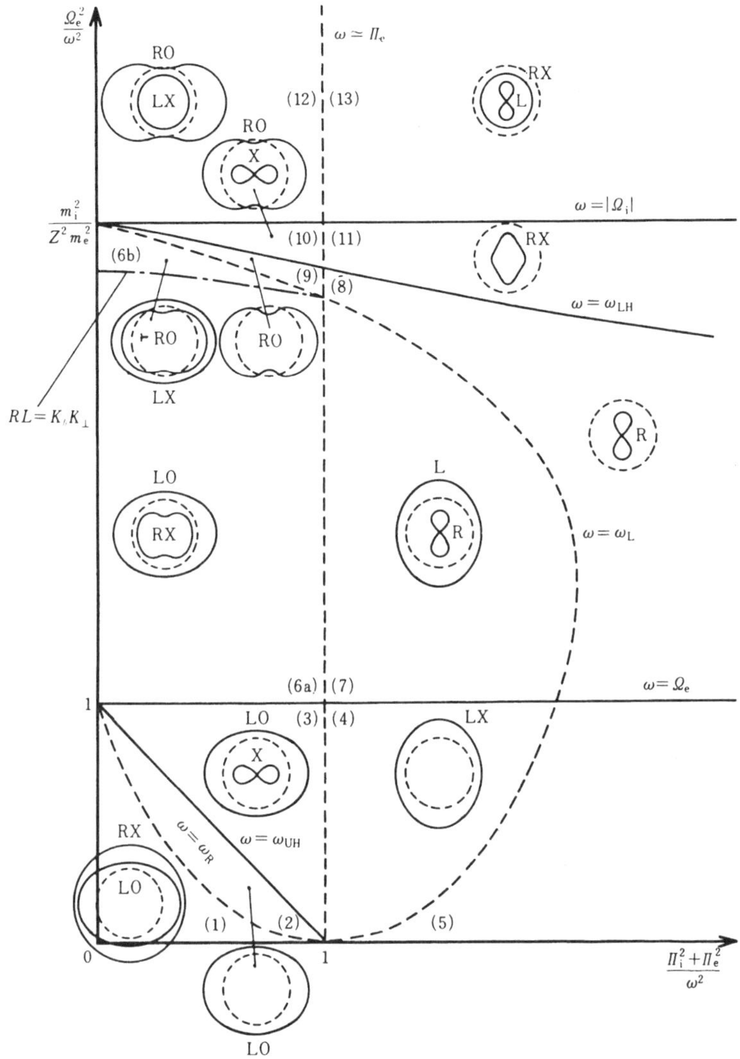


**Fig. 10.5.** CMA diagram of a two-component plasma. The surfaces of constant phase are drawn in each region. *Dotted circles* give the wave front in vacuum. The magnetic field is directed toward the top of the diagram

$c$. The steeper the gradient, the greater the phase velocity. The regions (in terms of $\omega$) of R and L waves at $\theta = 0$, and O and X waves at $\theta = \pi/2$, and F and S waves are shown in Fig. 10.4, for the case $\omega_L < \Pi_e < \Omega_e$.

We now explain the CMA diagram (Fig. 10.5), which was introduced by P.C. Clemmow and R.F. Mullaly and later modified by W.P. Allis [10.3].

The quantities $\Omega_e^2/\omega^2$ and $(\Pi_i^2+\Pi_e^2)/\omega^2$ are plotted along the vertical and horizontal axes, respectively. The cutoff conditions $R=0$ $(\omega=\omega_R)$, $L=0$ $(\omega=\omega_L)$, $K_\parallel=0$ $(\omega=\Pi_e)$ are shown by the dotted lines and the resonance conditions $R=\infty$ $(\omega=\Omega_e)$, $L=\infty$ $(\omega=\Omega_i)$, $K_\perp=0$ $(\omega=\Omega_{LH}, \omega=\Omega_{UH})$ are shown by solid lines. The cutoff and the resonance contours form the boundaries of the different regions. The boundary $RL=K_\parallel K_\perp$, at which O wave and X wave are exchanged, is also shown by broken and dotted lines in Fig. 10.5. The surfaces of constant phase for R, L and O, X waves are shown for the different regions. As the vertical and horizontal axes correspond to the magnitude of $B$ and the density $n_e$, one can easily assign waves to the corresponding regions simply by giving their frequencies $\omega$.

## 10.4 Various Waves

### 10.4.1 Alfven Wave

When the frequency $\omega$ is smaller than the ion cyclotron frequency ($\omega \ll |\Omega_i|$), the dielectric tensor $\boldsymbol{K}$ is expressed by

$$K_\perp = 1+\delta\,, \qquad K_\times = 0\,, \qquad K_\parallel = 1-\frac{\Pi_e^2}{\omega^2}\,, \tag{10.53}$$

where $\delta=\mu_0 n_i m_i c^2/B_0^2$. As $\Pi_e^2/\omega^2=(m_i/m_e)(\Omega_i^2/\omega^2)\delta$, we find $\Pi_e^2/\omega^2 \gg \delta$. Assuming that $\Pi_e^2/\omega^2 \gg 1$, we have $|K_\parallel| \gg |K_\perp|$. Then $A$, $B$, $C$ of (10.20) are given by

$$\begin{cases} A \approx -\dfrac{\Pi_e^2}{\omega^2}\cos^2\theta\,, \\ B \approx -\dfrac{\Pi_e^2}{\omega^2}(1+\delta)(1+\cos^2\theta)\,, \\ C \approx -\dfrac{\Pi_e^2}{\omega^2}(1+\delta)^2\,, \end{cases} \tag{10.54}$$

and the dispersion relations are

$$\frac{c^2}{N^2}=\frac{\omega^2}{k^2}=\frac{c^2}{1+\delta}=\frac{c^2}{1+\mu_0\rho_m c^2/B_0^2}\simeq\frac{B_0^2}{\mu_0\rho_m}\,, \tag{10.55}$$

$$\frac{c^2}{N^2}=\frac{\omega^2}{k^2}=\frac{c^2}{1+\delta}\cos^2\theta\,, \tag{10.56}$$

where $\rho_m$ is the mass density. The wave satisfying this dispersion relation is called the *Alfvén wave*. We define the *Alfvén velocity* by

$$v_A^2=\frac{c^2}{1+\delta}=\frac{c^2}{1+\mu_0\rho_m c^2/B_0^2}\simeq\frac{B_0^2}{\mu_0\rho_m}\,. \tag{10.57}$$

Equations (10.55) and (10.56) correspond to modes appearing in region (13) of the CMA diagram. Substituting (10.55) and (10.56) into (10.19) shows that $E_z = 0$ for either mode, whilst $E_x = 0$ for the mode (10.55) (R wave, F wave, X wave) and $E_y = 0$ for the mode (10.56) (L wave, S wave). From (10.6), we find for $\omega \ll |\Omega_i|$ that

$$\boldsymbol{E} + \boldsymbol{v}_i \times \boldsymbol{B}_0 = 0 \tag{10.58}$$

and $\boldsymbol{v}_i = (\boldsymbol{E} \times \boldsymbol{B}_0)/\boldsymbol{B}_0^2$, so that $\boldsymbol{v}_i$ of the mode (10.55) is

$$\boldsymbol{v}_i \approx \hat{\boldsymbol{x}} \cos(k_x x + k_z z - \omega t) , \tag{10.59}$$

and $\boldsymbol{v}_i$ of the mode (10.56) is

$$\boldsymbol{v}_i \approx \hat{\boldsymbol{y}} \cos(k_x x + k_z z - \omega t) , \tag{10.60}$$

where $\hat{\boldsymbol{x}}$, $\hat{\boldsymbol{y}}$ are unit vectors along the $x$ and $y$ axes, respectively. From these last equations, the fast mode (10.55) and (10.59) is called the *compressional mode* and the slow mode (10.56) and (10.60) is called the *torsional* or *shear mode*. The R wave (10.55) still exists, although it is deformed in the transition from region (13) to regions (11) and (8), but the L wave (10.56) disappears in these transitions.

It is clear from (10.58) that the plasma is frozen to the magnetic field. There is a tension $B^2/2\mu_0$ along the magnetic field lines and a pressure $B^2/2\mu_0$ exerted perpendicularly to the magnetic field. As the plasma, of mass density $\rho_m$, sticks to the field lines, the wave propagation speed in the direction of the field is $B_0^2/(\mu_0 \rho_m)$.

### 10.4.2 Ion Cyclotron Wave and Fast Wave

Let us consider the case where the frequency $\omega$ is shifted from low frequency toward the ion cyclotron frequency and $\Pi_e^2/\omega^2 \gg 1$. The corresponding waves are located in regions (13) and (11) of the CMA diagram. When $|\omega| \ll \Omega_e$, $\delta \gg 1$, and $\Pi_e^2/\omega^2 \gg 1$, the values of $K_\perp$, $K_\times$ and $K_\parallel$ are

$$K_\perp = -\frac{\delta \Omega_i^2}{\omega^2 - \Omega_i^2} , \qquad K_\times = -\frac{\delta \omega \Omega_i^2}{\omega^2 - \Omega_i^2} , \qquad K_\parallel = -\frac{\Pi_e^2}{\omega^2} . \tag{10.61}$$

Since $\Pi_e^2/\omega^2 = (m_i/m_e)(\Omega_i^2/\omega^2)\delta \gg \delta$, the coefficients $A$, $B$, $C$ are

$$\begin{cases} A = -\dfrac{\Pi_e^2}{\omega^2} \cos^2\theta , \\ B = \dfrac{\Pi_e^2}{\omega^2} \dfrac{\delta \Omega_i^2}{\omega^2 - \Omega_i^2} (1 + \cos^2\theta) , \\ C = \dfrac{\Pi_e^2}{\omega^2} \dfrac{\delta^2 \Omega_i^2}{\omega^2 - \Omega_i^2} . \end{cases} \tag{10.62}$$

The dispersion equation becomes ($\Pi_i^2 = \Omega_i^2 \delta$, $v_A^2 = c^2/\delta$)

$$N^4 \cos^2\theta - N^2 \frac{\delta \Omega_{\rm i}^2}{\Omega_{\rm i}^2 - \omega^2}(1 + \cos^2\theta) + \frac{\delta^2 \Omega_{\rm i}^2}{\Omega_{\rm i}^2 - \omega^2} = 0 \,. \tag{10.63}$$

Setting $N^2 \cos^2\theta = c^2 k_\parallel^2/\omega^2$, and $N^2 \sin^2\theta = c^2 k_\perp^2/\omega^2$, we may write (10.63) as

$$k_\perp^2 c^2 = \frac{\omega^4 \delta^2 \Omega_{\rm i}^2 - \omega^2(2\delta\Omega_{\rm i}^2 k_\parallel^2 c^2 + k_\parallel^4 c^4) + \Omega_{\rm i}^2 k_\parallel^4 c^4}{\omega^2(\delta\Omega_{\rm i}^2 + k_\parallel^2 c^2) - \Omega_{\rm i}^2 k_\parallel^2 c^2} \,, \tag{10.64a}$$

$$\frac{k_\perp^2}{k_\parallel^2} = \frac{(\omega/v_{\rm A}k_\parallel)^4 - (\omega/v_{\rm A}k_\parallel)^2 - (\omega/\Omega_{\rm i})^2 + 1}{(\omega/v_{\rm A}k_\parallel)^2 - (1 - \omega^2/\Omega_{\rm i}^2)} \tag{10.64b}$$

$$= \frac{\left[(\omega/v_{\rm A}k_\parallel)^2 - (1 - \omega/\Omega_{\rm i})\right]\left[(\omega/v_{\rm A}k_\parallel)^2 - (1 + \omega/\Omega_{\rm i})\right]}{(\omega/v_{\rm A}k_\parallel)^2 - (1 - \omega^2/\Omega_{\rm i}^2)} \,.$$

Therefore resonance occurs at

$$\omega^2 = \Omega_{\rm i}^2 \frac{k_\parallel^2 c^2}{k_\parallel^2 c^2 + \delta\Omega_{\rm i}^2} = \Omega_{\rm i}^2 \frac{k_\parallel^2 c^2}{k_\parallel^2 c^2 + \Pi_{\rm i}^2} \,, \tag{10.65a}$$

$$\left(\frac{\omega}{v_{\rm A}k_\parallel}\right)^2 = 1 - \left(\frac{\omega}{\Omega_{\rm i}}\right)^2 \,. \tag{10.65b}$$

When $|\omega|$ approaches $|\Omega_{\rm i}|$, the dispersion equation (10.63) approaches

$$N^2 \approx \frac{\delta}{1 + \cos^2\theta} \,, \tag{10.66}$$

$$N^2 \cos^2\theta \approx \delta(1 + \cos^2\theta)\Omega_{\rm i}^2 \Omega_{\rm i}^2 - \omega^2 \,. \tag{10.67}$$

The mode (10.66) corresponds to the compressional Alfvén mode (fast wave) and is not affected by the ion cyclotron resonance. The dispersion relation (10.67) is that of the *ion cyclotron wave*, and can be expressed by

$$\omega^2 = \Omega_{\rm i}^2 \left(1 + \frac{\Pi_{\rm i}^2}{k_\parallel^2 c^2} + \frac{\Pi_{\rm i}^2}{k_\parallel^2 c^2 + k_\perp^2 c^2}\right)^{-1} \,. \tag{10.68}$$

Note that here $\omega^2$ is always less than $\Omega_{\rm i}^2$.

The ions move in a left-circular motion (i.e., in the direction of the ion Larmor motion) at $\omega \simeq |\Omega_{\rm i}|$ for both waves [see (10.30)]. The mode (10.66) satisfies $\mathrm{i}E_x/E_y = 1$, i.e., it is circularly polarized, with the electric field rotating the opposite way to the ion Larmor motion.

The ion cyclotron wave satisfies

$$\frac{\mathrm{i}E_x}{E_y} \approx -\frac{\omega}{|\Omega_{\rm i}|}\frac{1}{1 + k_\perp^2/k_\parallel^2} \,, \tag{10.69}$$

i.e., the electric field is elliptically polarized, rotating in the same direction as the ion Larmor motion.

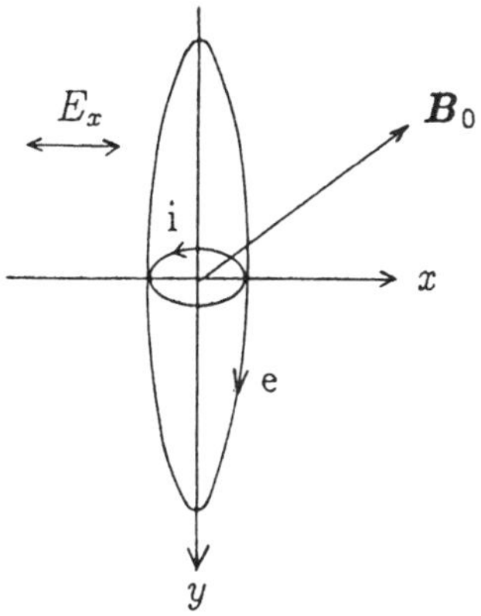


**Fig. 10.6.** Orbits of ions and electrons at lower hybrid resonance

### 10.4.3 Lower Hybrid Resonance

The frequency at lower hybrid resonance at $\theta = \pi/2$ is given by

$$\omega^2 = \omega_{\rm LH}^2 \,, \quad \frac{1}{\omega_{\rm LH}^2} = \frac{1}{\Omega_{\rm i}^2 + \Pi_{\rm i}^2} + \frac{1}{|\Omega_{\rm i}|\Omega_{\rm e}} \,, \quad \frac{\omega_{\rm LH}^2}{|\Omega_{\rm i}|\Omega_{\rm e}} = \frac{\Pi_{\rm i}^2 + \Omega_{\rm i}^2}{\Pi_{\rm i}^2 + |\Omega_{\rm i}|\Omega_{\rm e} + \Omega_{\rm i}^2} \,. \tag{10.70}$$

When the density is high and $\Pi_{\rm i}^2 \gg |\Omega_{\rm i}|\Omega_{\rm e}$, it follows that $\omega_{\rm LH} = (|\Omega_{\rm i}|\Omega_{\rm e})^{1/2}$. When $\Pi_{\rm i}^2 \ll |\Omega_{\rm i}|\Omega_{\rm e}$, then $\omega_{\rm LH}^2 = \Pi_{\rm i}^2 + \Omega_{\rm i}^2$. At lower hybrid resonance, we have $E_y = E_z = 0$ and $E_x \neq 0$.

When the density is high (i.e., $\Pi_{\rm i}^2 > |\Omega_{\rm i}|\Omega_{\rm e}$), then $|\Omega_{\rm i}| \ll \omega_{\rm LH} \ll \Omega_{\rm e}$ and the analysis of the motions of ions and electrons becomes simple. From (10.7), the velocity is given by

$$v_{k,x} = \frac{{\rm i}\epsilon_k E_x}{B_0} \frac{\omega|\Omega_k|}{\omega^2 - \Omega_k^2} \,, \tag{10.71}$$

and $v_{k,x} = {\rm d}x_k/{\rm d}t = -{\rm i}\omega x_k$ yields

$$x_k = \frac{-\epsilon_k E_x}{B_0} \frac{|\Omega_k|}{\omega^2 - \Omega_k^2} \,. \tag{10.72}$$

At $\omega^2 = |\Omega_{\rm i}|\Omega_{\rm e}$, we find that $x_{\rm i} \approx -E_x/B_0\Omega_{\rm e}$ and $x_{\rm e} \simeq -E_x/B_0\Omega_{\rm e}$, or $x_{\rm i} \approx x_{\rm e}$ (see Fig. 10.6). Consequently, charge separation does not occur and the lower hybrid wave can exist.

We have been discussing lower hybrid resonance at $\theta = \pi/2$. Let us consider the case in which $\theta$ is slightly different from $\theta = \pi/2$. The resonance condition is obtained from (10.24) as follows:

$$K_\perp \sin^2\theta + K_\parallel \cos^2\theta = 0\,. \tag{10.73}$$

$K_\perp$ is given by (10.46), (10.50) and (10.51), and (10.73) reduces to

$$\frac{(\omega^2-\omega_{\mathrm{LH}}^2)(\omega^2-\omega_{\mathrm{UH}}^2)}{(\omega^2-\Omega_{\mathrm{i}}^2)(\omega^2-\Omega_{\mathrm{e}}^2)}\sin^2\theta + \left(1-\frac{\Pi_{\mathrm{e}}^2}{\omega^2}\right)\cos^2\theta = 0\,. \tag{10.74}$$

When $\theta$ is near $\pi/2$ and $\omega$ is not much different from $\omega_{\mathrm{LH}}$, we find that

$$\begin{aligned}\omega^2-\omega_{\mathrm{LH}}^2 &= \frac{(\omega_{\mathrm{LH}}^2-\Omega_{\mathrm{e}}^2)(\omega_{\mathrm{LH}}^2-\Omega_{\mathrm{i}}^2)}{\omega_{\mathrm{LH}}^2-\omega_{\mathrm{UH}}^2}\frac{\Pi_{\mathrm{e}}^2-\omega_{\mathrm{LH}}^2}{\omega_{\mathrm{LH}}^2}\cos^2\theta \\ &\approx \frac{\Omega_{\mathrm{e}}^2\Pi_{\mathrm{e}}^2}{\omega_{\mathrm{UH}}^2}\left[1-\left(\frac{\Omega_{\mathrm{i}}}{\omega_{\mathrm{LH}}}\right)^2\right]\left[1-\left(\frac{\omega_{\mathrm{LH}}}{\Pi_{\mathrm{e}}}\right)^2\right]\cos^2\theta\,.\end{aligned}$$

Since $\omega_{\mathrm{UH}}^2\omega_{\mathrm{LH}}^2 = \Omega_{\mathrm{i}}^2\Omega_{\mathrm{e}}^2 + \Pi_{\mathrm{e}}^2|\Omega_{\mathrm{i}}|\Omega_{\mathrm{e}}$, it follows that $\omega^2$ is expressed by

$$\omega^2 = \omega_{\mathrm{LH}}^2\left\{1+\frac{m_{\mathrm{i}}}{Zm_{\mathrm{e}}}\cos^2\theta\frac{\left[1-\left(\frac{\Omega_{\mathrm{i}}}{\omega_{\mathrm{LH}}}\right)^2\right]\left[1-\left(\frac{\omega_{\mathrm{LH}}}{\Pi_{\mathrm{e}}}\right)^2\right]}{\left(1+\frac{|\Omega_{\mathrm{i}}|\Omega_{\mathrm{e}}}{\Pi_{\mathrm{e}}^2}\right)}\right\}\,. \tag{10.75}$$

When $\Pi_{\mathrm{e}}^2/|\Omega_{\mathrm{i}}|\Omega_{\mathrm{e}} \approx \delta = c^2/v_{\mathrm{A}}^2 \gg 1$, (10.75) becomes

$$\omega^2 = \omega_{\mathrm{LH}}^2\left(1+\frac{m_{\mathrm{i}}}{Zm_{\mathrm{e}}}\cos^2\theta\right)\,. \tag{10.76}$$

Even if $\theta$ differs from $\pi/2$ by only a slight amount $(Zm_{\mathrm{e}}/m_{\mathrm{i}})^{1/2}$, $\omega^2$ becomes $\omega^2 \approx 2\omega_{\mathrm{LH}}^2$, so that (10.76) holds only in the region very near $\theta = \pi/2$.

### 10.4.4 Upper Hybrid Resonance

The upper hybrid resonant frequency $\omega_{\mathrm{UH}}$ is given by

$$\omega_{\mathrm{UH}}^2 = \Pi_{\mathrm{e}}^2 + \Omega_{\mathrm{e}}^2\,. \tag{10.77}$$

Since this frequency is much larger than $|\Omega_{\mathrm{i}}|$, ion motion can be neglected.

### 10.4.5 Electron Cyclotron Wave

Let us consider high-frequency waves, so that ion motion can be neglected. When $\omega \gg |\Omega_{\mathrm{i}}|$, we find

$$K_\perp \approx 1-\frac{\Pi_{\mathrm{e}}^2}{\omega^2-\Omega_{\mathrm{e}}^2}\,, \qquad K_\times \approx -\frac{\Pi_{\mathrm{e}}^2}{\omega^2-\Omega_{\mathrm{e}}^2}\frac{\Omega_{\mathrm{e}}}{\omega}\,, \qquad K_\parallel = 1-\frac{\Pi_{\mathrm{e}}^2}{\omega^2}\,. \tag{10.78}$$

The solution of the dispersion equation $AN^4 - BN^2 + C = 0$, viz.,

$$N^2 = \frac{B \pm (B^2 - 4AC)^{1/2}}{2A} ,$$

may be modified to

$$N^2 - 1 = \frac{-2(A - B + C)}{2A - B \pm (B^2 - 4AC)^{1/2}} = \frac{-2\Pi_e^2(1 - \Pi_e^2/\omega^2)}{2\omega^2(1 - \Pi_e^2/\omega^2) - \Omega_e^2 \sin^2\theta \pm \Omega_e \Delta} , \tag{10.79}$$

$$\Delta = \left[ \Omega_e^2 \sin^4\theta + 4\omega^2 \left(1 - \frac{\Pi_e^2}{\omega^2}\right)^2 \cos^2\theta \right]^{1/2} . \tag{10.80}$$

The ordinary wave and extraordinary wave will be obtained by taking the plus and minus sign, respectively, in (10.79). In the case

$$\Omega_e^2 \sin^4\theta \gg 4\omega^2 \left(1 - \frac{\Pi_e^2}{\omega^2}\right)^2 \cos^2\theta , \tag{10.81}$$

we find

$$N^2 = \frac{1 - \Pi_e^2/\omega^2}{1 - (\Pi_e^2/\omega^2)\cos^2\theta} , \tag{10.82}$$

$$N^2 = \frac{(1 - \Pi_e^2/\omega^2)^2\omega^2 - \Omega_e^2 \sin^2\theta}{(1 - \Pi_e^2/\omega^2)\omega^2 - \Omega_e^2 \sin^2\theta} . \tag{10.83}$$

Equation (10.82) becomes $N^2 = K_\parallel = 1 - \Pi_e^2/\omega^2$ at $\theta \sim \pi/2$ and does not depend on the magnitude of the magnetic field. This wave is used for density measurements by microwave interferometry. In the case

$$\Omega_e^2 \sin^4\theta \ll 4\omega^2 \left(1 - \frac{\Pi_e^2}{\omega^2}\right)^2 \cos^2\theta , \tag{10.84}$$

with the additional condition

$$\Omega_e^2 \sin^2\theta \ll \left| 2\omega^2 \left(1 - \frac{\Pi_e^2}{\omega^2}\right) \right| , \tag{10.85}$$

the dispersion relations become

$$N^2 = 1 - \frac{\Pi_e^2}{(\omega + \Omega_e \cos\theta)\omega} , \tag{10.86}$$

$$N^2 = 1 - \frac{\Pi_e^2}{(\omega - \Omega_e \cos\theta)\omega} . \tag{10.87}$$

Equation (10.86) corresponds to the L wave and (10.87) to the R wave. R-wave resonance occurs near the electron cyclotron frequency. This wave can propagate in regions (7) and (8) of the CMA diagram, where the frequency is less than the plasma frequency. This wave is called the *electron cyclotron wave*. It should be noted that the assumptions (10.84) and (10.85) are not satisfied near $K_{\parallel} = 1 - \Pi_{\rm e}^2/\omega^2 \simeq 0$.

The electron cyclotron wave is also called the *whistler wave*. Electromagnetic disturbances initiated by lightning flashes propagate through the ionosphere along magnetic field lines. The frequency of a lightning-induced whistler wave falls in the audio region, and its group velocity increases with frequency, so that this wave is perceived as a whistle of descending tone. This is why it is called the whistler wave.

## 10.5 Conditions for Electrostatic Waves

When the electric field $\boldsymbol{E}$ can be expressed by an electrostatic potential $\phi$,

$$\boldsymbol{E} = -\nabla\phi = -\mathrm{i}\boldsymbol{k}\phi \, , \tag{10.88}$$

the resulting wave is called an *electrostatic wave*. The electric field $\boldsymbol{E}$ is always parallel to the propagation vector $\boldsymbol{k}$, so that the electrostatic wave is longitudinal. The magnetic field $\boldsymbol{B}_1$ of the electrostatic wave is always zero:

$$\boldsymbol{B}_1 = \boldsymbol{k} \times \boldsymbol{E}/\omega = 0 \, . \tag{10.89}$$

Alfv́en waves are not electrostatic waves. We will discuss here the conditions for electrostatic waves. Since the dispersion relation is

$$\boldsymbol{N} \times (\boldsymbol{N} \times \boldsymbol{E}) + \boldsymbol{K}\cdot\boldsymbol{E} = 0 \, ,$$

the scalar product with $\boldsymbol{N}$ becomes

$$\boldsymbol{N}\cdot\boldsymbol{K}\cdot(\boldsymbol{E}_{\parallel} + \boldsymbol{E}_{\perp}) = 0 \, ,$$

where $\boldsymbol{E}_{\parallel}$ and $\boldsymbol{E}_{\perp}$ are the components of the electric field parallel and perpendicular to $\boldsymbol{k}$. If $|\boldsymbol{E}_{\parallel}| \gg |\boldsymbol{E}_{\perp}|$, the wave is electrostatic and the dispersion relation becomes

$$\boldsymbol{N}\cdot\boldsymbol{K}\cdot\boldsymbol{N} = 0 \, . \tag{10.90}$$

Rewriting the general dispersion relation as

$$(N^2 - \boldsymbol{K})\cdot\boldsymbol{E}_{\perp} = \boldsymbol{K}\cdot\boldsymbol{E}_{\parallel} \, ,$$

it follows that $|\boldsymbol{E}_{\parallel}| \gg |\boldsymbol{E}_{\perp}|$ holds when

$$|N^2| \gg |K_{ij}| \tag{10.91}$$

is satisfied for all $K_{ij}$. The dispersion relation (10.90) for the electrostatic wave is then

$$k_x^2 K_{xx} + 2k_x k_z K_{xz} + k_z^2 K_{zz} = 0 \, . \tag{10.92}$$

The condition (10.91) for the electrostatic wave indicates that the phase velocity $\omega/k = c/N$ of this wave is low. The $K_{ij}$ have already been given by (10.9)–(10.12) for cold plasmas, and the general formula for hot plasma will be discussed in Chap. 12. We have stated that the magnetic field $\boldsymbol{B}_1$ of the electrostatic wave is zero. Disturbances of the magnetic field propagate with the Alfvén velocity $v_{\rm A} \simeq B_0^2/(\mu_0 n_{\rm i} m_{\rm i})$. If the phase velocity of the wave is much lower than $v_{\rm A}$, the disturbance of the magnetic field will be damped within a few cycles of the wave and the propagated magnetic field disturbance becomes zero. When the electron thermal velocity $v_{T_{\rm e}}$ is taken as a typical phase velocity for electrostatic waves, then the condition of $v_{\rm A} > v_{T_{\rm e}}$ reduces to

$$\frac{B_0^2}{\mu_0 n_{\rm i} m_{\rm i} v_{T_{\rm e}}^2} = \frac{2m_{\rm e}}{\beta_{\rm e} m_{\rm i}} > 1 , \qquad \beta_{\rm e} < \frac{2m_{\rm e}}{m_{\rm i}} .$$

This measures the extent to which a wave is electrostatic.

At resonance, the refractive index $N$ becomes infinite. As the $K_{ij}$ are finite for lower hybrid and upper hybrid resonance, the condition (10.91) is satisfied, and these hybrid waves are electrostatic. Since some of the $K_{ij}$ become infinite for the ion or electron cyclotron waves, these cyclotron waves are not always electrostatic.

# 11 Landau Damping and Cyclotron Damping

The existence of a damping mechanism by which plasma particles absorb wave energy even in a collisionless plasma was found by L.D. Landau, under the condition that the plasma is not cold and the velocity distribution is of finite extent. Energy-exchange processes between particles and waves are possible even in a collisionless plasma and play important roles in plasma heating by waves (wave absorption) and in the mechanism of instabilities (wave amplification). These important processes will be explained in terms of simplified physical models in this chapter, whilst in Chap. 12, they will be described more systematically. In hot plasma models, a pressure term and particle–wave interaction term appear in the dielectric tensor that are absent in the dielectric tensor for a cold plasma.

## 11.1 Landau Damping (Amplification)

Let us assume that many particles drift with different velocities in the direction of the lines of magnetic force. When an electrostatic wave (a longitudinal wave with $\boldsymbol{k} \parallel \boldsymbol{E}$) propagates along the lines of magnetic force, an interaction appears between the wave and a group of particles (see Fig. 11.1). Take the $z$ axis in the direction of the magnetic field and denote the unit vector in this direction by $\hat{\boldsymbol{z}}$. Then the electric field and the velocity $\boldsymbol{v} = v\hat{\boldsymbol{z}}$ satisfy

$$\boldsymbol{E} = \hat{\boldsymbol{z}} E \cos(kz - \omega t) , \tag{11.1}$$

$$m\frac{\mathrm{d}v}{\mathrm{d}t} = qE\cos(kz - \omega t) . \tag{11.2}$$

The electric field $\boldsymbol{E}$ is a quantity of the first order. The zeroth-order solution of (11.2) is

$$z = v_0 t + z_0$$

and the first-order equation is

$$m\frac{\mathrm{d}v_1}{\mathrm{d}t} = qE\cos(kz_0 + kv_0 t - \omega t) . \tag{11.3}$$

The solution of (11.3) for the initial condition $v_1 = 0$ at $t = 0$ is

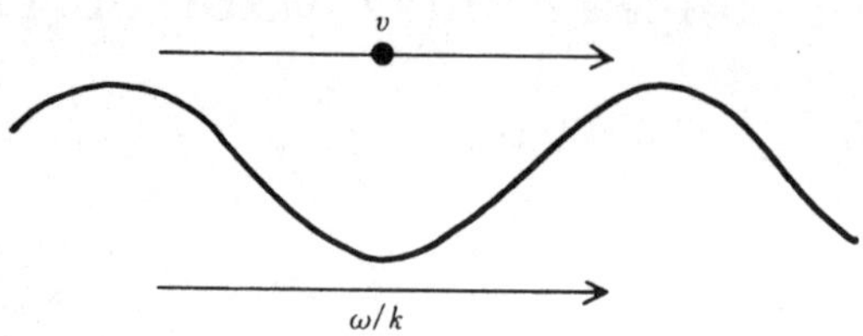


**Fig. 11.1.** Propagation of wave and motion of particles in the process of Landau damping

$$v_1 = \frac{qE}{m}\frac{\sin(kz_0 + kv_0t - \omega t) - \sin kz_0}{kv_0 - \omega} \ . \tag{11.4}$$

The kinetic energy of the particle becomes

$$\frac{\mathrm{d}}{\mathrm{d}t}\frac{mv^2}{2} = v\frac{\mathrm{d}}{\mathrm{d}t}mv = v_1\frac{\mathrm{d}}{\mathrm{d}t}mv_1 + v_0\frac{\mathrm{d}}{\mathrm{d}t}mv_2 + \cdots \ . \tag{11.5}$$

From (11.2) and (11.4), we have the relation

$$\begin{aligned} m\frac{\mathrm{d}(v_1+v_2)}{\mathrm{d}t} &= qE\cos\left[k(z_0 + v_0t + z_1) - \omega t\right] \\ &= qE\cos(kz_0 + \alpha t) - qE\sin(kz_0+\alpha t)kz_1 \ , \end{aligned}$$

$$z_1 = \int_0^t v_1 \, \mathrm{d}t = \frac{qE}{m}\left[\frac{-\cos(kz_0+\alpha t) + \cos kz_0}{\alpha^2} - \frac{t\sin kz_0}{\alpha}\right] ,$$

where

$$\alpha \equiv kv_0 - \omega \ .$$

Using these, we may put (11.5) into the form

$$\begin{aligned} \frac{\mathrm{d}}{\mathrm{d}t}\frac{mv^2}{2} &= \frac{q^2E^2}{m}\left[\frac{\sin(kz_0+\alpha t) - \sin kz_0}{\alpha}\right]\cos(kz_0+\alpha t) \\ &\quad - \frac{kv_0q^2E^2}{m}\left[\frac{-\cos(kz_0+\alpha t)+\cos kz_0}{\alpha^2} - \frac{t\sin kz_0}{\alpha}\right]\sin(kz_0+\alpha t) \ . \end{aligned}$$

The average of the foregoing quantity with respect to the initial position $z_0$ is

$$\left\langle \frac{\mathrm{d}}{\mathrm{d}t}\frac{mv^2}{2}\right\rangle_{z_0} = \frac{q^2E^2}{2m}\left(\frac{-\omega\sin\alpha t}{\alpha^2} + t\cos\alpha t + \frac{\omega t\cos\alpha t}{\alpha}\right) . \tag{11.6}$$

When we take the velocity average of (11.6) over $v_0$ with the weighting factor, i.e., distribution function $f(v_0)$ (defining $\alpha \equiv kv_0 - \omega$),

$$f(v_0) = f\left(\frac{\alpha + \omega}{k}\right) = g(\alpha) ,$$

the rate of increase of the kinetic energy of the particles is obtained. The distribution function is normalized:

$$\int_{-\infty}^{\infty} f(v_0)\,\mathrm{d}v_0 = \frac{1}{k}\int g(\alpha)\,\mathrm{d}\alpha = 1 .$$

The integral of the second term of (11.6), viz.,

$$\frac{1}{k}\int g(\alpha) t \cos\alpha t\,\mathrm{d}\alpha = \frac{1}{k}\int g\left(\frac{x}{t}\right)\cos x\,\mathrm{d}x , \tag{11.7}$$

approaches zero as $t \to \infty$. The integral of the third term of (11.6) becomes

$$\frac{\omega}{k}\int \frac{g(\alpha) t \cos\alpha t}{\alpha}\,\mathrm{d}\alpha = \frac{\omega}{k}\int \frac{t}{x} g\left(\frac{x}{t}\right)\cos x\,\mathrm{d}x . \tag{11.8}$$

The function $g(\alpha)$ can be considered to be the sum of an even and an odd function. The even function does not contribute to the integral. The contribution of the odd function approaches zero when $t \to \infty$ if $g(\alpha)$ is continuous at $\alpha = 0$. Therefore, only the contribution of the first term in (11.6) remains and we find

$$\left\langle \frac{\mathrm{d}}{\mathrm{d}t}\frac{mv^2}{2}\right\rangle_{z_0,v_0} = -\frac{\omega q^2 E^2}{2mk}\mathrm{P}\int \frac{g(\alpha)\sin\alpha t}{\alpha^2}\,\mathrm{d}\alpha , \tag{11.9}$$

where P denotes Cauchy's principal value. The main contribution to the integral comes from near $\alpha = 0$, so that $g(\alpha)$ may be expanded around $\alpha = 0$:

$$g(\alpha) = g(0) + \alpha g'(0) + \frac{\alpha^2}{2}g''(0) + \cdots .$$

As $\sin\alpha t/\alpha^2$ is an odd function, only the second term of the foregoing equation contributes to the integral and we find for large $t$ that

$$\begin{aligned}\left\langle \frac{\mathrm{d}}{\mathrm{d}t}\frac{mv^2}{2}\right\rangle_{z_0,v_0} &= -\frac{\omega q^2 E^2}{2m|k|}\int_{-\infty}^{\infty} \frac{g'(0)\sin\alpha t}{\alpha}\,\mathrm{d}\alpha \\ &= \frac{-\pi q^2 E^2}{2m|k|}\left(\frac{\omega}{k}\right)\left[\frac{\partial f(v_0)}{\partial v_0}\right]_{v_0=\omega/k} .\end{aligned} \tag{11.10}$$

If the number of particles moving slightly slower than the phase velocity of the wave is larger than the number moving slightly faster, i.e., if $v_0 \partial f_0/\partial v_0 < 0$, the group of particles as a whole gains energy from the wave and the wave is damped. On the contrary, when $v_0 \partial f_0/\partial v_0 > 0$ at $v_0 = \omega/k$, the particles gives their energy to the wave and the amplitude of the wave increases (Fig. 11.2). This mechanism is called *Landau damping* or *amplification* [11.1]. Experimental verification of Landau damping of waves in a

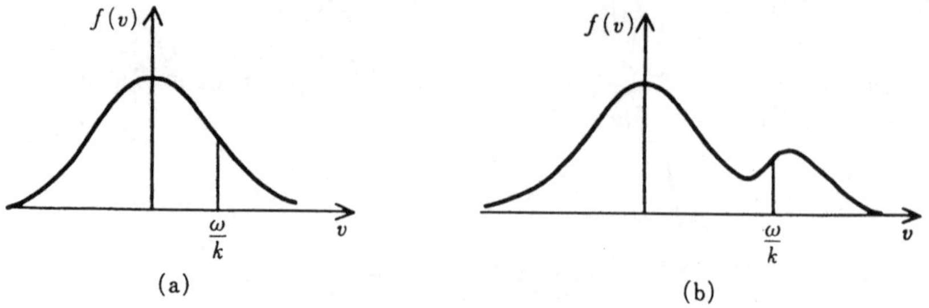


**Fig. 11.2.** (**a**) Landau damping and (**b**) Landau amplification

collisionless plasma was demonstrated by J.M. Malemberg and C.B. Wharton [11.2] in 1965, twenty years after Landau's prediction.

The growth rate (11.10) of the kinetic energy of the particles must be equal to the damping rate of the wave energy. Therefore the growth rate $\gamma$ of the amplitude of the wave field is obtained by ($\gamma < 0$ in the damping case)

$$n\left\langle \frac{\mathrm{d}}{\mathrm{d}t}\frac{mv^2}{2} \right\rangle_{z_0 v_0} = -2\gamma W \ ,$$

and the growth rate $\gamma$ is given by

$$\frac{\gamma}{\omega} = \frac{\pi}{2}\left(\frac{\Pi}{\omega}\right)^2 \left(\frac{\omega}{|k|}\right)\left[v_0 \frac{\partial f(v_0)}{\partial v_0}\right]_{v_0=\omega/k} , \qquad (11.11)$$

where $\Pi^2 = nq^2/\epsilon_0 m$, $W \approx 2\epsilon_0 E^2/4$, $\int f(v)\mathrm{d}v = 1$.

There is a restriction on the applicability of linear Landau damping. When this phenomenon runs its course before the particle orbit deviates from the linear-approximation solution, the reductions leading to linear Landau damping are justified. The period of oscillation in the potential well of the electric field of the wave gives the time for the particle orbit to deviate from the linear approximation ($\omega^2 \sim eEk/m$ from $m\omega^2 x = eE$). The period of oscillation is

$$\tau_{\mathrm{osc}} = \frac{1}{\omega_{\mathrm{osc}}} \approx \left(\frac{m}{ekE}\right)^{1/2} .$$

Consequently, the condition for the applicability of linear Landau damping is that the Landau damping time $1/\gamma$ should be shorter than $\tau_{\mathrm{osc}}$ or the collision time $1/\nu_{\mathrm{coll}}$ should be shorter than $\tau_{\mathrm{osc}}$:

$$|\gamma\tau_{\mathrm{osc}}| > 1 \ , \qquad (11.12)$$

$$|\nu_{\mathrm{coll}}\tau_{\mathrm{osc}}| > 1 \ . \qquad (11.13)$$

On the other hand, it was assumed that particles are collisionless. The condition that the collision time $1/\nu_{\mathrm{coll}}$ is longer than $\lambda/v_{\mathrm{rms}}$ is necessary for the asymptotic approximation of the integral (11.9) as $t \to \infty$, where $\lambda$ is the wavelength of the wave and $v_{\rm rms}$ is the spread in the velocity distribution:

$$\frac{1}{\nu_{\rm coll}} > \frac{2\pi}{kv_{\rm rms}} \,. \tag{11.14}$$

## 11.2 Transit Time Damping

We have already described the properties of Alfvén waves in cold plasmas. There are compressional and torsional modes. The compressional mode becomes magnetosonic in hot plasmas, as described in Chap. 5. In the low-frequency region, the magnetic moment $\mu_{\rm m}$ is conserved and the equation of motion along the field lines is

$$m\frac{{\rm d}v_z}{{\rm d}t} = -\mu_{\rm m}\frac{\partial B_{1z}}{\partial z} \,. \tag{11.15}$$

This equation is the same as that for Landau damping if $-\mu_{\rm m}$ and $\partial B_{1z}/\partial z$ are replaced by the electric charge and the electric field, respectively. The rate of change of the kinetic energy is derived similarly, and is equal to

$$\left\langle \frac{\rm d}{{\rm d}t}\frac{mv^2}{2}\right\rangle_{z_0,v_0} = -\frac{\pi\mu_{\rm m}^2|k|}{2m}|B_{1z}|^2\left(\frac{\omega}{k}\right)\left[\frac{\partial f(v_0)}{\partial v_0}\right]_{v_0=\omega/k} \,. \tag{11.16}$$

This phenomenon is called *transit time damping.*

## 11.3 Cyclotron Damping

The mechanism of cyclotron damping is different from that of Landau damping. Here the electric field of the wave is perpendicular to the direction of the magnetic field and the particle drift and accelerates the particle perpendicularly to the drift direction. Let us consider a simple case in which the thermal energy of particles perpendicular to the magnetic field is zero and the velocity of particles parallel to the magnetic field $\boldsymbol{B}_0 = B_0\hat{\boldsymbol{z}}$ is $V$. The equation of motion is

$$m\frac{\partial \boldsymbol{v}}{\partial t} + mV\frac{\partial \boldsymbol{v}}{\partial z} = q(\boldsymbol{E}_1 + \boldsymbol{v}\times\hat{\boldsymbol{z}}B_0 + V\hat{\boldsymbol{z}}\times\boldsymbol{B}_1) \,. \tag{11.17}$$

As our interest is in the perpendicular acceleration, we assume $(\boldsymbol{E}_1\cdot\hat{\boldsymbol{z}}) = 0$. $\boldsymbol{B}_1$ is given by $\boldsymbol{B}_1 = (\boldsymbol{k}\times\boldsymbol{E})/\omega$. With the definitions $v^{\pm} = v_x \pm {\rm i}v_y$ and $E^{\pm} = E_x \pm {\rm i}E_y$, the solution for the initial condition $\boldsymbol{v} = 0$ at $t = 0$ is

$$\begin{aligned} v^{\pm} &= \frac{{\rm i}qE^{\pm}(\omega - kV)\exp({\rm i}kz - {\rm i}\omega t)}{m\omega}\frac{1-\exp({\rm i}\omega t - {\rm i}kVt \pm {\rm i}\Omega t)}{\omega - kV \pm \Omega} \,, \\ \Omega &= \frac{-qB_0}{m} \,. \end{aligned} \tag{11.18}$$

The macroscopic value of $\boldsymbol{v}_\perp$ is obtained by taking the average weighted by the distribution function $f_0(V)$ as follows:

$$\langle \boldsymbol{v}_\perp \rangle = \frac{\mathrm{i}q\exp(\mathrm{i}kz - \mathrm{i}\omega t)}{2m}\left[(c^+ + c^-)\boldsymbol{E}_\perp + \mathrm{i}(c^+ - c^-)\boldsymbol{E}_\perp \times \hat{\boldsymbol{z}}\right] , \tag{11.19}$$

$$c^\pm = \alpha^\pm - \mathrm{i}\beta^\pm , \tag{11.20}$$

$$\alpha^\pm = \int_{-\infty}^{\infty} \mathrm{d}V \frac{f_0(V)(1-kV/\omega)\left[1-\cos(\omega - kV \pm \Omega)t\right]}{\omega - kV \pm \Omega} , \tag{11.21}$$

$$\beta^\pm = \int_{-\infty}^{\infty} \mathrm{d}V \frac{f_0(V)(1-kV/\omega)\sin(\omega - kV \pm \Omega)t}{\omega - kV \pm \Omega} . \tag{11.22}$$

As $t$ becomes large, we find that

$$\alpha^\pm \to P\int_{-\infty}^{\infty} \mathrm{d}V \frac{f_0(V)(1-kV/\omega)}{\omega - kV \pm \Omega} , \tag{11.23}$$

$$\beta^\pm \to \frac{\mp\pi\Omega}{\omega|k|} f_0\left(\frac{\omega \pm \Omega}{k}\right) . \tag{11.24}$$

When

$$t \gg \frac{2\pi}{kV_{\mathrm{rms}}} , \tag{11.25}$$

where $V_{\mathrm{rms}} = \langle V^2 \rangle^{1/2}$ is the spread of the velocity distribution, the approximations (11.19)–(11.24) are justified. The absorption of the wave energy by the plasma particles is given by

$$\left\langle \mathrm{Re}\left[q\boldsymbol{E}\exp(\mathrm{i}kz - \mathrm{i}\omega t)\right]\left(\mathrm{Re}\langle \boldsymbol{v}_\perp \rangle\right)\right\rangle_z = \frac{q^2}{4m}\left(\beta^+|E_x + \mathrm{i}E_y|^2 + \beta^-|E_x - \mathrm{i}E_y|^2\right) . \tag{11.26}$$

Let us consider the case of electrons ($\Omega_\mathrm{e} > 0$). As described in Sect. 10.2, the wave $N^2 = R$ propagating in the direction of the magnetic field ($\theta = 0$) satisfies $E_x + \mathrm{i}E_y = 0$, so that the absorption power becomes

$$P_\mathrm{e} = \frac{q^2}{4m}\beta^-|E_x - \mathrm{i}E_y|^2 .$$

When $\omega > 0$, (11.24) indicates that $\beta^- > 0$. When $\omega < 0$, $\beta^-$ is nearly zero since $f_0\left[(\omega - \Omega_\mathrm{e})/k\right] \ll 1$.

Let us consider the case of ions ($-\Omega_\mathrm{i} > 0$). In a similar way, we find that

$$P_\mathrm{i} = \frac{q^2}{4m}\beta^+|E_x + \mathrm{i}E_y|^2 .$$

When $\omega > 0$, (11.24) indicates that $\beta^+ > 0$. When $\omega < 0$, $\beta^+$ is nearly zero, since $f_0\left(\omega + \Omega_\mathrm{i}/k\right) \ll 1$.

The *cyclotron velocity* $V_c$ is defined so that the Doppler shifted frequency (the frequency of the wave felt by a particle running with the velocity $V$) is equal to the cyclotron frequency, that is,

$$\omega - kV_c \pm \Omega = 0\,, \qquad V_c = \frac{\omega}{k}\left(1 \pm \frac{\Omega}{\omega}\right)\,.$$

Accordingly, particles absorb the wave energy when the absolute value of the cyclotron velocity is smaller than the absolute value of the phase velocity of the wave ($\pm\Omega/\omega < 0$) [see (11.24)]. This phenomena is called *cyclotron damping*.

Let us consider the change in the kinetic energy of the particles in the case of cyclotron damping. Then the equation of motion is

$$m\frac{\mathrm{d}\boldsymbol{v}}{\mathrm{d}t} - q(\boldsymbol{v}\times\boldsymbol{B}_0) = q\boldsymbol{E}_\perp + q(\boldsymbol{v}\times\boldsymbol{B}_1)\,.$$

Since $\boldsymbol{B}_1 = (\boldsymbol{k}\times\boldsymbol{E})/\omega$ and $E_z = 0$, we have

$$m\frac{\mathrm{d}v_z}{\mathrm{d}t} = \frac{qk_z}{\omega}(\boldsymbol{v}_\perp\cdot\boldsymbol{E}_\perp)\,,$$

$$m\frac{\mathrm{d}\boldsymbol{v}_\perp}{\mathrm{d}t} - q(\boldsymbol{v}_\perp\times\boldsymbol{B}_0) = q\boldsymbol{E}_\perp\left(1 - \frac{k_z v_z}{\omega}\right)\,,$$

so that

$$m\boldsymbol{v}_\perp\cdot\frac{\mathrm{d}\boldsymbol{v}_\perp}{\mathrm{d}t} = q(\boldsymbol{v}_\perp\cdot\boldsymbol{E}_\perp)\left(1 - \frac{k_z v_z}{\omega}\right)\,.$$

Then

$$\frac{\mathrm{d}}{\mathrm{d}t}\left(\frac{mv_z^2}{2}\right) = \frac{k_z v_z}{\omega - k_z v_z}\frac{\mathrm{d}}{\mathrm{d}t}\left(\frac{mv_\perp^2}{2}\right)\,,$$

$$v_\perp^2 + \left(v_z - \frac{\omega}{k_z}\right)^2 = \text{const.}$$

In the analysis of cyclotron damping, we assumed that $v_z = V$ is constant. The condition for the validity of the linearized theory is [11.3]

$$\frac{k_z^2 q^2 E_\perp^2 |\omega - k_z v_z| t^3}{24\omega^2 m^2} < 1\,.$$

We have discussed the case in which the perpendicular thermal energy is zero. When the perpendicular thermal energy is larger than the parallel thermal energy, so-called cyclotron instability may occur. The mutual interaction between particles and wave will be discussed again in Chap. 12 in relation to heating and instabilities.

## 11.4 Quasi-Linear Theory of Evolution in the Distribution Function

It has been assumed that the perturbation is small and the zeroth-order terms do not change. Under these assumptions, the linearized equations for the perturbations can be analyzed. However, if the perturbations grow, the zeroth-order quantities may change and the growth rate of the perturbations may change due to the evolution of the zeroth-order quantities. Finally, the perturbations saturate (growth rate becomes zero) and shift to a steady state. Let us consider the simple case when $\boldsymbol{B} = 0$ and there is a one-dimensional electrostatic perturbation ($\boldsymbol{B}_1 = 0$). Ions are uniformly distributed. Then the electron distribution function $f(x, v, t)$ obeys the following Vlasov equation:

$$\frac{\partial f}{\partial t} + v\frac{\partial f}{\partial x} - \frac{e}{m}E\frac{\partial f}{\partial v} = 0 \, . \tag{11.27}$$

Let the distribution function $f$ be divided into two parts, viz.,

$$f(x, v, t) = f_0(v, t) + f_1(x, v, t) \, , \tag{11.28}$$

where $f_0$ is the slowly changing zeroth-order term and $f_1$ is the oscillatory first-order term. It is assumed that the time derivative of $f_0$ is the second-order term. When (11.28) is substituted into (11.27), the first and the second terms satisfy

$$\frac{\partial f_1}{\partial t} + v\frac{\partial f_1}{\partial x} = \frac{e}{m}E\frac{\partial f_0}{\partial v} \, , \tag{11.29}$$

$$\frac{\partial f_0}{\partial t} = \frac{e}{m}E\frac{\partial f_1}{\partial v} \, . \tag{11.30}$$

$f_1$ and $E$ may be expressed by Fourier integrals:

$$f_1(x, v, t) = \frac{1}{(2\pi)^{1/2}}\int f_k(v)\exp\{\mathrm{i}[kx - \omega(k)t]\}\mathrm{d}k \, , \tag{11.31}$$

$$E(x, t) = \frac{1}{(2\pi)^{1/2}}\int E_k\exp\{\mathrm{i}[kx - \omega(k)t]\}\mathrm{d}k \, . \tag{11.32}$$

Since $f_1$ and $E$ are real, $f_{-k} = f_k^*$, $E_{-k} = E_k^*$, $\omega(-k) = -\omega^*(k)$, where $\omega(k) = \omega_{\mathrm{r}}(k) + \mathrm{i}\gamma(k)$. Substituting (11.31) and (11.32) into (11.29) yields

$$f_k(v) = \frac{e}{m}\left[\frac{\mathrm{i}}{\omega(k) - kv}\right]E_k\frac{\partial f_0}{\partial v} \, . \tag{11.33}$$

If (11.32) and (11.33) are substituted into (11.30), we find

$$\begin{aligned}\frac{\partial f_0(v,t)}{\partial t} = \left(\frac{e}{m}\right)^2\frac{\partial}{\partial v}\Bigg\langle\frac{1}{2\pi}\int E_{k'}\exp\{\mathrm{i}[k'x - \omega(k')t]\}\mathrm{d}k' \\ \times\frac{\mathrm{i}}{\omega(k) - kv}E_k\frac{\partial f_0(v,t)}{\partial v}\exp\{\mathrm{i}[kx - \omega(k)t]\}\mathrm{d}k\Bigg\rangle \, .\end{aligned} \tag{11.34}$$

The statistical average of (11.34) (integration over $x$) reduces to

$$\frac{\partial f_0(v,t)}{\partial t} = \frac{\partial}{\partial v}\left[D_{\mathrm{v}}(v)\frac{\partial f_0(v,t)}{\partial v}\right] , \tag{11.35}$$

$$D_{\mathrm{v}}(v) = \left(\frac{e}{m}\right)^2 \int_{-\infty}^{\infty} \frac{\mathrm{i}|E_k|^2 \exp\left[2\gamma(k)t\right]}{\omega_{\mathrm{r}}(k) - kv + \mathrm{i}\gamma(k)}\mathrm{d}k$$
$$= \left(\frac{e}{m}\right)^2 \int_{-\infty}^{\infty} \frac{\gamma(k)|E_k|^2 \exp\left[2\gamma(k)t\right]}{\left[\omega_{\mathrm{r}}(k) - kv\right]^2 + \gamma(k)^2}\mathrm{d}k .$$

When $|\gamma(k)| \ll |\omega_{\mathrm{r}}(k)|$, the diffusion coefficient in velocity space is

$$D_{\mathrm{v}}(v) = \left(\frac{e}{m}\right)^2 \pi \int |E_k|^2 \exp\left[2\gamma(k)t\right]\delta\left[\omega_{\mathrm{r}}(k) - kv\right]\mathrm{d}k$$
$$= \left(\frac{e}{m}\right)^2 \frac{\pi}{|v|}|E_k|^2 \exp\left[2\gamma(k)t\right]\Big|_{\omega/k=v} . \tag{11.36}$$

From Poisson's equation and (11.33), the dispersion equation can be derived:

$$\nabla \cdot \boldsymbol{E} = -\frac{e}{\epsilon_0}\int f_1 \mathrm{d}v , \qquad \mathrm{i}kE_k = -\frac{e}{\epsilon_0}\int f_k \mathrm{d}v ,$$

$$1 + \frac{\Pi_{\mathrm{e}}^2}{k}\frac{1}{n}\int \left[\frac{1}{\omega(k) - kv}\right]\frac{\partial f_0}{\partial v}\mathrm{d}v = 0 . \tag{11.37}$$

Under the assumption that $|\gamma| \ll |\omega_{\mathrm{r}}|$ ($\omega = \omega_{\mathrm{r}} + \mathrm{i}\gamma$), the solution of (11.37) for $\gamma$ is given by the same equation as (11.11).

Equation (11.35) is the diffusion equation in the velocity space. When the electron distribution function is given by the profile shown in Fig. 11.2b, where $v\partial f/\partial v > 0$ has positive gradient near $v_1$, then waves with phase velocity $\omega/k \approx v_1$ grow due to Landau amplification and the amplitude of $|E_k|$ increases. The diffusion coefficient $D_{\mathrm{v}}$ in velocity space becomes large and anomalous diffusion takes place in velocity space. The positive gradient of $\partial f/\partial v$ near $\sim v_1$ decreases and the profile of the distribution function eventually becomes flat near $v \sim v_1$.

Let us consider the other case. When a wave is externally excited (by antenna) in a plasma with Maxwellian distribution function, as shown in Fig. 11.2a, the diffusion coefficient $D_{\mathrm{v}}$ at $v = \omega/k$ is increased. The gradient of the distribution function near $v = \omega/k$ becomes flat, as can be seen in Fig. 16.17 of Chap. 16.

# 12 Hot Plasma

When the temperature of a plasma becomes hot and the thermal velocities $v_T = (\kappa T/m)^{1/2}$ of electrons or ions become comparable to the phase velocity of waves, Landau damping or amplification may occur, as described in Chap. 11. In order to study the propagation and absorption (damping) or excitation (amplification) of waves and perturbation systematically, the dielectric tensor of a hot plasma must be used. The dielectric tensor of a hot plasma is explained and derived in Sects. 12.1–12.3 and 12.8.

We discuss wave heating, i.e., wave heating in the ion cyclotron frequency range (ICRF), in Sect. 12.4, lower hybrid heating (LHH) in Sect. 12.5, and electron cyclotron heating (ECH) in Sect. 12.6. The physical processes of wave heating are not simple and the interactions of waves and plasmas display a great deal of variety, so that various applications are possible depending on the development of wave heating methods.

Waves are excited in the plasma by antennas or waveguides located outside the plasma (*excitation of wave, antenna–plasma coupling*). When the electric field of the excited wave is parallel to the confining magnetic field of the plasma, the electron, which can move along the magnetic field, may cancel the electric field. However, if the frequency of the wave is higher than the plasma frequency the electron cannot follow the change in the electric field, and the wave then propagates through the plasma. When the electric field of the excited wave is perpendicular to the magnetic field, the electrons move in the direction of $\boldsymbol{E} \times \boldsymbol{B}$ (under the condition $\omega < \Omega_{\mathrm{e}}$) and so cannot cancel the electric field. In this case the wave can propagate through the plasma even if the wave frequency is lower than the plasma frequency. Excitation consists in pumping the high-frequency electromagnetic wave into the plasma through the coupling system. If the structure of the coupling system has the same periodicity as the eigenmode wave, the wave can be excited resonantly (*resonant excitation*). Electron cyclotron heating (and neutral beam injection) can be launched in vacuum and propagate directly into the plasma without attenuation or interaction with the edge. Consequently, the launching structures do not have to be in close proximity to the plasma and have an advantage against thermal load and erosion by the plasma.

Excited waves may propagate and pass through the plasma center without damping (heating) in some cases and may refract and turn back to the

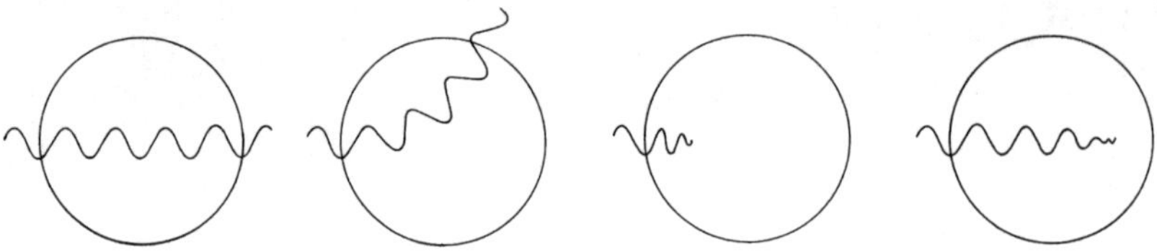

**Fig. 12.1.** Fate of excited wave: passing through, refraction and reflection, absorption near boundary, and absorption at center of plasma

external region without passing the plasma center, or may be reflected by the cutoff layer (see Fig. 12.1). The wave may be converted to another type by *mode conversion* (*wave propagation*).

The waves propagating in the plasma are absorbed and damped at the locations where Landau damping and cyclotron damping occur and heat the plasma. Therefore, the plasma center must be heated so that the waves can propagate into the plasma center without absorption and be absorbed when they reach the plasma center (*wave heating*).

When the velocity distribution function deviates from the stable Maxwell distribution, the plasma may be unstable due to Landau and cyclotron amplification. This type of instability is called a *velocity space instability*. Electron beam instability is described as an example in Sect. 12.7.

## 12.1 Energy Flow

Energy transport and the propagation of waves in the plasma medium are very important in the wave heating of plasmas. The equation of energy flow is derived by taking the difference between the scalar product of $\boldsymbol{H}$ with (10.15) and the scalar product of $\boldsymbol{E}$ with (10.16):

$$\nabla\cdot(\boldsymbol{E}\times\boldsymbol{H})+\boldsymbol{E}\cdot\frac{\partial\boldsymbol{D}}{\partial t}+\boldsymbol{H}\cdot\frac{\partial\boldsymbol{B}}{\partial t}=0\,. \tag{12.1}$$

$\boldsymbol{P}\equiv\boldsymbol{E}\times\boldsymbol{H}$ is called the *Poynting vector* and represents the energy flow of the electromagnetic field. This Poynting equation does not include the effect of electric resistivity by electron–ion collisions.

Plasmas are dispersive media and dielectric tensors are dependent on the propagation vector $\boldsymbol{k}$ and the frequency $\omega$. Denote the Fourier components of $\boldsymbol{E}(\boldsymbol{r},t)$ and $\boldsymbol{D}(\boldsymbol{r},t)$ by $\boldsymbol{E}_{k\,\omega}(\boldsymbol{k},\omega)$ and $\boldsymbol{D}_{k\,\omega}(\boldsymbol{k},\omega)$, respectively. Then we find

$$\boldsymbol{D}_{k\,\omega}=\frac{1}{(2\pi)^2}\int\boldsymbol{D}(\boldsymbol{r},t)\exp\big[-\mathrm{i}(\boldsymbol{k}\cdot\boldsymbol{r}-\omega t)\big]\,\mathrm{d}\boldsymbol{r}\,\mathrm{d}t\,,$$

$$\boldsymbol{E}_{k\,\omega}=\frac{1}{(2\pi)^2}\int\boldsymbol{E}(\boldsymbol{r},t)\exp\big[-\mathrm{i}(\boldsymbol{k}\cdot\boldsymbol{r}-\omega t)\big]\,\mathrm{d}\boldsymbol{r}\,\mathrm{d}t\,.$$

There is the following relation between them:

$$D_{k\,\omega}(\boldsymbol{k},\omega) = \epsilon_0 \boldsymbol{K}(\boldsymbol{k},\omega)\cdot \boldsymbol{E}_{k\,\omega}(\boldsymbol{k},\omega)\ ,$$

and we have

$$\boldsymbol{D}(\boldsymbol{r},t) = \frac{1}{(2\pi)^2}\epsilon_0 \int \boldsymbol{K}(\boldsymbol{k},\omega)\cdot \boldsymbol{E}_{k\,\omega}(\boldsymbol{k},\omega) \exp\left[\mathrm{i}(\boldsymbol{k}\cdot\boldsymbol{r} - \omega t)\right] \mathrm{d}\boldsymbol{k}\,\mathrm{d}\omega\ ,$$

$$\boldsymbol{E}(\boldsymbol{r},t) = \frac{1}{(2\pi)^2} \int \boldsymbol{E}_{\boldsymbol{k},\omega}(\boldsymbol{k},\omega) \exp\left[\mathrm{i}(\boldsymbol{k}\cdot\boldsymbol{r} - \omega t)\right] \mathrm{d}\boldsymbol{k}\,\mathrm{d}\omega\ .$$

From the formula for the Fourier integral:

$$\boldsymbol{D}(\boldsymbol{r},t) = \epsilon_0 \int \widehat{\boldsymbol{K}}(\boldsymbol{r}-\boldsymbol{r}', t-t')\cdot \boldsymbol{E}(\boldsymbol{r}',t')\,\mathrm{d}\boldsymbol{r}'\,\mathrm{d}t'\ ,$$

where $\widehat{\boldsymbol{K}}(\boldsymbol{r},t)$ is

$$\widehat{\boldsymbol{K}}(\boldsymbol{r},t) = \frac{1}{(2\pi)^4} \int \boldsymbol{K}(\boldsymbol{k},\omega) \exp\left[i(\boldsymbol{k}\cdot\boldsymbol{r} - \omega t)\right] \mathrm{d}\boldsymbol{k}\,\mathrm{d}\omega\ .$$

Therefore analysis of general electromagnetic fields in dispersive media is not simple. However, if the electric field consists of Fourier components in a narrow region near $\boldsymbol{k}_0$, $\omega_0$, and if $\boldsymbol{K}$ changes slowly as $\boldsymbol{k}$, $\omega$ change, then we can use the relation

$$\boldsymbol{D}(\boldsymbol{r},t) = \epsilon_0 \boldsymbol{K}(\boldsymbol{k}_0,\omega_0)\cdot \boldsymbol{E}(\boldsymbol{r},t).$$

From now on, we will discuss only this simple case. The relation between the magnetic induction $\boldsymbol{B}$ and the magnetic intensity $\boldsymbol{H}$ is $\boldsymbol{B} = \mu_0 \boldsymbol{H}$.

The quasi-periodic functions $A$, $B$ may be expressed by

$$A = A_0 \exp\left[-\mathrm{i}\int_{-\infty}^{t} (\omega_\mathrm{r} + \mathrm{i}\omega_\mathrm{i})\mathrm{d}t'\right] = A_0 \exp(-\mathrm{i}\phi_\mathrm{r} + \phi_\mathrm{i})\ ,$$

$$B = B_0 \exp\left[-\mathrm{i}\int_{-\infty}^{t} (\omega_\mathrm{r} + \mathrm{i}\omega_\mathrm{i})\mathrm{d}t'\right] = B_0 \exp(-\mathrm{i}\phi_\mathrm{r} + \phi_\mathrm{i})\ ,$$

where $\phi_\mathrm{r}$ and $\phi_\mathrm{i}$ are real. Denoting the average of the product of the real part of $A$ with the real part of $B$ by $\overline{AB}$, we find

$$\begin{aligned}\overline{AB} &= \frac{1}{2}\times\frac{1}{2}\Big\langle\left[A_0\exp(-\mathrm{i}\phi_\mathrm{r}+\phi_\mathrm{i}) + A_0^*\exp(\mathrm{i}\phi_\mathrm{r}+\phi_\mathrm{i})\right] \\ &\qquad\times\left[B_0\exp(-\mathrm{i}\phi_\mathrm{r}+\phi_\mathrm{i}) + B_0^*\exp(\mathrm{i}\phi_\mathrm{r}+\phi_\mathrm{i})\right]\Big\rangle \\ &= \frac{1}{4}(A_0B_0^* + A_0^*B_0)\exp(2\phi_\mathrm{i}) = \frac{1}{2}\mathrm{Re}(AB^*)\ . \end{aligned} \tag{12.2}$$

The average of the Poynting equation becomes

$$\nabla\cdot\boldsymbol{P}+\frac{\partial W}{\partial t}=0\,, \tag{12.3}$$

$$\boldsymbol{P}=\frac{1}{2\mu_0}\mathrm{Re}(\boldsymbol{E}_0\times\boldsymbol{B}_0^*)\exp 2\int_{-\infty}^{t}\omega_{\mathrm{i}}\mathrm{d}t'\,, \tag{12.4}$$

$$\begin{aligned}\frac{\partial W}{\partial t}&=\frac{1}{2}\mathrm{Re}\left[\left(\frac{\boldsymbol{B}^*}{\mu_0}\cdot\frac{\partial\boldsymbol{B}}{\partial t}\right)+\epsilon_0\boldsymbol{E}^*\cdot\frac{\partial}{\partial t}(\boldsymbol{K}\cdot\boldsymbol{E})\right]\\&=\frac{1}{2}\mathrm{Re}\left[-\mathrm{i}\omega\frac{\boldsymbol{B}^*\cdot\boldsymbol{B}}{\mu_0}+\epsilon_0(-\mathrm{i}\omega)\boldsymbol{E}^*\cdot\boldsymbol{K}\cdot\boldsymbol{E}\right]\\&=\frac{1}{2}\omega_{\mathrm{i}}\frac{\boldsymbol{B}\cdot\boldsymbol{B}^*}{\mu_0}+\frac{\epsilon_0}{2}\left[\omega_{\mathrm{i}}\mathrm{Re}(\boldsymbol{E}^*\cdot\boldsymbol{K}\cdot\boldsymbol{E})+\omega_{\mathrm{r}}\mathrm{Im}(\boldsymbol{E}^*\cdot\boldsymbol{K}\cdot\boldsymbol{E})\right]\,.\end{aligned} \tag{12.5}$$

From the relations

$$\boldsymbol{E}^*\cdot\boldsymbol{K}\cdot\boldsymbol{E}=\sum_i E_i^*\sum_j K_{ij}E_j,$$

$$\begin{aligned}\boldsymbol{E}\cdot\boldsymbol{K}^*\cdot\boldsymbol{E}^*&=\sum_i E_i\sum_j K_{ij}^*E_j^*=\sum_j E_j^*\sum_i (K_{ji}^{\mathrm{T}})^*E_i\\&=\sum_i E_i^*\sum_j (K_{ij}^{\mathrm{T}})^*E_j\,,\end{aligned}$$

we find

$$\mathrm{Re}(\boldsymbol{E}^*\cdot\boldsymbol{K}\cdot\boldsymbol{E})=\boldsymbol{E}^*\cdot\frac{\boldsymbol{K}+(\boldsymbol{K}^{\mathrm{T}})^*}{2}\cdot\boldsymbol{E}\,,$$

$$\mathrm{Im}(\boldsymbol{E}^*\cdot\boldsymbol{K}\cdot\boldsymbol{E})=\boldsymbol{E}^*\cdot\frac{(-\mathrm{i})\left[\boldsymbol{K}-(\boldsymbol{K}^{\mathrm{T}})^*\right]}{2}\cdot\boldsymbol{E}\,.$$

$(\boldsymbol{K}^{\mathrm{T}})^*$ is the complex conjugate of the transpose matrix $\boldsymbol{K}^{\mathrm{T}}$ of $\boldsymbol{K}$ (in which the columns and rows of components are exchanged), i.e., $K_{ij}^{\mathrm{T}}\equiv K_{ji}$. When the matrices $\boldsymbol{M}$ and $(\boldsymbol{M}^{\mathrm{T}})^*$ are equal, this kind of matrix is called a *Hermite matrix*. For the Hermite matrix, $(\boldsymbol{E}^*\cdot\boldsymbol{M}\cdot\boldsymbol{E})$ is always real. The dielectric tensor may be decomposed to

$$\boldsymbol{K}(\boldsymbol{k},\omega)=\boldsymbol{K}_{\mathrm{H}}(\boldsymbol{k},\omega)+\mathrm{i}\boldsymbol{K}_{\mathrm{I}}(\boldsymbol{k},\omega)\,.$$

As we shall explain in Sect. 12.3, $\boldsymbol{K}_{\mathrm{H}}$ and $\boldsymbol{K}_{\mathrm{I}}$ are Hermite when $\boldsymbol{k}$, $\omega$ are real. It will be proved that the term $\mathrm{i}\boldsymbol{K}_{\mathrm{I}}$ corresponds to Landau damping and cyclotron damping. When the imaginary part of $\omega$ is much smaller than the real part ($\omega=\omega_{\mathrm{r}}+\mathrm{i}\omega_{\mathrm{i}}$, $|\omega_{\mathrm{i}}|\ll|\omega_{\mathrm{r}}|$), we may write

$$\boldsymbol{K}(\boldsymbol{k},\omega_{\mathrm{r}}+i\omega_{\mathrm{i}})\approx\boldsymbol{K}_{\mathrm{H}}(\boldsymbol{k},\omega_{\mathrm{r}})+i\omega_{\mathrm{i}}\frac{\partial}{\partial\omega_{\mathrm{r}}}\boldsymbol{K}_{\mathrm{H}}(\boldsymbol{k},\omega_{\mathrm{r}})+i\boldsymbol{K}_{\mathrm{I}}(\boldsymbol{k},\omega_{\mathrm{r}}),$$

$$\frac{\boldsymbol{K}+(\boldsymbol{K}^{\mathrm{T}})^*}{2}=\boldsymbol{K}_{\mathrm{H}},\qquad\frac{-i[\boldsymbol{K}-(\boldsymbol{K}^{\mathrm{T}})^*]}{2}=\omega_{\mathrm{i}}\frac{\partial}{\partial\omega_{\mathrm{r}}}\boldsymbol{K}_{\mathrm{H}}+\boldsymbol{K}_{\mathrm{I}}.$$

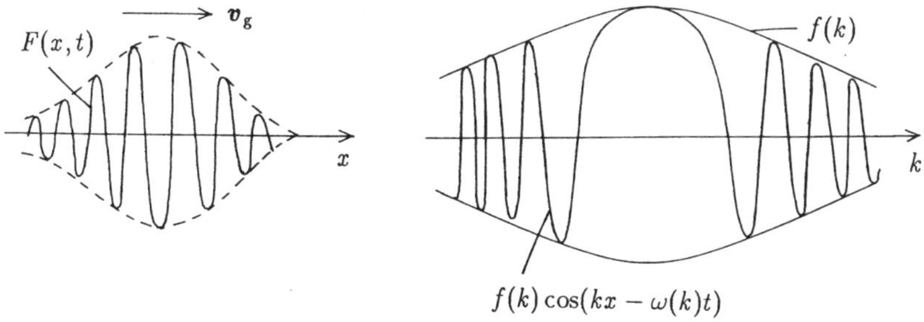


**Fig. 12.2.** $F(x,t)$ and $f(k)\cos\left[kx - w(k)t\right]$

When the Hermite component of $W$ (the term associated with $\boldsymbol{K}_{\rm H}$ in $W$) is denoted by $W_0$, this $W_0$ is given by

$$W_0 = \frac{1}{2}{\rm Re}\left[\frac{\boldsymbol{B}_0^*\cdot\boldsymbol{B}_0}{2\mu_0} + \frac{\epsilon_0}{2}\boldsymbol{E}_0^*\cdot\boldsymbol{K}_{\rm H}\cdot\boldsymbol{E}_0 + \frac{\epsilon_0}{2}\boldsymbol{E}_0^*\cdot\left(\omega_{\rm r}\frac{\partial}{\partial\omega_{\rm r}}\boldsymbol{K}_{\rm H}\right)\cdot\boldsymbol{E}_0\right]$$
$$= \frac{1}{2}{\rm Re}\left\{\frac{\boldsymbol{B}_0^*\cdot\boldsymbol{B}_0}{2\mu_0} + \frac{\epsilon_0}{2}\boldsymbol{E}_0^*\cdot\left[\frac{\partial}{\partial\omega}(\omega\boldsymbol{K}_{\rm H})\right]\cdot\boldsymbol{E}_0\right\}, \qquad (12.6)$$

and (12.3) and (12.5) yield

$$\frac{\partial W_0}{\partial t} = -\omega_{\rm r}\frac{1}{2}\epsilon_0\boldsymbol{E}_0^*\cdot\boldsymbol{K}_{\rm I}\cdot\boldsymbol{E}_0 - \nabla\cdot\boldsymbol{P}\,. \qquad (12.7)$$

The first term in (12.6) is the energy density of the magnetic field and the second term is the energy density of the electric field which includes the kinetic energy of coherent motion associated with the wave. Equation (12.6) gives the energy density of the wave in a dispersive media. The first term on the right-hand side of (12.7) represents the Landau and cyclotron dampings and the second term is the divergence of the flow of wave energy.

Let us consider the velocity of movement of the wave packet given by

$$F(\boldsymbol{r},t) = \int_{-\infty}^{\infty} f(\boldsymbol{k})\exp{\rm i}\left[\boldsymbol{k}\cdot\boldsymbol{r} - \omega(\boldsymbol{k})t\right]{\rm d}\boldsymbol{k}\,, \qquad (12.8)$$

when $\omega = \omega(\boldsymbol{k})$ is given. If $f(\boldsymbol{k})$ varies slowly, the position of the maximum of $F(\boldsymbol{r},t)$ at $t$ is the position of the stationary phase of

$$\frac{\partial}{\partial k_\alpha}\left[\boldsymbol{k}\cdot\boldsymbol{r} - \omega(\boldsymbol{k})t\right] = 0 \qquad (\alpha = x,y,z)\,,$$

because the main contribution to the integral (12.8) comes from the region near the stationary phase, as shown in Fig. 12.2. Consequently, the velocity of the wave packet is

$$\left( \frac{x}{t} = \frac{\partial \omega(\boldsymbol{k})}{\partial k_x}, \quad \frac{y}{t} = \frac{\partial \omega(\boldsymbol{k})}{\partial k_y}, \quad \frac{z}{t} = \frac{\partial \omega(\boldsymbol{k})}{\partial k_z} \right),$$

that is

$$\boldsymbol{v}_{\mathrm{g}} = \left( \frac{\partial \omega}{\partial k_x}, \frac{\partial \omega}{\partial k_y}, \frac{\partial \omega}{\partial k_z} \right). \tag{12.9}$$

This velocity is called the *group velocity* and represents the velocity of energy flow.

## 12.2 Ray Tracing

When the wavelength of waves in the plasma is much less than the characteristic length (typically the minor radius $a$), the WKB approximation (geometrical optics approximation) can be applied. Let the dispersion relation be $D(\boldsymbol{k}, \omega, \boldsymbol{r}, t) = 0$. The direction of wave energy flow is given by the group velocity $\boldsymbol{v}_{\mathrm{g}} = \partial\omega/\partial\boldsymbol{k} \equiv (\partial\omega/\partial k_x,\ \partial\omega/\partial k_y,\ \partial\omega/\partial k_z)$, so that the optical ray can be given by $\mathrm{d}\boldsymbol{r}/\mathrm{d}t = \boldsymbol{v}_{\mathrm{g}}$. Although the quantities $(\boldsymbol{k}, \omega)$ change according to the change in $\boldsymbol{r}$, they always satisfy $D = 0$. Then the optical ray can be obtained by

$$\frac{\mathrm{d}\boldsymbol{r}}{\mathrm{d}s} = \frac{\partial D}{\partial \boldsymbol{k}}, \qquad \frac{\mathrm{d}\boldsymbol{k}}{\mathrm{d}s} = -\frac{\partial D}{\partial \boldsymbol{r}}, \tag{12.10}$$

$$\frac{\mathrm{d}t}{\mathrm{d}s} = -\frac{\partial D}{\partial \omega}, \qquad \frac{\mathrm{d}\omega}{\mathrm{d}s} = \frac{\partial D}{\partial t}. \tag{12.11}$$

Here $s$ is a measure of the length along the optical ray. Along the optical ray the variation $\delta D$ becomes zero:

$$\delta D = \frac{\partial D}{\partial \boldsymbol{k}} \cdot \delta \boldsymbol{k} + \frac{\partial D}{\partial \omega} \delta\omega + \frac{\partial D}{\partial \boldsymbol{r}} \cdot \delta \boldsymbol{r} + \frac{\partial D}{\partial t} \delta t = 0, \tag{12.12}$$

and $D(\boldsymbol{k}, \omega, \boldsymbol{r}, t) = 0$ is satisfied. Equations (12.10) and (12.11) reduce to

$$\frac{\mathrm{d}\boldsymbol{r}}{\mathrm{d}t} = \frac{\mathrm{d}\boldsymbol{r}}{\mathrm{d}s} \left( \frac{\mathrm{d}t}{\mathrm{d}s} \right)^{-1} = -\frac{\partial D}{\partial \boldsymbol{k}} \left( \frac{\partial D}{\partial \omega} \right)^{-1} = \left( \frac{\partial \omega}{\partial \boldsymbol{k}} \right)_{\boldsymbol{r}, t=\mathrm{const.}} = \boldsymbol{v}_{\mathrm{g}}.$$

Equation (12.10) has the same form as the equation of motion with Hamiltonian $D$. When $D$ does not depend on $t$ explicitly, $D = \mathrm{const.} = 0$ corresponds to the energy conservation law. If the plasma medium does not depend on $z$, $k_z = \mathrm{const.}$ corresponds to the momentum conservation law and is the same as Snell's law, $N_{\parallel} = \mathrm{const.}$

When $\boldsymbol{k} = \boldsymbol{k}_{\mathrm{r}} + \mathrm{i}\boldsymbol{k}_{\mathrm{i}}$ is a solution of $D = 0$ for a given real $\omega$ and $|\boldsymbol{k}_{\mathrm{i}}| \ll |\boldsymbol{k}_{\mathrm{r}}|$ is satisfied, we have

$$D(\boldsymbol{k}_{\mathrm{r}} + \mathrm{i}\boldsymbol{k}_{\mathrm{i}}, \omega) = \mathrm{Re}D(\boldsymbol{k}_{\mathrm{r}}, \omega) + \frac{\partial \mathrm{Re}D(\boldsymbol{k}_{\mathrm{r}}, \omega)}{\partial \boldsymbol{k}_{\mathrm{r}}} \cdot \mathrm{i}\boldsymbol{k}_{\mathrm{i}} + \mathrm{i}\mathrm{Im}D(\boldsymbol{k}_{\mathrm{r}}, \omega) = 0,$$

hence,

$$\mathrm{Re}D(\boldsymbol{k}_\mathrm{r},\omega)=0\ ,$$

$$\boldsymbol{k}_\mathrm{i}\cdot\frac{\partial \mathrm{Re}D(\boldsymbol{k}_\mathrm{r},\omega)}{\partial \boldsymbol{k}_\mathrm{r}}=-\mathrm{Im}D(\boldsymbol{k}_\mathrm{r},\omega)\ . \tag{12.13}$$

Then the wave intensity $I(\boldsymbol{r})$ becomes

$$I(\boldsymbol{r})=I(\boldsymbol{r}_0)\exp\left(-2\int_{\boldsymbol{r}_0}^{\boldsymbol{r}}\boldsymbol{k}_\mathrm{i}\cdot\mathrm{d}\boldsymbol{r}\right)\ , \tag{12.14}$$

$$\int \boldsymbol{k}_\mathrm{i}\cdot\mathrm{d}\boldsymbol{r}=\int \boldsymbol{k}_\mathrm{i}\cdot\frac{\partial D}{\partial \boldsymbol{k}}\mathrm{d}s=-\int \mathrm{Im}D(\boldsymbol{k}_\mathrm{r},\omega)\mathrm{d}s=-\int\frac{\mathrm{Im}D(\boldsymbol{k}_\mathrm{r},\omega)}{|\partial D/\partial \boldsymbol{k}|}\mathrm{d}l\ , \tag{12.15}$$

where $\mathrm{d}l$ is the length along the optical ray. Therefore the wave absorption can be estimated from (12.14) and (12.15) by tracing many optical rays. The geometrical optical approximation can provide the average wave intensity with a space resolution of, say, two or three times the wavelength.

## 12.3 Dielectric Tensor of Hot Plasma

In the process of wave absorption by a hot plasma, Landau damping or cyclotron damping are the most important damping processes, as discussed in Chap. 11. These are due to the interaction between the wave and so-called *resonant particles* satisfying

$$\omega-k_z v_z-n\Omega=0\ ,\qquad n=0,\pm1,\pm2,\dots\ .$$

In coordinates moving with the same velocity, the electric field is static ($\omega=0$) or has a cyclotron harmonic frequency ($\omega=n\Omega$). The case $n=0$ corresponds to Landau damping, whilst $n=1$ corresponds to electron cyclotron damping and $n=-1$ corresponds to ion cyclotron damping ($\omega>0$ is assumed).

Although nonlinear or stochastic processes accompany wave heating in many cases, the experimental results of wave heating or absorption are usually well described by linear or quasi-linear theories. The basis of the linear theory is the dispersion relation with the dielectric tensor $\boldsymbol{K}$ of a finite-temperature plasma. The absorbed power $P^{\mathrm{ab}}$ per unit volume of plasma is given by the first term on the right-hand side of (12.7):

$$P^{\mathrm{ab}}=\omega_\mathrm{r}\left(\frac{\epsilon_0}{2}\right)\boldsymbol{E}^*\cdot\boldsymbol{K}_\mathrm{I}\cdot\boldsymbol{E}\ .$$

Since $\boldsymbol{K}_\mathrm{H}$, $\boldsymbol{K}_\mathrm{I}$ is a Hermite matrix for real $\boldsymbol{k}$, $\omega$, as will be shown later in this section, the absorbed power $P^{\mathrm{ab}}$ is given by

$$P^{\mathrm{ab}}=\omega_\mathrm{r}\left(\frac{\epsilon_0}{2}\right)\mathrm{Re}\big[\boldsymbol{E}^*\cdot(-\mathrm{i})\boldsymbol{K}\cdot\boldsymbol{E}\big]_{\omega=\omega_\mathrm{r}}\ . \tag{12.16}$$

It is clear from the expression (12.19) for $\boldsymbol{K}$ that the absorbed power $P^{\mathrm{ab}}$ reduces to

$$P^{\mathrm{ab}}=\omega\frac{\epsilon_0}{2}\Big[|E_x|^2\mathrm{Im}K_{xx}+|E_y|^2\mathrm{Im}K_{yy}+|E_z|^2\mathrm{Im}K_{zz} \tag{12.17}$$
$$+2\mathrm{Im}(E_x^*E_y)\mathrm{Re}K_{xy}+2\mathrm{Im}(E_y^*E_z)\mathrm{Re}K_{yz}+2\mathrm{Im}(E_x^*E_z)\mathrm{Re}K_{xz}\Big].$$

Since (10.3) gives $\boldsymbol{j}=-\mathrm{i}\omega\boldsymbol{P}=-\mathrm{i}\epsilon_0\omega(\boldsymbol{K}-\boldsymbol{I})\cdot\boldsymbol{E}$, (12.16) may be described by

$$P^{\mathrm{ab}}=\frac{1}{2}\mathrm{Re}(\boldsymbol{E}^*\cdot\boldsymbol{j})_{\omega=\omega_{\mathrm{r}}}. \tag{12.18}$$

The process driving the dielectric tensor $\boldsymbol{K}$ of a finite-temperature plasma is described in Sect. 12.8. When the plasma is bi-Maxwellian,

$$f_0(v_\perp,v_z)=n_0F_\perp(v_\perp)F_z(v_z), \qquad F_\perp(v_\perp)=\frac{m}{2\pi\kappa T_\perp}\exp\left(-\frac{mv_\perp^2}{2\kappa T_\perp}\right),$$

$$F_z(v_z)=\left(\frac{m}{2\pi\kappa T_z}\right)^{1/2}\exp\left[-\frac{m(v_z-V)^2}{2\kappa T_z}\right],$$

the dielectric tensor $\boldsymbol{K}$ is given by

$$\boldsymbol{K}=\boldsymbol{I}+ \tag{12.19}$$
$$\sum_{\mathrm{i,e}}\frac{\Pi^2}{\omega^2}\left[\sum_n\left\{\zeta_0Z(\zeta_n)-\left(1-\frac{1}{\lambda_T}\right)[1+\zeta_nZ(\zeta_n)]\right\}\mathrm{e}^{-b}\boldsymbol{X}_n+2\eta_0^2\lambda_T\boldsymbol{L}\right],$$

$$\boldsymbol{X}_n=\begin{pmatrix} n^2I_n/b & \mathrm{i}n(I_n'-I_n) & -(2\lambda_T)^{1/2}\eta_n\dfrac{n}{\alpha}I_n \\ -\mathrm{i}n(I_n'-I_n) & (n^2/b+2b)I_n-2bI_n' & \mathrm{i}(2\lambda_T)^{1/2}\eta_n\alpha(I_n'-I_n) \\ -(2\lambda_T)^{1/2}\eta_n\dfrac{n}{\alpha}I_n & -\mathrm{i}(2\lambda_T)^{1/2}\eta_n\alpha(I_n'-I_n) & 2\lambda_T\eta_n^2I_n \end{pmatrix}, \tag{12.20}$$

$$Z(\zeta)\equiv\frac{1}{\pi^{1/2}}\int_{-\infty}^{\infty}\frac{\exp(-\beta^2)}{\beta-\zeta}\,\mathrm{d}\beta,$$

where $I_n(b)$ is the $n$th modified Bessel function and

$$\eta_n\equiv\frac{\omega+n\Omega}{2^{1/2}k_zv_{T_z}}, \qquad \zeta_n\equiv\frac{\omega-k_zV+n\Omega}{2^{1/2}k_zv_{T_z}},$$

$$\lambda_T\equiv\frac{T_z}{T_\perp}, \qquad b\equiv\left(\frac{k_xv_{T\perp}}{\Omega}\right)^2, \qquad \alpha\equiv b^{1/2},$$

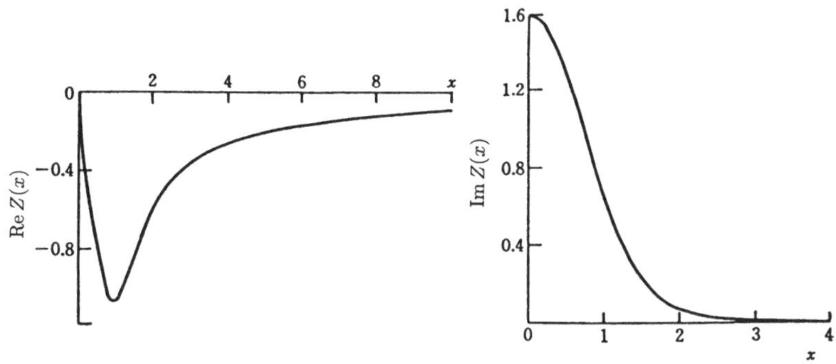


**Fig. 12.3.** Real part Re$Z(x)$ and imaginary part Im$Z(x)$ of $Z(x)$

$$v_{T_z}^2 \equiv \frac{\kappa T_z}{m}\,, \qquad v_{T_\perp}^2 \equiv \frac{\kappa T_\perp}{m}\,.$$

The components of the matrix $\boldsymbol{L}$ are zero except for $L_{zz} = 1$.

When the plasma is isotropic Maxwellian ($T_z = T_\perp$) and $V = 0$, then $\eta_n = \zeta_n$, $\lambda_T = 1$, and (12.19) reduces to

$$\boldsymbol{K} = \boldsymbol{I} + \sum_{\mathrm{i,e}} \frac{\Pi^2}{\omega^2}\left[\sum_{n=-\infty}^{\infty} \zeta_0 Z(\zeta_n)\mathrm{e}^{-b}\boldsymbol{X}_n + 2\zeta_0^2 \boldsymbol{L}\right]\,. \tag{12.21}$$

$Z(\zeta)$ is called the *plasma dispersion function*. When the imaginary part Im $\omega$ of $\omega$ is smaller than the real part Re $\omega$ in magnitude ($|\mathrm{Im}\,\omega| < |\mathrm{Re}\,\omega|$), the imaginary part of the plasma dispersion function is

$$\mathrm{Im}Z(\zeta) = \mathrm{i}\frac{k_z}{|k_z|}\pi^{1/2}\exp(-\zeta^2)\,.$$

The real part Re$Z(x)$ ($x$ is real) is shown in Fig. 12.3.

The real part of $Z(x)$ is

$$\mathrm{Re}Z(x) = -2x\left[1 - (2/3)x^2 + \cdots\right]\,,$$

when $x \ll 1$ (hot plasma) and

$$\mathrm{Re}Z(x) = -x^{-1}\left[1 + (1/2)x^{-2} + (3/4)x^{-4} + \cdots\right]\,,$$

when $x \gg 1$ (cold plasma) [12.1–12.3]. The imaginary part of $Z(\zeta)$ represents the Landau damping and cyclotron damping terms, as described later in this section.

When $T \to 0$, that is, $\zeta_n \to \pm\infty$, $b \to 0$, the dielectric tensor of the hot plasma reduces to the dielectric tensor (10.9)–(10.13) of a cold plasma.

When $b = (k_x \rho_\Omega)^2 \ll 1$, where $\rho_\Omega = v_{T\perp}/\Omega$ is the Larmor radius, $\mathrm{e}^{-b}\boldsymbol{X}_{\mathrm{n}}$ can be expanded in powers of $b$ using

$$I_n(b) = \left(\frac{b}{2}\right)^n \sum_{l=0}^{\infty} \frac{1}{l!(n+l)!} \left(\frac{b}{2}\right)^{2l}$$

$$= \left(\frac{b}{2}\right)^n \left[\frac{1}{n!} + \frac{1}{1!(n+1)!}\left(\frac{b}{2}\right)^2 + \frac{1}{2!(n+2)!}\left(\frac{b}{2}\right)^4 + \cdots\right].$$

The expansion in powers of $b$ and the inclusion of terms up to the second harmonics in $\boldsymbol{K}$ gives

$$K_{xx} = 1 + \sum_j \left(\frac{\Pi_j}{\omega}\right)^2 \zeta_0 \left[(Z_1 + Z_{-1})\left(\frac{1}{2} - \frac{b}{2} + \cdots\right) + (Z_2 + Z_{-2})\left(\frac{b}{2} - \frac{b^2}{2} + \cdots\right) + \cdots\right]_j,$$

$$K_{yy} = 1 + \sum_j \left(\frac{\Pi_j}{\omega}\right)^2 \zeta_0 \left[Z_0(2b + \cdots) + (Z_1 + Z_{-1})\left(\frac{1}{2} - \frac{3b}{2} + \cdots\right) + (Z_2 + Z_{-2})\left(\frac{b}{2} - b^2 + \cdots\right) + \cdots\right]_j,$$

$$K_{zz} = 1 - \sum_j \left(\frac{\Pi_j}{\omega}\right)^2 \zeta_0 \left[2\zeta_0 W_0(1 - b + \cdots) + (\zeta_1 W_1 + \zeta_{-1} W_{-1})(b + \cdots) + (\zeta_2 W_2 + \zeta_{-2} W_{-2})\left(\frac{b^2}{4} + \cdots\right) + \cdots\right]_j,$$

$$K_{xy} = \mathrm{i} \sum_j \left(\frac{\Pi_j}{\omega}\right)^2 \zeta_0 \left[(Z_1 - Z_{-1})\left(\frac{1}{2} - b + \cdots\right) + (Z_2 - Z_{-2})\left(\frac{b}{2} + \cdots\right) + \cdots\right]_j,$$

$$K_{xz} = 2^{1/2} \sum_j \left(\frac{\Pi_j}{\omega}\right)^2 b^{1/2} \zeta_0 \left[(W_1 - W_{-1})\left(\frac{1}{2} + \cdots\right) + (W_2 - W_{-2})\left(\frac{b}{4} + \cdots\right) + \cdots\right]_j,$$

$$K_{yz} = -2^{1/2}\mathrm{i}\sum_j \left(\frac{\Pi_j}{\omega}\right)^2 b^{1/2}\zeta_0 \left[W_0\left(-1+\frac{3}{2}b+\cdots\right)\right. \quad (12.22)$$

$$\left. +(W_1+W_{-1})\left(\frac{1}{2}+\cdots\right)+(W_2-W_{-2})\left(\frac{b}{4}+\cdots\right)+\cdots\right]_j ,$$

$$K_{yx} = -K_{xy}\ , \qquad K_{zx} = K_{xz}\ , \qquad K_{zy} = -K_{zy}\ ,$$

where

$$Z_{\pm n} \equiv Z(\zeta_{\pm n})\ , \qquad W_n \equiv -\big[1+\zeta_n Z(\zeta_n)\big]\ , \qquad \zeta_n = \frac{\omega + n\Omega}{2^{1/2}k_z(\kappa T_z/m)^{1/2}}\ .$$

When $x \gg 1$,

$$\mathrm{Re}W(x) = (1/2)x^{-2}\big[1+(3/2)x^{-2}+\cdots\big]\ .$$

The absorbed power by Landau damping (including transit time damping) may be estimated from the terms associated with the imaginary part $G_0$ of $\zeta_0 Z(\zeta_0)$ in (12.22) of $K_{ij}$:

$$G_0 \equiv \mathrm{Im}\zeta_0 Z(\zeta_0) = (k_z/|k_z|)\pi^{1/2}\zeta_0\exp(-\zeta_0^2)\ .$$

Since

$$(\mathrm{Im}K_{yy})_0 = \left(\frac{\Pi_j}{\omega}\right)^2 2bG_0\ , \qquad (\mathrm{Im}K_{zz})_0 = \left(\frac{\Pi_j}{\omega}\right)^2 2\zeta_0^2 G_0\ ,$$

$$(\mathrm{Re}K_{yz})_0 = \left(\frac{\Pi_j}{\omega}\right)^2 2^{1/2}b^{1/2}\zeta_0 G_0\ ,$$

the contribution of these terms to the absorption power (12.17) is

$$P_0^{\mathrm{ab}} = \omega\left(\frac{\Pi_j}{\omega}\right)^2 G_0\epsilon_0\left[|E_y|^2 b + |E_z|^2\zeta_0^2 + \mathrm{Im}(E_y^* E_z)(2b)^{1/2}\zeta_0\right]\ . \quad (12.23)$$

The first term is the transit time damping, equal to (11.16). The second term is the Landau damping, equal to (11.10). The third arises from interference between the first two terms.

The absorption power due to cyclotron damping and harmonic cyclotron damping is obtained by the contribution from the terms

$$G_{\pm n} \equiv \mathrm{Im}\zeta_0 Z_{\pm n} = \frac{k_z}{|k_z|}\pi^{1/2}\zeta_0\exp(-\zeta_{\pm n}^2)\ ,$$

and for the case $b \ll 1$,

$$(\mathrm{Im}K_{xx})_{\pm n} = (\mathrm{Im}K_{yy})_{\pm n} = \left(\frac{\Pi_j}{\omega}\right)^2 G_{\pm n}\alpha_n\ ,$$

$$(\mathrm{Im}K_{zz})_{\pm n} = \left(\frac{\Pi_j}{\omega}\right)^2 2\zeta_{\pm n}^2 G_{\pm n} b\alpha_n n^{-2} ,$$

$$(\mathrm{Re}K_{xy})_{\pm n} = -\left(\frac{\Pi_j}{\omega}\right)^2 G_{\pm n}(\pm\alpha_n) ,$$

$$(\mathrm{Re}K_{yz})_{\pm n} = -\left(\frac{\Pi_j}{\omega}\right)^2 (2b)^{1/2}\zeta_{\pm n} G_{\pm n}\alpha_n n^{-1} ,$$

$$(\mathrm{Im}K_{xz})_{\pm n} = -\left(\frac{\Pi_j}{\omega}\right)^2 (2b)^{1/2}\zeta_{\pm n} G_{\pm n}(\pm\alpha_n) n^{-1} ,$$

$$\alpha_n = n^2(2\times n!)^{-1}\left(\frac{b}{2}\right)^{n-1} .$$

The contribution of these terms to the absorbed power (12.17) is

$$P_{\pm n}^{\mathrm{ab}} = \omega\left(\frac{\Pi_j}{\omega}\right)^2 G_n\left(\frac{\epsilon_0}{2}\right)\alpha_n|E_x \pm \mathrm{i}E_y|^2 . \tag{12.24}$$

Since

$$\zeta_n = \frac{\omega + n\Omega_{\mathrm{i}}}{2^{1/2}k_z v_{T_{\mathrm{i}}}} = \frac{\omega - n|\Omega_{\mathrm{i}}|}{2^{1/2}k_z v_{T_{\mathrm{i}}}} ,$$

the term in $+n$ is dominant for *ion cyclotron damping* ($\omega > 0$), and since

$$\zeta_{-n} = \frac{\omega - n\Omega_{\mathrm{e}}}{2^{1/2}k_z v_{T_{\mathrm{e}}}} ,$$

the term in $-n$ is dominant for *electron cyclotron damping* ($\omega > 0$). The relative strength of $\boldsymbol{E}$ components can be estimated from the following equations:

$$\begin{aligned}
(K_{xx} - N_\parallel^2)E_x + K_{xy}E_y + (K_{xz} + N_\perp N_\parallel)E_z &= 0 ,\\
-K_{xy}E_x + (K_{yy} - N_\parallel^2 - N_\perp^2)E_y + K_{yz}E_z &= 0 ,\\
(K_{xz} + N_\perp N_\parallel)E_x - K_{yz}E_y + (K_{zz} - N_\perp^2)E_z &= 0 .
\end{aligned} \tag{12.25}$$

For cold plasmas,

$$K_{xx} \to K_\perp ,\quad K_{yy} \to K_\perp ,\quad K_{zz} \to K_\parallel ,\quad K_{xy} \to -\mathrm{i}K_\times ,$$

$$K_{xz} \to 0 ,\quad K_{yz} \to 0 ,$$

can be substituted into (12.25), and the ratio is

$$E_x : E_y : E_z = (K_\perp - N^2)\times(K_\parallel - N_\perp^2) : -\mathrm{i}K_x(K_\parallel - N_\perp^2) : -N_\parallel N_\perp(K_\perp - N^2) .$$

In order to obtain the magnitude of the electric field, one must solve the Maxwell equations with the dielectric tensor of (12.19). In this case the density, the temperature, and the magnetic field are functions of the coordinates. Therefore the simplified model must be used for analytical solutions. Otherwise numerical calculations are required to derive the wave field.

## 12.4 Wave Heating in the Ion Cyclotron Frequency Range

The dispersion relation of waves in the ion cyclotron frequency range (ICRF) is given by (10.64a) and (10.64c) and reduces to

$$N_\parallel^2 = \frac{N_\perp^2}{2[1-(\omega/\Omega_\mathrm{i})^2]}\left[-\left[1-\left(\frac{\omega}{\Omega_\mathrm{i}}\right)^2\right]+\frac{2\omega^2}{k_\perp^2 v_\mathrm{A}^2}\right.$$
$$\left.\pm\left\{\left[1-\left(\frac{\omega}{\Omega_\mathrm{i}}\right)^2\right]^2+4\left(\frac{\omega}{\Omega_\mathrm{i}}\right)^2\left(\frac{\omega}{k_\perp v_\mathrm{A}}\right)^4\right\}^{1/2}\right] .$$

The plus sign corresponds to the slow wave (L wave, ion cyclotron wave), and the minus sign corresponds to the fast wave (R wave, extraordinary wave). When $1-\omega^2/\Omega_\mathrm{i}^2 \ll 2(\omega/k_\perp v_\mathrm{A})^2$, the dispersion relation becomes

$$k_z^2 = 2\left(\frac{\omega^2}{v_\mathrm{A}^2}\right)\left(1-\frac{\omega^2}{\Omega_\mathrm{i}^2}\right)^{-1} \qquad \text{(slow wave)} ,$$

$$k_z^2 = -\frac{k_\perp^2}{2}+\frac{\omega^2}{2v_\mathrm{A}^2} \qquad \text{(fast wave)} .$$

Since the externally excited waves usually have propagation vectors with $0 < k_z^2 < (\pi/a)^2$, $k_\perp^2 > (\pi/a)^2$ usually, there are constraints for slow wave, viz.,

$$\frac{\omega^2}{v_\mathrm{A}^2}\frac{2}{(1-\omega^2/\Omega_\mathrm{i}^2)} < \left(\frac{\pi}{a}\right)^2 , \qquad n_{20}a^2 < 1.3\times10^{-3}\frac{A}{Z^2}\frac{\Omega_\mathrm{i}^2}{\omega^2}\left(1-\frac{\omega^2}{\Omega_\mathrm{i}^2}\right) ,$$

and for fast wave [12.4], viz.,

$$\frac{\omega^2}{2v_\mathrm{A}^2} > \left(\frac{\pi}{a}\right)^2 , \qquad n_{20}a^2 > 0.5\times10^{-2}\frac{A}{Z^2}\frac{\Omega_\mathrm{i}^2}{\omega^2}$$

where $n_{20}$ is the ion density in $10^{20}\,\mathrm{m}^{-3}$, $a$ is the plasma radius in meters, and $A$ is the atomic number.

An ion cyclotron wave (slow wave) can be excited by a Stix coil [12.1] and can propagate and heat ions in a low-density plasma. But it cannot propagate in a high-density plasma like that of a tokamak.

The fast wave is an extraordinary wave in this frequency range and can be excited by a loop antenna, which generates a high-frequency electric field perpendicular to the magnetic field (see Sect. 10.2). The fast wave can propagate in a high-density plasma. The fast wave in a plasma with a single ion species has $E_x + \mathrm{i}E_y = 0$ at $\omega = |\Omega_\mathrm{i}|$ in the cold plasma approximation, so that it is not absorbed by ion cyclotron damping. However, the electric field of the fast wave in a plasma with two ion species is $E_x + \mathrm{i}E_y \neq 0$, so that the fast wave can be absorbed, i.e., the fast wave can heat the ions in this case.

Let us consider the heating of a plasma with two ion species, M and m, by a fast wave. The masses, charge numbers, and densities of the M and m ions are denoted by $m_{\mathrm{M}}$, $Z_{\mathrm{M}}$, $n_{\mathrm{M}}$ and $m_{\mathrm{m}}$, $Z_{\mathrm{m}}$, $n_{\mathrm{m}}$, respectively. When we use

$$\eta_{\mathrm{M}} \equiv \frac{Z_{\mathrm{M}}^2 n_{\mathrm{M}}}{n_{\mathrm{e}}}, \qquad \eta_{\mathrm{m}} \equiv \frac{Z_{\mathrm{m}}^2 n_{\mathrm{m}}}{n_{\mathrm{e}}},$$

we have $\eta_{\mathrm{M}}/Z_{\mathrm{M}} + \eta_{\mathrm{m}}/Z_{\mathrm{m}} = 1$ since $n_{\mathrm{e}} = Z_{\mathrm{M}} n_{\mathrm{M}} + Z_{\mathrm{m}} n_{\mathrm{m}}$. Since $(\Pi_{\mathrm{e}}/\omega)^2 \gg 1$ in an ICRF wave, the dispersion relation in the cold plasma approximation is given by (10.2) as follows:

$$N_\perp^2 = \frac{(R - N_\parallel^2)(L - N_\parallel^2)}{K_\perp - N_\parallel^2},$$

$$R = -\frac{\Pi_{\mathrm{i}}^2}{\omega^2}\left[\frac{(m_{\mathrm{M}}/m_{\mathrm{m}})\eta_{\mathrm{m}}\omega}{\omega + |\Omega_{\mathrm{m}}|} + \frac{\eta_{\mathrm{M}}\omega}{\omega + |\Omega_{\mathrm{M}}|} - \frac{\omega}{|\Omega_{\mathrm{M}}|/Z_{\mathrm{M}}}\right],$$

$$L = -\frac{\Pi_{\mathrm{i}}^2}{\omega^2}\left[\frac{(m_{\mathrm{M}}/m_{\mathrm{m}})\eta_{\mathrm{m}}\omega}{\omega - |\Omega_{\mathrm{m}}|} + \frac{\eta_{\mathrm{M}}\omega}{\omega - |\Omega_{\mathrm{M}}|} + \frac{\omega}{|\Omega_{\mathrm{M}}|/Z_{\mathrm{M}}}\right],$$

$$K_\perp = -\frac{\Pi_{\mathrm{i}}^2}{\omega^2}\left[\frac{(m_{\mathrm{M}}/m_{\mathrm{m}})\eta_{\mathrm{m}}\omega^2}{\omega^2 - \Omega_{\mathrm{m}}^2} + \frac{\eta_{\mathrm{M}}\omega^2}{\omega^2 - \Omega_{\mathrm{M}}^2}\right],$$

$$\Pi_{\mathrm{i}}^2 \equiv \frac{n_{\mathrm{e}} e^2}{\epsilon_0 m_{\mathrm{M}}}.$$

Therefore ion–ion hybrid resonance occurs at $K_\perp - N_\parallel^2 = 0$, that is,

$$\frac{\eta_{\mathrm{m}}(m_{\mathrm{M}}/m_{\mathrm{m}})\omega^2}{\omega^2 - \Omega_{\mathrm{m}}^2} + \frac{\eta_{\mathrm{M}}\omega^2}{\omega^2 - \Omega_{\mathrm{M}}^2} \approx -\frac{\omega^2}{\Pi_{\mathrm{i}}^2} N_\parallel^2 \approx 0,$$

$$\omega^2 \approx \omega_{\mathrm{IH}} \equiv \frac{\eta_{\mathrm{M}} + \eta_{\mathrm{m}}(\mu^2/\mu')}{\eta_{\mathrm{M}} + \eta_{\mathrm{m}}/\mu'}\Omega_{\mathrm{m}}^2, \qquad \mu' \equiv \frac{m_{\mathrm{m}}}{m_{\mathrm{M}}}, \qquad \mu \equiv \frac{\Omega_{\mathrm{M}}}{\Omega_{\mathrm{m}}} = \frac{m_{\mathrm{m}} Z_{\mathrm{M}}}{m_{\mathrm{M}} Z_{\mathrm{m}}}.$$

Figure 12.4 shows the ion–ion hybrid resonance layer $K_\perp - N_\parallel^2 = 0$, the L cutoff layer $L - N_\parallel^2 = 0$, and the R cutoff layer $R - N_\parallel^2 = 0$ of a tokamak plasma with two ion species $D^+$ (M ion) and $H^+$ (m ion).

Since the $K_{zz}$ component of the dielectric tensor is much larger than the other component, even in a hot plasma, the dispersion relation of a hot plasma is [12.5]

$$\begin{vmatrix} K_{xx} - N_\parallel^2 & K_{xy} \\ -K_{xy} & K_{yy} - N_\parallel^2 - N_\perp^2 \end{vmatrix} = 0. \tag{12.26}$$

When we use the relation $K_{yy} \equiv K_{xx} + \Delta K_{yy}$, $|\Delta K_{yy}| \ll |K_{xx}|$,

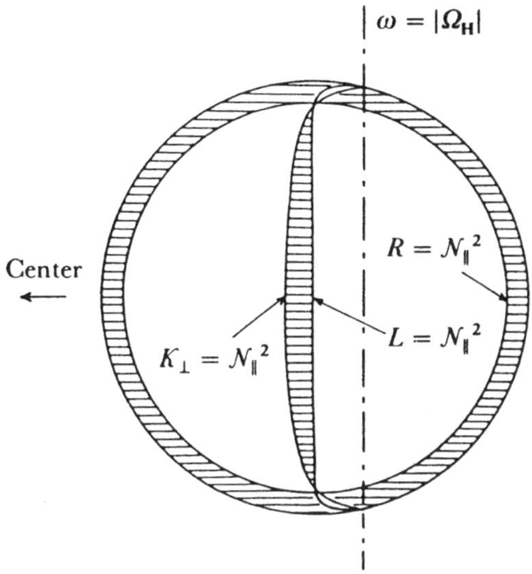


**Fig. 12.4.** L cutoff layer ($L = N_\parallel^2$), R cutoff layer ($R = N_\parallel^2$), and ion–ion hybrid resonance layer ($K_\perp = N_\parallel^2$) of an ICRF wave in a tokamak with two ion components $D^+$, $H^+$. The *shaded area* is the region $N_\perp^2 < 0$

$$N_\perp^2 = \frac{(K_{xx} - N_\parallel^2)(K_{xx} + \Delta K_{yy} - N_\parallel^2) + K_{xy}^2}{K_{xx} - N_\parallel^2}$$

$$\approx \frac{(K_{xx} + \mathrm{i}K_{xy} - N_\parallel^2)(K_{xx} - \mathrm{i}K_{xy} - N_\parallel^2)}{K_{xx} - N_\parallel^2} \, .$$

When $\omega^2$ is near $\omega_{\mathrm{IH}}^2$, $K_{xx}$ is given by [see (12.22)]

$$K_{xx} = -\frac{\Pi_{\mathrm{i}}^2}{\omega^2}\left[\frac{m_{\mathrm{M}}}{2m_{\mathrm{m}}}\eta_{\mathrm{m}}\zeta_0 Z(\zeta_1) + \frac{\eta_{\mathrm{M}}\omega^2}{\omega^2 - \Omega_{\mathrm{M}}^2}\right] .$$

The resonance condition is $K_{xx} = N_\parallel^2$. The value of $Z(\zeta_1)$ that appears in the dispersion equation is finite and $0 > Z(\zeta_1) > -1.08$. The condition

$$\eta_{\mathrm{m}} \geq \eta_{\mathrm{cr}} \equiv \frac{2}{1.08}\frac{m_{\mathrm{m}}}{m_{\mathrm{M}}} 2^{1/2} N_\parallel \frac{v_{T_{\mathrm{i}}}}{c}\left(\frac{\eta_{\mathrm{M}}\omega^2}{\omega^2 - \Omega_{\mathrm{M}}^2} + N_\parallel^2 \frac{\omega^2}{\Pi_{\mathrm{i}}^2}\right)$$

is required to obtain the resonance condition. This differs from the cold plasma dispersion equation (note the difference between $K_{xx}$ and $K_\perp$).

It can be deduced from the dispersion equation (12.26) that the mode conversion [12.5] from the fast wave to the ion Bernstein wave occurs at the resonance layer when $\eta_{\mathrm{m}} \geq \eta_{\mathrm{cr}}$. When the L cutoff layer and the ion–ion hybrid resonance layer are close to each other, as shown in Fig. 12.4, the fast wave propagating from the outside torus penetrates the L cutoff layer partly by the tunneling effect and is converted to the ion Bernstein wave. The mode-converted wave is absorbed by ion cyclotron damping or electron Landau damping. The theory of mode conversion is described in Chap. 10 of [12.1]. ICRF experiments related to this topic were carried out in TFR.

When $\eta_{\rm m} < \eta_{\rm cr}$, $K_\perp = N_\parallel^2$ cannot be satisfied and the ion–ion hybrid resonance layer disappears. In this case a fast wave excited by the loop antenna outside the torus can pass through the R cutoff region (because the width is small), to be reflected by the L cutoff layer and bounced back and forth in the region surrounded by $R = N_\parallel^2$ and $L = N_\parallel^2$. In this region, there is a layer satisfying $\omega = |\Omega_{\rm m}|$, and the minority m ions are heated by the fundamental ion cyclotron damping. The majority M ions are heated by Coulomb collisions with m ions. If the mass of M ions is $l$ times the mass of m ions, the M ions are also heated by the $l$ th harmonic ion cyclotron damping. This type of experiment was carried out in PLT with good heating efficiency. This is called *minority heating*. The absorption power $P_{\rm e0}$ due to electron Landau damping per unit volume is given by (12.23), and it is important only when $\zeta_0 \leq 1$. In this case we have $E_y/E_z \approx K_{zz}/K_{yz} \approx 2\zeta_0^2/[2^{1/2}b^{1/2}\zeta_0(-{\rm i})]$ and $P_{\rm e0}$ is [12.6]

$$P_{\rm e0} = \frac{\omega\epsilon_0}{4}|E_y|^2\left(\frac{\Pi_{\rm e}}{\omega}\right)^2\left(\frac{k_\perp v_{T_{\rm e}}}{\Omega_{\rm e}}\right)^2 2\zeta_{0\rm e}\pi^{1/2}\exp(-\zeta_{0\rm e}^2)\,. \tag{12.27}$$

The absorption power $P_{{\rm i}n}$ by $n$ th harmonic ion cyclotron damping is given by (12.24) as follows:

$$P_{{\rm i}n} = \frac{\omega\epsilon_0}{2}|E_x + {\rm i}E_y|^2\left(\frac{\Pi_{\rm i}}{\omega}\right)^2\left(\frac{n^2}{2\times n!}\right)\left(\frac{b}{2}\right)^{n-1}$$
$$\times\frac{\omega}{2^{1/2}k_z v_{T_{\rm i}}}\pi^{1/2}\exp\left[-\frac{(\omega - n|\Omega_{\rm i}|)^2}{2(k_z v_{T_{\rm i}})^2}\right]\,. \tag{12.28}$$

The absorption power due to second harmonic cyclotron damping is proportional to the beta value of the plasma. In order to evaluate the absorption power from (12.27) and (12.28), we need the spatial distributions of $E_x$ and $E_y$ and it is possible to calculate these distributions numerically [12.7].

In the range of the higher harmonic ion cyclotron frequencies, i.e., $\omega \sim 2\Omega_{\rm i}$, $3\Omega_{\rm i}$, the direct excitation of the ion Bernstein wave has been studied by an external antenna or waveguide, which generates a high-frequency electric field parallel to the magnetic field [12.8].

## 12.5 Lower Hybrid Heating

Since $|\Omega_{\rm i}| \ll \Pi_{\rm i}$ in a tokamak plasma ($n_{\rm e} \geq 10^{13}\,{\rm cm}^{-3}$), the lower hybrid resonance frequency becomes

$$\omega_{\rm LH}^2 = \frac{\Pi_{\rm i}^2 + \Omega_{\rm i}^2}{1 + \Pi_{\rm e}^2/\Omega_{\rm e}^2 + Zm_{\rm e}/m_{\rm i}} \approx \frac{\Pi_{\rm i}^2}{1 + \Pi_{\rm e}^2/\Omega_{\rm e}^2} \,.$$

We have the relations $\Omega_{\rm e} \gg \omega_{\rm LH} \gg |\Omega_{\rm i}|$, $\Pi_{\rm i}^2/\Pi_{\rm e}^2 = |\Omega_{\rm i}|/\Omega_{\rm e}$. For a given frequency $\omega$, lower hybrid resonance $\omega = \omega_{\rm LH}$ occurs at the position where the electron density satisfies the condition

$$\frac{\Pi_{\rm e}^2(x)}{\Omega_{\rm e}^2} = \frac{\Pi_{\rm res}^2}{\Omega_{\rm e}^2} \equiv p \,, \qquad p = \frac{\omega^2}{\Omega_{\rm e}|\Omega_{\rm i}| - \omega^2} \,.$$

When the dispersion equation (10.20) of cold plasma is solved for $N_\perp^2$ using $N^2 = N_\parallel^2 + N_\perp^2$, we have

$$N_\perp^2 = \frac{K_\perp \widetilde{K}_\perp - K_\times^2 + K_\parallel \widetilde{K}_\perp}{2K_\perp} \pm \left[ \left( \frac{K_\perp \widetilde{K}_\perp - K_\times^2 + K_\parallel \widetilde{K}_\perp}{2K_\perp} \right)^2 + \frac{K_\parallel}{K_\perp}(K_\times^2 - \widetilde{K}_\perp^2) \right]^{1/2} ,$$

where $\widetilde{K}_\perp = K_\perp - N_\parallel^2$. The relations

$$h(x) \equiv \frac{\Pi_{\rm e}^2(x)}{\Pi_{\rm res}^2} \,, \quad K_\perp = 1 - h(x) \,, \quad K_\times = ph(x)\frac{\Omega_{\rm e}}{\omega} \,, \quad K_\parallel = 1 - \beta_\Pi h(x) \,,$$

$$\beta_\Pi \equiv \frac{\Pi_{\rm res}^2}{\omega^2} \sim O(m_{\rm i}/m_{\rm e}) \,, \quad \alpha \equiv \frac{\Pi_{\rm res}^2}{\omega \Omega_{\rm e}} \sim O(m_{\rm i}/m_{\rm e})^{1/2} \,, \quad \beta_\Pi h \gg 1 \,,$$

reduce this to

$$N_\perp^2(x) = \frac{\beta_\Pi h}{2(1-h)} \left[ N_\parallel^2 - (1 - h + ph) \pm \left\{ \left[ N_\parallel^2 - (1-h+ph) \right]^2 - 4(1-h)ph \right\}^{1/2} \right] . \tag{12.29}$$

The slow wave corresponds to the plus sign in (12.29). In order for the slow wave to propagate from the plasma edge with low density ($h \ll 1$) to the plasma center with high density ($\Pi_{\rm e}^2 = \Pi_{\rm res}^2$, $h = 1$), $N_\perp(x)$ must be real. We thus require the condition

$$N_\parallel > (1-h)^{1/2} + (ph)^{1/2} \,.$$

The right-hand side of the inequality has maximum value $(1+p)^{1/2}$ in the range $0 < h < 1$, so that the condition of accessibility of the lower hybrid wave to the resonant region becomes

$$N_\parallel^2 > N_{\parallel,\rm cr}^2 = 1 + p = 1 + \frac{\Pi_{\rm res}^2}{\Omega_{\rm e}^2} \,. \tag{12.30}$$

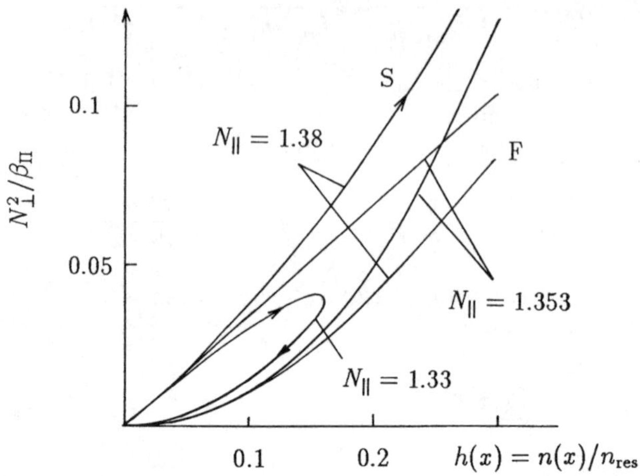


**Fig. 12.5.** Plot of lower hybrid wave in $N_\perp^2$–$h(x)$ $(= \Pi_{\rm e}^2(x)/\Pi_{\rm res}^2)$ diagram for the case $p = 0.353$, $N_{\parallel\rm cr}^2 = 1+p = 1.353$. This corresponds to an $\rm H^+$ plasma in $B = 3\,\rm T$, and $f = \omega/2\pi = 10^9\,\rm Hz$. The electron density for the parameter $\beta_\Pi = 7.06 \times 10^3$ $(= \Pi_{\rm res}^2/\omega^2)$ is $n_{\rm res} = 0.31 \times 10^{20}\,\rm m^{-3}$

If this condition is not satisfied, the externally excited slow wave propagates into the position where the square root term in (12.29) becomes zero and transforms to the fast wave there. Then the fast wave returns to the low-density region (see Fig. 12.5). The slow wave that satisfies the accessibility condition can approach the resonance region and $N_\perp$ can become large, so that the dispersion relation of a hot plasma must be used to examine the behavior of this wave. Near the lower hybrid resonance region, the electrostatic wave approximation (12.35) is applicable. Since $|\Omega_{\rm i}| \ll \omega \ll \Omega_{\rm e}$, the ion and electron contribution terms are given by (12.37) and (12.39), respectively, i.e.,

$$1 + \frac{\Pi_{\rm e}^2}{k^2}\frac{m_{\rm e}}{T_{\rm e}}\left[1 + I_0 {\rm e}^{-b}\zeta_0 Z(\zeta_0)\right] + \Pi_{\rm i}^2 k^2 \frac{m_{\rm i}}{T_{\rm i}}\left[1 + \zeta Z(\zeta)\right] = 0\,,$$

where

$$\zeta_0 = \frac{\omega}{2^{1/2} k_z v_{T_{\rm e}}} \qquad \text{and} \qquad \zeta = \frac{\omega}{2^{1/2} k v_{T_{\rm i}}} \approx \frac{\omega}{2^{1/2} k_\perp v_{T_{\rm i}}}\,.$$

Since

$$I_0 {\rm e}^{-b} \approx 1 - b + \frac{3}{4}b^2\,, \quad \zeta_0 \gg 1\,, \quad \zeta \gg 1\,, \quad 1 + \zeta Z(\zeta) \approx -\frac{1}{2}\zeta^{-2} - \frac{3}{4}\zeta^{-4}\,,$$

we have

$$\left(\frac{3\Pi_{\rm i}^2}{\omega^4}\frac{\kappa T_{\rm i}}{m_{\rm i}} + \frac{3}{4}\frac{\Pi_{\rm e}^2}{\Omega_{\rm e}^4}\frac{\kappa T_{\rm e}}{m_{\rm e}}\right)k_\perp^4 - \left(1 + \frac{\Pi_{\rm e}^2}{\Omega_{\rm e}^2} - \frac{\Pi_{\rm i}^2}{\omega^2}\right)k_\perp^2 - \left(1 - \frac{\Pi_{\rm e}^2}{\omega^2}\right)k_z^2 = 0\,. \tag{12.31}$$

Using the notation $\rho_{\rm i} = v_{T_{\rm i}}/|\Omega_{\rm i}|$ and

$$s^2 \equiv 3\left(\frac{|\Omega_{\rm e}\Omega_{\rm i}|}{\omega^2} + \frac{1}{4}\frac{T_{\rm e}}{T_{\rm i}}\frac{\omega^2}{|\Omega_{\rm e}\Omega_{\rm i}|}\right) = 3\left(\frac{1+p}{p} + \frac{1}{4}\frac{T_{\rm e}}{T_{\rm i}}\frac{p}{1+p}\right) ,$$

we have

$$\frac{3\Pi_{\rm i}^2}{\omega^4}\frac{\kappa T_{\rm i}}{m_{\rm i}} + \frac{3}{4}\frac{\Pi_{\rm e}^2}{\Omega_{\rm e}^4}\frac{\kappa T_{\rm e}}{m_{\rm e}} = \frac{\Pi_{\rm i}^2}{\omega^2}\frac{m_{\rm e}}{m_{\rm i}}\frac{v_{T_{\rm i}}^2 s^2}{\Omega_{\rm i}} ,$$

$$1 + \frac{\Pi_{\rm e}^2}{\Omega_{\rm e}^2} - \frac{\Pi_{\rm i}^2}{\omega^2} = \frac{1}{1+p}\frac{1-h}{h}\frac{\Pi_{\rm i}^2}{\omega^2} .$$

Then the dimensionless form of (12.31) is

$$(k_\perp \rho_{\rm i})^4 - \frac{1-h}{h}\frac{m_{\rm i}}{m_{\rm e}}\frac{1}{(1+p)s^2}(k_\perp\rho_{\rm i})^2 + \left(\frac{m_{\rm i}}{m_{\rm e}}\right)^2 \frac{1}{s^2}(k_z\rho_{\rm i})^2 = 0 . \quad (12.32)$$

This dispersion equation has two solutions. One corresponds to the slow wave in a cold plasma and the other to the plasma wave in a hot plasma. The slow wave transforms to the plasma wave at the location where (12.31) or (12.32) has equal roots [12.9–12.11]. The condition of zero discriminant is $1/h = 1 + 2k_z\rho_{\rm i}(1+p)s$ and

$$\frac{\Pi_{\rm e}^2(x)}{\Omega_{\rm e}^2} = \frac{\Pi_{\rm M.C.}^2}{\Omega_{\rm e}^2} \equiv \frac{p}{1 + 2k_z\rho_{\rm i}(1+p)s} .$$

Accordingly, mode conversion occurs at the position satisfying

$$\frac{\omega^2}{\Pi_{\rm i}^2} = \left(1 - \frac{\omega^2}{|\Omega_{\rm i}|\Omega_{\rm e}}\right) + \frac{N_\parallel v_{T_{\rm e}} 2\sqrt{3}}{c}\left[\frac{T_{\rm i}}{T_{\rm e}} + \frac{1}{4}\left(\frac{\omega^2}{\Omega_{\rm i}\Omega_{\rm e}}\right)^2\right]^{1/2} ,$$

and the value of $k_\perp^2\rho_{\rm i}^2$ at this position becomes

$$k_\perp^2\rho_{\rm i}^2\big|_{\rm M.C.} = \frac{m_{\rm i}}{m_{\rm e}}\frac{k_z\rho_{\rm i}}{s} .$$

If the electron temperature is high enough at the plasma center to satisfy $v_{T_{\rm e}} > (1/3)c/N_\parallel$, the wave is absorbed by electrons due to electron Landau damping.

After mode conversion, the value $N_\perp$ becomes large so that $c/N_\perp$ becomes comparable to the ion thermal velocity ($c/N_\perp \sim v_{T_{\rm i}}$). Since $\omega \gg |\Omega_{\rm i}|$, the ion motion is not affected by the magnetic field within the time scale $\omega^{-1}$. Therefore the wave with phase velocity $c/N$ is absorbed by ions due to ion Landau damping. When ions have velocity $v_{\rm i}$ larger than $c/N_\perp$ ($v_{\rm i} > c/N_\perp$), they are accelerated or decelerated at each time satisfying $v_{\rm i}\cos(\Omega_{\rm i}t) \approx c/N_\perp$ and are subjected to stochastic heating.

The wave is excited by the array of waveguides, as shown in Fig. 12.6, with an appropriate phase difference to provide the necessary parallel index

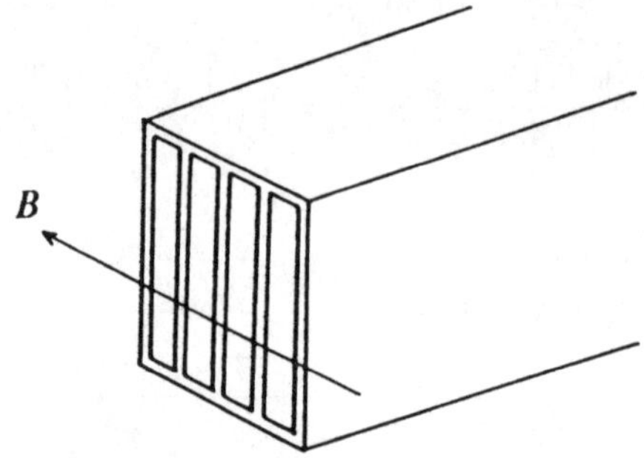


**Fig. 12.6.** Array of waveguides to excite a lower hybrid wave (slow wave)

$N_{\parallel} = k_z c/\omega = 2\pi c/(\lambda_z \omega)$. In the low-density region at the plasma boundary, the component of the electric field parallel to the magnetic field is larger for the slow wave than for the fast wave. Therefore the waveguide directions are arranged to excite the electric field parallel to the line of magnetic force. The coupling of waves to plasmas is discussed in detail in [12.12] and the experiments relating to lower hybrid heating (LHH) are reviewed in [12.13].

For a current to be driven by the lower hybrid wave, the accessibility condition (12.30) and $c/N_{\parallel} \gg v_{T_e}$ are required. If the electron temperature is high and $\kappa T_e \sim 10\,\text{keV}$, then $v_{T_e}/c$ is already $\sim 1/7$. Even if $N_{\parallel}$ is chosen to be small under the accessibility condition (12.30), the wave is subjected to absorption by electron damping in the outer part of the plasma, and the wave cannot be expected to propagate into the central part of the plasma.

When the value of $N_{\parallel}$ is chosen to be $N_{\parallel} \sim (1/3)(c/v_{T_e})$, electron heating can be expected and has indeed been observed experimentally. Under conditions where mode conversion can occur, ion heating can be expected. However, the experimental results are less clear than those for electron heating.

## 12.6 Electron Cyclotron Heating

The dispersion relation of waves in the electron cyclotron range of frequency in a cold plasma is given by (10.79). The plus and minus signs in (10.79) correspond to ordinary and extraordinary waves, respectively. The ordinary wave can only propagate when $\omega^2 > \Pi_e^2$, as is clear from (10.86) (when $\theta = \pi/2$). This wave can be excited by an array of waveguides, like the one used for lower hybrid waves (Fig. 12.6), which emits an electric field parallel to the magnetic field. The phase of each waveguide is selected to provide the appropriate value of the parallel index $N_{\parallel} = k_z c/\omega = 2\pi c/(\omega \lambda_z)$.

The dispersion relation of the extraordinary wave is given by (10.87). When $\theta = \pi/2$, it is given by (10.52). One must satisfy $\omega_{UH}^2 > \omega^2 > \omega_L^2, \omega_{LH}^2$.

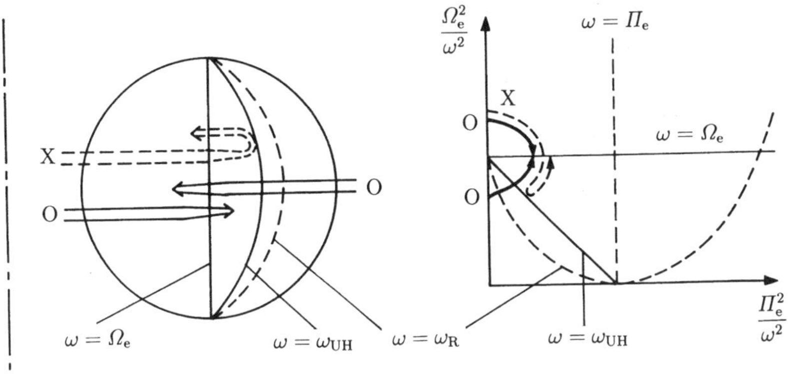


**Fig. 12.7.** Locations of electron cyclotron resonance ($\omega = \Omega_e$), upper hybrid resonance ($\omega = \omega_{LH}$), and R cut off ($\omega = \omega_R$), when $\Omega_{e0} > \Pi_{e0}$ in tokamak configuration, where $\Omega_{e0}$ and $\Pi_{e0}$ are the electron cyclotron resonance frequency and the plasma frequency at the plasma center, respectively (*left*). CMA diagram near the electron cyclotron frequency region (*right*)

As can be seen from the CMA diagram in Fig. 10.5, the extraordinary wave can access the plasma center from the high magnetic field side (see Fig. 12.7) but cannot access from the low field side because of the $\omega = \omega_R$ cutoff. The extraordinary wave can be excited by the waveguide, which emits an electric field perpendicular to the magnetic field (see Sect. 10.2).

The ion's contribution to the dielectric tensor is negligible. When relations $b \ll 1$, $\zeta_0 \gg 1$ are satisfied for the electron, the dielectric tensor of a hot plasma is

$$K_{xx} = K_{yy} = 1 + X\zeta_0 Z_{-1}/2\,, \qquad K_{zz} = 1 - X + N_\perp^2 \chi_{zz}\,,$$

$$K_{xy} = -\mathrm{i}X\zeta_0 Z_{-1}/2\,, \qquad K_{xz} = N_\perp \chi_{xz}\,, \qquad K_{yz} = \mathrm{i}N_\perp \chi_{yz}\,,$$

$$\chi_{xz} \approx \chi_{yz} \approx 2^{-1/2} X Y^{-1} \frac{v_T}{c} \zeta_0 (1 + \zeta_{-1} Z_{-1})\,,$$

$$\chi_{zz} \approx X Y^{-2} \left(\frac{v_T}{c}\right)^2 \zeta_0 \zeta_{-1} (1 + \zeta_{-1} Z_{-1})\,,$$

$$X \equiv \frac{\Pi_e^2}{\omega^2}\,, \qquad Y \equiv \frac{\Omega_e}{\omega}\,, \qquad \zeta_{-1} = \frac{\omega - \Omega_e}{2^{1/2} k_z v_T}\,, \qquad N_\perp = \frac{k_\perp c}{\omega}\,.$$

The Maxwell equations are

$$(K_{xx} - N_\parallel^2) E_x + K_{xy} E_y + N_\perp (N_\parallel + \chi_{xz}) E_z = 0\,,$$

$$-K_{xy} E_x + (K_{yy} - N_\parallel^2 - N_\perp^2) E_y + \mathrm{i} N_\perp \chi_{yz} E_z = 0\,,$$

$$N_\perp (N_\parallel + \chi_{xz}) E_x - \mathrm{i} N_\perp \chi_{yz} E_y + \left[1 - X - N_\perp^2 (1 - \chi_{zz})\right] E_z = 0\,.$$

The solution is

$$\frac{E_x}{E_z} = -\frac{\mathrm{i}N_\perp^2\chi_{xz}(N_\parallel + \chi_{xz}) + K_{xy}\left[1 - X - N_\perp^2(1-\chi_{zz})\right]}{N_\perp\left[\mathrm{i}\chi_{xz}(K_{xx} - N_\parallel^2) + K_{xy}(N_\parallel + \chi_{xz})\right]} ,$$

$$\frac{E_y}{E_z} = -\frac{N_\perp^2(N_\parallel + \chi_{xz})^2 - (K_{xx} - N_\parallel^2)\left[1 - X - N_\perp^2(1-\chi_{zz})\right]}{N_\perp\left[\mathrm{i}\chi_{xz}(K_{xx} - N_\parallel^2) + K_{xy}(N_\parallel + \chi_{xz})\right]} .$$

The absorption power $P_{-1}$ per unit volume is given by (12.24) as follows:

$$P_{-1} = \omega X \zeta_0 \frac{\pi^{1/2}}{2} \exp\left[-\frac{(\omega - \Omega_\mathrm{e})^2}{2k_z^2 v_{T_\mathrm{e}}^2}\right] \frac{\epsilon_0}{2}|E_x - \mathrm{i}E_y|^2 .$$

When $\omega = \Omega_\mathrm{e}$, then

$$\zeta_{-1} = 0 , \quad Z_{-1} = \mathrm{i}\pi^{1/2} , \quad K_{xx} = 1 + \mathrm{i}h , \quad K_{xy} = h ,$$

$$\chi_{yz} = \chi_{xz} = 2^{1/2} X \frac{v_{T_\mathrm{e}}}{c}\zeta_0 = \frac{X}{2N_\parallel} , \quad \chi_{zz} = 0 , \quad h \equiv \pi^{1/2}\frac{\zeta_0 X}{2} .$$

Therefore the dielectric tensor $\boldsymbol{K}$ becomes

$$\boldsymbol{K} = \begin{pmatrix} 1+\mathrm{i}h & h & N_\perp\chi_{xz} \\ -h & 1+\mathrm{i}h & \mathrm{i}N_\perp\chi_{xz} \\ N_\perp\chi_{xz} & -\mathrm{i}N_\perp\chi_{xz} & 1-X \end{pmatrix} .$$

For the ordinary wave (O wave), we have

$$\frac{E_x - \mathrm{i}E_y}{E_z} = \frac{\mathrm{i}N_\perp^2(\mathrm{O})N_\parallel(N_\parallel + \chi_{xz}) - \mathrm{i}(1 - N_\parallel^2)\left[1 - X - N_\perp^2(\mathrm{O})\right]}{N_\perp(\mathrm{O})\left[N_\parallel h + \mathrm{i}\chi_{xz}(1 - N_\parallel^2)\right]} .$$

When $N_\parallel \ll 1$ and the incident angle is nearly perpendicular, (10.82) gives $1 - X - N_\perp^2(\mathrm{O}) = (1-X)N_\parallel^2$. Since $\chi_{xz} = X/2N_\parallel$, then $\chi_{xz} \gg N_\parallel$. Therefore the foregoing equation reduces to

$$\frac{E_x - \mathrm{i}E_y}{E_z} = \frac{\mathrm{i}N_\perp(\mathrm{O})N_\parallel\chi_{xz}}{N_\parallel h + \mathrm{i}\chi_{xz}} .$$

For the extraordinary wave (X wave), we have

$$\frac{E_x - \mathrm{i}E_y}{E_y} = -\frac{\mathrm{i}N_\perp^2(\mathrm{X})N_\parallel(N_\parallel + \chi_{xz}) - \mathrm{i}(1 - N_\parallel^2)\left[1 - X - N_\perp^2(\mathrm{X})\right]}{N_\perp^2(\mathrm{X})(N_\parallel + \chi_{xz})^2 - (K_{xx} - N_\parallel^2)\left[1 - X - N_\perp^2(\mathrm{X})\right]} .$$

When $N_\parallel \ll 1$ and $\omega = \Omega_\mathrm{e}$, (10.83) gives $1 - X - N_\perp^2(\mathrm{X}) \approx -1 + N_\parallel^2$. Since $\chi_{xz}^2 = (2\pi)^{-1/2}(v_{T_\mathrm{e}}/cN_\parallel)Xh \ll h$, the foregoing equation reduces to

$$\frac{E_x - \mathrm{i}E_y}{E_y} = \frac{-\left[1 + N_\perp^2(\mathrm{X})N_\parallel(N_\parallel + \chi_{xz})\right]}{h - \mathrm{i}\left[1 + N_\perp^2(\mathrm{X})(N_\parallel + \chi_{xz})^2\right]} \sim -\frac{1}{h}\,.$$

The absorption power per unit volume [12.14] at $\omega = \Omega_\mathrm{e}$ for the ordinary wave is

$$P_{-1}(\mathrm{O}) \approx \frac{\omega\epsilon_0}{2}|E_z|^2 \frac{hN_\perp^2(\mathrm{O})N_\parallel^2\chi_{xz}^2}{(N_\parallel h)^2 + \chi_{xz}^2}\exp(-\zeta_{-1}^2) \tag{12.33}$$

$$\approx \frac{\omega\epsilon_0}{2}|E_z|^2 \frac{1}{(2\pi)^{1/2}}\left(\frac{\Pi_\mathrm{e}}{\omega}\right)^2\left(\frac{v_{T_\mathrm{e}}}{cN_\parallel}\right)\frac{N_\perp^2(\mathrm{O})N_\parallel^2}{N_\parallel^2 + (v_{T_\mathrm{e}}/c)^2(2/\pi)}\,,$$

and the absorption power per unit volume for the extraordinary wave is

$$P_{-1}(\mathrm{X}) \sim \frac{\omega\epsilon_0}{2}|E_y|^2\frac{1}{h} = \frac{\omega\epsilon_0}{2}|E_y|^2 2\left(\frac{2}{\pi}\right)^{1/2}\left(\frac{\Pi_\mathrm{e}}{\omega}\right)^{-2}\left(\frac{N_\parallel v_{T_\mathrm{e}}}{c}\right)\,. \tag{12.34}$$

Since $P(\mathrm{O}) \propto n_\mathrm{e}T_\mathrm{e}^{1/2}/N_\parallel$, $P(\mathrm{X}) \propto N_\parallel T_\mathrm{e}^{1/2}/n_\mathrm{e}$, the ordinary wave is absorbed to a greater extent in higher density and for perpendicular incidence, whereas the extraordinary wave has the opposite tendency.

Experiments with electron cyclotron heating (ECH) have been carried out by T-10, ISX-B, JFT-2, D-IIID, and so on, and the good heating efficiency of ECH has been demonstrated. Heating and current drive by electron cyclotron waves are reviewed in [12.15].

## 12.7 Velocity Space Instabilities (Electrostatic Waves)

Besides the magnetohydrodynamic instabilities discussed in Chap. 8, there is another type of instability caused by deviations of the velocity space distribution function from the stable Maxwell form. Instabilities which depend on the shape of the velocity distribution function are called *velocity space instabilities* or *microscopic instabilities*. However, the distinction between microscopic and macroscopic or MHD instabilities is not always clear.

### 12.7.1 Dispersion Equation of Electrostatic Wave

In this section, we describe the characteristics of the perturbation of electrostatic waves. In this case the electric field can be expressed by $\boldsymbol{E} = -\boldsymbol{\nabla}\phi = -\mathrm{i}\boldsymbol{k}\phi$. The dispersion equation of the electrostatic wave is give by (Sect. 10.5)

$$k_x^2K_{xx} + 2k_xk_zK_{xz} + k_z^2K_{zz} = 0\,. \tag{12.35}$$

When the zeroth-order distribution function is expressed by

$$f_0(v_\perp, v_z) = n_0F_\perp(v_\perp)F_z(v_z)\,, \qquad F_\perp(v_\perp) = \frac{m}{2\pi\kappa T_\perp}\exp\left(-\frac{mv_\perp^2}{2\kappa T_\perp}\right)\,,$$

$$F_z(v_z) = \left(\frac{m}{2n\kappa T_\perp}\right)^{1/2} \exp\left[-\frac{m(v_z - V)^2}{2\kappa T_z}\right] ,$$

the dispersion equation is given by substituting $K_{xx}$, $K_{xz}$, $K_{zz}$ expressed by (12.19) and (12.20), into (12.35),

$$k_x^2 + k_z^2 + \sum_{\mathrm{i,e}} \Pi^2 \frac{m}{\kappa T_z} \left\{1 + \sum_{n=-\infty}^{\infty} \left[1 + \frac{T_z}{T_\perp}\frac{(-n\Omega)}{\omega_n}\right] \zeta_n Z(\zeta_n) I_n(b) \mathrm{e}^{-b}\right\} = 0 , \tag{12.36}$$

derived in Sect. 12.8, where

$$\zeta_n \equiv \frac{\omega_n}{2^{1/2} k_z v_{T_z}} , \qquad \omega_n \equiv \omega - k_z V + n\Omega ,$$

$$b = \left(\frac{k_x v_{T_\perp}}{\Omega}\right)^2 , \qquad v_{T_z}^2 = \frac{\kappa T_z}{m} , \qquad v_{T_\perp}^2 = \frac{\kappa T_\perp}{m} ,$$

$I_n(b)$ is the $n$th modified Bessel function, and $Z(\zeta)$ is the plasma dispersion function. When the frequency of the wave is much higher than the cyclotron frequency ($|\omega| \gg |\Omega|$) or the magnetic field is very weak ($B \to 0$), then we find $\zeta_n \to \zeta_0$, $n\Omega/\omega_n \to 0$, $\sum I_n(b)\mathrm{e}^{-b} = 1$, so that the dispersion equation reduces to

$$k_x^2 + k_z^2 + \sum_{\mathrm{i,e}} \Pi^2 \frac{m}{\kappa T_z} [1 + \zeta_0 Z(\zeta_0)] = 0 \qquad (|\omega| \gg |\Omega|) . \tag{12.37}$$

When $B = 0$, the dispersion equation is given by

$$k^2 + \sum_{\mathrm{i,e}} \Pi^2 \frac{m}{\kappa T} [1 + \zeta Z(\zeta)] = 0 \qquad \left(\zeta = \frac{\omega - kV}{2^{1/2} k v_T} , \quad B = 0\right) . \tag{12.38}$$

When the frequency of the wave is much lower than the cyclotron frequency, i.e., $|\omega| \ll |\Omega|$, then we find $\zeta_n \to \infty\ (n \neq 0)$, $\zeta_n Z_n \to -1$ and $\sum I_n(b)\mathrm{e}^{-b} = 1$:

$$k_x^2 + k_z^2 + \sum_{\mathrm{i,e}} \Pi^2 \frac{m}{\kappa T_z} \left\{I_0 \mathrm{e}^{-b} [1 + \zeta_0 Z(\zeta_0)] + \frac{T_z}{T_\perp}\left(1 - I_0 \mathrm{e}^{-b}\right)\right\} = 0$$
$$(|\omega| \ll |\Omega|) . \tag{12.39}$$

When the frequency of the wave is much higher than the cyclotron frequency or the magnetic field is very small, the dispersion equation (12.35) with (12.71) reduces to

$$k_x^2 + k_z^2 + \sum_{\mathrm{i,e}} \Pi^2 \frac{m}{\kappa T_z} \left[k_z \frac{\kappa T_z}{m} \int \frac{\partial (f/n_0)/\partial v_z}{\omega - k_z v_z} \mathrm{d}v_z\right] = 0 .$$

Partial integration gives

$$\frac{k_x^2 + k_z^2}{k_z^2} = \sum_{\mathrm{i,e}} \Pi^2 \int \frac{f/n_0}{(\omega - k_z v_z)^2} \mathrm{d}v_z . \tag{12.40}$$

### 12.7.2 Electron Beam Instability

Let us consider the interaction of a weak electron beam of velocity $V_0$ with a plasma which consists of cold ions and hot electrons. The dispersion equation (12.40) of an electrostatic wave with $k_x = 0$ ($E_x = E_y = 0$, $E_z \neq 0$, $\boldsymbol{B}_1 = 0$) is given by

$$K_{zz} = K_\parallel - \frac{\Pi_\mathrm{b}^2}{(\omega - kV_0)^2}$$

$$= 1 - \frac{\Pi_\mathrm{i}^2}{\omega^2} - \Pi_\mathrm{e}^2 \int_{-\infty}^{\infty} \frac{f_\mathrm{e}(v_z)/n_0}{(\omega - kv_z)^2}\,\mathrm{d}v_z - \frac{\Pi_\mathrm{b}^2}{(\omega - kV_0)^2} = 0\,. \qquad (12.41)$$

For the limit of a weak beam ($\Pi_\mathrm{b}^2 \to 0$), the dispersion equation reduces to $K_\parallel(\omega, k) \approx 0$, if $\omega \neq kV_0$. The dispersion equation including the effect of a weak beam must be in the form

$$\omega - kV_0 = \delta_\omega(k) \qquad \left[\delta_\omega(k) \ll kV_0\right]\,.$$

Using $\delta_\omega^2$, (12.41) reduces to

$$\frac{\Pi_\mathrm{b}^2}{\delta_\omega^2} = K_\parallel(\omega = kV_0, k) + \left(\frac{\partial K_\parallel}{\partial \omega}\right)_{\omega = kV_0} \delta_\omega\,.$$

If $\omega = kV_0$ does not satisfy $K_\parallel = 0$, then $K_\parallel \neq 0$ holds and the second term on the right-hand side of the foregoing equation can be neglected:

$$\frac{\Pi_\mathrm{b}^2}{\delta_\omega^2} = K_\parallel(\omega = kV_0, k)\,.$$

The expression for $K_\parallel(\omega = kV_0, k)$ is

$$K_\parallel(\omega_\mathrm{r}) = K_\mathrm{R}(\omega_\mathrm{r}) + \mathrm{i}K_\mathrm{I}(\omega_\mathrm{r})\,.$$

The $K_\mathrm{I}$ term is the Landau damping term (see Sect. 12.3).

When the condition $\omega = kV_0$ is in a region where Landau damping is ineffective, then $|K_\mathrm{I}| \ll |K_\mathrm{R}|$ and the dispersion equation becomes

$$\frac{\Pi_\mathrm{b}^2}{(\omega - kV_0)^2} = K_\mathrm{R}\,. \qquad (12.42)$$

Therefore if the condition

$$K_\mathrm{R} < 0 \qquad (12.43)$$

is satisfied, $\delta_\omega$ is imaginary and the wave is unstable. When the dielectric constant is negative, electric charges are likely to be bunched and we can predict the occurrence of this instability.

If $\omega = kV_0$ is in a region where Landau damping is effective, the condition of instability is that the wave energy density $W_0$ in a dispersive medium (12.6) should be negative, i.e.,

$$W_0 = \frac{\epsilon_0}{2} E_z^* \frac{\partial}{\partial \omega}(\omega K_{zz}) E_z < 0 \, .$$

In other words, a *negative energy wave* is the condition for instability, because the absolute value of $W_0$ increases if $\partial W_0/\partial t$ is negative:

$$\frac{\partial W_0}{\partial t} = \frac{\partial}{\partial t}\left[\frac{\epsilon_0}{2} E_z^* \frac{\partial}{\partial \omega}(\omega K_{zz}) E_z\right] = -\frac{\omega_{\mathrm{r}}}{2}\epsilon_0 E_z^* K_{\mathrm{I}} E_z < 0 \, .$$

When energy is lost from the wave by Landau damping, the amplitude of the wave increases because the wave energy density is negative. Readers may refer to [12.16] for a more detailed analysis of the beam–plasma interaction.

### 12.7.3 Various Velocity Space Instabilities

In Sect. 12.7.2, a simple case of electron beam instability was described. There are various velocity space instabilities.

The distribution function of a plasma confined in a mirror field is zero for the loss cone region $(v_\perp/v)^2 < 1/R_{\mathrm{M}}$, where $R_{\mathrm{M}}$ is the mirror ratio (see Sect. 2.5). The instability associated with this is called *loss-cone instability* [12.17].

Plasmas heated by ICRF have higher ion temperature in the perpendicular direction than in the parallel direction. In this case instabilities with higher harmonic ion cyclotron frequencies may occur. This type of instability is called a *Harris instability* [12.18,12.19]. A Harris instability is electrostatic and is analyzed by the dispersion relation (12.79).

In general plasmas are hot and dense in the center and cold and low density on the outside. The instabilities driven by temperature gradient and density gradient are called *drift instabilities.* The electrostatic drift instability of an inhomogeneous plasma can be analyzed by the dispersion equation (12.84) and (12.85) for the inhomogeneous plasma [12.3,12.20].

In a toroidal field, trapped particles always exist on the outside, where the magnetic field is weak. Instabilities induced by trapped particles are called *trapped particle instabilities* [12.21].

## 12.8 Derivation of Dielectric Tensor in Hot Plasma

### 12.8.1 Formulation of Dispersion Relation in Hot Plasma

The dispersion relation of a cold plasma was derived in Chap. 10. In the unperturbed state, both the electrons and the ions are motionless in a cold plasma. However, in a hot plasma, electrons and ions move along spiral trajectories, even in the unperturbed state. The motion of charged particles in a uniform magnetic field $\boldsymbol{B}_0 = B_0\hat{\boldsymbol{z}}$ may be described by

$$\frac{\mathrm{d}\boldsymbol{r}'}{\mathrm{d}t'} = \boldsymbol{v}' , \qquad \frac{\mathrm{d}\boldsymbol{v}'}{\mathrm{d}t'} = \frac{q}{m}\boldsymbol{v}' \times \boldsymbol{B}_0 . \tag{12.44}$$

Assuming that $\boldsymbol{r}' = \boldsymbol{r}$, $\boldsymbol{v}' = \boldsymbol{v} = (v_\perp \cos\theta, v_\perp \sin\theta, v_z)$ at $t' = t$, the solution of (12.44) is obtained as follows:

$$v_x'(t') = v_\perp \cos\left[\theta + \Omega(t'-t)\right] , \quad v_y'(t') = v_\perp \sin\left[\theta + \Omega(t'-t)\right] , \quad v_z'(t') = v_z , \tag{12.45}$$

$$\begin{aligned} x'(t') &= x + \frac{v_\perp}{\Omega}\Big\{ \sin\left[\theta + \Omega(t'-t)\right] - \sin\theta \Big\} , \\ y'(t') &= y - \frac{v_\perp}{\Omega}\Big\{ \cos\left[\theta + \Omega(t'-t)\right] - \cos\theta \Big\} , \\ z'(t') &= z + v_z(t'-t) , \end{aligned} \tag{12.46}$$

where $\Omega = -qB_0/m$ and $v_x = v_\perp \cos\theta$, $v_y = v_\perp \sin\theta$. The analysis of the behavior due to a perturbation of this system must be based on Boltzmann's equation. The distribution function $f_k(\boldsymbol{r}, \boldsymbol{v}, t)$ for the $k$th kind of particles is

$$\frac{\partial f_k}{\partial t} + \boldsymbol{v}\cdot\boldsymbol{\nabla}_{\boldsymbol{r}} f_k + \frac{q_k}{m_k}(\boldsymbol{E} + \boldsymbol{v} \times \boldsymbol{B})\cdot\boldsymbol{\nabla}_{\boldsymbol{v}} f_k = 0 . \tag{12.47}$$

Maxwell's equation are

$$\boldsymbol{\nabla}\cdot\boldsymbol{E} = \frac{1}{\epsilon_0}\sum_k q_k \int \boldsymbol{v} f_k \mathrm{d}\boldsymbol{v} , \tag{12.48}$$

$$\frac{1}{\mu_0}\boldsymbol{\nabla} \times \boldsymbol{B} = \epsilon_0 \frac{\partial \boldsymbol{E}}{\partial t} + \sum_k q_k \int \boldsymbol{v} f_k \mathrm{d}\boldsymbol{v} , \tag{12.49}$$

$$\boldsymbol{\nabla} \times \boldsymbol{E} = -\frac{\partial \boldsymbol{B}}{\partial t} , \tag{12.50}$$

$$\boldsymbol{\nabla}\cdot\boldsymbol{B} = 0 . \tag{12.51}$$

As usual, we indicate zeroth-order quantities (the unperturbed state) by a subscript 0 and first-order perturbation terms by a subscript 1. The first-order terms are expressed in the form $\exp \mathrm{i}(\boldsymbol{k}\cdot\boldsymbol{r} - \omega t)$. Using

$$f_k = f_{k0}(\boldsymbol{r}, \boldsymbol{v}) + f_{k1} , \tag{12.52}$$

$$\boldsymbol{B} = \boldsymbol{B}_0 + \boldsymbol{B}_1 , \tag{12.53}$$

$$\boldsymbol{E} = 0 + \boldsymbol{E}_1 , \tag{12.54}$$

we can linearize (12.47)–(12.51) as follows:

$$\boldsymbol{v}\cdot\boldsymbol{\nabla}_{\boldsymbol{r}} f_{k0} + \frac{q_k}{m_k}(\boldsymbol{v}\times\boldsymbol{B}_0)\cdot\boldsymbol{\nabla}_{\boldsymbol{v}} f_{k0} = 0\,, \tag{12.55}$$

$$\sum_k q_k \int f_{k0}\mathrm{d}\boldsymbol{v} = 0\,, \tag{12.56}$$

$$\frac{1}{\mu_0}\boldsymbol{\nabla}\times\boldsymbol{B}_0 = \sum_k q_k \int \boldsymbol{v} f_{k0}\mathrm{d}\boldsymbol{v} = \boldsymbol{j}_0\,, \tag{12.57}$$

$$\frac{\partial f_{k1}}{\partial t} + \boldsymbol{v}\cdot\boldsymbol{\nabla}_{\boldsymbol{r}} f_{k1} + \frac{q_k}{m_k}(\boldsymbol{v}\times\boldsymbol{B}_0)\cdot\boldsymbol{\nabla}_{\boldsymbol{v}} f_{k1} = -\frac{q_k}{m_k}(\boldsymbol{E}_1 + \boldsymbol{v}\times\boldsymbol{B}_1)\cdot\boldsymbol{\nabla}_{\boldsymbol{v}} f_{k0}\,, \tag{12.58}$$

$$\mathrm{i}\boldsymbol{k}\cdot\boldsymbol{E}_1 = \frac{1}{\epsilon_0}\sum_k q_k \int f_{k1}\mathrm{d}\boldsymbol{v}\,, \tag{12.59}$$

$$\frac{1}{\mu_0}\boldsymbol{k}\times\boldsymbol{B}_1 = -\omega\left(\epsilon_0\boldsymbol{E}_1 + \frac{\mathrm{i}}{\omega}\sum_k q_k \int \boldsymbol{v} f_{k1}\mathrm{d}\boldsymbol{v}\right)\,, \tag{12.60}$$

$$\boldsymbol{B}_1 = \frac{1}{\omega}(\boldsymbol{k}\times\boldsymbol{E}_1)\,. \tag{12.61}$$

The right-hand side of (12.58) is a linear equation in $\boldsymbol{E}_1$, as is clear from (12.61), so that $f_{k1}$ is given as a linear function in $\boldsymbol{E}_1$. The electric tensor of the hot plasma is defined by $\boldsymbol{K}$ ($\boldsymbol{D} = \epsilon_0\boldsymbol{K}\cdot\boldsymbol{E}$), given by

$$\boldsymbol{E}_1 + \frac{\mathrm{i}}{\epsilon\omega}\boldsymbol{j} = \boldsymbol{E}_1 + \frac{\mathrm{i}}{\epsilon_0\omega}\sum_k q_k \int \boldsymbol{v} f_{k1}\mathrm{d}\boldsymbol{v} \equiv \boldsymbol{K}\cdot\boldsymbol{E}_1\,. \tag{12.62}$$

The linear relation for $\boldsymbol{E}_1$ is derived from (12.60) and (12.61):

$$\boldsymbol{k}\times(\boldsymbol{k}\times\boldsymbol{E}_1) + \frac{\omega^2}{c^2}\boldsymbol{K}\cdot\boldsymbol{E}_1 = 0\,, \tag{12.63}$$

and the dispersion relation is obtained by equating the determinant of the coefficient matrix of the linear equation to zero. Consequently, if $f_{k1}$ can be solved from (12.58), then $\boldsymbol{K}$ can be obtained. As for cold plasmas, the properties of waves in hot plasmas can be studied using the dispersion relation of a hot plasma.

### 12.8.2 Solution of Linearized Vlasov Equation

When the right-hand side of (12.58) is time-integrated along the particle orbit (12.45) and (12.46) in the unperturbed state, we find

$$f_{k1}(\boldsymbol{r},\boldsymbol{v},t) = -\frac{q_k}{m_k}\int_{-\infty}^{t}\left\{\boldsymbol{E}_1\big(\boldsymbol{r}'(t'),t'\big) + \frac{1}{\omega}\boldsymbol{v}'(t')\times\big[\boldsymbol{k}\times\boldsymbol{E}_1\big(\boldsymbol{r}'(t'),t'\big)\big]\right\}$$
$$\cdot\boldsymbol{\nabla}'_{\boldsymbol{v}} f_{k0}\big(\boldsymbol{r}'(t'),\boldsymbol{v}'(t')\big)\mathrm{d}t'\,. \tag{12.64}$$

Substituting (12.64) into (12.58) yields

$$-\frac{q_k}{m_k}\left[\boldsymbol{E}_1+\frac{1}{\omega}\boldsymbol{v}\times(\boldsymbol{k}\times\boldsymbol{E}_1)\right]\cdot\nabla_{\boldsymbol{v}}f_{k0}$$
$$-\frac{q_k}{m_k}\int_{-\infty}^{t}\left[\frac{\partial}{\partial t}+\boldsymbol{v}\cdot\nabla_{\boldsymbol{r}}+\frac{q_k}{m_k}(\boldsymbol{v}\times\boldsymbol{B}_0)\cdot\nabla_{\boldsymbol{v}}\right][\text{integrand of (12.64)}]\,\mathrm{d}t'$$
$$=-\frac{q_k}{m_k}(\boldsymbol{E}_1+\boldsymbol{v}\times\boldsymbol{B}_1)\cdot\nabla_{\boldsymbol{v}}f_{k0}\,. \tag{12.65}$$

Therefore if it is proven that the second term on the left-hand side of (12.65) is zero, (12.64) is confirmed to be the solution of (12.58). When the variables $(\boldsymbol{r},\boldsymbol{v},t)$ are changed to $(\boldsymbol{r}',\boldsymbol{v}',t')$ using (12.45) and (12.46), the differential operators in the second term on the left-hand side of (12.65) reduce to

$$\begin{aligned}\frac{\partial}{\partial t}&=\frac{\partial t'}{\partial t}\frac{\partial}{\partial t'}+\frac{\partial \boldsymbol{r}'}{\partial t}\cdot\nabla'_{\boldsymbol{r}}+\frac{\partial \boldsymbol{v}'}{\partial t}\cdot\nabla'_{\boldsymbol{v}}\\&=\frac{\partial(t'-t)}{\partial t}\left[\frac{\partial \boldsymbol{r}'}{\partial(t'-t)}\cdot\nabla'_{\boldsymbol{r}}+\frac{\partial \boldsymbol{v}'}{\partial(t'-t)}\cdot\nabla'_{\boldsymbol{v}}\right]\\&=-\boldsymbol{v}'\cdot\nabla'_{\boldsymbol{r}}-\frac{q_k}{m_k}(\boldsymbol{v}'\times\boldsymbol{B}_0)\cdot\nabla'_{\boldsymbol{v}}\,,\end{aligned}$$

$$\boldsymbol{v}\cdot\nabla_{\boldsymbol{r}}=\boldsymbol{v}\cdot\nabla'_{\boldsymbol{r}}\,,$$

$$\begin{aligned}\frac{\partial}{\partial v_x}&=\frac{\partial \boldsymbol{r}'}{\partial v_x}\cdot\nabla'_{\boldsymbol{r}}+\frac{\partial \boldsymbol{v}'}{\partial v_x}\cdot\nabla'_{\boldsymbol{v}}\\&=\frac{1}{\Omega}\left\{\sin\Omega(t'-t)\frac{\partial}{\partial x'}+\left[-\cos\Omega(t'-t)+1\right]\frac{\partial}{\partial y'}\right\}\\&\quad+\left[\cos\Omega(t'-t)\frac{\partial}{\partial v'_x}+\sin\Omega(t'-t)\frac{\partial}{\partial v'_y}\right]\,,\end{aligned}$$

$$\begin{aligned}\frac{\partial}{\partial v_y}&=\frac{1}{\Omega}\left\{\left[\cos\Omega(t'-t)-1\right]\frac{\partial}{\partial x'}+\sin\Omega(t'-t)\frac{\partial}{\partial y'}\right\}\\&\quad+\left[-\sin\Omega(t'-t)\frac{\partial}{\partial v'_x}+\cos\Omega(t'-t)\frac{\partial}{\partial v'_y}\right]\,,\end{aligned}$$

$$\begin{aligned}\frac{q}{m}(\boldsymbol{v}\times\boldsymbol{B}_0)\cdot\nabla_{\boldsymbol{v}}&=-\Omega\left(v_y\frac{\partial}{\partial v_x}-v_x\frac{\partial}{\partial v_y}\right)\\&=v'_x\frac{\partial}{\partial x'}+v'_y\frac{\partial}{\partial y'}-\left(v_x\frac{\partial}{\partial x'}+v_y\frac{\partial}{\partial y'}\right)-\Omega\left(v'_y\frac{\partial}{\partial v'_x}-v'_x\frac{\partial}{\partial v'_y}\right)\\&=(\boldsymbol{v}'-\boldsymbol{v})\cdot\nabla'_{\boldsymbol{r}}+\frac{q}{m}(\boldsymbol{v}'\times\boldsymbol{B}_0)\cdot\nabla'_{\boldsymbol{v}}\,.\end{aligned}$$

Therefore the second term on the left-hand side of (12.65) is zero.

Since the first-order terms vary as $\exp(-\mathrm{i}\omega t)$, the integral (12.64) converges when the imaginary part of $\omega$ is positive. When the imaginary part of $\omega$ is negative, the solution can be given by analytic continuation from the region of the positive imaginary part.

### 12.8.3 Dielectric Tensor of Hot Plasma

The zeroth-order distribution function $f_0$ must satisfiy (12.55), or

$$f_0(\boldsymbol{r}, \boldsymbol{v}) = f(v_\perp, v_z) \,, \qquad v_\perp^2 = v_x^2 + v_y^2 \,.$$

Let us consider

$$\boldsymbol{E}_1(\boldsymbol{r}', t') = \boldsymbol{E} \exp \mathrm{i}(\boldsymbol{k}\cdot\boldsymbol{r}' - \omega t') \,.$$

The $z$ axis is taken along the $\boldsymbol{B}_0$ direction and the $x$ axis is taken in the plane spanned by $\boldsymbol{B}_0$ and the propagation vector $\boldsymbol{k}$, so that the $y$ component of the propagation vector is zero ($k_y = 0$), i.e.,

$$\boldsymbol{k} = k_x \hat{\boldsymbol{x}} + k_z \hat{\boldsymbol{z}} \,.$$

Then (12.64) reduces to

$$f_1(\boldsymbol{r}, \boldsymbol{v}, t) = -\frac{q}{m} \exp \mathrm{i}(k_x x + k_z z - \omega t) \int_\infty^t \left[ \left( 1 - \boldsymbol{k}\cdot\boldsymbol{v}'\omega \right) \boldsymbol{E} + (\boldsymbol{v}'\cdot\boldsymbol{E}) \frac{\boldsymbol{k}}{\omega} \right] \cdot \nabla'_{\boldsymbol{v}} f_0$$

$$\times \exp\left\{ \mathrm{i} \frac{k_x v_\perp}{\Omega} \sin\left[\theta + \Omega(t' - t)\right] - \mathrm{i} \frac{k_x v_\perp}{\Omega} \sin\theta + \mathrm{i}(k_z v_z - \omega)(t' - t) \right\} \mathrm{d}t' \,.$$

We introduce $\tau = t' - t$ and use the following formulas for the Bessel function:

$$\exp(\mathrm{i}a \sin\theta) = \sum_{m=-\infty}^{\infty} J_m(a) \exp \mathrm{i}m\theta \,, \qquad J_{-m}(a) = (-1)^m J_m(a) \,,$$

$$\exp\Big(\ \Big) = \sum_{m=-\infty}^{\infty} \sum_{n=-\infty}^{\infty} J_m \exp(-\mathrm{i}m\theta) J_n \exp\left[\mathrm{i}n(\theta + \Omega\tau)\right] \exp \mathrm{i}(k_z v_z - \omega)\tau \,.$$

Since

$$\left[ \left( 1 - \frac{\boldsymbol{k}\cdot\boldsymbol{v}'}{\omega} \right) \boldsymbol{E} + (\boldsymbol{v}'\cdot\boldsymbol{E}) \frac{\boldsymbol{k}}{\omega} \right] \cdot \nabla'_{\boldsymbol{v}} f_0 = \frac{\partial f_0}{\partial v_z} \left[ \left( 1 - \frac{k_x v'_x}{\omega} \right) E_z + (v'_x E_x + v'_y E_y) \frac{k_z}{\omega} \right]$$

$$+ \frac{\partial f_0}{p v_\perp} \left[ \left( 1 - \frac{k_z v'_z}{\omega} \right) \left( E_x \frac{v'_x}{v_\perp} + E_y \frac{v'_y}{v_\perp} \right) + v_z E_z \frac{k_x}{\omega} \frac{v'_x}{v_\perp} \right]$$

$$= \left[ \frac{\partial f_0}{\partial v_\perp} \left( 1 - \frac{k_z v_z}{\omega} \right) + \frac{\partial f_0}{\partial v_z} \frac{k_z v_\perp}{\omega} \right]$$

$$\times \left\{ \frac{E_x}{2} \left[ \mathrm{e}^{\mathrm{i}(\theta + \Omega\tau)} + \mathrm{e}^{-\mathrm{i}(\theta + \Omega\tau)} \right] + \frac{E_y}{2\mathrm{i}} \left[ \mathrm{e}^{\mathrm{i}(\theta + \Omega\tau)} - \mathrm{e}^{-\mathrm{i}(\theta + \Omega\tau)} \right] \right\}$$

$$+ \left( \frac{\partial f_0}{\partial v_\perp} \frac{k_x v_z}{\omega} - \frac{\partial f_0}{\partial v_z} \frac{k_x v_\perp}{\omega} \right) \frac{E_z}{2} \left[ \mathrm{e}^{\mathrm{i}(\theta + \Omega\tau)} + \mathrm{e}^{-\mathrm{i}(\theta + \Omega\tau)} \right] + \frac{\partial f_0}{\partial v_z} E_z \,,$$

we find

$$f_1(\boldsymbol{r},\boldsymbol{v},t) = -\frac{q}{m}\exp \mathrm{i}(k_x x + k_z z - \omega t)$$
$$\times \sum_{m,n}\left[U\left(\frac{J_{n-1}+J_{n+1}}{2}\right)E_x - \mathrm{i}U\left(\frac{J_{n-1}-J_{n+1}}{2}\right)E_y\right.$$
$$\left. + \left(W\frac{J_{n-1}+J_{n+1}}{2} + \frac{\partial f_0}{\partial v_z}J_n\right)E_z\right]\frac{J_m(a)\exp\left[-\mathrm{i}(m-n)\theta\right]}{\mathrm{i}(k_z v_z - \omega + n\Omega)}\,,$$

where

$$U = \left(1 - \frac{k_z v_z}{\omega}\right)\frac{\partial f_0}{\partial v_\perp} + \frac{k_z v_\perp}{\omega}\frac{\partial f_0}{\partial v_z}\,, \tag{12.66}$$

$$W = \frac{k_x v_z}{\omega}\frac{\partial f_0}{\partial v_\perp} - \frac{k_x v_\perp}{\omega}\frac{\partial f_0}{\partial v_z}\,, \tag{12.67}$$

$$a = \frac{k_x v_\perp}{\Omega}\,, \qquad \Omega = -\frac{qB}{m}\,, \tag{12.68}$$

and

$$\frac{J_{n-1}(a) + J_{n+1}(a)}{2} = \frac{nJ_n(a)}{a}\,, \qquad \frac{J_{n-1}(a) - J_{n+1}(a)}{2} = \frac{\mathrm{d}}{\mathrm{d}a}J_n(a)\,.$$

Since $f_1$ is obtained, the dielectric tensor $\boldsymbol{K}$ of a hot plasma reduces from (12.62) to

$$(\boldsymbol{K} - \boldsymbol{I})\cdot\boldsymbol{E} = \frac{\mathrm{i}}{\epsilon_0\omega}\sum_j q_j\int \boldsymbol{v} f_{j1}\mathrm{d}\boldsymbol{v}\,. \tag{12.69}$$

Since $v_x = v_\perp\cos\theta$, $v_y = v_\perp\sin\theta$, $v_z = v_z$, only the terms in $\mathrm{e}^{\mathrm{i}(m-n)\theta} = \mathrm{e}^{\pm\mathrm{i}\theta}$ in $f_{j1}$ can contribute to the $x$, $y$ components of the integral (12.69) and only the term in $\mathrm{e}^{\mathrm{i}(m-n)\theta} = 1$ in $f_{j1}$ can contribute to the $z$ component of the integral (12.69). Hence,

$$\boldsymbol{K} = \boldsymbol{I} - \sum_j \frac{\Pi_j^2}{\omega n_{j0}}\sum_{n=-\infty}^{\infty}\int \mathrm{d}\boldsymbol{v}\frac{S_{jn}}{k_z v_z - \omega + n\Omega_j}\,, \tag{12.70}$$

where

$$S_{jn} = \begin{pmatrix} v_\perp\left(n\dfrac{J_n}{a}\right)^2 U & -\mathrm{i}v_\perp\left(n\dfrac{J_n}{a}\right)J_n' U & v_\perp\left(n\dfrac{J_n}{a}\right)J_n\left(\dfrac{\partial f_0}{\partial v_z} + \dfrac{n}{a}W\right) \\ \mathrm{i}v_\perp J_n'\left(n\dfrac{J_n}{a}\right)U & v_\perp (J_n')^2 U & \mathrm{i}v_\perp J_n' J_n\left(\dfrac{\partial f_0}{\partial v_z} + \dfrac{n}{a}W\right) \\ v_z J_n\left(n\dfrac{J_n}{a}\right)U & -\mathrm{i}v_z J_n J_n' U & v_z J_n^2\left(\dfrac{\partial f_0}{\partial v_z} + \dfrac{n}{a}W\right) \end{pmatrix}$$

and

$$\Pi_j^2 = \frac{n_j q_j^2}{\epsilon_0 m_j} .$$

When we use the relations

$$\frac{v_z U - v_\perp \left( \frac{\partial f_0}{\partial v_z} + \frac{n\Omega}{k_x v_\perp} W \right)}{k_z v_z - \omega + n\Omega} = -\frac{v_z}{\omega}\frac{\partial f_0}{\partial v_\perp} + \frac{v_\perp}{\omega}\frac{\partial f_0}{\partial v_z} ,$$

$$\sum_{n=-\infty}^{\infty} J_n^2 = 1 , \quad \sum_{n=-\infty}^{\infty} J_n J_n' = 0 , \quad \sum_{n=-\infty}^{\infty} n J_n^2 = 0 , \quad J_{-n} = (-1)^n J_n ,$$

and replace $n$ by $-n$, then (12.70) reduces to

$$\boldsymbol{K} = \boldsymbol{I} - \sum_j \frac{\Pi_j^2}{\omega} \sum_{n=-\infty}^{\infty} \int T_{jn} \frac{v_\perp^{-1} U_j n_{j0}^{-1}}{k_z v_z - \omega - n\Omega_j} \mathrm{d}\boldsymbol{v}$$
$$- \boldsymbol{L} \sum_j \frac{\Pi_j^2}{\omega^2} \left( 1 + \frac{1}{n_{j0}} \int \frac{v_z^2}{v_\perp} \frac{\partial f_{j0}}{\partial v_\perp} \mathrm{d}\boldsymbol{v} \right) ,$$

where

$$T_{jn} = \begin{pmatrix} v_\perp^2 \left( n\frac{J_n}{a} \right)\left( n\frac{J_n}{a} \right) & \mathrm{i} v_\perp^2 \left( n\frac{J_n}{a} \right) J_n' & -v_\perp v_z \left( n\frac{J_n}{a} \right) J_n \\ -\mathrm{i} v_\perp^2 J_n' \left( n\frac{J_n}{a} \right) & v_\perp^2 J_n' J_n' & \mathrm{i} v_\perp v_z J_n' J_n \\ -v_\perp v_z J_n \left( n\frac{J_n}{a} \right) & -\mathrm{i} v_\perp v_z J_n J_n' & v_z^2 J_n J_n \end{pmatrix}$$

and all the components of the matrix $\boldsymbol{L}$ are zero except $L_{zz} = 1$. From the relations

$$\frac{U_j}{k_z v_z - \omega - n\Omega_j} = -\frac{1}{\omega}\frac{\partial f_{j0}}{\partial v_\perp} + \frac{-n\Omega_j \frac{\partial f_{j0}}{\partial v_\perp} + k_z v_\perp \frac{\partial f_{j0}}{\partial v_z}}{\omega(k_z v_z - \omega - n\Omega_j)} ,$$

$$\sum_{n=-\infty}^{\infty} (J_n')^2 = \frac{1}{2} , \quad \sum_{n=-\infty}^{\infty} \frac{n^2 J_n^2(a)}{a^2} = \frac{1}{2} ,$$

another expression for the dielectric tensor is obtained:

$$\boldsymbol{K} = \left( 1 - \frac{\Pi_j^2}{\omega^2} \right) \boldsymbol{I} - \sum_{j,n} \frac{\Pi_j^2}{\omega^2} \int \frac{T_{jn}}{k_z v_z - \omega - n\Omega_j} \left( \frac{-n\Omega_j}{v_\perp}\frac{\partial f_{j0}}{\partial v_\perp} + k_z \frac{\partial f_{j0}}{\partial v_z} \right) \frac{1}{n_{j0}} \mathrm{d}\boldsymbol{v} . \tag{12.71}$$

Using $\boldsymbol{N} \equiv c\boldsymbol{k}/\omega$, (12.63) becomes

$$(K_{xx} - N_\parallel^2) E_x + K_{xy} E_y + (K_{xz} + N_\perp N_\parallel) E_z = 0 ,$$

$$K_{yx}E_x + (K_{yy} - N^2)E_y + K_{yz}E_z = 0 ,$$

$$(K_{zx} + N_\perp N_\parallel)E_x + K_{zy}E_y + (K_{zz} - N_\perp^2)E_z = 0 ,$$

where $N_\parallel$ is the $z$ component of $\boldsymbol{N}$ (parallel to $\boldsymbol{B}$ ) and $N_\perp$ is the $x$ component of $\boldsymbol{N}$ (perpendicular to $\boldsymbol{B}$ ). The dispersion relation is given by equating the determinant of the coefficient matrix to zero.

### 12.8.4 Dielectric Tensor of Bi-Maxwellian Plasma

When the zeroth-order distribution function is bi-Maxwellian,

$$f_0(v_\perp, v_z) = n_0 F_\perp(v_\perp) F_z(v_z) , \tag{12.72}$$

$$F_\perp(v_\perp) = \frac{m}{2\pi\kappa T_\perp} \exp\left(-\frac{mv_\perp^2}{2\kappa T_\perp}\right) , \tag{12.73}$$

$$F_z(v_z) = \left(\frac{m}{2\pi\kappa T_z}\right)^{1/2} \exp\left[-\frac{m(v_z - V)^2}{2\kappa T_z}\right] , \tag{12.74}$$

we find

$$\left(-\frac{n\Omega_j}{v_\perp}\frac{\partial f_0}{\partial v_\perp} + k_z \frac{\partial f_0}{\partial v_z}\right)\frac{1}{n_0} = m\left[\frac{n\Omega_j}{\kappa T_\perp} - \frac{k_z(v_z - V)}{\kappa T_z}\right] F_\perp(v_\perp)F_z(v_z) .$$

The integration over $v_z$ can be carried out using the *plasma dispersion function* $Z(\zeta)$. The plasma dispersion function $Z(\zeta)$ is defined by

$$Z(\zeta) \equiv \frac{1}{\pi^{1/2}} \int_{-\infty}^{\infty} \frac{\exp(-\beta^2)}{\beta - \zeta} \mathrm{d}\beta . \tag{12.75}$$

Using the relations

$$\int_{-\infty}^{\infty} \frac{F_z}{k_z(v_z - V) - \omega_n} \mathrm{d}v_z = \frac{1}{\omega_n}\zeta_n Z(\zeta_n) ,$$

$$\int_{-\infty}^{\infty} \frac{k_z(v_z - V)F_z}{k_z(v_z - V) - \omega_n} \mathrm{d}v_z = 1 + \zeta_n Z(\zeta_n) ,$$

$$\int_{-\infty}^{\infty} \frac{\left[k_z(v_z - V)\right]^2 F_z}{k_z(v_z - V) - \omega_n} \mathrm{d}v_z = \omega_n \left[1 + \zeta_n Z(\zeta_n)\right] ,$$

$$\int_{-\infty}^{\infty} \frac{\left[k_z(v_z - V)\right]^3 F_z}{k_z(v_z - V) - \omega_n} \mathrm{d}v_z = \frac{k_z^2(\kappa T_z)}{m} + \omega_n^2 \left[1 + \zeta_n Z(\zeta_n)\right] ,$$

$$\omega_n \equiv \omega - k_z V + n\Omega , \qquad \zeta_n \equiv \frac{\omega - k_z V + n\Omega}{k_z(2\kappa T_z/m)^{1/2}} ,$$

$$\int_0^{\infty} J_n^2(b^{1/2}x) \exp\left(-\frac{x^2}{2\alpha}\right) x \mathrm{d}x = \alpha I_n(\alpha b) \mathrm{e}^{-b\alpha} ,$$

$$\sum_{n=-\infty}^{\infty} I_n(b) = \mathrm{e}^b \,, \qquad \sum_{n=-\infty}^{\infty} n I_n(b) = 0 \,, \qquad \sum_{n=-\infty}^{\infty} n^2 I_n(b) = b\mathrm{e}^b \,,$$

where $I_n(x)$ is the $n$ th modified Bessel function, the formula for the dielectric tensor of a bi-Maxwellian plasma is [12.3]:

$$\boldsymbol{K} = \boldsymbol{I} + \sum_{\mathrm{i,e}} \frac{\Pi^2}{\omega^2} \left\{ \sum_n \left[ \zeta_0 Z(\zeta_n) - \left(1 - \frac{1}{\lambda_T}\right) [1 + \zeta_n Z(\zeta_n)] \right] \mathrm{e}^{-b} \boldsymbol{X}_n \right.$$

$$\left. + 2\eta_0^2 \lambda_T \boldsymbol{L} \right\} , \tag{12.76}$$

where

$$\boldsymbol{X}_n = \begin{pmatrix} n^2 I_n/b & \mathrm{i}n(I_n' - I_n) & -(2\lambda_T)^{1/2}\eta_n \frac{n}{\alpha} I_n \\ -\mathrm{i}n(I_n' - I_n) & (n^2/b + 2b)I_n - 2bI_n' & \mathrm{i}(2\lambda_T)^{1/2}\eta_n \alpha (I_n' - I_n) \\ -(2\lambda_T)^{1/2}\eta_n \frac{n}{\alpha} I_n & -\mathrm{i}(2\lambda_T)^{1/2}\eta_n \alpha (I_n' - I_n) & 2\lambda_T \eta_n^2 I_n \end{pmatrix} , \tag{12.77}$$

$$\eta_n \equiv \frac{\omega + n\Omega}{2^{1/2} k_z v_{T_z}} \,, \qquad \lambda_T \equiv \frac{T_z}{T_\perp} \,, \qquad b \equiv \left(\frac{k_x v_{T_\perp}}{\Omega}\right)^2 \,,$$

$$\alpha \equiv \frac{k_x v_{T_\perp}}{\Omega} \,, \qquad v_{T_z}^2 \equiv \frac{\kappa T_z}{m} \,, \qquad v_{T_\perp}^2 \equiv \frac{\kappa T_\perp}{m} \,,$$

and the matrix components of $\boldsymbol{L}$ are all zero except for $L_{zz} = 1$.

### 12.8.5 Dispersion Relation of Electrostatic Wave

When the electric field $\boldsymbol{E}$ of the waves is expressed by an electrostatic potential $\boldsymbol{E} = -\boldsymbol{\nabla}\phi$, the waves are called electrostatic waves. The dispersion relation of the electrostatic wave in a hot plasma reduces from (10.92) to

$$k_x^2 K_{xx} + 2k_x k_z K_{xz} + k_z^2 K_{zz} = 0 \,. \tag{12.78}$$

When $K_{xx}$, $K_{xz}$, and $K_{zz}$ of (12.76) and (12.77) are substituted into (12.78), we find

$$0 = k_x^2 + k_z^2 + \sum_{\mathrm{i,e}} \frac{\Pi^2}{\omega^2} \left\{ k_z^2 2\eta_0^2 \lambda_T \right.$$

$$+ \sum_{n=-\infty}^{\infty} \left[ \frac{n^2 I_n}{b} k_x^2 - (2\lambda_T)^{1/2} \eta_n \frac{n}{b^{1/2}} I_n 2k_x k_z + 2\lambda_T \eta_n^2 I_n k_z^2 \right]$$

$$\left. \times \left[ \eta_0 Z(\zeta_n) - \left(1 - \frac{1}{\lambda_T}\right) [1 + \zeta_n Z(\zeta_n)] \right] \mathrm{e}^{-b} \right\} .$$

Using $\sum_{n=-\infty}^{\infty} I_n(b)\mathrm{e}^{-b} = 1$, we have

$$0 = k_x^2 + k_z^2 + \sum_{\mathrm{i,e}} \frac{\Pi^2}{\omega^2}\left\{\frac{m\omega^2}{\kappa T_\perp}\right.$$

$$\left. + \sum_{n=-\infty}^{\infty} \frac{m\omega^2}{\kappa T_\perp} I_n \left[\zeta_0 Z(\zeta_n) - \left(1 - \frac{1}{\lambda_T}\right)[1 + \zeta_n Z(\zeta_n)]\right] \mathrm{e}^{-b}\right\} = 0 \,,$$

$$0 = k_x^2 + k_z^2 + \sum_{\mathrm{i,e}} \Pi^2 \frac{m}{\kappa T_z} \left\{1 + \sum_{n=-\infty}^{\infty} \left[1 + \frac{T_z}{T_\perp}\left(\frac{-n\Omega}{\omega_n}\right)\right] \zeta_n Z(\zeta_n) I_n \mathrm{e}^{-b}\right\} . \tag{12.79}$$

When the density and temperature of the zeroth-order state change in the direction of $y$, we must resort to (12.59) and (12.64). Since the electrostatic wave $\boldsymbol{E} = -\nabla\phi_1$, $\boldsymbol{B}_1 = 0$ is considered in this section, (12.59) and (12.64) reduce to

$$-\nabla^2\phi_1 = \frac{1}{\epsilon_0}\sum_k q_k \int f_{k1} \mathrm{d}\boldsymbol{v} \,, \tag{12.80}$$

$$f_{k1} = \frac{q_k}{m_k}\int_{-\infty}^{t} \nabla'_{\boldsymbol{r}}\phi_1(\boldsymbol{r}',t')\cdot\nabla'_{\boldsymbol{v}} f_{k0}(\boldsymbol{r}',\boldsymbol{v}')\mathrm{d}t' \,. \tag{12.81}$$

The zeroth-order distribution function $f_{k0}$ must satisfy (12.55), i.e.,

$$v_y\frac{\partial f_0}{\partial y} - \Omega\left(v_y\frac{\partial}{\partial v_x} - v_x\frac{\partial}{\partial v_y}\right) f_0 = 0 \,. \tag{12.82}$$

Since $v_\perp^2 = \alpha$, $(v_z - V)^2 = \beta$, and $y + v_x/\Omega = \gamma$ are the solutions of the equation for the particle motion, $f_0(\alpha,\beta,\gamma)$ satisfies (12.82) and we adopt the following zeroth-order distribution function:

$$f_0\left(v_\perp^2, (v_z - V)^2, y + \frac{v_x}{\Omega}\right) \tag{12.83}$$

$$= n_0\left\{1 + \left[-\epsilon + \delta_\perp\frac{v_\perp^2}{2v_{T_\perp}^2} + \delta_z\frac{(v_z - V)^2}{2v_{T_z}^2}\right]\left(y + \frac{v_x}{\Omega}\right)\right\}$$

$$\times\left(\frac{1}{2\pi v_{T_\perp}^2}\right)\left(\frac{1}{2\pi v_{T_z}^2}\right)^{1/2}\exp\left[-\frac{v_\perp^2}{2v_{T_\perp}^2} - \frac{(v_z - V)^2}{2v_{T_z}^2}\right] .$$

The density gradient and temperature gradient of this distribution function are

$$\frac{1}{n_0}\frac{\mathrm{d}n_0}{\mathrm{d}y} = -\epsilon + \delta_\perp + \frac{\delta_z}{2} \,, \qquad \frac{1}{T_\perp}\frac{\mathrm{d}T_\perp}{\mathrm{d}y} = \delta_\perp \,, \qquad \frac{1}{T_z}\frac{\mathrm{d}T_z}{\mathrm{d}y} = \delta_z \,.$$

Let us consider the perturbation $\phi_1(\boldsymbol{r},t) = \phi_1(y)\exp(\mathrm{i}k_x x + \mathrm{i}k_z z - \mathrm{i}\omega t)$ and assume $|(k_x^2 + k_z^2)\phi_1| \gg |\partial^2\phi_1/\partial y^2|$. Following the same approach as in Sect. 12.8.3, (12.80) reduces to the following dispersion relation for the electrostatic wave in an inhomogeneous plasma [12.3, 12.20]:

$$0 = (k_x^2 + k_z^2) - \sum_j \Pi_j^2 \frac{1}{n_{0j}} \int\int\int [\ ]_j \mathrm{d}\theta \mathrm{d}v_\perp \mathrm{d}v_\perp \mathrm{d}v_z \ ,$$

$$0 = k_x^2 + k_z^2 + \sum_j \Pi_j^2 \Bigg\{ \frac{1}{v_{T_z}^2} + \sum_{n=-\infty}^{\infty} I_n(b)\mathrm{e}^{-b} \Bigg[ \left( \frac{1}{v_{T_z}^2} - \frac{1}{v_{T_\perp}^2} \frac{n\Omega}{\omega_n} \right) \zeta_n Z(\zeta_n)$$
$$- \frac{1}{v_{T_\perp}^2} \frac{n}{k_x} \Bigg\{ [\epsilon + \delta_\perp - f_n(b)\delta_\perp] \left[ 1 + \frac{n\Omega}{\omega_n} \zeta_n Z(\zeta_n) \right]$$
$$- \frac{\delta_z}{2} \left[ 1 + \frac{n\Omega\omega_n}{k_z^2 v_{T_z}^2} [1 + \zeta_n Z(\zeta_n)] \right] \Bigg\}$$
$$+ \frac{1}{v_{T_z}^2} \frac{n}{k_x} \Bigg\{ [\epsilon + \delta_z - f_n(b)\delta_\perp][1 + \zeta_n Z(\zeta_n)]$$
$$- \frac{\delta_z}{2} \left[ 1 + \frac{\omega_n^2}{k_z^2 v_{T_z}^2} [1 + \zeta_n Z(\zeta_n)] \right] \Bigg\}$$
$$+ \frac{k_x}{\Omega} \left\{ [\epsilon - f_n(b)\delta_\perp] \frac{\zeta_n}{\omega_n} Z(\zeta_n) - \frac{\delta_z}{2} \frac{\omega_n}{k_z^2 v_{T_z}^2} [1 + \zeta_n Z(\zeta_n)] \right\} \Bigg]_n \Bigg\}_j \ , \quad (12.84)$$

where $f_n(b) \equiv (1-b) + bI_n'(b)/I_n(b)$ and we have used the relation

$$\int_{-\infty}^{\infty} J_n^2(b^{1/2}x) \exp\left(-\frac{x^2}{2}\right) \frac{x^2}{2} x \mathrm{d}x = f_n(b) I_n(b) \mathrm{e}^{-b} \ .$$

For low frequencies ($\omega \ll |\Omega|$), we have relations $\zeta_n \gg 1$ ($n \neq 0$), $\zeta_n Z(\zeta_n) \to -1$ ($n \neq 0$) and $1 + \zeta_n Z(\zeta_n) \to -(1/2)\zeta_n^{-2}$ ($n \neq 0$), and (12.84) reduces to

$$0 = k_x^2 + k_z^2 + \sum_j \Pi_j^2 \Bigg\{ \frac{1}{v_{T_z}^2} + I_0(b)\mathrm{e}^{-b} \Bigg[ \frac{1}{v_{T_z}^2} [1 + \zeta_0 Z(\zeta_0)] - \frac{1}{v_{T_\perp}^2}$$
$$+ \frac{k_x}{\Omega\omega_0} [\epsilon - f_0(b)\delta_\perp] \zeta_0 Z(\zeta_0) - \frac{k_x}{\Omega\omega_0} \delta_z \zeta_0^2 [1 + \zeta_0 Z(\zeta_0)] \Bigg] \Bigg\}_j \ , \quad (12.85)$$

using $\sum_{-\infty}^{\infty} I_n(b)\mathrm{e}^{-b} = 1$.

When an isotropic plasma with $v_{T_\perp} = v_{T_z} = v_T$ has no temperature gradient $\delta_\perp = \delta_z = 0$ and $V = 0$, we have the familiar dispersion relation of a drift wave due to the density gradient as follows:

$$0 = k_x^2 + k_z^2 + \sum_j \Pi_j^2 \left\{ \frac{1}{v_T^2} + I_0(b)\mathrm{e}^{-b} \left[ \frac{1}{v_T^2}\zeta_0 Z(\zeta_0) + \frac{k_x}{\Omega\omega_0}\epsilon\zeta_0 Z(\zeta_0) \right] \right\}_j .$$

We can usually assume $b_\mathrm{e} = 0$ for electrons. Then it reduces to

$$0 = (k_x^2 + k_z^2)\frac{v_{T_\mathrm{e}}^2}{\Pi_\mathrm{e}^2} + 1 + \zeta_\mathrm{e} Z(\zeta_\mathrm{e}) \left(1 - \frac{\omega_\mathrm{e}^*}{\omega}\right)$$
$$+ \frac{ZT_\mathrm{e}}{T_\mathrm{i}} \left[ 1 + I_0(b)\mathrm{e}^{-b}\zeta_0 Z(\zeta_0) \left(1 - \frac{\omega_\mathrm{i}^*}{\omega}\right) \right] , \qquad (12.86)$$

where

$$\omega_\mathrm{e}^* = -k_x\epsilon\frac{v_{T_\mathrm{e}}^2}{\Omega_\mathrm{e}} = -k_x\epsilon\frac{T_\mathrm{e}}{eB} \qquad \text{and} \qquad \omega_\mathrm{i}^* = -k_x\epsilon\frac{v_{T_\mathrm{i}}^2}{\Omega_\mathrm{i}} = k_x\epsilon\frac{T_\mathrm{i}}{ZeB} .$$

Note that the $x$ direction is opposite to the electron drift velocity $\boldsymbol{v}_\mathrm{de}$, $y$ is in the direction of the negative density gradient and $z$ is in the direction of the magnetic field.

# 13 Instabilities Driven by Energetic Particles

Sustained ignition of thermonuclear plasma depends on heating by highly energetic alpha particles produced from fusion reactions. Excess loss of the energetic particles may be caused by fishbone instability and toroidal Alfvén eigenmodes induced by energetic particles. Such losses do not only reduce the alpha particle heating efficiency, but may also lead to excess heat loading and damage to plasma-facing components. These problems have been studied in experiments and analyzed theoretically. In this chapter, basic aspects of theories on collective instabilities by energetic particles are described.

## 13.1 Fishbone Instability

Fishbone oscillations were first observed in PDX experiments with nearly perpendicular neutral beam injection. The poloidal magnetic field fluctuations associated with these instabilities have a characteristic skeletal signature on the Mirnov coils (see Sect. 16.1), which is what suggested the name of fishbone oscillations. Particle bursts corresponding to loss of energetic beam ions are correlated with fishbone events, reducing the beam heating efficiency. The structure of the mode was identified as an $m = 1$, $n = 1$ internal kink mode, with a precursor oscillation frequency close to the thermal ion diamagnetic frequency as well as the fast ion magnetic toroidal precessional frequency.

### 13.1.1 Formulation

The theoretical analysis of the *fishbone instability* described here is mainly due to Chen, White and Rosenbluth [13.1]. The core plasma is treated using the ideal MHD analysis and the hot component is treated using a gyrokinetic description. The first-order equation for the displacement $\boldsymbol{\xi}$ is [see (8.24)]

$$\rho_{\rm m}\gamma^2\boldsymbol{\xi} = \boldsymbol{j}\times\delta\boldsymbol{B} + \delta\boldsymbol{j}\times\boldsymbol{B} - \nabla\delta p_{\rm c} - \nabla\delta p_{\rm h}\ , \tag{13.1}$$

where $\delta p_{\rm c}$ is the first-order pressure disturbance of the core plasma ($\nabla\delta p_{\rm c} = -\boldsymbol{\xi}\cdot\nabla p_{\rm c} - \gamma_{\rm s} p\nabla\cdot\boldsymbol{\xi}$) and $\delta p_{\rm h}$ is the first-order pressure disturbance of the hot component. The following ideal MHD relations hold:

$$\delta\boldsymbol{E}_\perp = \gamma\boldsymbol{\xi}\times\boldsymbol{B}\ ,\quad \delta\boldsymbol{E}_\parallel = 0\ ,\quad \delta\boldsymbol{B} = \nabla\times(\boldsymbol{\xi}\times\boldsymbol{B})\ ,\quad \delta\boldsymbol{j} = \nabla\cdot\delta\boldsymbol{B}\ .$$

Multiplying (13.1) by $\int \mathrm{d}\boldsymbol{r}\boldsymbol{\xi}^*$ and assuming a fixed conducting boundary, we have

$$\delta W_{\mathrm{MHD}} + \delta W_{\mathrm{K}} + \delta I = 0 , \tag{13.2}$$

where

$$\delta I = \frac{\gamma^2}{2} \int \rho_{\mathrm{m}} |\boldsymbol{\xi}|^2 \mathrm{d}\boldsymbol{r} , \tag{13.3}$$

$$\delta W_{\mathrm{K}} = \frac{1}{2} \int \boldsymbol{\xi} \cdot \boldsymbol{\nabla} \delta p_{\mathrm{h}} \mathrm{d}\boldsymbol{r} , \tag{13.4}$$

and $\delta W_{\mathrm{MHD}}$ is the potential energy of the core plasma associated with the displacement $\boldsymbol{\xi}$, discussed in Sect. 8.3.2 and given by (8.80). $\delta W_{\mathrm{K}}$ is the contribution from the hot component.

### 13.1.2 MHD potential Energy

Let us consider the MHD term of $\delta W_{\mathrm{MHD}}$, which consists of the contribution $\delta W^{\mathrm{s}}_{\mathrm{MHD}}$ from the singular region near the rational surface and the contribution $\delta W^{\mathrm{ext}}_{\mathrm{MHD}}$ from the external region. The external contribution $\delta W^{\mathrm{ext}}_{\mathrm{MHD}}$ from the cylindrical circular plasma is already given by (8.93):

$$\frac{\delta W^{\mathrm{ext}}_{\mathrm{MHDcycl}}}{2\pi R} = \frac{\pi}{2\mu_0} \int_0^a \left( f \left| \frac{\mathrm{d}\xi_r}{\mathrm{d}r} \right|^2 + g |\xi_r|^2 \right) \mathrm{d}r , \tag{13.5}$$

where $f$ and $g$ are given by (8.94) and (8.96). When $r/R \ll 1$ is assumed, $f$ and $g$ of the $(-m, n)$ mode are

$$f = \frac{r^3}{R^2} B_z^2 \left( \frac{1}{q} - \frac{n}{m} \right)^2 \left[ 1 - \left( \frac{nr}{mR} \right)^2 \right] ,$$

$$g = \frac{r}{R} B_z^2 \left( \left( \frac{1}{q} - \frac{n}{m} \right)^2 \left( (m^2 - 1) + \frac{n^2 r^2}{R^2} \right) \left( 1 - \left( \frac{nr}{mR} \right)^2 \right) + 2 \left( \frac{1}{q^2} - \left( \frac{n}{m} \right)^2 \right) \left( \frac{n}{m} \right) \left( \frac{r}{R} \right)^2 \right) ,$$

where $q(r) \equiv rB_z/RB_\theta(r)$ is the safety factor. Let us consider the $m = 1$ perturbation with the singular radius $r = r_{\mathrm{s}}$, $q(r_{\mathrm{s}}) = m/n$. In this case the displacement is $\xi_r = \mathrm{const.}$ for $0 < r < r_{\mathrm{s}}$ and $\xi_r = 0$ for $r_{\mathrm{s}} < r < a$ (see Sect. 8.3.2). Then $\delta W^{\mathrm{ext}}_{\mathrm{MHDcycl}}$ reduces to [13.2]

$$\frac{\delta W^{\mathrm{ext}}_{\mathrm{MHDcycl}}}{2\pi R} = \frac{\pi B^2_{\theta\mathrm{s}}}{2\mu_0} |\xi_{\mathrm{s}}|^2 \left( \frac{r_{\mathrm{s}}}{R} \right)^2 \left[ -\beta_{\mathrm{p}} - \int_0^1 \rho^3 \left( \frac{1}{q^2} + \frac{1}{q} - 3 \right) \mathrm{d}\rho \right] , \tag{13.6}$$

where $\rho = r/r_{\mathrm{s}}$, $\beta_{\mathrm{p}} \equiv \langle p \rangle_{\mathrm{s}} / (B^2_{\theta\mathrm{s}}/2\mu_0)$ and $B_{\theta\mathrm{s}} \equiv \left[ r_{\mathrm{s}}/Rq(r_{\mathrm{s}}) \right] B_z$ is the poloidal field at $r = r_{\mathrm{s}}$. The pressure $\langle p \rangle_{\mathrm{s}}$ is defined by

$$\langle p\rangle_s = -\int_0^{r_s}\left(\frac{r}{r_s}\right)^2 \frac{dp}{dr}dr = \frac{1}{r_s^2}\int_0^{r_s}(p-p_s)2r dr \,. \tag{13.7}$$

The MHD potential energy $\delta W_{\rm MHDtor}^{\rm ext}/2\pi R$ per unit length of toroidal plasma with circular cross-section is given by [13.3]

$$\frac{\delta W_{\rm MHDtor}^{\rm ext}}{2\pi R} = \left(1-\frac{1}{n^2}\right)\frac{\delta W_{\rm MHDcycl}^{\rm ext}}{2\pi R} + \frac{\pi B_{\theta s}^2}{2\mu_0}|\xi_s|^2 \delta\hat{W}_{\rm T} \,,$$

$$\delta\hat{W}_{\rm T} = \pi\left(\frac{r_s}{R}\right)^2 3(1-q_0)\left(\frac{13}{144}-\beta_{\rm ps}^2\right) \,. \tag{13.8}$$

For $m=1$ and $n=1$, $\delta W_{\rm MHDtor}^{\rm ext}/2\pi R$ reduces to just the $\delta\hat{W}_{\rm T}$ term.

Let us consider the contribution from the singular region. In this case we must solve for the displacement $\xi_r$ in the singular region near the rational surface. The equation of motion in the singular surface was treated in Sect. 9.1 on tearing instability. From (9.13) and (9.9), we have (in the limit $x \ll 1$)

$$\mu_0\rho_{\rm m}\gamma^2\frac{\partial^2\xi_r}{\partial x^2} = {\rm i}F\frac{\partial^2 B_{1r}}{\partial x^2} \,, \tag{13.9}$$

$$\gamma B_{1r} = {\rm i}F\gamma\xi_r + \frac{\eta}{\mu_0 r_s^2}\frac{\partial^2}{\partial x^2}B_{1r} \,, \tag{13.10}$$

where

$$F \equiv (\boldsymbol{k}\cdot\boldsymbol{B}) = -\frac{m}{r}B_\theta + \frac{n}{R}B_z = \frac{B_\theta}{r}(-m+nq) = \frac{B_\theta n}{r}\frac{dq}{dr}\Delta r = \frac{B_\theta ns}{r_s}x \,,$$

$$x \equiv \frac{r-r_s}{r_s} \,, \qquad s \equiv r_s\frac{dq}{dr}\bigg|_{r_s} \,.$$

With the normalizations

$$\psi \equiv \frac{{\rm i}B_{1r}r_s}{B_{\theta,s}sn} \,, \quad \tau_{{\rm A}\theta} \equiv \frac{r_s}{(B_\theta^2/\mu_0\rho_m)^{1/2}} \,, \quad \tau_{\rm R} \equiv \frac{\mu_0 r_s^2}{\eta} \,, \quad S_{\rm R} \equiv \frac{\tau_{\rm R}}{\tau_{{\rm A}\theta}} \,,$$

we have

$$\gamma^2\left(\frac{\tau_{{\rm A}\theta}}{ns}\right)^2\xi_r'' = x\psi'' \,, \qquad \psi = -x\xi_r + \frac{1}{\gamma\tau_{{\rm A}\theta}S_{\rm R}}\psi'' \,. \tag{13.11}$$

In the limit $S_{\rm R}\to\infty$, (13.11) yields

$$\left[\left(\gamma\frac{\tau_{{\rm A}\theta}}{ns}\right)^2 + x^2\right]\xi_r'' + 2x\xi_r' = 0 \,,$$

and the solution is [13.4]

$$\xi_r' = \frac{(\xi_0/\pi)(\gamma\tau_{A\theta}/ns)^{-1}}{x^2/(\gamma\tau_{A\theta}/ns)^2+1}\,, \qquad \xi_r(x) = \xi_\infty - \frac{\xi_0}{\pi}\tan^{-1}\left[\frac{x}{\gamma\tau_{A\theta}/(ns)}\right]\,. \tag{13.12}$$

Since the external solution of $m = 1$ is $\xi_r = \xi_s$ as $x \to -\infty$ and $\xi_r = 0$ as $x \to \infty$, the matching conditions for the external solution yield $\xi_\infty = \xi_s/2$ and $\xi_0 = \xi_s$. The term $\delta W^s_{MHD}$ for the singular region is

$$\frac{\delta W^s_{MHD}}{2\pi R} = \frac{\pi}{2\mu_0}\int_{r_s-\Delta}^{r_s+\Delta} r^3(\boldsymbol{k}\cdot\boldsymbol{B})^2\left(\frac{\partial\xi_r}{\partial r}\right)^2 dr = \frac{\pi}{2\mu_0}\frac{B^2_{\theta s}}{2\pi}sn\gamma\tau_{A\theta}|\xi_s|^2\,. \tag{13.13}$$

Equation (13.13) is the expression in the case of a cylindrical plasma. For the toroidal plasma, $\tau_{A\theta}$ is replaced by $3^{1/2}r_s/(B^2_\theta/\mu_0\rho)^{1/2}$, where $3^{1/2}$ is the standard toroidal factor $(1+2q^2)^{1/2}$ [13.5]. Therefore the total sum of MHD contributions with $m = 1$, $n = 1$ is (assuming $\gamma\tau_{A\theta} \ll 1$)

$$\begin{aligned}\delta W_{MHD} + \delta I &= 2\pi R\frac{B^2_{\theta s}}{2\mu_0}|\xi_s|^2\left(\delta\hat{W}^T + \gamma\tau_{A\theta}\frac{s}{2} + \pi\gamma^2\tau^2_{A\theta}\right)\\ &\approx 2\pi R\frac{B^2_{\theta s}}{2\mu_0}|\xi_s|^2\left(\delta\hat{W}^T + \gamma\tau_{A\theta}\frac{s}{2}\right)\,. \end{aligned}\tag{13.14}$$

### 13.1.3 Kinetic Integral of Hot Component

The perturbed distribution $\delta F_h$ of the hot ion component is given by the gyrokinetic equation in the case of low beta and zero gyro-radius approximation as follows [13.6]:

$$\delta F_h \equiv \frac{e}{m}\delta\phi\frac{\partial}{\partial E}F_{0h} + \delta H_h\,, \qquad \left[v_\parallel\frac{\partial}{\partial l} - i(\omega - \hat{\omega}_{dh})\right]\delta H_h = i\frac{e}{m}Q(\delta\phi - v_\parallel\delta A_\parallel)\,, \tag{13.15}$$

where $\partial/\partial l \equiv \boldsymbol{b}\cdot\boldsymbol{\nabla}$, $\delta A_\parallel = (-i/\omega)\partial\delta\phi/\partial l$ due to $E_\parallel = 0$ [see (13.45)] and

$$E \equiv \frac{v^2}{2}\,, \quad \mu \equiv \frac{v_\perp^2}{2B}\,, \quad \omega_c \equiv \frac{eB}{m}\,, \quad Q \equiv \left(\omega\frac{\partial}{\partial E} + \hat{\omega}_{*h}\right)F_{0h}\,,$$

$$\hat{\omega}_{dh} \equiv -i\boldsymbol{v}_{dh}\cdot\boldsymbol{\nabla}\,, \quad \boldsymbol{v}_{dh} \equiv \left(v_\parallel^2 + \frac{v_\perp^2}{2}\right)\frac{m}{eB}(\boldsymbol{b}\times\boldsymbol{\kappa})\,,$$

$$\hat{\omega}_{*h} \equiv -i\omega_c^{-1}\frac{\boldsymbol{b}\times\nabla F_{0h}}{F_{0h}}\cdot\boldsymbol{\nabla} \approx -\frac{m}{eBr}\frac{1}{F_{0h}}\frac{\partial}{\partial r}\,.$$

$\boldsymbol{v}_{dh}$ is the magnetic drift velocity and $|\hat{\omega}_{dh}|$ is the diamagnetic drift frequency of the hot ions. $\boldsymbol{\kappa} = (\boldsymbol{b}\cdot\boldsymbol{\nabla})\boldsymbol{b}$ is the vector toward the center of curvature of the magnetic field line and the magnitude is $R^{-1}$ (see Sect. 2.4). $\delta\phi$ is the scalar potential such that $\boldsymbol{\nabla}\delta\phi = -i\omega\boldsymbol{\xi}\times\boldsymbol{B}$. When we set

$$\delta H_h = -\frac{1}{\omega}\frac{e}{m}Q\delta\phi + \delta G_h\,, \tag{13.16}$$

we have

$$v_\parallel \frac{\partial \delta H_\mathrm{h}}{\partial l} = \mathrm{i}(\omega - \hat{\omega}_\mathrm{dh})\delta G_\mathrm{h} + \mathrm{i}\frac{\hat{\omega}_\mathrm{dh}}{\omega}\frac{e}{m}Q\delta\phi - \frac{1}{\omega^2}\frac{e}{m}v_\parallel \frac{\partial \delta\phi}{\partial l} \ .$$

Taking the average $\bar{A} \equiv \oint (A/v_\parallel)\mathrm{d}l / \oint \mathrm{d}l/v_\parallel$ of both sides of the foregoing equation yields

$$\delta G_\mathrm{h} = -\frac{e}{m}Q\frac{1}{\omega - \overline{\hat{\omega}_\mathrm{dh}}}\frac{\overline{\hat{\omega}_\mathrm{dh}\delta\phi}}{\omega} \tag{13.17}$$

and

$$\begin{aligned}\frac{\hat{\omega}_\mathrm{dh}\delta\phi}{\omega} &= -\frac{\mathrm{i}}{\omega}\frac{m(v_\parallel^2 + v_\perp^2/2)}{eB}(\boldsymbol{b}\times\boldsymbol{\kappa})\cdot\boldsymbol{\nabla}\delta\phi \\ &= -\frac{1}{\omega}\frac{m(v_\parallel^2 + v_\perp^2/2)}{eB}(\boldsymbol{b}\times\boldsymbol{\kappa})\cdot\omega(\boldsymbol{\xi}\times\boldsymbol{B}) \\ &= \frac{m(v_\parallel^2 + v_\perp^2/2)}{e}(\boldsymbol{\kappa}\cdot\boldsymbol{\xi}) = -\frac{m}{e}Jv^2 = -\frac{m}{e}J2E \ ,\end{aligned}$$

where

$$\begin{aligned}J \equiv -\frac{1}{v^2}(v_\parallel^2 + v_\perp^2/2)(\boldsymbol{\kappa}\cdot\boldsymbol{\xi}) &\approx \frac{v_\parallel^2 + v_\perp^2/2}{v^2}\frac{(\cos\theta\xi_r + \sin\theta\xi_\theta)}{R} \\ &\sim \frac{v_\parallel^2 + v_\perp^2/2}{v^2}\frac{\mathrm{e}^{-\mathrm{i}\theta}\xi_r}{R}\end{aligned} \tag{13.18}$$

($\boldsymbol{\xi}$ is incompressible).One notes that frequencies $\omega$, $\omega_\mathrm{dh}$ are much smaller than the hot ion transit and bounce frequencies $v_\parallel/R$, $\epsilon^{1/2}v/qR$. For untrapped particles ($\delta G_\mathrm{hu}$) and trapped particles ($\delta G_\mathrm{ht}$), we have

$$\delta G_\mathrm{hu} \approx 0 \ , \qquad \delta G_\mathrm{ht} \approx 2QE\frac{\bar{J}}{\omega - \overline{\hat{\omega}_\mathrm{dh}}} \ . \tag{13.19}$$

The perturbed pressure tensor due to the hot ion component is

$$\delta P_\mathrm{h} = -\boldsymbol{\xi}_\perp\cdot\boldsymbol{\nabla}\left[P_\perp I + (P_\parallel - P_\perp)\boldsymbol{bb}\right] + \delta P_\perp I + (\delta P_\parallel - \delta P_\perp)\boldsymbol{bb} \ , \tag{13.20}$$

where

$$\delta P_\perp = \int \frac{mv_\perp^2}{2}\delta F_\mathrm{h} 2\pi v_\perp \mathrm{d}v_\perp \mathrm{d}v_\parallel \ , \qquad \delta P_\parallel = \int mv_\parallel^2 \delta F_\mathrm{h} 2\pi v_\perp \mathrm{d}v_\perp \mathrm{d}v_\parallel \ .$$

The first term on the right-hand side of (13.20) has similar form to the pressure term of the core plasma. Since the beta of the hot ion component $\beta_\mathrm{h}$ is much smaller than $\beta_\mathrm{c}$ for the core plasma, the first term in (13.20) can be neglected. Since $E = v^2/2$, $\mu = v_\perp^2/2B$ and $\alpha \equiv \mu/E$ are defined, we have

$$v_\parallel^2 = 2E(1 - \alpha B) \ , \qquad v_\perp^2 = 2B\alpha E \ ,$$

$$2\pi v_\perp \mathrm{d}v_\perp \mathrm{d}v_\parallel = 2^2\pi \frac{BE}{v_\parallel}\mathrm{d}E\mathrm{d}\alpha = 2^{3/2}\pi B \frac{E^{1/2}}{(1-\alpha B)^{1/2}}\mathrm{d}\alpha\,\mathrm{d}E\,.$$

Then the perturbed pressure of the hot ion component reduces to

$$\begin{aligned}\delta P_\perp &= 2^{3/2}\pi B \int \frac{E^{1/2}}{(1-\alpha B)^{1/2}}\mathrm{d}\alpha\,\mathrm{d}E\, m\alpha BE\delta F_\mathrm{h} \\ &= 2^{5/2} mB \int_{B_\mathrm{max}^{-1}}^{B^{-1}} \mathrm{d}\alpha (1-\alpha B)^{1/2} \int_0^E \mathrm{d}EE^{3/2} \frac{\alpha B}{2(1-\alpha B)}\delta F_\mathrm{h}\,, \end{aligned} \tag{13.21}$$

$$\begin{aligned}\delta P_\parallel &= 2^{3/2}\pi B \int \frac{E^{1/2}}{(1-\alpha B)^{1/2}}\mathrm{d}\alpha\,\mathrm{d}E\, m2E(1-\alpha B)\delta F_\mathrm{h} \\ &= 2^{5/2} mB \int_{B_\mathrm{max}^{-1}}^{B^{-1}} \mathrm{d}\alpha (1-\alpha B)^{1/2} \int_0^E \mathrm{d}EE^{3/2}\delta F_\mathrm{h}\,. \end{aligned} \tag{13.22}$$

The divergence of the second pressure term on the right-hand side of (13.20) is

$$\begin{aligned}(\nabla\delta P_\mathrm{h})_\beta &= \sum_\alpha \frac{\partial \delta P_\perp}{\partial x_\alpha}\delta_{\alpha\beta} + \sum_\alpha \frac{\partial(\delta P_\parallel - \delta P_\perp)}{\partial x_\alpha} b_\alpha b_\beta + (\delta P_\parallel - \delta P_\perp)\sum_\alpha \frac{\partial(b_\alpha b_\beta)}{\partial x_\alpha} \\ &= \frac{\partial \delta P_\perp}{\partial x_\beta} + b_\beta(\boldsymbol{b}\cdot\nabla)(\delta P_\parallel - \delta P_\perp) + (\delta P_\parallel - \delta P_\perp)\left[(\boldsymbol{b}\cdot\nabla)b_\beta + b_\beta(\nabla\cdot\boldsymbol{b})\right]\,,\end{aligned}$$

$$(\nabla\delta P_\mathrm{h})_\perp = \nabla_\perp\delta P_\perp + (\delta P_\parallel - \delta P_\perp)(\boldsymbol{b}\cdot\nabla)\boldsymbol{b} = \nabla_\perp\delta P_\perp + (\delta P_\parallel - \delta P_\perp)\boldsymbol{\kappa}\,, \tag{13.23}$$

$$(\nabla\delta P_\mathrm{h})_\parallel = \nabla_\parallel\delta P_\perp + (\boldsymbol{b}\cdot\nabla)(\delta P_\parallel - \delta P_\perp) + (\delta P_\parallel - \delta P_\perp)\nabla\cdot\boldsymbol{b}\,. \tag{13.24}$$

The kinetic integral $\delta W_\mathrm{K}$ is

$$\begin{aligned}\delta W_\mathrm{K} &= \frac{1}{2}\int \boldsymbol{\xi}_\perp^*\cdot\nabla\delta P \mathrm{d}\boldsymbol{r} = \frac{1}{2}\int \boldsymbol{\xi}_\perp^*\cdot\left[\nabla_\perp\delta P_\perp + (\delta P_\parallel - \delta P_\perp)\boldsymbol{\kappa}\right]\mathrm{d}\boldsymbol{r} \\ &= -\frac{1}{2}\int \left[\nabla\cdot\boldsymbol{\xi}_\perp^*\delta P_\perp - (\delta P_\parallel - \delta P_\perp)\boldsymbol{\xi}_\perp^*\cdot\boldsymbol{\kappa}\right]\mathrm{d}\boldsymbol{r} \\ &= -2^{3/2}\pi m \int \mathrm{d}\boldsymbol{r} B \int \mathrm{d}\alpha\mathrm{d}E \frac{E^{3/2}}{(1-\alpha B)^{1/2}} \\ &\quad \times \left[\nabla\cdot\boldsymbol{\xi}_\perp^* \frac{v_\perp^2}{2v^2} - (\boldsymbol{\xi}_\perp^*\cdot\boldsymbol{\kappa})\frac{v_\parallel^2 - v_\perp^2/2}{v^2}\right]\delta F_\mathrm{h}\,.\end{aligned}$$

Since $\nabla\cdot\boldsymbol{\xi}_\perp + 2(\boldsymbol{\xi}\cdot\boldsymbol{k}) \approx 0$ [see (8.116)], the term in square brackets in the integrand is

$$-\frac{v_\parallel^2 + v_\perp^2/2}{v^2}(\boldsymbol{\xi}_\perp^* \cdot \boldsymbol{\kappa}) \approx -\frac{1}{2}(\boldsymbol{\xi}_\perp^* \cdot \boldsymbol{\kappa}) .$$

$\delta W_K$ reduces to

$$\begin{aligned}
\frac{\delta W_K}{2\pi R} &= -\frac{2^{3/2}\pi}{2\pi R} m_h \int \mathrm{d}\boldsymbol{r} B \int \mathrm{d}\alpha\, \mathrm{d}E \frac{E^{3/2}}{(1-\alpha B)^{1/2}} \frac{\bar{J}^* Q E \bar{J}2}{\omega - \overline{\hat{\omega}_{dh}}} \\
&= -2^{5/2}\pi^2 m_h \int \mathrm{d}r\, r \frac{1}{2\pi} \int \mathrm{d}\theta B \int \mathrm{d}\alpha\, \mathrm{d}E \frac{E^{3/2}}{(1-\alpha B)^{1/2}} \frac{\bar{J}^* Q E \bar{J}2}{\omega - \overline{\hat{\omega}_{dh}}} \\
&= -2^{7/2}\pi^2 m_h \int_0^{r_s} \mathrm{d}r\, r \int_{(1-r/R)}^{(1+r/R)} \mathrm{d}(\alpha B) \int \mathrm{d}E K_b E^{5/2} \frac{\overline{\bar{J}^* Q E \bar{J}}}{\omega - \overline{\hat{\omega}_{dh}}} \\
&\approx 2^{3/2}\pi^2 m_h \frac{|\boldsymbol{\xi}|^2}{R^2} \int_0^{r_s} \mathrm{d}r\, r \int \mathrm{d}(\alpha B) \int \mathrm{d}E E^{5/2} \frac{K_2^2}{K_b} \frac{Q}{\overline{\hat{\omega}_{dh}} - \omega} \\
&\equiv \frac{B_{\theta s}^2}{2\mu_0} |\boldsymbol{\xi}_s|^2 \delta \hat{W}_K ,
\end{aligned} \tag{13.25}$$

where

$$K_b = \oint \frac{\mathrm{d}\theta}{2\pi} \frac{1}{(1-\alpha B)^{1/2}} = \oint \frac{\mathrm{d}\theta}{2\pi} \frac{v}{v_\parallel} , \qquad K_2 = \oint \frac{\mathrm{d}\theta}{2\pi} \frac{\cos\theta}{(1-\alpha B)^{1/2}} .$$

Therefore the dispersion relation (13.2) reduces to

$$-\frac{\mathrm{i}\omega}{\omega_A} + \delta \hat{W}^T + \delta \hat{W}_K = 0 , \tag{13.26}$$

where $\omega_A \equiv (\tau_A s/2)^{-1}$ and $\gamma$ is replaced by $-\mathrm{i}\omega$.

### 13.1.4 Growth Rate of Fishbone Instability

Let us assume a model distribution for slowing down hot ions with initial velocity $v_{mx}^2/2 = E_{mx}$:

$$F_{h0} = c_0 \frac{\delta(\alpha - \alpha_0)}{E^{3/2}} \qquad (E < E_{mx}) . \tag{13.27}$$

Then the pressure $p_h$ and the density $n_h$ of hot ions are

$$\begin{aligned}
p_h &= \int 2^{3/2}\pi B \frac{E^{1/2}}{(1-\alpha B)^{1/2}} \mathrm{d}\alpha\, \mathrm{d}E (mE) F_{h0} \\
&= c_0 \int_0^{E_{mx}} 2^{3/2}\pi B \frac{m}{(1-\alpha_0 B)^{1/2}} \mathrm{d}E = c_0 2^{3/2}\pi B m K_b E_{mx} ,
\end{aligned} \tag{13.28}$$

$$c_0 = \frac{p_h}{2^{3/2}\pi B m K_b E_{mx}} , \tag{13.29}$$

$$n_{\rm h} = \int_{T_{\rm c}}^{E_{\rm mx}} 2^{3/2}\pi B \frac{E^{1/2}}{(1-\alpha B)^{1/2}} {\rm d}\alpha {\rm d}E F_{\rm h0}$$

$$= c_0 2^{3/2}\pi B m K_{\rm b} E_{\rm mx} = p_{\rm h} \frac{\ln(E_{\rm mx}/T_{\rm c})}{E_{\rm mx}} . \tag{13.30}$$

The kinetic integral is

$$\begin{aligned}
\frac{\delta W_{\rm K}}{2\pi R} &= \frac{r_{\rm s}^2}{R^2}|\boldsymbol{\xi}_{\rm s}|^2 \frac{1}{r_{\rm s}^2}\int_0^{r_{\rm s}} {\rm d}r\, r \int {\rm d}E 2^{3/2}\pi^2 m B E^{5/2}\left(\frac{K_2^2}{K_{\rm b}}\right) \\
&\quad \times \left[\frac{-(3/2)\omega c_0 E^{-5/2} - (\partial c_0/\partial r)(m/eBr)E^{-3/2}}{mE/(2eBRr)-\omega}\right] \\
&= \frac{r_{\rm s}^2}{R^2}\frac{B^2}{2\mu_0}|\boldsymbol{\xi}_{\rm s}|^2 \frac{1}{r_{\rm s}^2}\int_0^{r_{\rm s}} {\rm d}r\, r\pi \frac{K_2^2}{K_{\rm b}^2}\frac{1}{E_{\rm mx}} \\
&\quad \times \int_0^{E_{\rm mx}} {\rm d}E \frac{-(3/2)\beta_{\rm h} - 2(\partial\beta_{\rm h}/\partial r)R(mE/2eBRr\omega)}{mE/(2eBRr\omega)-1} \\
&= \frac{r_{\rm s}^2}{R^2}\frac{B^2}{2\mu_0}|\boldsymbol{\xi}_{\rm s}|^2 \frac{1}{r_{\rm s}^2}\int_0^{r_{\rm s}} {\rm d}r\, r\pi \frac{K_2^2}{K_{\rm b}^2}\frac{\omega}{\omega_{\rm dh,mx}} \\
&\quad \times \left(-\frac{3}{2}\beta_{\rm h}\int_0^{y_{\rm mx}} \frac{{\rm d}y}{y-1} - 2\frac{\partial\beta_{\rm h}}{\partial r}R\int\frac{y{\rm d}y}{y-1}\right) \\
&= \frac{B_{\theta{\rm s}}^2}{2\mu_0}|\boldsymbol{\xi}_{\rm s}|^2 \frac{\pi}{2}\frac{K_2^2}{K_{\rm b}^2}\left\{ -\frac{3}{2}\langle\beta_{\rm h}\rangle \frac{\omega}{\omega_{\rm dh,mx}}\ln\left(1-\frac{\omega_{\rm dh,mx}}{\omega}\right)\right. \\
&\quad \left. -2\left\langle\frac{\partial\beta_{\rm h}}{\partial r}\right\rangle R\left[1+\frac{\omega}{\omega_{\rm dh,mx}}\ln\left(1-\frac{\omega_{\rm dh,mx}}{\omega}\right)\right]\right\} . 
\end{aligned} \tag{13.31}$$

As the second term of $\langle(\partial\beta_{\rm h}/\partial r)\rangle R$ is dominant, the dispersion relation reduces to

$$-{\rm i}\Omega\frac{\omega_{\rm dh,mx}}{\omega_{\rm A}} + \delta\hat{W}^{\rm T} + \pi\frac{K_2^2}{K_{\rm b}^2}\left\langle-\frac{\partial\beta_{\rm h}}{\partial r}\right\rangle R\left[1+\Omega\ln\left(1-\frac{1}{\Omega}\right)\right] = 0 , \tag{13.32}$$

where

$$\Omega \equiv \frac{\omega}{\omega_{\rm dh,mx}} , \quad \omega_{\rm dh,mx} \equiv \frac{mv_{\rm mx}^2/2}{eBRr} , \quad \beta_{\rm h} \equiv \frac{p_{\rm h}}{B^2/2\mu_0} .$$

Let us consider the case $\delta\hat{W}^{\rm T} = 0$. Then (13.32) is

$$-{\rm i}\alpha_{\rm h}\Omega + \Omega\ln\left(1-\frac{1}{\Omega}\right) + 1 = 0 , \tag{13.33}$$

where

$$\alpha_{\rm h} \equiv \frac{\omega_{\rm dh,mx}}{\omega_{\rm A}}\left(\pi K_2^2 K_{\rm b}^2\left\langle-\frac{\partial\beta_{\rm h}}{\partial r}\right\rangle R\right)^{-1} .$$

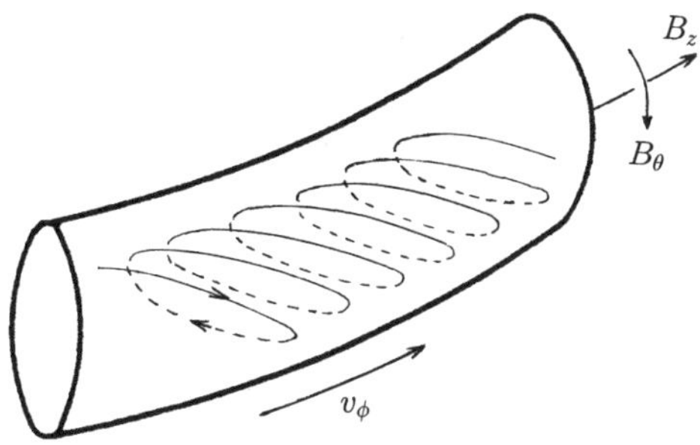


**Fig. 13.1.** Toroidal precession of banana orbit of trapped ions

Under the assumption $(1 - 1/\Omega_r) < 0$ and $|\Omega_i| \ll |\Omega_r|$, (13.33) reduces to

$$-i\alpha_h(\Omega_r + i\Omega_i) + (\Omega_r + i\Omega_i)\left[\ln\left(\frac{1}{\Omega_r} - 1\right) + \pi i - \frac{\Omega_i}{(1/\Omega_r - 1)\Omega_r^2} i\right] + 1 = 0 . \tag{13.34}$$

From the real and imaginary parts of (13.34), we have

$$\Omega_i = \frac{\pi - \alpha_h}{-\ln(1/\Omega_r - 1) + \left[1 - (1/\Omega_r)\right]}\Omega_r , \tag{13.35}$$

$$\Omega_r = \frac{1 - (\pi - \alpha_h)\Omega_i}{-\ln(1/\Omega_r - 1)} . \tag{13.36}$$

In the case of a marginally unstable state $\pi = \alpha_h$, that is, $\Omega_i = 0$, $\Omega_r$ is given by

$$\Omega_r = \frac{1}{-\ln(1/\Omega_r - 1)} \quad \longrightarrow \quad \Omega_r = \frac{1}{1 + \exp(-1/\Omega_r)} = \frac{1}{2}\left(1 + \tanh\frac{1}{2\Omega_r}\right) ,$$

and $\Omega_r \approx 0.75$. A necessary condition for the excitation of a fishbone instability is $\Omega_i > 0$, that is, $\alpha_h < \pi$ and

$$\left\langle -\frac{\partial \beta_h}{\partial r}\right\rangle r_s > \frac{r_s}{R}\frac{\omega_{dh,mx}}{\omega_A}\frac{1}{\pi^2}\frac{K_b^2}{K_2^2} .$$

There is a threshold for $\langle|\partial\beta_h/\partial r|\rangle r_s$ for the instability.

Banana orbits of trapped ions drift in the toroidal direction, as shown in Fig. 13.1. The toroidal precession velocity and frequency are[1]

[1] The toroidal vertical drift velocity is $v_d = mv_\perp^2/2eBR$, so that the poloidal displacement of particles between bounces is $r\delta\theta \sim v_d\tau_d$, $\tau_d$ being the bounce period. Since $d\phi/d\theta = q$ along the magnetic field line, the associated toroidal displacement between bounces is $Rd\phi = Rqv_d\tau_d/r$, $q = 1$. Thus the toroidal precession velocity is given by (13.37).

$$v_\phi = \frac{mv_\perp^2/2}{eBr}\,, \qquad \omega_\phi = \frac{mv_\perp^2/2}{eBRr}\,. \tag{13.37}$$

Therefore $\omega_{\rm dh,mx}$ is equal to the toroidal precession frequency of trapped ions with the initial (maximum) velocity. It seems that the fishbone instability is due to an interaction between energetic particles and the $m = 1$, $n = 1$ MHD perturbation. The interaction of resonant type is characterized by Landau damping. The resonance is between the toroidal wave velocity of the instability and the toroidal precession of trapped energetic particles.

## 13.2 Toroidal Alfvén Eigenmode

Alfvén waves in a homogeneous magnetic field in an infinite plasma were discussed in Sect. 5.4. Shear Alfvén waves and fast and slow magnetosonic waves appear. In the case of an incompressible plasma ($\nabla\cdot\boldsymbol{\xi} = 0$ or specific heat ratio $\gamma \to \infty$), only the shear Alfvén wave can exist.

In the case of a cylindrical plasma in an axisymmetric magnetic field, the displacement of the MHD perturbation $\boldsymbol{\xi}(r,\theta,z) = \boldsymbol{\xi}(r)\exp{\rm i}(-m\theta+kz-\omega t)$ is given by the Hain–Lüst equations (8.103)–(8.106), as discussed in Sect. 8.4. In the case of an incompressible plasma, the Hain–Lüst equation (8.106) reduces to [recalling that the perturbation is assumed to be $\boldsymbol{\xi}(r)\exp{\rm i}(m\theta + kz - \omega t)$ in Sect. 8.4]

$$\frac{\rm d}{{\rm d}r}\left(\frac{F^2-\mu_0\rho_{\rm m}\omega^2}{m^2/r^2+k^2}\right)\frac{1}{r}\frac{\rm d}{{\rm d}r}(r\xi_r) + \Bigg\{-(F^2-\mu_0\rho_{\rm m}\omega^2) + 2B_\theta\frac{\rm d}{{\rm d}r}\left(\frac{B_\theta}{r}\right)$$
$$+\frac{4k^2B_\theta^2F^2}{r^2(m^2/r^2+k^2)(F^2-\mu_0\rho_{\rm m}\omega^2)} + 2r\frac{\rm d}{{\rm d}r}\left[\frac{(m/r)FB_\theta}{r^2(m^2/r^2+k^2)}\right]\Bigg\}\xi_r = 0\,, \tag{13.38}$$

where

$$F = (\boldsymbol{k}\cdot\boldsymbol{B}) = \left[-\frac{m}{r}B_\theta(r) + \frac{n}{R}B_z(r)\right] = \frac{B_z}{R}\left[n - \frac{m}{q(r)}\right]\,, \qquad q(r) = \frac{R}{r}\frac{B_z}{B_\theta}\,.$$

The position at which $F^2 - \mu_0\rho_{\rm m}\omega^2 = 0 \to \omega^2 = \boldsymbol{k}_\parallel^2 v_{\rm A}^2$, $v_{\rm A}^2 \equiv B^2/\mu_0\rho_{\rm m}$ holds is the singular radius. It was shown by Hasegawa and Chen [13.7] that, at this singular radius (resonant layer), the shear Alfvén wave is mode-converted to the kinetic Alfvén wave and absorbed by Landau damping. The Alfvén wave is therefore stable in the cylindrical plasma.

Alfvén waves were also treated in Sects. 10.4.1 and 10.4.2 using the cold plasma model. The dispersion relation in a homogeneous infinite plasma is given by (10.64c), showing that Alfvén resonance occurs at $\omega^2 \approx k_\parallel^2 v_{\rm A}^2$ and that the compressional Alfvén wave and shear Alfvén wave cutoffs occur at $\omega^2 = k_\parallel^2 v_{\rm A}^2(1+\omega/\Omega_{\rm i})$ and $\omega^2 = k_\parallel^2 v_{\rm A}^2(1-\omega/\Omega_{\rm i})$, respectively.

### 13.2.1 Toroidicity-Induced Alfvén Eigenmode

Let us consider shear Alfvén waves in a toroidal plasma and the perturbation of the $(-m, n)$ mode given by

$$\phi(r, \theta, z, t) = \phi(r) \exp \mathrm{i} \left( -m\theta + n\frac{z}{R} - \omega t \right) , \tag{13.39}$$

where $R$ is the major radius of the torus and

$$k_\parallel = \frac{\boldsymbol{k} \cdot \boldsymbol{B}}{B} = \frac{1}{R} \left[ n - \frac{m}{q(r)} \right] .$$

The resonant conditions of the $m$ and $m+1$ modes in a linear cylindrical plasma are

$$\frac{\omega^2}{v_\mathrm{A}^2} - k_{\parallel m}^2 = 0 , \qquad \frac{\omega^2}{v_\mathrm{A}^2} - k_{\parallel m+1}^2 = 0 .$$

However, the $m$ mode wave can couple with $m \pm 1$ in a toroidal plasma since the magnitude of the toroidal field changes as $B_z = B_{z0}\big[1 - (r/R)\cos\theta\big]$, as will be shown later in this section. Then the resonant condition of $m$ and $m+1$ modes in a toroidal plasma becomes

$$\begin{vmatrix} \dfrac{\omega^2}{v_\mathrm{A}^2} - k_{\parallel m}^2 & \alpha\epsilon \dfrac{\omega^2}{v_\mathrm{A}^2} \\ \alpha\epsilon \dfrac{\omega^2}{v_\mathrm{A}^2} & \dfrac{\omega^2}{v_\mathrm{A}^2} - k_{\parallel m+1}^2 \end{vmatrix} = 0 ,$$

where $\epsilon = r/R$, and $\alpha$ is a constant of order of 1. Then the solutions are

$$\frac{\omega_\pm^2}{v_\mathrm{A}^2} = \frac{k_{\parallel m}^2 + k_{\parallel m+1}^2 \pm \left[ (k_{\parallel m}^2 - k_{\parallel m+1}^2)^2 + 4\alpha\epsilon^2 k_{\parallel m}^2 k_{\parallel m+1}^2 \right]^{1/2}}{2(1 - \alpha^2\epsilon^2)} . \tag{13.40}$$

The resonant condition (13.40) is plotted in Fig. 13.2.

At the radius satisfying $k_{\parallel m}^2 = k_{\parallel m+1}^2$, the difference of $\omega_\pm$ becomes minimum and the radius is given by

$$\frac{1}{R}\left[ n - \frac{m}{q(r)} \right] = -\frac{1}{R}\left[ n - \frac{m+1}{q(r)} \right] , \quad q(r_0) = \frac{m + 1/2}{n} , \tag{13.41}$$

$$k_{\parallel m} = -k_{\parallel m+1} = \frac{1}{2q(r_0)R} .$$

$q(r_0) = 1.5$ for the case $m = 1$ and $n = 1$. Therefore Alfvén resonance does not exist in the frequency gap $\omega_- < \omega < \omega_+$.

The continuum Alfvén waves correspond to excitation of shear Alfvén waves on a given flux surface where the mode frequency is resonant $\omega^2 = k_{\parallel m}^2 v_\mathrm{A}^2(r)$, and such a resonance leads to wave damping. However, frequencies

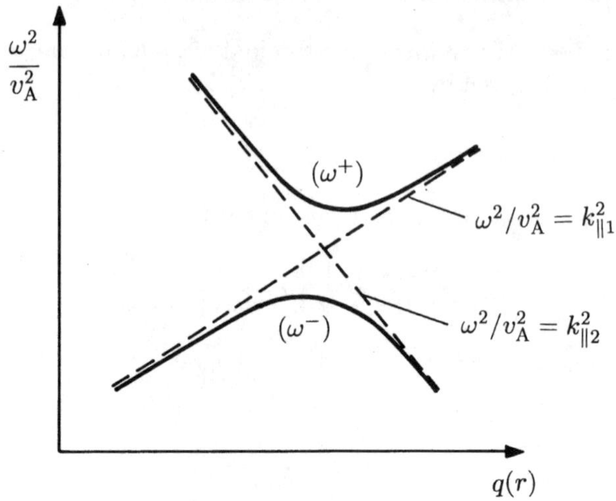


**Fig. 13.2.** Alfvén resonance frequency $\omega$ of toroidally coupled $m$ and $m+1$ modes

excited within the spectral gaps are not resonant with the continuum and hence will not damp in the gap region. This allows a discrete eigenfrequency of the *toroidicity-induced Alfvén eigenmode* or *toroidal Alfvén eigenmode* (TAE) to be established. This TAE can easily be destabilized by the kinetic effect of energetic particles.

The equations of the TAE will be described here according to Berk, Van Dam, Guo, and Lindberg [13.8]. The equations of the first-order perturbations are

$$\nabla \cdot \boldsymbol{j}_1 = 0 \, , \qquad \rho \frac{\mathrm{d}\boldsymbol{v}_1}{\mathrm{d}t} = (\boldsymbol{j} \times \boldsymbol{B})_1 \, , \tag{13.42}$$

$$\boldsymbol{E}_1 = \nabla \phi_1 - \frac{\partial \boldsymbol{A}_1}{\partial t} \, , \qquad \boldsymbol{B}_1 = \nabla \times \boldsymbol{A}_1 \, . \tag{13.43}$$

For ideal, low beta MHD waves, we have the relations

$$\boldsymbol{E}_{\parallel} = 0 \, , \qquad \boldsymbol{B}_{\parallel 1} = 0 \, , \qquad \boldsymbol{A}_1 = A_{\parallel 1} \boldsymbol{b} \, , \tag{13.44}$$

so that

$$\mathrm{i}\omega A_{\parallel 1} = \boldsymbol{b} \cdot \nabla \phi_1 \, , \qquad \boldsymbol{v}_1 = \frac{\boldsymbol{E}_1 \times \boldsymbol{b}}{B} \, . \tag{13.45}$$

From (13.42), we have

$$\nabla \cdot \boldsymbol{j}_{\perp 1} + \nabla \cdot (j_{\parallel 1} \boldsymbol{b}) = 0 \tag{13.46}$$

and

$$-\mathrm{i}\omega\rho(\boldsymbol{v}_1 \times \boldsymbol{b}) = (\boldsymbol{j}_{\perp 1} \times \boldsymbol{B}) \times \boldsymbol{b} + (\boldsymbol{j} \times \boldsymbol{B}_1) \times \boldsymbol{b} \, ,$$

$$\boldsymbol{j}_{\perp 1} = -\frac{\mathrm{i}\omega\rho}{B^2} \boldsymbol{E}_{\perp 1} + \frac{j_{\parallel}}{B} \boldsymbol{B}_{\perp 1} \, . \tag{13.47}$$

Equations (13.43)–(13.45) yield

$$\boldsymbol{B}_{\perp 1} = \boldsymbol{\nabla} \times (A_{\parallel 1}\boldsymbol{b}) = \boldsymbol{\nabla}\left(\frac{A_{\parallel 1}}{B}\right) \times \boldsymbol{B} + \frac{A_{\parallel 1}}{B}\boldsymbol{\nabla} \times \boldsymbol{B}$$

$$\approx -\frac{\mathrm{i}}{\omega}\boldsymbol{\nabla}\left(\frac{\boldsymbol{b}\cdot\boldsymbol{\nabla}\phi_1}{B}\right) \times \boldsymbol{B}\,, \tag{13.48}$$

$$j_{\parallel 1} = \boldsymbol{b}\cdot\boldsymbol{j}_1 = \frac{1}{\mu_0}\boldsymbol{b}\cdot\boldsymbol{\nabla} \times \boldsymbol{B}_{\perp 1}$$

$$= -\frac{\mathrm{i}}{\omega\mu_0}\boldsymbol{b}\cdot\boldsymbol{\nabla} \times \left\{B^2\boldsymbol{\nabla}_\perp\left[\frac{(\boldsymbol{B}\cdot\boldsymbol{\nabla})\phi_1}{B^2}\right] \times \frac{\boldsymbol{B}}{B^2}\right\}$$

$$= \frac{\mathrm{i}}{\omega\mu_0}\left(\boldsymbol{b}\cdot\frac{\boldsymbol{B}}{B^2}\right)\boldsymbol{\nabla}\cdot\left[B^2\boldsymbol{\nabla}_\perp\left(\frac{\boldsymbol{B}\cdot\boldsymbol{\nabla}\phi_1}{B^2}\right)\right]$$

$$= \frac{\mathrm{i}}{\omega\mu_0 B}\boldsymbol{\nabla}\cdot\left[B^2\boldsymbol{\nabla}_\perp\left(\frac{\boldsymbol{B}\cdot\boldsymbol{\nabla}\phi_1}{B^2}\right)\right]\,. \tag{13.49}$$

Then (13.46)–(13.49) yield

$$0 = \boldsymbol{\nabla}\cdot\left(\mathrm{i}\frac{\omega}{\mu_0}\frac{1}{v_{\mathrm{A}}^2}\boldsymbol{\nabla}_\perp\phi_1\right) + \boldsymbol{\nabla}\left(\frac{j_\parallel}{B}\boldsymbol{B}_{\perp 1}\right) + \boldsymbol{\nabla}\left(\frac{j_{\parallel 1}}{B}\boldsymbol{B}\right)\,,$$

$$0 = \boldsymbol{\nabla}\cdot\left(\frac{\omega^2}{v_{\mathrm{A}}^2}\boldsymbol{\nabla}_\perp\phi_1\right) + \mu_0\boldsymbol{\nabla}\left(\frac{j_\parallel}{B}\right)\cdot\boldsymbol{B} \times \boldsymbol{\nabla}\left[\frac{(\boldsymbol{B}\cdot\boldsymbol{\nabla})\phi_1}{B^2}\right]$$

$$+(\boldsymbol{B}\cdot\boldsymbol{\nabla})\left\{\frac{1}{B^2}\boldsymbol{\nabla}\cdot\left[B^2\boldsymbol{\nabla}_\perp\cdot\left(\frac{\boldsymbol{B}\cdot\boldsymbol{\nabla}\phi_1}{B^2}\right)\right]\right\}\,. \tag{13.50}$$

When coordinates $(R, \varphi, Z)$ and $(r, \theta, \zeta)$ are introduced by

$$R = R_0 + r\cos\theta\,, \qquad Z = r\sin\theta\,, \qquad \varphi = -\frac{\zeta}{R}\,,$$

with the notation

$$\phi_1(r, \theta, \zeta, t) = \sum_m \phi_m(r)\exp\mathrm{i}(-m\theta + n\varphi - \omega t)\,,$$

$$(\boldsymbol{b}\cdot\boldsymbol{\nabla})\phi_m = \frac{\mathrm{i}}{R_0}\left[n - \frac{m}{q(r)}\right]\phi_m = \mathrm{i}k_{\parallel m}\phi_m\,,$$

$$k_{\parallel m} = \frac{1}{R_0}\left[n - \frac{m}{q(r)}\right]\,, \qquad E_m \equiv \frac{\phi_m}{R}\,.$$

equation (13.50) reduces to [13.8]

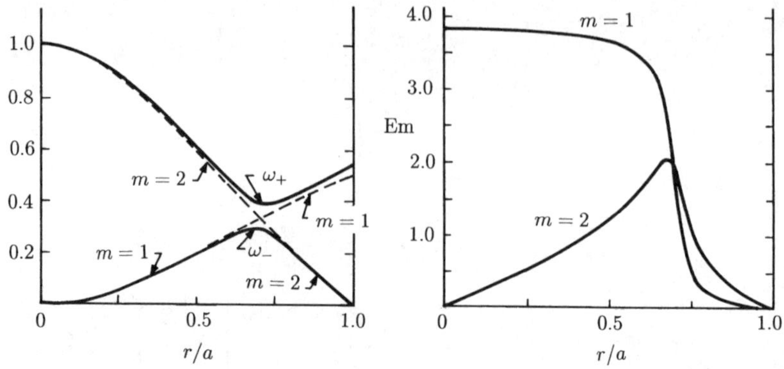


**Fig. 13.3.** *Left*: toroidal shear Alfvén resonance frequencies $\Omega$ corresponding to $(n = 1, m = 1)$ and $(n = 1, m = 2)$, $q(r) = 1 + (r/a)^2$, $a/R = 0.25$, $\Omega \equiv \omega/\left[v_{\rm A}(0)/R_0\right]$. *Right*: structure of the global mode amplitude as a function of radius [13.8]

$$0 = \frac{\mathrm{d}}{\mathrm{d}r}\left[r^3\left(\frac{\omega^2}{v_{\rm A}^2} - k_{\parallel m}^2\right)\frac{\mathrm{d}E_m}{\mathrm{d}r}\right] + r^2 E_m \frac{\mathrm{d}}{\mathrm{d}r}\left(\frac{\omega}{v_{\rm A}}\right)^2$$
$$-(m^2-1)\left(\frac{\omega^2}{v_{\rm A}^2} - k_{\parallel m}^2\right) r E_m + \frac{\mathrm{d}}{\mathrm{d}r}\left[r^3\left(\frac{\omega}{v_{\rm A}}\right)^2 \frac{2r}{R_0}\left(\frac{\mathrm{d}E_{m+1}}{\mathrm{d}r} + \frac{\mathrm{d}E_{m-1}}{\mathrm{d}r}\right)\right] . \tag{13.51}$$

As can be seen from Fig. 13.3, the mode structure has a sharp transition of $m = 1$ and $m = 2$ components at the gap location. Therefore $m$ and $m + 1$ modes near the gap location reduce to

$$\left(\frac{\omega^2}{v_{\rm A}^2} - k_{\parallel m}^2\right)\frac{\mathrm{d}E_m}{\mathrm{d}r} + \frac{2r}{R_0}\left(\frac{\omega}{v_{\rm A}}\right)^2 \frac{\mathrm{d}E_{m+1}}{\mathrm{d}r} \approx 0 ,$$

$$\left(\frac{\omega^2}{v_{\rm A}^2} - k_{\parallel m+1}^2\right)\frac{\mathrm{d}E_{m+1}}{\mathrm{d}r} + \frac{2r}{R_0}\left(\frac{\omega}{v_{\rm A}}\right)^2 \frac{\mathrm{d}E_m}{\mathrm{d}r} \approx 0 ,$$

so that the toroidal shear Alfvén resonance frequency is given by

$$\begin{vmatrix} \left(\frac{\omega^2}{v_{\rm A}^2} - k_{\parallel m}^2\right) & 2\epsilon\left(\frac{\omega}{v_{\rm A}}\right)^2 \\ 2\epsilon\left(\frac{\omega}{v_{\rm A}}\right)^2 & \left(\frac{\omega^2}{v_{\rm A}^2} - k_{\parallel m+1}^2\right) \end{vmatrix} = 0 . \tag{13.52}$$

When the Shafranov shift is included in the coordinates $(R, \varphi, Z)$ and $(r, \theta, \zeta)$, the coupling constant becomes $2.5\epsilon$ instead of $2\epsilon$ [13.8].

The energy integral from (13.51) without the coupling term of the $m \pm 1$ modes reduces to the following equation by partial integration:

$$G(\omega, E_m) \equiv \mathrm{P}\int_0^a \mathrm{d}r\, r\Biggl\{\left[r^2\left(\frac{\mathrm{d}E_m}{\mathrm{d}r}\right)^2 + (m^2-1)E_m^2\right]\left(\frac{\omega^2}{v_\mathrm{A}^2} - k_{\parallel m}^2\right) - \omega^2 r E_m^2 \frac{\mathrm{d}}{\mathrm{d}r}\frac{1}{v_\mathrm{A}^2}\Biggr\}$$
$$= E_m(r_\mathrm{s}^-)C_m(r_\mathrm{s}^-) - E_m(r_\mathrm{s}^+)C_m(r_\mathrm{s}^+)\,, \qquad (13.53)$$

where

$$C_m(r) = \left(\frac{\omega^2}{v_\mathrm{A}^2} - k_{\parallel m}^2\right) r^2 \frac{\mathrm{d}E_m}{\mathrm{d}r}\,, \quad E_m(a) = 0\,.$$

The radius $r = r_\mathrm{s}$ is the singular radius at which $(\omega^2/v_\mathrm{A})^2 - k_{\parallel m}^2 = 0$ and P denotes the principal value of the integral. From this formulation, it is possible to estimate the damping rate of TAE [13.8]:

$$\frac{\delta\omega}{\omega} = -\mathrm{i}\pi \frac{\mathrm{sgn}(\omega_0) C_m(r_\mathrm{s})^2}{r_\mathrm{s}^3 \left|\dfrac{\partial}{\partial r}\left(\dfrac{\omega^2}{v_\mathrm{A}^2} - k_{\parallel m}^2\right)\right| \omega_0 \dfrac{\partial G}{\partial \omega_0}}\,. \qquad (13.54)$$

Since $\omega_0 \partial G/\partial\omega_0 > 0$, we have $\mathrm{Im}(\delta\omega) < 0$. This is called *continuum damping*.

### 13.2.2 Instability of TAE Driven by Energetic Particles

The dynamics of energetic particles must be treated by kinetic theory. The basic equations are those due to Betti and Freidberg [13.9]:

$$\frac{\partial f_j}{\partial t} + \boldsymbol{v}\cdot f_j + \frac{q_j}{m_j}(\boldsymbol{E} + \boldsymbol{v}\times\boldsymbol{B})\cdot\nabla_v f_j = 0\,, \qquad (13.55)$$

$$\frac{\partial n_j}{\partial t} + \nabla\cdot(n_j \boldsymbol{u}_j) = 0\,, \qquad (13.56)$$

$$m_j \frac{\partial}{\partial t}(n_j \boldsymbol{u}_j) + \nabla\cdot P_j = q_j n_j (\boldsymbol{E} + \boldsymbol{u}_j \times \boldsymbol{B})\,, \qquad (13.57)$$

$$P_j = m_j \int \boldsymbol{v}\boldsymbol{v} f_j \mathrm{d}\boldsymbol{v}\,, \qquad (13.58)$$

$$\boldsymbol{B}_1 = \nabla\times(\boldsymbol{\xi}_\perp \times \boldsymbol{B})\,, \qquad (13.59)$$

$$\mu_0 \boldsymbol{j}_1 = \nabla \boldsymbol{B}_1 = \nabla\times\nabla\times(\boldsymbol{\xi}_\perp\times\boldsymbol{B})\,, \qquad (13.60)$$

$$\boldsymbol{j}_1 \times \boldsymbol{B} + \boldsymbol{j}\times\boldsymbol{B}_1 = \sum_j \left[\nabla P_{1j} - \mathrm{i}\omega m_j (n_{1j}\boldsymbol{u}_j + n_j \boldsymbol{u}_{1j})\right]$$
$$\approx \sum_j \left(\nabla P_{1j} - \rho\omega^2 \boldsymbol{\xi}_{\perp j}\right)\,. \qquad (13.61)$$

$F_j$ is the equilibrium distribution function of the axisymmetric torus. $F_j(\epsilon, p_\varphi)$ is assumed to be a function of the constants of motion $\varepsilon$ and $p_\varphi$, where

$$\epsilon = \frac{m_j}{2}v^2 + q_j\phi \,, \qquad p_\varphi = m_j R v_\varphi + q_j\psi \,, \qquad \psi = RA_\varphi \,, \tag{13.62}$$

$$RB_Z = \frac{\partial\psi}{\partial R} \,, \qquad RB_R = -\frac{\partial\psi}{\partial Z} \,.$$

The first-order perturbation $f_{1j}$ of the distribution function is

$$\frac{\partial f_{1j}}{\partial t} + \boldsymbol{v}\cdot f_{1j} + \frac{q_j}{m_j}(\boldsymbol{v}\times\boldsymbol{B})\cdot\boldsymbol{\nabla}_v f_{1j} = -\frac{q_j}{m_j}(\boldsymbol{E}+\boldsymbol{v}\times\boldsymbol{B}_1)\cdot\boldsymbol{\nabla}_v F_j \tag{13.63}$$

and

$$\boldsymbol{\nabla}_v F_j = \hat{\varphi}\frac{\partial p_\varphi}{\partial v_\varphi}\frac{\partial F_j}{\partial p_\varphi} + (\boldsymbol{\nabla}_v\epsilon)\frac{\partial F_j}{\partial\epsilon} = \hat{\varphi} m_j R\frac{\partial F_j}{\partial p_\varphi} + m_j\boldsymbol{v}\frac{\partial F_j}{\partial\epsilon}. \tag{13.64}$$

The solution is obtained by integrating along the particle orbit:

$$f_{1j} = -\frac{q_j}{m_j}\int_{-\infty}^{t}(\boldsymbol{E}+\boldsymbol{v}\times\boldsymbol{B}_1)\cdot\boldsymbol{\nabla}_v F_j \mathrm{d}t' \,. \tag{13.65}$$

It is assumed that perturbations have the form

$$Q_1 = Q_1(R,Z)\exp\mathrm{i}(n\varphi - \omega t) \,.$$

The second term $m_j\boldsymbol{v}(\partial F_j/\partial\epsilon)$ on the right-hand side of (13.64) contributes to the integral

$$-\frac{q_j}{m_j}\int_{-\infty}^{t}(\boldsymbol{E}+\boldsymbol{v}\times\boldsymbol{B}_1)\cdot m_j\boldsymbol{v}\frac{\partial F_j}{\partial\epsilon}\mathrm{d}t' = -q_j\frac{\partial F_j}{\partial\epsilon}\int_{-\infty}^{t}\boldsymbol{E}\cdot\boldsymbol{v}\mathrm{d}t' \,.$$

The contribution from the first term $m_j R(\partial F_j/\partial p_\varphi)$ is

$$\begin{aligned}
&-\frac{q_j}{m_j}\left[\int_{-\infty}^{t}E_\varphi m_j R\frac{\partial F_j}{\partial p_\varphi}\mathrm{d}t' + \int_{-\infty}^{t}m_j R(\boldsymbol{v}\times\boldsymbol{B}_1)_\varphi\frac{\partial F_j}{\partial p_\varphi}\mathrm{d}t'\right] \\
&= -q_j\frac{\partial F_j}{\partial p_\varphi}\left\{\int_{-\infty}^{t}E_\varphi R\mathrm{d}t' + \int_{-\infty}^{t}\frac{R}{\mathrm{i}\omega}\left[\boldsymbol{v}\times(\boldsymbol{\nabla}\times\boldsymbol{E})\right]_\varphi\mathrm{d}t'\right\} \\
&= -q_j\frac{\partial F_j}{\partial p_\varphi}\left\{\int_{-\infty}^{t}\frac{1}{-\mathrm{i}\omega}\frac{\partial(E_\varphi R)}{\partial t}\mathrm{d}t' + \int_{-\infty}^{t}\left[\frac{n}{\omega}(\boldsymbol{v}\cdot\boldsymbol{E}) - \frac{1}{\mathrm{i}\omega}(\boldsymbol{v}\cdot\boldsymbol{\nabla})(E_\varphi R)\right]\mathrm{d}t'\right\} \\
&= -q_j\frac{\partial F_j}{\partial p_\varphi}\left[\int_{-\infty}^{t}\frac{1}{-\mathrm{i}\omega}\frac{\mathrm{d}(E_\varphi R)}{\mathrm{d}t}\mathrm{d}t' + \int_{-\infty}^{t}\frac{n}{\omega}(\boldsymbol{v}\cdot\boldsymbol{E})\mathrm{d}t'\right] .
\end{aligned}$$

The solution is

$$f_{1j} = -\frac{q_j}{\omega}\left[\mathrm{i}\frac{\partial F_j}{\partial p_\varphi}RE_\varphi + \left(\omega\frac{\partial F_j}{\partial \epsilon} + n\frac{\partial F_j}{\partial p_\varphi}\right)\int_{-\infty}^{t}(\boldsymbol{E}\cdot\boldsymbol{v})\mathrm{d}t'\right] . \tag{13.66}$$

Since

$$E_{1\|} = 0\,, \quad -\mathrm{i}\omega\boldsymbol{\xi}_\perp = \frac{\boldsymbol{E}_\perp \times \boldsymbol{B}}{B^2}\,, \quad \boldsymbol{E}_\perp = \mathrm{i}\omega(\boldsymbol{\xi}_\perp \times \boldsymbol{B})\,,$$

$$RE_\varphi = \mathrm{i}\omega(\boldsymbol{\xi}_\perp \times \boldsymbol{B})_\varphi R = \mathrm{i}\omega(\xi_{\perp R}B_Z - \xi_{\perp Z}B_R)R = -\mathrm{i}\omega(\boldsymbol{\xi}\cdot\boldsymbol{\nabla}\psi)\,,$$

$$\begin{aligned}\boldsymbol{E}\cdot\boldsymbol{v} &= \mathrm{i}\omega(\boldsymbol{\xi}_\perp \times \boldsymbol{B})\cdot\boldsymbol{v} = -\mathrm{i}\omega\boldsymbol{\xi}_\perp\cdot(\boldsymbol{v}\times\boldsymbol{B}) = -\mathrm{i}\omega\boldsymbol{\xi}_\perp\cdot\frac{m_j}{q_j}\frac{\mathrm{d}\boldsymbol{v}}{\mathrm{d}t}\\ &= -\mathrm{i}\omega\frac{m_j}{q_j}\boldsymbol{\xi}_\perp\cdot\frac{\mathrm{d}\boldsymbol{v}}{\mathrm{d}t} = -\mathrm{i}\omega\frac{m_j}{q_j}\left[\frac{\mathrm{d}(\boldsymbol{\xi}_\perp\cdot\boldsymbol{v})}{\mathrm{d}t} - \boldsymbol{v}\cdot\frac{\mathrm{d}\boldsymbol{\xi}_\perp}{\mathrm{d}t}\right],\end{aligned}$$

$f_{1j}$ becomes

$$\begin{aligned}f_{1j} &= -q_j\frac{\partial F_j}{\partial p_\varphi}(\boldsymbol{\xi}\cdot\boldsymbol{\nabla}\psi) + \mathrm{i}m_j\left(\omega\frac{\partial F_j}{\partial \epsilon} + n\frac{\partial F_j}{\partial p_\varphi}\right)\left(\boldsymbol{\xi}_\perp\cdot\boldsymbol{v} - \int_{-\infty}^{t}\boldsymbol{v}\cdot\frac{\mathrm{d}\boldsymbol{\xi}_\perp}{\mathrm{d}t}\mathrm{d}t'\right)\\ &= -q_j\frac{\partial F_j}{\partial \psi} + \mathrm{i}m_j(\omega-\omega_{*j})\frac{\partial F_j}{\partial \epsilon}(\boldsymbol{\xi}_\perp\cdot\boldsymbol{v} - s_j)\,, \end{aligned} \tag{13.67}$$

where

$$s_j \equiv \int_{-\infty}^{t}\boldsymbol{v}\cdot\frac{\mathrm{d}\boldsymbol{\xi}_\perp}{\mathrm{d}t}\mathrm{d}t'\,, \qquad \omega_{*j} \equiv -\frac{n\partial F_j/\partial p_\varphi}{\partial F_j/\partial\epsilon}\,.$$

$s_j$ reduces to

$$s_j = \int_{-\infty}^{t}\left[\frac{v_\perp^2}{2}\boldsymbol{\nabla}\cdot\boldsymbol{\xi}_\perp + \left(\frac{v_\perp^2}{2} - v_\|^2\right)\boldsymbol{\xi}\cdot\boldsymbol{\kappa}\right]\mathrm{d}t'\,, \tag{13.68}$$

as will be shown at the end of this section. The perturbed pressure tensor is

$$P_{1j} = \int m_j\boldsymbol{v}\boldsymbol{v}f_{1j}\mathrm{d}\boldsymbol{v} = P_{1\perp j}I + (P_{1\|j} - P_{1\perp j})\boldsymbol{b}\boldsymbol{b}\,, \tag{13.69}$$

and $\boldsymbol{\nabla}P_{1j}$ is given by (13.23) and (13.24). Then the equation of motion is

$$-\rho\omega^2\boldsymbol{\xi}_\perp = \boldsymbol{F}_\perp(\boldsymbol{\xi}_\perp) + \mathrm{i}\boldsymbol{D}_\perp(\boldsymbol{\xi}_\perp)\,, \tag{13.70}$$

$$\boldsymbol{F}_\perp(\boldsymbol{\xi}_\perp) = \boldsymbol{j}_1\times\boldsymbol{B} + \boldsymbol{j}\times\boldsymbol{B}_1 + \boldsymbol{\nabla}(\boldsymbol{\xi}_\perp\cdot\boldsymbol{\nabla}P_1)\,, \tag{13.71}$$

$$\boldsymbol{D}_\perp(\boldsymbol{\xi}_\perp) = m_j\int\left[\frac{v_\perp^2}{2}\boldsymbol{\nabla}_\perp + \left(v_\|^2 - \frac{v_\perp^2}{2}\right)\boldsymbol{\kappa}\right]m_j(\omega-\omega_{*j})\frac{\partial F_j}{\partial\epsilon}s_j\mathrm{d}\boldsymbol{v}\,. \tag{13.72}$$

$\boldsymbol{F}_\perp(\boldsymbol{\xi}_\perp)$ is the ideal MHD force operator for incompressible displacement. $\boldsymbol{D}_\perp(\boldsymbol{\xi}_\perp)$ contains the contribution of energetic particles. Equations (13.70)–(13.72) describe the low frequency, finite wave number stability of energetic particle Alfvén waves in an axisymmetric torus.

The energy integral of (13.70) consists of the plasma kinetic energy normalization $K_{\mathrm{M}}$, the ideal MHD perpendicular potential energy $\delta W_{\mathrm{MHD}}$, and the kinetic contribution to the energy integral $\delta W_{\mathrm{K}}$:

$$\omega^2 K_{\mathrm{M}} = \delta W_{\mathrm{MHD}} + \delta W_{\mathrm{K}} \,, \tag{13.73}$$

where

$$K_{\mathrm{M}} = \frac{1}{2}\int \rho|\boldsymbol{\xi}_\perp|^2 \mathrm{d}\boldsymbol{r} \,, \qquad \delta W_{\mathrm{MHD}} = -\frac{1}{2}\int \boldsymbol{\xi}_\perp^* \boldsymbol{F}_\perp(\boldsymbol{\xi}_\perp)\mathrm{d}\boldsymbol{r} \,,$$

$$\delta W_{\mathrm{K}} = -\frac{\mathrm{i}}{2}\int \boldsymbol{\xi}_\perp^* \boldsymbol{D}_\perp(\boldsymbol{\xi}_\perp)\mathrm{d}\boldsymbol{r} \,.$$

After a simple integration by parts, $\delta W_{\mathrm{K}}$ can be written as

$$\delta W_{\mathrm{K}} = \frac{\mathrm{i}}{2}\sum_j \int (\omega - \omega_{*j})\frac{\partial F_j}{\partial \epsilon} s_j \frac{\mathrm{d}s_j^*}{\mathrm{d}t}\mathrm{d}\boldsymbol{v}\mathrm{d}\boldsymbol{r} \,, \tag{13.74}$$

since

$$\frac{\mathrm{d}s_j^*}{\mathrm{d}t} = m_j\left[\frac{v_\perp^2}{2}\boldsymbol{\nabla}_\perp\cdot\boldsymbol{\xi}^* + \left(\frac{v_\perp^2}{2} - v_\parallel\right)\boldsymbol{\xi}_\perp\cdot\boldsymbol{\kappa}\right] .$$

On the other hand, $\mathrm{d}s_j^*/\mathrm{d}t$ is given by

$$\frac{\mathrm{d}s_j^*}{\mathrm{d}t} = \mathrm{i}\omega^* s_j^* + D s_j^* \,, \qquad D \equiv (\boldsymbol{v}\cdot\boldsymbol{\nabla}) + \frac{q_j}{m_j}(\boldsymbol{v}\times\boldsymbol{B})\cdot\boldsymbol{\nabla}_v \,.$$

With the notation $s_j \equiv a_j + \mathrm{i}c_j$ ($a_j$ and $c_j$ real), we have

$$s_j\frac{\mathrm{d}s_j^*}{\mathrm{d}t} = \mathrm{i}\omega^*|\mathrm{s}_j|^2 + \mathrm{i}(c_j D a_j - a_j D c_j) + \frac{1}{2}D(a_j^2 + c_j^2) \,.$$

The contribution of the last term to the integral (13.74) by $\mathrm{d}\boldsymbol{r}\mathrm{d}\boldsymbol{v}$ is zero, since $F_j$ and $\omega_{*j}$ are functions of the constants of motion $\epsilon$ and $p_\varphi$, and

$$\delta W_{\mathrm{K}} = \frac{1}{2}\sum_j \int (\omega - \omega_{*j})\frac{\partial F_j}{\partial \varepsilon}(\mathrm{i}\omega_{\mathrm{i}}|s_j|^2 + R_j)\mathrm{d}\boldsymbol{v}\mathrm{d}\boldsymbol{r} \,,$$

$$R_j = c_j D a_j - a_j D c_j - \omega_{\mathrm{r}}|s_j|^2 \,.$$

The desired expression for the growth rate is obtained by setting the real and imaginary parts of (13.73) equal to zero:

$$\omega_{\mathrm{r}}^2 = \frac{\delta W_{\mathrm{MHD}}}{K_{\mathrm{M}}} + O(\beta) \,. \tag{13.75}$$

$O(\beta)$ is the contribution of the $R_j$ term. In the limit $\omega_{\mathrm{i}} \ll \omega_{\mathrm{r}}$, the imaginary part yields

$$\omega_{\mathrm{i}} \approx \frac{W_{\mathrm{K}}}{K_{\mathrm{M}}} , \qquad W_{\mathrm{K}} \equiv \lim_{\omega_{\mathrm{i}} \to 0} \left[ \frac{1}{4\omega_{\mathrm{r}}} \sum_j \int (\omega - \omega_{*j}) \frac{\partial F_j}{\partial \epsilon} \omega_{\mathrm{i}} |s_j|^2 \mathrm{d}\boldsymbol{v} \mathrm{d}\boldsymbol{r} \right] . \tag{13.76}$$

Let us estimate (13.76). Since $\nabla\cdot\boldsymbol{\xi}_\perp + 2\boldsymbol{\xi}_\perp\cdot\boldsymbol{\kappa} \approx 0$ [see (8.116)], $s_j$ is

$$s_j = -m_j \int_{-\infty}^{t} \left( v_\parallel^2 + \frac{v_\perp^2}{2} \right) (\boldsymbol{\kappa}\cdot\boldsymbol{\xi}_\perp) \mathrm{d}t' = m_j \int_{-\infty}^{t} \left( v_\parallel^2 + \frac{v_\perp^2}{2} \right) \frac{\xi_R}{R} \mathrm{d}t' ,$$

where

$$\xi_R = \xi_r \cos\theta - \xi_\theta \sin\theta = \xi_r \frac{\mathrm{e}^{\mathrm{i}\theta} + \mathrm{e}^{-\mathrm{i}\theta}}{2} - \xi_\theta \frac{\mathrm{e}^{\mathrm{i}\theta} - \mathrm{e}^{-\mathrm{i}\theta}}{2\mathrm{i}} .$$

Recalling that

$$\nabla\cdot\boldsymbol{\xi} = \frac{1}{r}\frac{\partial (r\xi_r)}{\partial r} - \mathrm{i}\frac{m}{r}\xi_\theta \approx 0 ,$$

$\xi_r$ and $\xi_\theta$ are

$$\xi_r = \sum_m \xi_m(r) \mathrm{e}^{-\mathrm{i}m\theta} , \qquad \xi_\theta = -\mathrm{i} \sum_m \frac{[r\xi_m(r)]'}{m} \mathrm{e}^{-\mathrm{i}m\theta} .$$

Since the leading-order guiding centers of orbits of energetic particles are given by

$$r(t') = r(t) , \quad \theta(t') = \frac{v_\parallel B_\theta}{r B_\varphi}(t' - t) + \theta(t) , \quad \varphi(t') = \frac{v_\parallel}{r}(t' - t) + \varphi(t) ,$$

perturbations along the orbit become

$$\begin{aligned} &\exp \mathrm{i}\big[ -m\theta(t') + n\varphi(t') - \omega t' \big] \\ &\quad = \exp\left[ \mathrm{i} \left( -\frac{m B_\theta}{r B_\varphi} v_\parallel + \frac{n v_\parallel}{R} - \omega \right) (t' - t) \right] \exp \mathrm{i}\big[ -m\theta(t) + n\varphi(t) - \omega t \big] \\ &\quad = \exp\big[ -\mathrm{i}(\omega - \omega_m)(t' - t) \big] \exp \mathrm{i}\big[ -m\theta(t) + n\varphi(t) - \omega t \big] , \end{aligned}$$

where

$$\omega_m = \frac{v_\parallel}{R}\left( n - \frac{m}{q} \right)$$

and

$$\begin{aligned} s_j &= \frac{m_j (v_\parallel^2 + v_\perp^2/2)}{R} \frac{1}{2} \sum_m \big[ \xi_{m-1} + \xi_{m+1} - \mathrm{i}\xi_{\theta(m-1)} + \mathrm{i}\xi_{\theta(m+1)} \big] \\ &\qquad \times \exp \mathrm{i}(-m\theta + n\varphi - \omega t) \int_{\infty}^{0} \exp\big[ -\mathrm{i}(\omega - \omega_m) t'' \big] \mathrm{d}t'' \\ &= \mathrm{i}\frac{m_j}{2R}\left( v_\parallel^2 + \frac{v_\perp^2}{2} \right) \sum_m \left[ \xi_{m-1} + \xi_{m+1} - \mathrm{i}\frac{(r\xi_{m-1})'}{(m-1)} + \mathrm{i}(r\xi_{m+1})'(m+1) \right] \\ &\qquad \times \left[ \frac{\exp \mathrm{i}(-\mathrm{i}m\theta + n\varphi - \omega t)}{(\omega - \omega_m)} \right] . \end{aligned} \tag{13.77}$$

It is assumed that the perturbation consists primarily of two toroidally coupled harmonics $\boldsymbol{\xi}_m$ and $\boldsymbol{\xi}_{m+1}$ and that all other harmonics are essentially zero. Strong coupling occurs in a narrow region of thickness $\sim \epsilon a$ localized about the surface $r = r_0$ corresponding to $q(r_0) = (2m+1)/2n = q_0$. The mode localization implies that $\xi'_{m\pm1}$ terms dominate in (13.77). Substituting these results into the expression for $s_j$ and retaining only those terms which do not average to zero in $\theta$ leads to the expression

$$|s_j|^2 = \frac{m_j^2 r_0^2}{4R}\left(v_\parallel^2 + \frac{v_\perp^2}{2}\right)^2 \left[\frac{|\xi'_{(m+1)}|^2}{(m+1)^2} + \frac{|\xi'_m|^2}{m^2}\right]\left(\frac{1}{|\omega-\omega_m|^2} + \frac{1}{|\omega-\omega_{m-1}|^2}\right),$$

since $\omega_{m+1} = -\omega_m$ and $\omega_{m+2} = -\omega_{m-1}$. $K_\mathrm{M}$ is given by

$$K_\mathrm{M} = \frac{r_0^2\rho_0^2}{2}\int\left[\frac{|\xi'_m|^2}{m^2} + \frac{|\xi'_{m+1}|^2}{(m+1)^2}\right]\mathrm{d}\boldsymbol{r}\,. \tag{13.78}$$

Using the relations $\omega_\mathrm{r} \approx k_\parallel v_\mathrm{A}$, $k_\parallel = 1/(2q_0R)$, $q_0 = 2m + 1/2n$, we obtain the following expression for the growth rate:

$$\frac{\omega_\mathrm{i}}{k_\parallel v_\mathrm{A}} = \lim_{\omega_\mathrm{i}\to 0}\sum_j \frac{\mu_0 m_j^2 q_0^2}{2B^2}\int\left(v_\parallel^2 + \frac{v_\perp^2}{2}\right)^2\left(\omega_\mathrm{r}\frac{\partial F_j}{\partial\epsilon} + \frac{n}{q_j}\frac{\partial F_j}{\partial\psi}\right)$$
$$\times\left(\frac{\omega_\mathrm{i}}{|\omega-\omega_m|^2} + \frac{\omega_\mathrm{i}}{|\omega-\omega_{m-1}|^2}\right)\mathrm{d}\boldsymbol{v}\,. \tag{13.79}$$

Using the formula $\lim_{\varepsilon\to0}\int_{-\infty}^{\infty}\varepsilon/(x^2+\varepsilon^2)\mathrm{d}x = \pi$, a short calculation to find the integral with respect to $v_\parallel$ yields

$$\frac{\omega_\mathrm{i}}{k_\parallel v_\mathrm{A}} = \sum_j \frac{2\pi^2\mu_0 m_j^2 R q_0^3}{2B^2}\int\left(v_\parallel^2 + \frac{v_\perp^2}{2}\right)^2\left(\omega_\mathrm{r}\frac{\partial F_j}{\partial\epsilon} + \frac{n}{q_j}\frac{\partial F_j}{\partial\psi}\right)v_\perp \mathrm{d}v_\perp\Bigg|_{v_\parallel = v_\mathrm{A}}$$
$$+\cdots\Bigg|_{v_\parallel = v_\mathrm{A}/3}\,. \tag{13.80}$$

Note that $\omega_m = v_\parallel/(2q_0R)$, $\omega_{m-1} = 3v_\parallel/(2q_0R)$, $\omega_\mathrm{r} = v_\mathrm{A}/2q_0R$. Equation (13.80) gives the TAE growth rate for an arbitrary distribution function $F_j(\epsilon,\psi)$. The second term of (13.80) is due to sideband resonance.

The growth rate is easily evaluated for a Maxwellian distribution

$$F_j = n_j\left(\frac{m_j}{2\pi T_j}\right)^{3/2}\exp\left(-\frac{m_j v^2}{2T_j}\right).$$

Here $n_j = n_j(\psi)$ and $T_j = T_j(\psi)$. Some straightforward calculation leads to

$$\left(\frac{\omega_\mathrm{i}}{k_\parallel v_\mathrm{A}}\right)_j = -q_0^2\beta_j\left(G_{mj}^\mathrm{T} - nq_0\delta_j\frac{H_{mj}^\mathrm{T} + \eta_j J_{mj}^\mathrm{T}}{1+\eta_j}\right), \tag{13.81}$$

where

$$\beta_j = \frac{n_j T_j}{B^2/2\mu_0}\,, \quad \delta_j = -r_{\mathrm{L}p_j}\frac{\mathrm{d}p_j/\mathrm{d}r}{p_j}\,, \quad r_{\mathrm{L}p_j} \equiv \frac{m_j v_{T_j}}{q_j B_\mathrm{p}}\,, \quad \eta_j \equiv \frac{\mathrm{d}\ln T_j}{\mathrm{d}\ln n_j}\,.$$

Each of these quantities is evaluated at $r = r_0$. The functions $G^\mathrm{T}_{mj}$, $H^\mathrm{T}_{mj}$ and $J^\mathrm{T}_{mj}$ are functions of a single parameter $\lambda_j \equiv v_\mathrm{A}/v_{T_j}$ ($v_{T_j} \equiv 2T_j/m_j$) and are given by

$$G^\mathrm{T}_{mj} = g^\mathrm{T}_{mj}(\lambda_j) + g^\mathrm{T}_{mj}(\lambda_j/3)\,, \quad g^\mathrm{T}_{mj}(\lambda_j) = \frac{\pi^{1/2}}{2}\lambda_j(1 + 2\lambda_j^2 + 2\lambda_j^4)\mathrm{e}^{-\lambda_j^2}\,,$$

$$H^\mathrm{T}_{mj} = h^\mathrm{T}_{mj}(\lambda_j)+h^\mathrm{T}_{mj}(\lambda_j/3)\,, \quad h^\mathrm{T}_{mj}(\lambda_j) = \frac{\pi^{1/2}}{2}(1+2\lambda_j^2+2\lambda_j^4)\mathrm{e}^{-\lambda_j^2}\,, \quad (13.82)$$

$$J^\mathrm{T}_{mj} = j^\mathrm{T}_{mj}(\lambda_j) + j^\mathrm{T}_{mj}(\lambda_j/3)\,, \quad j^\mathrm{T}_{mj}(\lambda_j) = \frac{\pi^{1/2}}{2}(3/2 + 2\lambda_j^2 + \lambda_j^4 + 2\lambda_j^6)\mathrm{e}^{-\lambda_j^2}\,.$$

For alpha particles it is more reasonable to assume a slowing down distribution

$$F_\alpha = \frac{A}{(v^2 + v_0^2)^{3/2}}\,, \qquad \left(0 < v < v_\alpha\,, \quad \frac{m_\alpha v_\alpha^2}{2} = 3.5\,\mathrm{MeV}\right)\,. \qquad (13.83)$$

$A$ and $v_0$ are related to the density and pressure as follows:

$$A \approx \frac{n_\alpha}{4\pi\ln(v_\alpha/v_0)}, \qquad p_\alpha \approx \frac{n_\alpha m_\alpha v_\alpha^2/2}{3\ln(v_\alpha/v_0)}, \qquad \frac{m_j v_0^2}{2} \approx \kappa T_j.$$

After another straightforward calculation we obtain an analogous expression for the alpha particle contribution to the growth rate:

$$\left(\frac{\omega_\mathrm{i}}{k_\parallel v_\mathrm{A}}\right)_\alpha = -q_0^2\beta_\alpha\left(G^\mathrm{T}_{\mathrm{s}\alpha} - nq_0\delta_\alpha H^\mathrm{T}_{\mathrm{s}\alpha}\right)\,, \qquad (13.84)$$

where

$$\beta_\alpha = \frac{p_\alpha}{B^2/2\mu_0}\,, \qquad \delta_\alpha = -\frac{2}{3}r_{\mathrm{L}\alpha}\frac{\mathrm{d}p_\alpha/\mathrm{d}r}{p_\alpha}\,, \qquad r_{\mathrm{L}p\alpha} = \frac{m_\alpha v_\alpha}{q_\alpha B_\mathrm{p}}\,.$$

The functions $G^\mathrm{T}_{\mathrm{s}\alpha}$ and $H^\mathrm{T}_{\mathrm{s}\alpha}$ are functions of the parameter $\lambda_\alpha \equiv v_\mathrm{A}/v_\alpha$ and are given by

$$G^\mathrm{T}_{\mathrm{s}\alpha} = g^\mathrm{T}_\mathrm{s}(\lambda_\alpha) + g^\mathrm{T}_\mathrm{s}(\lambda_\alpha/3)\,, \quad g^\mathrm{T}_\mathrm{s}(\lambda_\alpha) = \frac{3\pi}{16}\lambda_\alpha(3 + 4\lambda_\alpha - 6\lambda_\alpha^2 - \lambda_\alpha^4)H(1-\lambda_\alpha)\,,$$

$$H^\mathrm{T}_\mathrm{s} = h^\mathrm{T}_\mathrm{s}(\lambda_\alpha) + h^\mathrm{T}_\mathrm{s}(\lambda_\alpha/3)\,, \quad h^\mathrm{T}_{\mathrm{s}\alpha}(\lambda_\alpha) = \frac{3\pi}{16}(1 + 6\lambda_\alpha^2 - 4\lambda_\alpha^3 - 3\lambda_\alpha^4)H(1-\lambda_\alpha)\,. \qquad (13.85)$$

$H(1-\lambda_\alpha)$ is the Heaviside step function $[H(x)=1$ for $x>0$, $H(x)=0$ for $x<0]$. The final form of the growth rate is obtained by combining the contributions of ions and electrons in the core plasma and $\alpha$ particles:

$$\frac{\omega_{\rm i}}{k_\parallel v_{\rm A}} = -q_0^2\left[\beta_{\rm i}G^{\rm T}_{m{\rm i}} + \beta_{\rm e}G^{\rm T}_{m{\rm e}} + \beta_\alpha(G^{\rm T}_{{\rm s}\alpha} - nq_0\delta_\alpha H^{\rm T}_{{\rm s}\alpha})\right] , \tag{13.86}$$

where $\beta_{\rm i}$, $\beta_{\rm e}$ and $\beta_\alpha$ are $\beta_j \equiv n_jT_j/B^2/2\mu_0$ of ions and electrons of the core plasma and $\alpha$ particles. The contribution of ions and electrons in the core plasma is to Landau damping. The marginal condition for excitation of TAE is

$$\beta_\alpha > \frac{\beta_{\rm i}G^{\rm T}_{\rm i}(\lambda_{\rm i})}{nq_0\delta_\alpha H^{\rm T}_{{\rm s}\alpha} - G^{\rm T}_{{\rm s}\alpha}} , \qquad \delta_\alpha > \frac{G_{{\rm s}\alpha}}{nq_0H^{\rm T}_{{\rm s}\alpha}} . \tag{13.87}$$

Equation (13.68) is derived as follows:

$$\boldsymbol{v}\cdot\frac{{\rm d}\boldsymbol{\xi}}{{\rm d}t} = \sum_i v_i\frac{{\rm d}\xi_i}{{\rm d}t} = \sum_i v_i\frac{\partial\xi_i}{\partial t} + \sum_i v_i(\boldsymbol{v}\cdot\boldsymbol{\nabla})\xi_i = \sum_i -{\rm i}\omega v_i\xi_i + \sum_{i,j} v_iv_j\frac{\partial\xi_i}{x_j} ,$$

$$\boldsymbol{v} = v_\parallel\boldsymbol{b} + v_\perp\cos(\Omega t)\hat{\mathbf{e}}_\perp - v_\perp\sin(\Omega t)(\boldsymbol{b}\times\hat{\mathbf{e}}_\perp) ,$$

$$\begin{aligned}\overline{\boldsymbol{vv}} &= v_\parallel^2\boldsymbol{bb} + v_\perp^2\overline{\cos(\Omega t)^2}\hat{\mathbf{e}}_\perp\hat{\mathbf{e}}_\perp + v_\perp^2\overline{\sin(\Omega t)^2}(\boldsymbol{b}\times\hat{\mathbf{e}}_\perp)(\boldsymbol{b}\times\hat{\mathbf{e}}_\perp)\\ &= (v_\parallel^2 - v_\perp^2/2)\boldsymbol{bb} + (v_\perp^2/2)\left[\boldsymbol{bb} + \hat{\mathbf{e}}_\perp\hat{\mathbf{e}}_\perp + (\boldsymbol{b}\times\hat{\mathbf{e}}_\perp)(\boldsymbol{b}\times\hat{\mathbf{e}}_\perp)\right]\\ &= (v_\parallel^2 - v_\perp^2/2)\boldsymbol{bb} + (v_\perp^2/2)I ,\end{aligned}$$

$$\overline{\boldsymbol{v}\cdot\frac{{\rm d}\boldsymbol{\xi}}{{\rm d}t}} = -{\rm i}\omega\overline{v_\parallel\xi_\parallel} + (v_\perp^2/2)\boldsymbol{\nabla}\cdot\boldsymbol{\xi} + (v_\parallel^2 - v_\perp^2/2)\sum_{i,j} b_ib_j\frac{\partial\xi_i}{\partial x_j} ,$$

$$\sum_{i,j}\left(b_ib_j\frac{\partial\xi_i}{\partial x_j} + b_j\xi_i\frac{\partial b_i}{\partial x_j}\right) = \sum_{i,j} b_j\frac{\partial}{\partial x_j}(\xi_ib_i) ,$$

$$\sum_{i,j} b_ib_j\frac{\partial\xi_i}{\partial x_j} = -\xi\cdot(\boldsymbol{b}\cdot\boldsymbol{\nabla})\boldsymbol{b} + (\boldsymbol{b}\cdot\boldsymbol{\nabla})(\boldsymbol{\xi}\cdot\boldsymbol{b}) = -\boldsymbol{\kappa}\cdot\boldsymbol{\xi} + (\boldsymbol{b}\cdot\boldsymbol{\nabla})\xi_\parallel ,$$

$$\overline{\boldsymbol{v}\cdot\frac{{\rm d}\boldsymbol{\xi}}{{\rm d}t}} = (v_\perp^2/2)\boldsymbol{\nabla}\cdot\boldsymbol{\xi} + (v_\perp^2/2 - v_\parallel^2)\boldsymbol{\kappa}\cdot\boldsymbol{\xi}_\perp - {\rm i}\omega\overline{v_\parallel\xi_\parallel} + (v_\parallel^2 - v_\perp^2/2)\frac{\partial\xi_\parallel}{\partial l} .$$

Since $|\xi_\parallel| \ll |\boldsymbol{\xi}_\perp|$, we obtain

$$\boldsymbol{v}\cdot\frac{{\rm d}\boldsymbol{\xi}}{{\rm d}t} = (v_\perp^2/2)\boldsymbol{\nabla}\cdot\boldsymbol{\xi} + (v_\perp^2/2 - v_\parallel^2)\boldsymbol{\kappa}\cdot\boldsymbol{\xi}_\perp + a_1{\rm e}^{-{\rm i}\Omega t} + \cdots .$$

The third term is rapidly oscillating and the contribution to (13.68) is small.

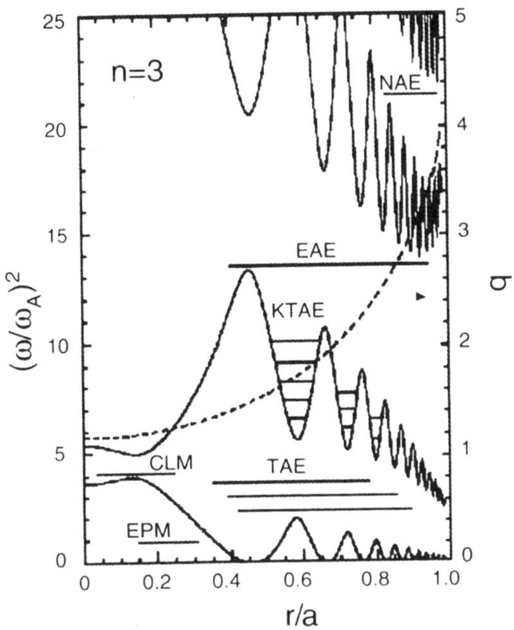


**Fig. 13.4.** Representative shear Alfvén frequency continuum curves as a function of minor radius $r$ [13.15]. *Horizontal lines* indicate the approximate radial location and mode width for the toroidal Alfvén eigenmode (TAE), kinetic TAE mode (KTAE), core-localized TAE mode (CLM), ellipticity Alfvén eigenmode (EAE), non-circular triangularity Alfvén eigenmode (NAE), and energetic particle continuum mode (EPM)

### 13.2.3 Various Alfvén Modes

In the last section we discussed the excitation of weakly damped low-$n$ TAE by super-Alfvénic energetic particles. High-$n$ TAE is analyzed in [13.10]. There are various Alfvén modes.

In high-temperature plasmas, non-ideal effects such as the finite Larmor radius of the core plasma become important in the gap region and cause the Alfvén continuum to split into a series of kinetic Alfvén eigenmodes (KTAE) at closely spaced frequencies above the ideal TAE frequency [13.11].

In the central region of the plasma, a low-shear version of TAE can arise, called the the core-localized mode (CLM) [13.12].

Non-circular shaping of the plasma poloidal cross-section creates other gaps in the Alfvén continuum, at high frequency. Ellipticity creats a gap at about twice the TAE frequency, within which there are ellipticity-induced Alfvén eigenmodes (EAE) [13.9]; likewise for triangularity-induced Alfvén eigenmodes (NAE) [13.9] at about three times the TAE frequency.

The ideal and kinetic TAEs are 'cavity' modes, whose frequencies are determined by the bulk plasma. In addition, a 'beam mode' can arise that is not a natural eigenmode of the plasma, but is supported by the presence of a population of energetic particles, and also destabilized by them. This so-called energetic particle mode (EPM) [13.13], which can also exist outside the TAE gaps, has a frequency related to the toroidal precession frequency and poloidal transit/bounce frequency of the fast ions. The beta-induced Alfvén eigenmode (BAE) [13.14] exists in the beta-induced gap. The schematic in Fig. 13.4 illustrates these different modes.

Close interaction between theory and experiment has led to many new discoveries concerning Alfvén eigenmodes in toroidal plasmas. A great deal of theoretical work has been carried out on energetic particle drive and competing damping mechanisms, such as continuum and radiative damping (mode conversion and Landau damping), ion Landau damping for both thermal and fast ions, electron damping and trapped electron collisional damping. For modes with low to moderate toroidal mode numbers $n$, continuum damping and ion Landau damping usually dominate, whereas high-$n$ modes, trapped collisional damping and radiative damping are strong stabilizing mechanisms. There are excellent reviews on toroidal Alfvén eigenmodes in [13.15, 13.16].

# 14 Computer Simulation

> What a plasma really is is a collection of a very large number of individual charged particles, all interacting with each other through mutual Coulomb forces and through the electric currents associated with their motion, while at the same time interacting with (and thereby modifying) any electromagnetic fields of external origin. It is this property of collective long-range interactions, of plasma with itself and of plasma with its electromagnetic environment, that gives rise to the great complexity of its behavior. (From R.F. Post [14.1])

The use of computers as an aid to understanding the complex nonlinear behavior of turbulence, such as mode–mode coupling, nonlinear saturation and self-organization, has continued to grow and has become a dominant factor in the theoretical component of plasma research. The evaluation of the saturation level of perturbations is essential in the study of plasma transport, which is very difficult to estimate in an analytical way.

However, computer simulation of plasma presents many difficulties. It is a numerically tough problem, since plasma behavior contains a wide range of length and time scales (see Sect. 2.9):

- resistive diffusion time $\tau_{\mathrm{R}} \sim 10^3\mathrm{s}$,
- Coulomb collision time $\tau_{\mathrm{ei}} \sim 0.1\,\mathrm{ms}$,
- MHD Alfvén transit time $\tau_{\mathrm{H}} \sim 0.1\,\mu\mathrm{s}$,
- ion and electron cyclotron periods $\tau_{\mathrm{ci}} \sim 30\,\mathrm{ns}$, $\tau_{\mathrm{ce}} \sim 7\,\mathrm{ps}$,
- electron plasma oscillation period $\tau_{\mathrm{p}} \sim 10\,\mathrm{ps}$,
- mean free path $\lambda_{\mathrm{ei}} \sim 10\,\mathrm{km}$,
- plasma radius $a \sim 1\,\mathrm{m}$,
- Larmor radii $\rho_{\mathrm{ci}} \sim 3\,\mathrm{mm}$, $\rho_{\mathrm{ce}} \sim 50\,\mu\mathrm{m}$,
- Debye length $\lambda_{\mathrm{D}} \sim 70\,\mu\mathrm{m}$ for typical fusion grade plasma.

Owing to these difficulties, simulations including all the relevant physics are not possible even with the help of the most advanced computers, at least in the near future. It is therefore necessary to use numerically feasible approximate models while retaining the essential features of the relevant plasma dynamics.

In this chapter, we provide a brief introduction to MHD models, the linearized kinetic model, the gyrofluid or gyro-Landau-fluid model, the gyroki-

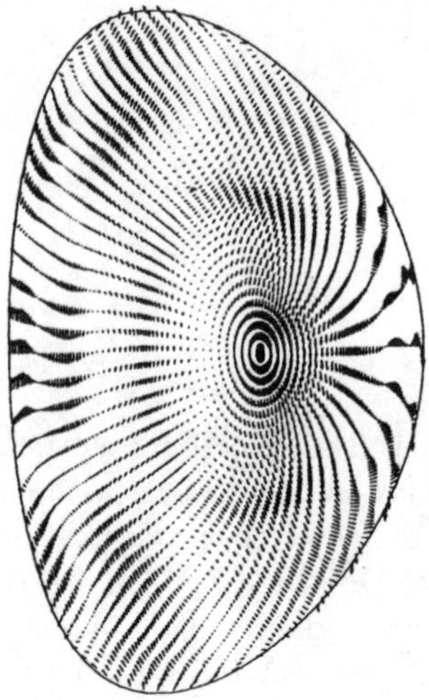

**Fig. 14.1.** Unstable global $n = 1$ mode [14.6]. The singular surfaces $q = 2, 3$ and 4 are visible because of the peaked shear velocity on them. $q_0 = 1.35$, $\beta = 3\%$

netic particle model, and the full orbit particle model. Numerical algorithms for simulations are not discussed here. Readers should refer to the excellent textbooks [14.2–14.4] and references therein.

## 14.1 MHD model

The first requirement for confined plasma is an ideal MHD stability. To study the linear growth rate of MHD instability, a variational method with the energy integral

$$L = \int \boldsymbol{\xi}\cdot\widehat{K}\boldsymbol{\xi}\mathrm{d}\boldsymbol{r} - \omega^2 \int \rho_{\mathrm{m}}\boldsymbol{\xi}^2\mathrm{d}\boldsymbol{r} \tag{14.1}$$

is used to evaluate the eigenvalue of $\omega^2$. The energy integral is described in (8.45)–(8.48) in Sect. 8.2. The linear growth rate is $\gamma = (-\omega^2)^{1/2}$ when $\omega^2 < 0$ $[\boldsymbol{\xi}(\boldsymbol{r}, t) = \boldsymbol{\xi}(\boldsymbol{r})\exp(-\mathrm{i}\omega t)]$.

To solve the eigenvalue problem, the regular finite element method is used in ERATO code [14.5] to avoid the introduction of spurious modes of numerical origin. Troyon et al. [14.6] evaluate the maximum growth rates of MHD instabilities in many cases for an elongated tokamak plasma and derive the beta scaling on the upper limit of the stable beta value $\beta_{\mathrm{c}}(\%) = \beta_{\mathrm{N}} I_{\mathrm{p}}\left[\mathrm{MA}/a(\mathrm{m})B_{\mathrm{t}}(\mathrm{T})\right]$ (see Sect. 16.4). Figure 14.1 shows the poloidal plasma flow associated with an unstable mode which develops when $\beta$ exceeds the limit.

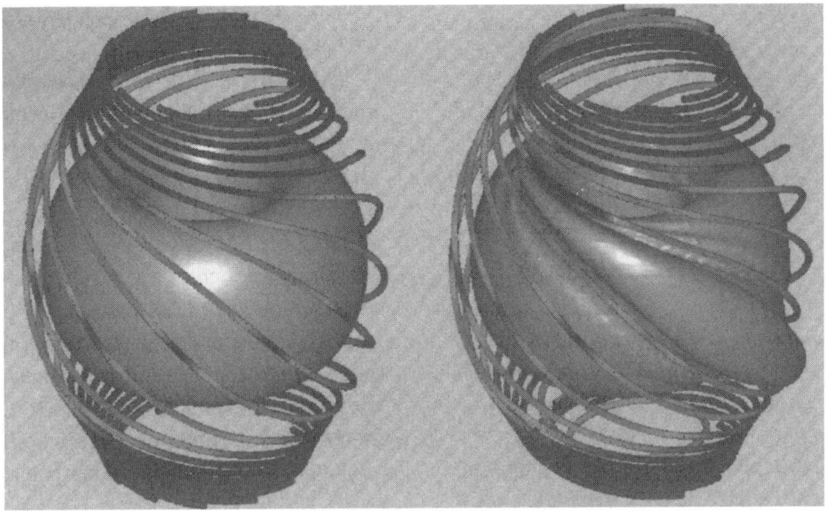

**Fig. 14.2.** Time development of the 3D profiles of the plasma pressure and magnetic field lines in the simulation of internal reconnection events in a spherical tokamak at the initial time (*left*) and $t = 197\tau_A$ (*right*) [14.11]

In the Princeton Plasma Physics Laboratory, the PEST code [14.7] has been developed, in which $\boldsymbol{\xi}$ is approximated by a linear superposition of finite linearly independent expansion functions. With this expansion, the calculus of variations for the energy integral reduces to a matrix eigenvalue problem.

The other method to study the MHD behavior of plasmas is to solve the MHD equations of motion under the appropriate initial and boundary conditions. An example of the full nonlinear MHD equations is:

$$\frac{\partial \rho}{\partial t} = -\nabla\cdot(\rho \boldsymbol{v}) ,$$

$$\frac{\partial \rho \boldsymbol{v}}{\partial t} = -\nabla\cdot(\rho \boldsymbol{v}\boldsymbol{v}) - \nabla p + \boldsymbol{j} \times \boldsymbol{B} + \mu \left[\nabla^2 \boldsymbol{v} + \frac{1}{3}\nabla(\nabla\cdot\boldsymbol{v})\right] , \qquad (14.2)$$

$$\frac{\partial \boldsymbol{B}}{\partial t} = -\nabla \times \boldsymbol{E} , \quad \frac{\partial p}{\partial t} = -\nabla\cdot(p\boldsymbol{v}) - (\gamma - 1)p\nabla\cdot\boldsymbol{v} + (\gamma - 1)(\eta j^2 + \Phi) ,$$

$$\boldsymbol{j} = \frac{1}{\mu_0}\nabla \times \boldsymbol{B} , \quad \boldsymbol{E} = -\boldsymbol{v} \times \boldsymbol{B} + \eta \boldsymbol{j} , \quad \Phi \equiv 2\mu \left[e_{ij}e_{ji} - \frac{1}{3}(\nabla\cdot\boldsymbol{v})^2\right] ,$$

$$e_{ij} \equiv \frac{1}{2}\left(\frac{\partial v_i}{\partial x_j} + \frac{\partial v_j}{\partial x_i}\right) .$$

The $\mu$ terms in the equations represent the viscosity effect and $\Phi$ is the viscous heating source term. The other notation is the same as in Sect. 5.3.

The phemonena of the dynamo and the reconnection of reversed pinch plasma (RFP) are analyzed in [14.8, 14.9]. High beta disruption in tokamaks is studied using the MH3D code [14.10] and nonlinear behavior of internal reconnection events in spherical tokamaks (very low aspect tokamaks, see Sect. 15.4) is simulated in [14.11]. Internal reconnection in a spherical tokamak is visualized in Fig. 14.2. In this simulation, the numerical grid consists typically of 128 × 64 × 128 points in $r, \phi, z$ coordinates. The time step is $\Delta t = \tau_{\mathrm{A}}/2\pi$, $\tau_{\mathrm{A}}$ being the Alfvén transit time encircling the magnetic axis.

## 14.2 Linearized Kinetic Model

The fundamental equations of the linearized kinetic model are the Maxwell equations with the dielectric tensor of a hot plasma:

$$\nabla \times \boldsymbol{E} = \mathrm{i}\omega \boldsymbol{B} \,, \qquad \epsilon_0 \nabla\cdot(\boldsymbol{K}\,\boldsymbol{E}) = \rho_{\mathrm{ext}} \,,$$

$$\nabla \times \boldsymbol{B} = -\frac{\mathrm{i}\omega}{c^2}\boldsymbol{K}\cdot\boldsymbol{E} + \mu_0 \boldsymbol{j}_{\mathrm{ext}} \,, \qquad \nabla\cdot\boldsymbol{B} = 0 \,,$$

$$\nabla \times \nabla \times \boldsymbol{E} - \omega^2 \boldsymbol{K}\,\boldsymbol{E} = \mathrm{i}\omega\mu_0 \boldsymbol{j}_{\mathrm{ext}} \,, \tag{14.3}$$

where $\boldsymbol{K}$ is the dielectric tensor given in Sect. 12.3. The vector $\mathrm{i}\boldsymbol{k}$ appearing in the tensor $\boldsymbol{K}$ should be replaced by the operator $\nabla$.

In terms of vector and scalar potentials $(\boldsymbol{A}, \phi)$, the Maxwell equations are

$$\boldsymbol{B} = \nabla \times \boldsymbol{A} \,, \quad \boldsymbol{E} = -\nabla\phi + \mathrm{i}\omega \boldsymbol{A} \,,$$

$$-\nabla \times \nabla \times \boldsymbol{A} + \frac{\omega^2}{c^2}\boldsymbol{K}\,\boldsymbol{A} + \frac{\mathrm{i}\omega}{c^2}\boldsymbol{K}\,\nabla\phi = -\mu_0 \boldsymbol{j}_{\mathrm{ext}} \,, \tag{14.4a}$$

$$\nabla\cdot(\boldsymbol{K}\,\nabla\phi) - \mathrm{i}\omega\nabla\cdot(\boldsymbol{K}\,\boldsymbol{A}) = -\frac{1}{\epsilon_0}\rho_{\mathrm{ext}} \,, \tag{14.4b}$$

$$\nabla\cdot\boldsymbol{A} = 0 \,.$$

It is possible to treat the phenomena associated with Landau damping or amplification and mode conversion using this analysis.

Ion–ion hybrid resonance in ICRF (see Sect. 12.4) is analyzed using the LION code [14.12], applying the finite hybrid element method to the variational form of (14.3). Alfvén waves are analyzed using the PENN code [14.13], applying the standard finite element method to the variational form of (14.4a) and (14.4b) without introducing spurious modes of numerical origin. Hydrogen minority heating in deuterium majority plasmas is studied using the PION code [14.14]. Two-ion hybrid resonance heating is studied using the TASK/WM code [14.15].

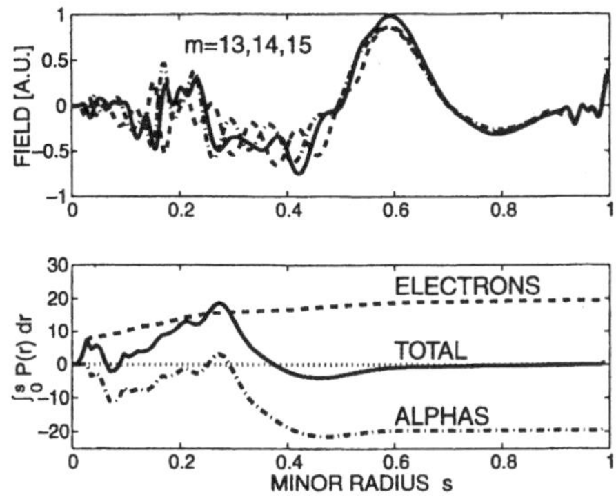


**Fig. 14.3.** Wave field of $n = 3$ (toroidal mode number) kinetic Alfvén waves and the integrated power transfer from AEs to particles in the region from $r = 0$ to $r = s$ for an optimized tokamak reactor with deeply reversed magnetic shear and a large central safety factor $q_0 = 4.5$ [14.16]. A negative value of the integrated power transfer means that AEs are excited by the particles. The critical energetic particle pressure for marginal stability is calculated here to be $\beta_{\text{fast}} < 0.1\%$

## 14.3 Modeling Bulk Plasma and Energetic Particles

Alfvén eigenmodes (AEs) can be excited by energetic particles such as fusion-produced alpha particles or energetic ions due to NBI or ICRF (see Sect. 13.2).

When analyzing Alfvén eigenmodes with the PENN code [14.16], the bulk plasma of the tokamak is treated using the kinetic wave equations as described in the last section and the energetic ions are treated using the gyrokinetic Vlasov equation. The power transfer is calculated perturbatively from a global eigenmode wave field $\boldsymbol{E}$ and the current density $\boldsymbol{j}_{nl}$ estimated from the non-adiabatic part of the perturbed distribution function. The power transfer $P$ is obtained from

$$P = \frac{1}{2}\text{Re} \int \boldsymbol{j}_{nl} \cdot \boldsymbol{E}^* \text{d}V \ .$$

Figure 14.3 shows the wave field of toroidal mode number $n = 3$ kinetic Alfvén waves and the integrated power transfer from AEs to particles in the region from $r = 0$ to $r = s$ for an optimized reactor with deeply reversed magnetic shear and a large central safety factor $q_0 = 4.5$.

In the TASK/WM code, the bulk plasma is treated using the kinetic wave equations and the effects of energetic ions are calculated from the drift kinetic equation [14.17].

In the NOVA-K code [14.18], thermal ions and electrons of the bulk plasma are treated using the MHD or moment equations of the gyrokinetic Vlasov equation. The perturbed pressure due to energetic ions is calculated

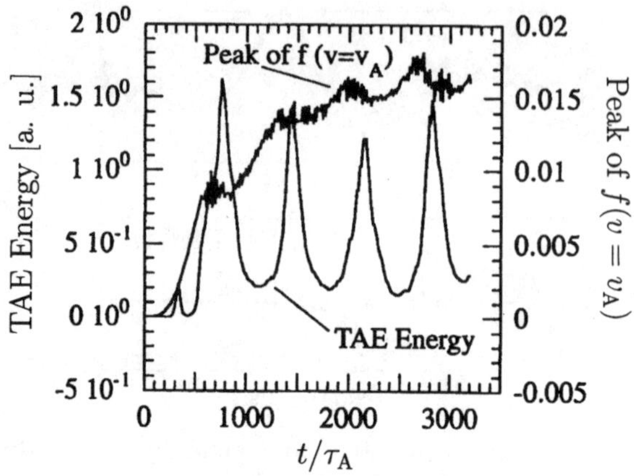


**Fig. 14.4.** Time evolution of total TAE (toroidal Alfvén eigenmode) energy and energetic ion distribution peak value for $v = v_A$ [14.19]

using the gyrokinetic equation and combined with the perturbed pressure terms from the bulk plasma.

A kinetic–MHD hybrid model [14.19] describes the bulk plasma using the full nonlinear MHD equations and includes the effect of energetic ions through the current $\boldsymbol{j}'_\alpha$ due to the energetic ions:

$$\boldsymbol{j}'_\alpha = \int (v_\parallel \boldsymbol{b} + \boldsymbol{v}_\mathrm{d}) f_\alpha \mathrm{d}^3\boldsymbol{v} + \boldsymbol{\nabla} \times \boldsymbol{M} \,, \qquad \boldsymbol{M} = -\int \mu \boldsymbol{b} f_\alpha \mathrm{d}^3\boldsymbol{v} \,,$$

where $\mu$ is the magnetic moment, $\boldsymbol{v}_\mathrm{d}$ is the curvature and $\boldsymbol{\nabla}B$ drift velocity. The distribution function of energetic ions $f_\alpha$ is calculated from the 4D $(R, \varphi, z, v)$ Fokker–Planck equation, where $v$ is the parallel component of $\boldsymbol{v}$. The momentum equation for bulk plasma (14.2) reduces to

$$\rho \left[ \frac{\partial \boldsymbol{v}}{\partial t} + (\boldsymbol{v}\cdot\boldsymbol{\nabla})\boldsymbol{v} \right] = \left( \frac{1}{\mu_0} \boldsymbol{\nabla} \times \boldsymbol{B} - \boldsymbol{j}'_\alpha \right) \times \boldsymbol{B} - \boldsymbol{\nabla} p \,.$$

This simulation reproduces some aspects of the experimentally observed periodic burst of Alfvén eigenmodes, as shown in Fig. 14.4.

## 14.4 Gyrofluid/Gyro-Landau-Fluid Models

In the derivation of the gyrokinetic equation, the fast time scale associated with the gyromotion of charged particles is asymptotically removed from the kinetic equation. The resulting gyrokinetic Vlasov equation is therefore simpler to solve than the full kinetic Vlasov equation, since it is independent of the gyroangle phase space coordinate. By multiplying the gyrokinetic Vlasov equation by an arbitrary gyrocenter phase space function and integrating over the velocity phase space, we obtain formulas for the gyrofluid model. The electrostatic gyrokinetic Vlasov equation for the distribution function $F(\boldsymbol{R}, v_\parallel, \mu, t)$ of the gyrocenter position $\boldsymbol{R}$, the parallel velocity $v_\parallel$, and the magnetic moment $\mu = v_\perp^2/2B$ is

$$BC(F) = \frac{\partial}{\partial t}(FB) + \boldsymbol{\nabla}\cdot\Big[FB(v_\parallel \boldsymbol{b} + \boldsymbol{v}_E + \boldsymbol{v}_\mathrm{d})\Big] + \frac{\partial}{\partial v_\parallel}\left\{FB\left[-\frac{e}{m}\boldsymbol{b}\cdot\boldsymbol{\nabla} J_0 \Phi \mu \boldsymbol{b}\cdot\boldsymbol{\nabla} B + v_\parallel(\boldsymbol{b}\cdot\boldsymbol{\nabla}\boldsymbol{b})\cdot\boldsymbol{v}_E\right]\right\} , \quad (14.5)$$

where $C(F)$ is the collision term for the test species with other species. The magnitude $B$ of the magnetic field is the Jacobian of the transformation from the $(v_\parallel, v_\perp)$ variable to $(v_\parallel, \mu)$. Since finite Larmor radius effects are retained, the particles feel the gyroaveraged $\boldsymbol{E}\times\boldsymbol{B}$ drift $\boldsymbol{v}_E = (\boldsymbol{b}\times J_0\Phi)/B$, where $J_0$ is the Bessel function of $(k_\perp v_\perp/\Omega)$. $\boldsymbol{v}_\mathrm{d}$ is the curvature and $\boldsymbol{\nabla} B$ drift velocity. The gyrokinetic Vlasov equation (14.5) reduces to [14.20]

$$\begin{aligned}0 = {} & \frac{\partial}{\partial t}(FB) + B\boldsymbol{\nabla}_\parallel\left(\frac{FBv_\parallel}{B}\right) + \boldsymbol{v}_\Phi\cdot(FBJ_0) + 2FBJ_0\frac{e}{T}\mathrm{i}\omega_\mathrm{d}\Phi \\ & + \frac{e}{T}\mathrm{i}\omega_\mathrm{d}(FBJ_1\Phi k_\perp v_\perp/2\Omega) + \frac{\mathrm{i}\omega_\mathrm{d}}{v_\mathrm{T}^2}FB(v_\parallel^2 + \mu B) \\ & - \frac{e}{m}\boldsymbol{\nabla}_\parallel\left(J_0\Phi B\frac{\partial F_0}{\partial v_\parallel}\right) + \frac{e}{m}J_0\Phi B\frac{\partial F_0}{\partial v_\parallel}\left(\frac{\mu B}{v_\mathrm{T}^2} - 1\right)\boldsymbol{\nabla}_\parallel \ln B \\ & - \mu B\frac{\partial}{\partial v_\parallel}(FB)\boldsymbol{\nabla}_\parallel \ln B - \frac{\partial}{\partial v_\parallel}(FBJ_0 v_\parallel)\frac{e}{T}\mathrm{i}\omega_\mathrm{d}\Phi , \qquad (14.6)\end{aligned}$$

where $\boldsymbol{v}_\Phi \equiv \boldsymbol{b}\times\boldsymbol{\nabla}\Phi/B$ and $\mathrm{i}\omega_\mathrm{d} \equiv (v_\mathrm{T}^2/\Omega^2)\boldsymbol{B}\times\boldsymbol{\nabla} B\cdot\boldsymbol{\nabla}$.

Taking an integral of the form $n\langle A\rangle = 2\pi\int \mathrm{d}v_\parallel \mathrm{d}\mu FBA$ leads to moment equations for 1, $v_\parallel$, $m(v_\parallel - u_\parallel)^2$, $mv_\perp^2/2$, $m(v_\parallel - u_\parallel)^3$, and $mv_\perp^2(v_\parallel - u_\parallel)/2$ for the density $n$, particle flux $nu_\parallel$, pressures $p_\parallel, p_\perp$, and energy fluxes $q_\parallel, q_\perp$, respectively, with the help of the appropriate closure approximation. The quasi-neutrality constraint is used to solve for $\Phi$.

The starting equation for the gyro-Landau-fluid model [14.21] is the gyrokinetic Vlasov equation [14.22, 14.23]

$$\begin{aligned}\left[\frac{\partial}{\partial t} + (v_\parallel \boldsymbol{b} + \boldsymbol{v}_E + \boldsymbol{v}_\mathrm{d}\cdot\boldsymbol{\nabla})\right]h &= \frac{q}{T}\left(\frac{\partial}{\partial t} + \frac{\mathrm{i}T}{qB}\boldsymbol{k}_\perp\times\boldsymbol{b}\cdot\boldsymbol{\nabla}\right)\Phi J_0 F_0 \\ &= \frac{q}{T}\left(\frac{\partial}{\partial t} + \mathrm{i}\omega_{*v}\right)\Phi J_0 F_0 , \qquad (14.7)\end{aligned}$$

where $h(\boldsymbol{R}, v_\parallel, \mu, t)$ is the nonadiabatic part of the perturbed distribution function $f = -(q/T)\Phi F_0 + h$ and $\omega_{*v} = 1 + \eta(v^2/2v_T^2 - 3/2)\omega_*$, with

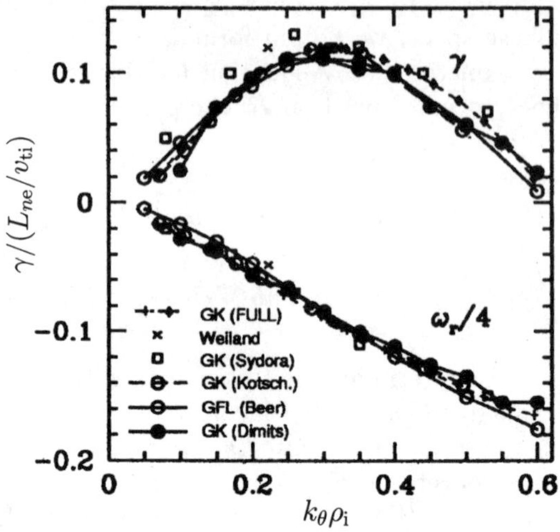


**Fig. 14.5.** Comparison of the frequency $\omega$ and linear growth rate $\gamma$ of toroidal ion-temperature-gradient driven (ITG) turbulence for the Sydora (global) and Dimits (flux-tube) nonlinear gyrokinetic codes, for the Kotschenreuther and Rewoldt (Full) linear gyrokinetic codes, the Beer nonlinear gyrofluid code, and the Weiland fluid calculation [14.24]. $L_{ne}$ is the scale of the electron density gradient and $v_{\rm ti}$ is the ion thermal velocity. Good agreement is observed between the gyrofluid code and the gyrokinetic codes

$\omega_* = k_\theta T/eBL_n$ the diamagnetic frequency. Here $L_n^{-1} = -{\rm d}\ln n/{\rm d}r$ and $\eta = L_n/L_T$. $\boldsymbol{v}_E$ is the $\boldsymbol{E} \times \boldsymbol{B}$ drift velocity, and $\boldsymbol{v}_{\rm d}$ is the curvature and $\boldsymbol{\nabla} B$ drift velocity. The moment equations of the density $N$, parallel flow velocity $U$, and pressures $P_\parallel$, $P_\perp$ are derived. This model is further generalized to the moment equations of the ion density $N_{\rm i}$, parallel ion flow velocity $U_{\rm i}$, ion pressures $P_\parallel$, $P_\perp$, trapped electron and untrapped electron densities $N_{\rm e}^{\rm t}$, $N_{\rm e}^{\rm u}$, trapped electron pressure $P_{\rm e}^{\rm t}$ and parallel magnetic potential $A$.

Figure 14.5 compares the calculated frequency and linear growth rate of toroidal ion-temperture-gradient driven (ITG) turbulence by the gyrokinetic codes and the gyrofluid code. Good agreement is observed between them [14.24]. Figure 14.6 compares the ion thermal diffusion coefficient $\chi_{\rm i}$ of ITG turbulence in a tokamak in nonlinear phase versus $R/L_T$ from the gyrofluid codes and from the gyrokinetic particle codes ($L_T^{-1} = -{\rm d}\ln T/{\rm d}r$, where $R$ is the major radius). The values of the threshold $R/L_{T_{\rm crit}}$ of the gyrofluid codes, from which the ion thermal diffusion coefficient $\chi_{\rm i}$ increases rapidly, differ somewhat from that of the LLNL gyrokinetic results [14.24].

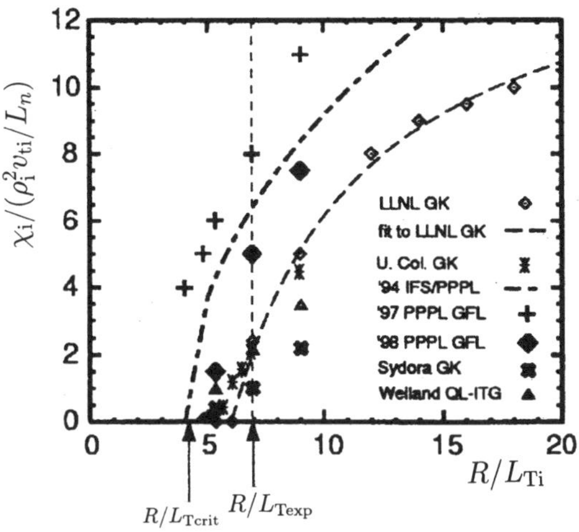


**Fig. 14.6.** Ion thermal diffusion coefficient $\chi_i$ of ITG turbulence versus $R/L_T$ from the gyrofluid codes 94IFS/PPPl, 97PPPL GFL, 98PPPL GFL and from the gyrokinetic particle codes of LLNL and U. Colorado flux-tube and UCLA (Sydora) global codes [14.24]. $R/L_{T_{\rm exp}} = 7$, $\chi_i/(\rho_i^2 v_{ti}/L_n) = 0.16$ is the experimental value from the DIII-D data base. The values of the threshold $R/L_{T_{\rm crit}}$ of the gyrofluid code differ somewhat from that of the LLNL gyrokinetic results

## 14.5 Gyrokinetic Particle Model

The Vlasov equation for a distribution function $F(\boldsymbol{x}, \boldsymbol{v}, t)$ in the space $(\boldsymbol{x}, \boldsymbol{v})$ is given by

$$\frac{\partial F}{\partial t} + \boldsymbol{v}\cdot\frac{\partial F}{\partial \boldsymbol{x}} + \frac{q}{m}(\boldsymbol{E} + \boldsymbol{v}\times\boldsymbol{B})\cdot\frac{\partial F}{\partial \boldsymbol{v}} = 0\,.$$

Let us apply a gyrokinetic change of variables from $(\boldsymbol{x}, \boldsymbol{v})$ to $(\boldsymbol{R}, \mu, v_\parallel, \varphi)$, where $\boldsymbol{R}$ is the gyrocenter coordinate, $\mu \equiv v_\perp^2/2B$ the magnetic moment, $v_\parallel$ the parallel component of velocity, $\varphi$ the phase angle, and

$$\boldsymbol{x} = \boldsymbol{R} + \boldsymbol{\rho}\,, \quad \boldsymbol{\rho} = \frac{\boldsymbol{b}\times\boldsymbol{v}_\perp}{\Omega}\,, \quad \Omega \equiv \frac{qB}{m}\,, \quad \boldsymbol{b} \equiv \frac{\boldsymbol{B}}{B}\,.$$

The distribution function $F(\boldsymbol{x}, \boldsymbol{v}, t)$ can be expressed by the sum of the zeroth-order term $f$, which is independent of the gyrophase, and the first-order term $g$, which does depend on the gyrophase:

$$F(\boldsymbol{x}, \boldsymbol{v}, t) = f(\boldsymbol{R}, \mu, v_\parallel, t) + g(\boldsymbol{R}, \mu, v_\parallel, \varphi, t)\,.$$

For an electrostatic perturbation $\boldsymbol{E} = -\boldsymbol{\nabla}\Phi$, $f$ and $g$ are given by [14.25]

$$0 = \frac{\partial f}{\partial t} + \left(v_\parallel \boldsymbol{b} + \boldsymbol{v}_{\rm d} - \frac{1}{B}\frac{\partial \Psi}{\partial \boldsymbol{R}}\times\boldsymbol{b}\right)\cdot\frac{\partial f}{\partial \boldsymbol{R}} + \left(-\frac{q}{m}\frac{\partial \Psi}{\partial \boldsymbol{R}}\cdot\boldsymbol{b} - \mu\boldsymbol{b}\cdot\frac{\partial B}{\partial \boldsymbol{R}}\right)\frac{\partial f}{\partial v_\parallel}\,, \tag{14.8}$$

$$g = \frac{q}{mB}\frac{\partial f}{\partial \mu}\left[\Phi(\boldsymbol{x}) - \langle\Phi\rangle\right] ,$$

where $\boldsymbol{v}_\mathrm{d}$ is the curvature and $\boldsymbol{\nabla} B$ drift, i.e., $\boldsymbol{v}_\mathrm{d} = \boldsymbol{b}\times\left[\mu\cdot\boldsymbol{\nabla}B + v_\parallel^2(\boldsymbol{b}\cdot\boldsymbol{\nabla})\boldsymbol{b}\right]/\Omega$, and $\langle\ \ \rangle = \oint \mathrm{d}\varphi/2\pi$ denotes the gyrophase average. $\Phi(\boldsymbol{x})$ and $\langle\Phi\rangle$ are expressed by

$$\Phi(\boldsymbol{x}) = \sum_k \phi(\boldsymbol{k}) \exp \mathrm{i}\boldsymbol{k}\cdot\boldsymbol{x} = \sum_k \phi(\boldsymbol{k}) \exp(\boldsymbol{k}\cdot\boldsymbol{R}) \exp \mathrm{i}\boldsymbol{k}\cdot\boldsymbol{\rho} ,$$

$$\langle\Phi\rangle \equiv \langle\Phi(\boldsymbol{R}+\boldsymbol{\rho})\rangle = \sum_k \phi(\boldsymbol{k}) J_0\left(\frac{\boldsymbol{k}_\perp v_\perp}{\Omega}\right) \exp(\mathrm{i}\boldsymbol{k}\cdot\boldsymbol{R}) .$$

When $f$ is Maxwellian in $v_\perp$, i.e., $f \propto \exp(-v_\perp^2/2v_T^2)/2\pi v_T^2$, we find

$$\frac{q}{mB}\frac{\partial f}{\partial \mu} = -\frac{q}{T} , \qquad \frac{T}{m} = v_T^2 ,$$

and $\Psi$ is given by

$$\Psi(\boldsymbol{R}) \equiv \langle\Phi\rangle + \frac{1}{2}\frac{q}{T}\left(\langle\Phi\rangle^2 - \langle\Phi^2\rangle\right) \simeq \langle\Phi\rangle - \frac{1}{2}\frac{q}{T}\frac{v_\perp^2}{\Omega^2}\left|\frac{\partial\langle\Phi\rangle}{\partial\boldsymbol{R}}\right|^2 ,$$

where we have used the relation

$$\langle\Phi^2\rangle = \sum_k\left[\sum_{k'}\phi(\boldsymbol{k}')\phi(\boldsymbol{k}-\boldsymbol{k}')\right] J_0(k_\perp v_\perp/\Omega) \exp \mathrm{i}\boldsymbol{k}\cdot\boldsymbol{R} .$$

Then the distribution function $F$ in $(\boldsymbol{x},\boldsymbol{v})$ space becomes

$$F(\boldsymbol{x},\boldsymbol{v},t) = \langle f(\boldsymbol{R},\mu,v_\parallel,t)\rangle\left\{1 - \frac{q}{T}\left[\Phi(\boldsymbol{x}) - \langle\langle\Phi\rangle(\boldsymbol{R})\rangle\right]\right\} , \qquad \boldsymbol{R} = \boldsymbol{x} - \boldsymbol{\rho} ,$$

and Poisson's equation is given by

$$\nabla^2\Phi(\boldsymbol{x}) = -\frac{e}{\epsilon_0}\int (F_\mathrm{i} - F_\mathrm{e})\mathrm{d}\boldsymbol{v} . \tag{14.9}$$

As $f$ is expressed by

$$f(\boldsymbol{R},\boldsymbol{v}_\perp,v_\parallel,t) = \sum_k \bar{f}(\boldsymbol{k},v_\parallel,t)\frac{1}{2\pi v_T^2}\exp\left(-\frac{v_\perp^2}{2v_T^2}\right)\exp \mathrm{i}\boldsymbol{k}\cdot(\boldsymbol{x}-\boldsymbol{\rho}) ,$$

it follows that $\langle f\rangle$ is

$$\langle f(\boldsymbol{R},\boldsymbol{v}_\perp,v_\parallel,t)\rangle = \sum_k \bar{f}(\boldsymbol{k},v_\parallel,t)\frac{1}{2\pi v_T^2}\exp\left(-\frac{v_\perp^2}{2v_T^2}\right)\exp(\mathrm{i}\boldsymbol{k}\cdot\boldsymbol{x}) J_0(k_\perp v_\perp/\Omega) .$$

We define $\tilde{f}(\boldsymbol{x},v_\parallel,t)$ by

$$\tilde{f}(\boldsymbol{x}, v_\parallel, t) \equiv \int \langle f(\boldsymbol{R}, \boldsymbol{v}_\perp, v_\parallel, t)\rangle 2\pi v_\perp \mathrm{d}v_\perp$$

$$= \sum_k \bar{f}(\boldsymbol{k}, v_\parallel, t) \exp(-b/2) \exp \mathrm{i}\boldsymbol{k}\cdot\boldsymbol{x} \ ,$$

where $\rho_T = v_T/\Omega$, $b \equiv (k_\perp \rho_T)^2$. Furthermore, $\tilde{\Phi}(\boldsymbol{x})$ is defined by

$$\tilde{f}(\boldsymbol{x}, v_\parallel, t)\tilde{\Phi}(\boldsymbol{x}) \equiv \int \langle\langle \Phi \rangle\rangle \langle f \rangle 2\pi v_\perp \mathrm{d}v_\perp \ .$$

Then we obtain

$$\tilde{\Phi}(\boldsymbol{x}) \simeq \sum_k \phi(\boldsymbol{k}) \Gamma_0(b) \exp \mathrm{i}\boldsymbol{k}\cdot\boldsymbol{x} \ , \qquad \Gamma_0(b) \equiv I_0(b) \exp(-b) \ .$$

The term $\int F \mathrm{d}\boldsymbol{v}$ appearing on the right-hand side of Poisson's equation (14.9) is expressed by

$$\int F \mathrm{d}\boldsymbol{v} = \hat{n}(\boldsymbol{x}) - \frac{q}{T}(\Phi - \tilde{\Phi})\hat{n}(\boldsymbol{x}) \ , \qquad \hat{n}(\boldsymbol{x}) \equiv \int \tilde{f}(\boldsymbol{x}, v_\parallel, t) \mathrm{d}v_\parallel \ .$$

Therefore (14.9) reduces to

$$\nabla^2 \Phi(\boldsymbol{x}) = -\frac{e}{\epsilon_0}(\hat{n}_\mathrm{i} - \hat{n}_\mathrm{e}) + \frac{T_\mathrm{e}}{T_\mathrm{i}} \frac{1}{\lambda_\mathrm{D}^2} \frac{\hat{n}_\mathrm{i}}{n_0}(\Phi - \tilde{\Phi}) \ . \tag{14.10}$$

Here we have assumed a zero electron Larmor radius. $\lambda_\mathrm{D} = (\epsilon_0 T_\mathrm{e}/n_0 e^2)^{1/2}$ is the Debye length. The second term on the right-hand side of (14.10) is

$$\frac{T_\mathrm{e}}{T_\mathrm{i}} \frac{1}{\lambda_\mathrm{D}^2}(\Phi - \tilde{\Phi}) \simeq \frac{T_\mathrm{e}}{T_\mathrm{i}} \frac{(k_\perp \rho_\mathrm{i})^2}{\lambda_\mathrm{D}^2} \Phi \approx -\frac{\Pi_\mathrm{i}^2}{\Omega_\mathrm{i}^2} \nabla_\perp^2 \Phi \ .$$

Usually, $\Pi_\mathrm{i}^2/\Omega_\mathrm{i}^2 = (T_\mathrm{e}/T_\mathrm{i})(\rho_\mathrm{i}/\lambda_\mathrm{D})^2$ is much larger than 1.

We shall now discuss numerical schemes for solving the gyrokinetic Vlasov–Poisson systems (14.8) and (14.10). Applying the discrete representation for the distribution function of $N$ particles, viz.,

$$f(\boldsymbol{R}, \mu, v_\parallel, t) = \sum_{j=1}^{N} \delta\big(\boldsymbol{R} - \boldsymbol{R}_j(t)\big) \delta(\mu - \mu_j) \delta\big(v_\perp - v_{\perp j}(t)\big) \ ,$$

to (14.8), the equations of motion in the gyrocenter coordinates for the $j$ th gyrokinetic particle must satisfy

$$\frac{\mathrm{d}\boldsymbol{R}_j}{\mathrm{d}t} = v_{\parallel j}\boldsymbol{b} + \boldsymbol{v}_\mathrm{d} - \frac{1}{B}\left(\frac{\partial \Psi}{\partial \boldsymbol{R}} \times \boldsymbol{b}\right)\bigg|_{R_j \mu_j} \ , \tag{14.11}$$

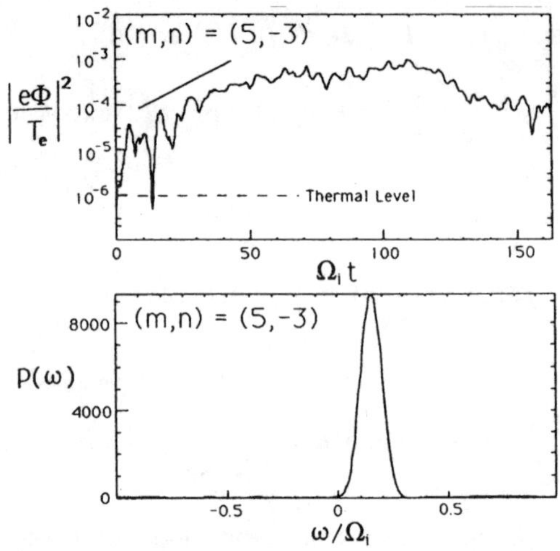


**Fig. 14.7.** *Upper*: temporal evolution of the mode amplitude of the drift-wave-trapped-electron mode $(m, n) = (5, -3)$. The *solid line* is the linear theory growth rate. *Lower*: power spectrum density versus frequency [14.26]

$$\frac{\mathrm{d}v_{\parallel j}}{\mathrm{d}t} = -\frac{q}{m}\left(\frac{\partial \Psi}{\partial \boldsymbol{R}}\cdot\boldsymbol{b}\right)\bigg|_{R_j\mu_j} - \mu_j \boldsymbol{B}\cdot\boldsymbol{\nabla}B\,. \tag{14.12}$$

Then the function $\tilde{f}$ is given by

$$\tilde{f}(\boldsymbol{x}, v_\parallel, t) = \int \langle f(\boldsymbol{R}, \boldsymbol{v}_\perp, v_\parallel, t)\rangle 2\pi v_\perp \mathrm{d}v_\perp = \sum_{j=1}^{N} \langle \delta(\boldsymbol{x} - \boldsymbol{R}_j - \boldsymbol{\rho}_j)\rangle \delta(v_\parallel - v_{\parallel j})\,.$$

In the first term on the right-hand side of (14.10), $\hat{n}_\mathrm{i}$ and $\hat{n}_\mathrm{e}$ are

$$\hat{n}_\mathrm{i} = \int \tilde{f}(\boldsymbol{x}, v_\parallel, t)\mathrm{d}v_\parallel = \sum_{j=1}^{N} \langle \delta(\boldsymbol{x} - \boldsymbol{R}_j^\mathrm{i} - \boldsymbol{\rho}_j^\mathrm{i})\rangle\,, \quad \hat{n}_\mathrm{e} = \sum_{j=1}^{N} \delta(\boldsymbol{x} - \boldsymbol{R}_j^\mathrm{e})\,.$$

When $k_\perp \rho_\mathrm{i} < 1$, the 4-point average by $\varphi = 0, \pi/2, \pi, 3\pi/2$ is a good approximation to the gyrophase average $\langle\ \rangle$.

The drift-wave-trapped-electron mode in a tokamak is studied in [14.26]. Simulation parameters used there include the system size $L_x \times L_y \times L_z = 64\Delta \times 64\Delta \times 32\Delta_z$, where $\Delta = \rho_\mathrm{i}$ and $\Delta_z = 25\rho_\mathrm{i}$. The aspect ratio was $R_0/a = 4$ and $n_0 = \langle n\rangle$=4 particles/cell (total ions $\sim 0.5\times 10^6$). $\langle T_\mathrm{e}\rangle/T_\mathrm{i} = 4$, $m_\mathrm{i}/m_\mathrm{e} = 1836$, $\Omega_\mathrm{i}\Delta t = 0.1$, $\omega_\mathrm{e}^*/\Omega_\mathrm{i} = 0.1m$, with $m$ the poloidal mode number, where $\omega_\mathrm{e}^* = k_\theta T_\mathrm{e}/L_n e B_0$. Results from the simulation of the drift-wave-trapped-electron mode with $\eta_\mathrm{e} \equiv \mathrm{d}(\ln T_\mathrm{e})/\mathrm{d}(\ln n_\mathrm{e})$=1 are shown in Fig. 14.7. The time evolution of the mode amplitude for $(m, n) = (5, -3)$ is shown in the upper figure. The saturation amplitude reaches $e\Phi/T_\mathrm{e} \approx 0.035$. The linear growth

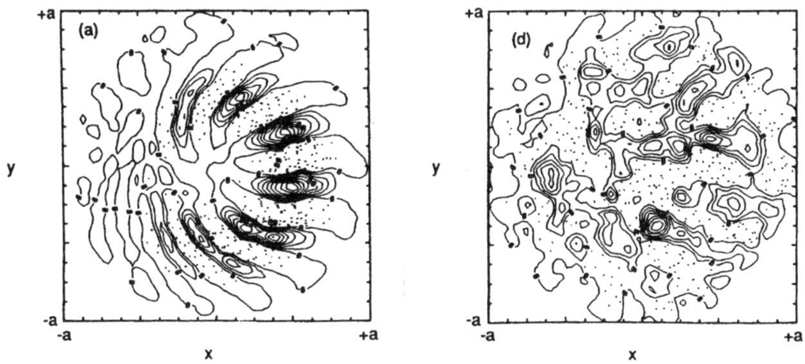


**Fig. 14.8.** Plots of the electrostatic potential of ITG turbulence in a poloidal cross-section during the linear phase (*left*) and non-linearly saturated steady state (*right*) [14.27]

rate in the initial phase and real frequency agree well with the theoretical linear eigenmode analysis.

The ion temperature gradient (ITG) mode is studied in [14.27]. In this analysis, the $\delta f/f$ method is used [14.28]. The electrons are assumed to be adiabatic ($\delta n_e/n_0 = e\phi/T_e$) and the total of number ions is $\sim 10^6$. Figure 14.8 plots the electrostatic potential during the linear phase and non-linearly saturated steady state.

The ion thermal diffusivity of ITG turbulence in a tokamak is studied in [14.29]. The number of particles in the simulation is in the range $5\times10^5$ to $1.34\times10^8$. For $10^6$ or more (2 particles per cell), $\chi_i$ at the later time does not appear to change with increasing particle number. The thermal diffusivity is defined formally as $\chi_i = 1.5L_n\langle\tilde{v}_r\tilde{T}_i\rangle/T_i$, where $\tilde{v}_r$ and $\tilde{T}_i$ are the fluctuating components of the radial ion velocity and ion temperature. The dependence of $\chi_i$ on $R/L_T$ is scanned and the fit can be expressed by an offset linear dependence on $R/L_T$ (see Fig. 14.6).

$$\frac{\chi_i}{\rho_i^2 v_{ti}/L_n} = 15.4\left(1.0 - 6.0\frac{L_T}{R}\right) .$$

We have discussed the gyrokinetic particle model of electrostatic perturbations. The formulation of the gyrokinetic particle model of electromagnetic perturbations has also been developed [14.30], in which the effect of magnetic field fluctuations is included as well as the effect of electric field fluctuations.

## 14.6 Full Orbit Particle Model

The fundamental equations of the full orbit particle model are simple and are given as follows [14.2]:

$$\frac{\mathrm{d}\boldsymbol{r}_{\mathrm{s}j}(t)}{\mathrm{d}t} = \boldsymbol{v}_{\mathrm{s}j}(t) , \quad \frac{\mathrm{d}\boldsymbol{v}_{\mathrm{s}j}(t)}{\mathrm{d}t} = \frac{q_{\mathrm{s}}}{m_{\mathrm{s}}}\Big[\boldsymbol{E}\big(\boldsymbol{r}_{\mathrm{s}j}(t),t\big) + \boldsymbol{v}_{\mathrm{s}j}(t) \times \boldsymbol{B}\big(\boldsymbol{r}_{\mathrm{s}j}(t),t\big)\Big] ,$$

$$\rho = \sum_{\mathrm{s=e,i}} q_{\mathrm{s}} \sum_j \delta\big(\boldsymbol{r} - \boldsymbol{r}_{\mathrm{s}j}(t)\big) , \quad \boldsymbol{j} = \sum_{\mathrm{s=e,i}} q_{\mathrm{s}} \sum_j \boldsymbol{v}_{\mathrm{s},j}(t)\delta\big(\boldsymbol{r} - \boldsymbol{r}_{\mathrm{s}j}(t)\big) ,$$

$$\epsilon_0 \boldsymbol{\nabla}\cdot\boldsymbol{E} = \rho , \quad \boldsymbol{\nabla}\cdot\boldsymbol{B} = 0 ,$$

$$\boldsymbol{\nabla} \times \boldsymbol{E} = -\frac{\partial \boldsymbol{B}}{\partial t} , \quad \boldsymbol{\nabla} \times \boldsymbol{B} = \mu_0 \boldsymbol{j} + \frac{1}{c^2}\frac{\partial \boldsymbol{E}}{\partial t} ,$$

where $m_{\mathrm{s}}$ and $q_{\mathrm{s}}$ are the mass and charge of species s. The other notation is as usual. However, the number of particles of real plasma is far beyond $10^7$–$10^8$, which is the limit of the most advanced supercomputer, at least in the near future. In the full orbit particle model, the concept of superparticle with finite size is introduced instead. Let us consider the system with volume of $V = L_x L_y L_z$, which contains $N_{\mathrm{e}}$ electrons and $N_{\mathrm{i}}$ ions. A number $\Lambda$ of particles are put together into one superparticle with mass and charge ($\Lambda \gg 1$)

$$m_{\mathrm{s}}^{\mathrm{sp}} = \Lambda m_{\mathrm{s}} , \qquad q_{\mathrm{s}}^{\mathrm{sp}} = \Lambda q_{\mathrm{s}} .$$

The average values of the density and temperature of superparticles are

$$n_{\mathrm{s}}^{\mathrm{sp}} = \frac{n_{\mathrm{s}}}{\Lambda} , \qquad T_{\mathrm{s}}^{\mathrm{sp}} = \Lambda T_{\mathrm{s}} .$$

Then the plasma frequency $\Pi_{\mathrm{s}}^{\mathrm{sp}}$, cyclotron frequency $\Omega_{\mathrm{s}}^{\mathrm{sp}}$, thermal velocity $v_T^{\mathrm{sp}}$, Alfvén velocity $v_{\mathrm{A}}^{\mathrm{sp}}$, Debye length $\lambda_{\mathrm{Ds}}^{\mathrm{sp}}$, Larmor radius $\rho_{\mathrm{cs}}^{\mathrm{sp}}$, and beta ratio $\beta^{\mathrm{sp}}$ are the same as in the original system. However, the Coulomb collision frequency is greatly enhanced, i.e.,

$$\nu_{\mathrm{ei}}^{\mathrm{sp}} \sim 0.4\frac{\Pi_{\mathrm{e}}^{\mathrm{sp}}}{n_{\mathrm{e}}^{\mathrm{sp}}(\lambda_{\mathrm{De}}^{\mathrm{sp}})^3} = 0.4\frac{\Lambda \Pi_{\mathrm{e}}}{n_{\mathrm{e}}(\lambda_{\mathrm{De}})^3} = \Lambda \nu_{\mathrm{ei}} .$$

The Coulomb collision frequency of superparticles with zero size can be comparable to 1/10–1/100 of the electron plasma frequency, when $N_{\mathrm{e}}^{\mathrm{sp}} \equiv n_{\mathrm{e}}^{\mathrm{sp}}(\lambda_{\mathrm{De}}^{\mathrm{sp}})^3$ is selected to be of order $10^0$–$10^1$. Low frequency waves will therefore be masked by Coulomb collision. To avoid this effect, a charge density distribution of finite size with shape factor of $S(\boldsymbol{r})$ is introduced:

$$q_{\mathrm{s}}^{\mathrm{sp}}\delta(\boldsymbol{r} - \boldsymbol{r}_{\mathrm{s}j}) \longrightarrow q_{\mathrm{s}}^{\mathrm{sp}} S(\boldsymbol{r} - \boldsymbol{r}_{\mathrm{s}j}) , \qquad \int S(\boldsymbol{r} - \boldsymbol{r}_{\mathrm{s}j})\mathrm{d}\boldsymbol{r} = 1 .$$

When the effective radius of the shape factor is $R$, the effective Coulomb collision frequency is reduced by one to three orders of magnitude depending on the size $R \sim (1$–$5)\lambda_{\mathrm{D}}$ and $n_{\mathrm{e}}(\lambda_{\mathrm{De}})^3 \sim (10$–$10^3)$ [14.31]. The dispersion relation with $k < R^{-1}$ is barely affected. Therefore the simulation using superparticles with a proper finite size can reproduce wave phenomena with wavelength greater than $R$. The equations of motion of superparticles are

$$\frac{\mathrm{d}\boldsymbol{r}_{sj}^{\mathrm{sp}}(t)}{\mathrm{d}t} = \boldsymbol{v}_{sj}^{\mathrm{sp}}(t) \;, \qquad \frac{\mathrm{d}\boldsymbol{v}_{sj}^{\mathrm{sp}}(t)}{\mathrm{d}t} = \frac{q_{\mathrm{s}}^{\mathrm{sp}}}{m_{\mathrm{s}}^{\mathrm{sp}}}\Big[\boldsymbol{E}^*\big(\boldsymbol{r}_{sj}(t),t\big) + \boldsymbol{v}_{sj}^{\mathrm{sp}}(t)\times\boldsymbol{B}^*\big(\boldsymbol{r}_{sj}(t),t\big)\Big] \;,$$

where $\boldsymbol{E}^*$ and $\boldsymbol{B}^*$ are the fields felt by finite-size superparticles, i.e.,

$$\boldsymbol{E}^*(\boldsymbol{r},t) \equiv \int \boldsymbol{E}(\boldsymbol{r}',t)S(\boldsymbol{r}'-\boldsymbol{r})\mathrm{d}\boldsymbol{r}' \;, \qquad \boldsymbol{B}^*(\boldsymbol{r},t) \equiv \int \boldsymbol{B}(\boldsymbol{r}',t)S(\boldsymbol{r}'-\boldsymbol{r})\mathrm{d}\boldsymbol{r}' \;,$$

$$\rho = \sum_{\mathrm{s=e,i}} q_{\mathrm{s}}^{\mathrm{sp}} \sum_{j=1}^{N_{\mathrm{s}}^{\mathrm{sp}}} S\big(\boldsymbol{r}-\boldsymbol{r}_{sj}(t)\big) \;, \qquad \boldsymbol{j} = \sum_{\mathrm{s=e,i}} q_{\mathrm{s}}^{\mathrm{sp}} \sum_{j=1}^{N_{\mathrm{s}}^{\mathrm{sp}}} \boldsymbol{v}_{\mathrm{s},j}(t) S\big(\boldsymbol{r}-\boldsymbol{r}_{sj}(t)\big) \;,$$

$$\epsilon_0 \boldsymbol{\nabla}\cdot\boldsymbol{E} = \rho \;, \qquad \boldsymbol{\nabla}\cdot\boldsymbol{B} = 0 \;,$$

$$\boldsymbol{\nabla}\times\boldsymbol{E} = -\frac{\partial \boldsymbol{B}}{\partial t} \;, \qquad \boldsymbol{\nabla}\times\boldsymbol{B} = \mu_0 \boldsymbol{j} + \frac{1}{c^2}\frac{\partial \boldsymbol{E}}{\partial t} \;.$$

When time and length are normalized by

$$\tilde{t} = t\Pi_{\mathrm{e}} \;, \qquad \tilde{\boldsymbol{r}} = \boldsymbol{r}\Delta^{-1} \;,$$

and $\boldsymbol{E}$, $\boldsymbol{B}$, $\rho$ and $\boldsymbol{j}$ are normalized by

$$\tilde{\boldsymbol{E}} = \frac{\boldsymbol{E}}{m_{\mathrm{e}}\Delta\Pi_{\mathrm{e}}^2/e} \;, \qquad \tilde{\boldsymbol{B}} = \frac{\boldsymbol{B}}{m_{\mathrm{e}}\Pi_{\mathrm{e}}/e} \;,$$

$$\tilde{\rho} = \frac{\rho}{en_{\mathrm{e}}} \;, \qquad \tilde{\boldsymbol{j}} = \frac{\boldsymbol{j}}{en_{\mathrm{e}}\Delta\Pi_{\mathrm{e}}} \;, \qquad \Delta = \text{grid size} \;,$$

then the dimensionless form of the above equations is [14.32]

$$\frac{\mathrm{d}\tilde{\boldsymbol{r}}_{sj}(\tilde{t})}{\mathrm{d}\tilde{t}} = \tilde{\boldsymbol{v}}_{sj}(\tilde{t}) \;, \tag{14.13}$$

$$\frac{\mathrm{d}\tilde{\boldsymbol{v}}_{sj}(\tilde{t})}{\mathrm{d}\tilde{t}} = \frac{Q_{\mathrm{s}}}{M_{\mathrm{s}}}\Big[\tilde{\boldsymbol{E}}^*\big(\tilde{\boldsymbol{r}}_{sj}(\tilde{t}),\tilde{t}\big) + \tilde{\boldsymbol{v}}_{sj}(\tilde{t})\times\tilde{\boldsymbol{B}}^*\big(\tilde{\boldsymbol{r}}_{sj}(\tilde{t}),\tilde{t}\big)\Big] \;, \tag{14.14}$$

$$\tilde{\rho} = \frac{\tilde{L}_x\tilde{L}_y\tilde{L}_z}{N_{\mathrm{e}}^{\mathrm{sp}}} \sum_{\mathrm{s=e,i}} Q_{\mathrm{s}} \sum_{j=1}^{N_{\mathrm{s}}^{\mathrm{sp}}} \tilde{S}\big(\tilde{\boldsymbol{r}}-\tilde{\boldsymbol{r}}_{sj}(\tilde{t})\big) \;, \tag{14.15}$$

$$\tilde{\boldsymbol{j}} = \frac{\tilde{L}_x\tilde{L}_y\tilde{L}_z}{N_{\mathrm{e}}^{\mathrm{sp}}} \sum_{\mathrm{s=e,i}} Q_{\mathrm{s}} \sum_{j=1}^{N_{\mathrm{s}}^{\mathrm{sp}}} \tilde{\boldsymbol{v}}_{\mathrm{s},j}(\tilde{t})\tilde{S}\big(\tilde{\boldsymbol{r}}-\tilde{\boldsymbol{r}}_{sj}(\tilde{t})\big) \;, \tag{14.16}$$

$$\tilde{\boldsymbol{\nabla}}\cdot\tilde{\boldsymbol{E}} = \tilde{\rho} \;, \qquad \tilde{\boldsymbol{\nabla}}\cdot\tilde{\boldsymbol{B}} = 0 \;, \tag{14.17}$$

$$\tilde{\boldsymbol{\nabla}}\times\tilde{\boldsymbol{E}} = -\frac{\partial\tilde{\boldsymbol{B}}}{\partial\tilde{t}} \;, \qquad \tilde{c}^2\tilde{\boldsymbol{\nabla}}\times\tilde{\boldsymbol{B}} = \tilde{\boldsymbol{j}} + \frac{\partial\tilde{\boldsymbol{E}}}{\partial\tilde{t}} \;, \tag{14.18}$$

$$\tilde{c} = \frac{c}{\Delta\Pi_{\mathrm{e}}} \;, \qquad \tilde{S}\big(\tilde{\boldsymbol{r}}-\tilde{\boldsymbol{r}}_{sj}(\tilde{t})\big) = \Delta^3 S\big(\boldsymbol{r}-\boldsymbol{r}_{sj}(t)\big) \;, \tag{14.19}$$

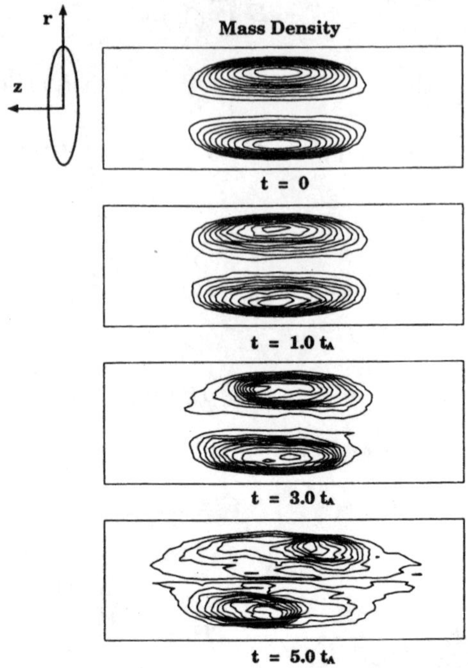


**Fig. 14.9.** Contour plots of mass density in the poloidal cross-section of the field-reversed configuration (FRC) at periods $t/t_A = 0.0$, 1.0, 3.0 and 5.0 [14.33]

where $c$ is the speed of light and $Q_s \equiv q_s/e$ ($Q_e = -1, Q_i = q_i/e$), $M_s \equiv m_s/m_e$ ($M_e = 1, M_i = m_i/m_e$).

The tilt stability of a field-reversed configuration (FRC) is studied by full orbit particle ions and electrons in [14.33]. The grid number is $49 \times 49 \times 32$ and the total number of superparticles is $10^6$. Furthermore, $\Pi_e \Delta t = 1.5$, $c\Delta t/\Delta < 1$. The ion and electron mass ratio is set to be $m_i/m_e = 50$ and $\Omega_e \sim \Pi_e/5$. Figure 14.9 shows contour plots of mass density in the poloidal cross-section at periods of $t/\tau_A = 0.0$, 1.0, 3.0 and 5.0. $\tau_A$ is the Alfvén transit time. In this simulation, $|\Omega_i| < \tau_A^{-1}$. The stabilizing effect on the tilting by cycling ions which cross the separatrix is discussed. Tilt stability of FRC is also discussed using the hybrid model of full orbit particle ions and fluid electrons [14.34]. The stabilizing effect due to finite ion Larmor radius is analyzed.

When $|\Omega_i| \gg \tau_A^{-1}$, full orbit particle simulation of electromagnetic perturbation is very difficult due to the excess amount of computer run time. The toroidal particle code (TPC) has been developed for electrostatic turbulence [14.35]. TPC solves Poisson's equation

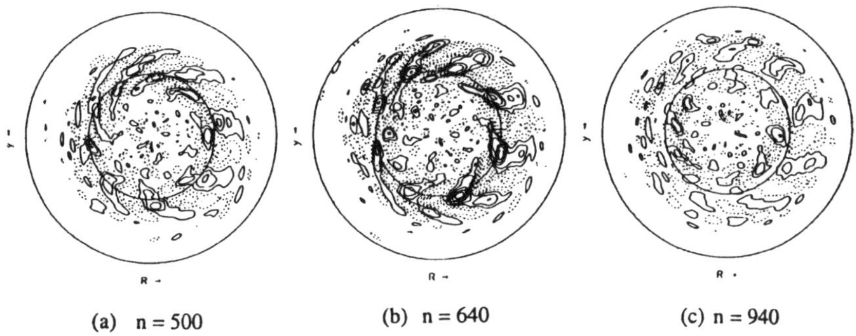


**Fig. 14.10.** Potential structure in poloidal cross-section at three different times for a reversed magnetic shear configuration of a tokamak [14.36]. In the quasi-steady state (**c**), the discontinuity of the potential structure across the $q_{\mathrm{min}}$ surface is recovered

$$\nabla^2 \Phi = -\frac{1}{\epsilon_0}\rho = -\frac{e}{\epsilon_0}\sum_j \left[\frac{q_{\mathrm{i}}}{e} S(\boldsymbol{r} - \boldsymbol{r}_j^{\mathrm{i}}) - S(\boldsymbol{r} - \boldsymbol{r}_j^{\mathrm{e}})\right] . \tag{14.20}$$

The ion motion is given by the Lorentz equation

$$\frac{\mathrm{d}\boldsymbol{r}^{\mathrm{i}}}{\mathrm{d}t} = \boldsymbol{v}^{\mathrm{i}} , \qquad \frac{\mathrm{d}\boldsymbol{v}^{\mathrm{i}}}{\mathrm{d}t} = \frac{q^{\mathrm{i}}}{m^{\mathrm{i}}}\left(\boldsymbol{E} + \boldsymbol{v}^{\mathrm{i}} \times \boldsymbol{B}\right) , \tag{14.21}$$

and the electron motion is given by the drift equation

$$\frac{\mathrm{d}v_{\parallel}^{\mathrm{e}}}{\mathrm{d}t} = -\frac{e}{m_{\mathrm{e}}} E_{\parallel} - \mu(\boldsymbol{b}\cdot\boldsymbol{\nabla})B . \tag{14.22}$$

In [14.36], the ion temperature gradient (ITG) turbulences in the tokamak configuration are analyzed using TPC. In these simulations, electrons are treated as an adiabatic fluid ($\delta n_{\mathrm{e}}/n_{\mathrm{e}} = e\Phi/T_{\mathrm{e}}$). The effects of the reversed magnetic shear configuration on ITG turbulence are studied (see Sect. 16.7) and discontinuities and/or gaps in the structure of the ITG perturbed potential across the $q_{\mathrm{min}}$ surface are observed, as shown in Fig. 14.10.

Part II

# Controlled Nuclear Fusion

# 15 Development of Fusion Research

The major research effort in the area of controlled nuclear fusion focuses on the confinement of hot plasmas by means of strong magnetic fields. Magnetic confinements are classified into toroidal and open end configurations. Confinement in a linear mirror field may have advantages over toroidal confinement with respect to stability and anomalous diffusion across the magnetic field. However, the end loss due to particles leaving along magnetic lines of force is determined solely by diffusion in the velocity space, i.e., the confinement time cannot be improved by increasing the intensity of the magnetic field or the plasma size. Ways must be found to suppress end loss.

Toroidal magnetic confinements have no open ends. In the simple toroidal field, ions and electrons drift in opposite directions due to the gradient of the magnetic field. This gradient $B$ drift causes the charge separation that induces the electric field $\boldsymbol{E}$ directed parallel to the major axis of the torus. The subsequent $\boldsymbol{E} \times \boldsymbol{B}$ drift tends to carry the plasma ring outward. In order to reduce the $\boldsymbol{E} \times \boldsymbol{B}$ drift, the upper and lower parts of the plasma must be connected by magnetic lines of force and the separated charges short-circuited along these field lines. Accordingly, a poloidal component of the magnetic field is essential to the equilibrium of toroidal plasmas, and toroidal devices may be classified according to the method used to generate the poloidal field. The tokamak (Chap. 16) and the reversed field (Sect. 17.1) pinch devices use the plasma current along the toroid, whereas the toroidal stellarator (Sect. 17.2) has helical conductors or equivalent winding outside the plasma to produce the appropriate rotational transform angles.

Besides the study of magnetic confinement systems, inertial confinement approaches are being actively investigated. If a very dense and hot plasma could be produced within a very short time, it might be possible to complete the nuclear fusion reaction before the plasma starts to expand. An extreme example is a hydrogen bomb. This type of confinement is called inertial confinement. In laboratory experiments, high-power laser beams or particle beams are focused onto small solid deuterium and tritium targets, thereby producing very dense, hot plasma within a short time. Because of the development of the technologies of high-power energy drivers, the approaches along this line have some foundation in reality. Inertial confinement will be discussed briefly in Chap. 18.

The various kinds of approach that are actively investigated in controlled thermonuclear fusion are classified as follows:

- Magnetic confinement
  - Toroidal system
    - Axially symmetric
      - Tokamak
      - Reversed field pinch
      - Spheromak
    - Axially asymmetric
      - Stellarator system
      - Heliac
      - Bumpy torus
  - Open end system
    - Mirror, Tandem mirror
    - Field reversal configuration
    - Cusp
- Inertial confinement
  - Laser
  - Ion beam, Electron beam

## 15.1 From Secrecy to International Collaboration

Basic research into controlled thermonuclear fusion probably began right after World War II in the United States, the Soviet Union, and the United Kingdom in strict secrecy. There are on record many speculations about research into controlled thermonuclear fusion even in the 1940s. The United States program, called Project Sherwood, has been described in detail by Bishop [15.1]. Bishop states that Z pinch experiments for linear and toroidal configurations at the Los Alamos Scientific Laboratory were carried out in an attempt to overcome sausage and kink instabilities. The astrophysicist L. Spitzer Jr. started the figure-eight toroidal stellarator project at Princeton University in 1951. At the Lawrence Livermore National Laboratory, mirror confinement experiments were conducted. At the Atomic Energy Research Establishment in Harwell, United Kingdom, the Zeta experiment was started [15.2] and at the I.V. Kurchatov Institute of Atomic Energy in the Soviet Union, experiments on a mirror called Ogra and on tokamaks were carried out [15.3].

The first United Nations International Conference on the Peaceful Uses of Atomic Energy was held in Geneva in 1955. Although this conference was concerned with peaceful applications of nuclear fission, the chairman, H.J. Bhabha, hazarded the prediction that ways of controlling fusion energy that would render it industrially usable would be found in less than two decades. However, as we have seen, research into controlled nuclear fusion encountered serious and unexpected difficulties. It was soon recognized that the realization of a practical fusion reactor was a long way off and that basic research on plasma physics and the international exchange of scientific information were absolutely necessary. From around that time articles on controlled nuclear fusion started appearing regularly in academic journals. Lawson's paper on the conditions for fusion was published in January 1957 [15.4] and several important theories on MHD instabilities had by that time begun to appear [15.5, 15.6]. Experimental results of the Zeta [15.7] (Zero Energy Thermonuclear Assembly) and Stellarator [15.8] projects were made public in January 1958. In the fusion sessions of the second United Nations International Conference on the Peaceful Uses of Atomic Energy, held in Geneva, 1–13 September 1958 [15.9, 15.10], many results of research that had proceeded in secrecy were revealed.

L.A. Artsimovich expressed his impression of this conference as "something that might be called a display of ideas." The second UN conference marks the beginning of open rather than secret international cooperation and competition in fusion research.

In Japan, controlled fusion research started in the Japan Atomic Energy Institute (JAERI) under the ministry of science and technology and in the Institute of Plasma Physics, Nagoya University, under the ministry of education and culture, in the early 1960s [15.11].

The First International Conference on Plasma Physics and Controlled Nuclear Fusion Research was held in Salzburg in 1961 under the auspices of the International Atomic Energy Agency (IAEA). At the Salzburg conference [15.12] the big projects were fully discussed. Among these were Zeta, Alpha, Stellarator C, Ogra, and DCX. Theta pinch experiments (Scylla, Thetatron, etc.) appeared to be more popular than linear pinches. The papers on the large scale experimental projects such as Zeta or Stellarator C all reported struggles with various instabilities. L.A. Artsimovich said in his summary of experimental results: "Our original beliefs that the doors into the desired regions of ultra-high temperature would open smoothly [...] have proved as unfounded as the sinner's hope of entering Paradise without passing through Purgatory." The importance of the PR-2 experiments of M.S. Ioffe and others was soon widely recognized [15.12, Vol. 3, p. 1045]. These experiments demonstrated that the plasma confined in a minimum-$B$ configuration is MHD stable.

The Second International Conference on Plasma Physics and Controlled Nuclear Fusion Research was held at Culham in 1965 [15.13]. The stabilizing effect of minimum-$B$ configurations was confirmed by many experiments. An absolute minimum-$B$ field cannot be realized in a toroidal configuration. Instead of this, the average minimum-$B$ concept was introduced [15.13, Vol. 1, pp. 103, 145]. Ohkawa and others succeeded in confining plasmas for much longer than the Bohm time with toroidal multipole configurations [15.13, Vol. 2, p. 531] and demonstrated the effectiveness of the average minimum-$B$ configuration. Artsimovich and others reported on a series of tokamak experiments [15.13, T-5, Vol. 2, p. 577; T-3, p. 595; T-2, p. 629; TM-2, p. 647; TM-1, p. 659]. Further experiments with Zeta and Stellarator C were also reported. However, the confinement times for these big devices were only of the order of the Bohm time, and painful examinations of loss mechanisms had to be carried out. Theta pinch experiments were still the most actively pursued. The ion temperatures produced by means of theta pinches were several hundred eV to several keV, and confinement times were limited only by end losses. One of the important goals of the theta pinch experiments had thus been attained, and it marked the turning point from linear theta pinch to toroidal pinch experiments.

In this conference, the effectiveness of minimum-$B$, average minimum-$B$, and shear configurations was thus confirmed. Many MHD instabilities were seen to be well understood experimentally as well as theoretically. Methods of stabilizing against MHD instabilities seemed to be becoming gradually clearer. The importance of velocity-space instabilities due to the non-Maxwellian distribution function of the confined plasma was recognized. There had been and were subsequently to be reports on loss-cone instabilities [15.17], Harris instability [15.18] (1959), drift instabilities [15.19] (1963, 1965), etc. The experiment by J.M. Malmberg and C.B. Wharton [15.13, Vol. 1, p. 485] was the first experimental verification of Landau damping.

L. Spitzer Jr. concluded in his summary talk at Culham that: "most of the serious obstacles have been overcome, sometimes after years of effort by a great number of scientists. We can be sure that there will be many obstacles ahead but we have good reason to hope that these will be surmounted by the cooperative efforts of scientists in many nations."

## 15.2 Artsimovich Era

The Third International Conference [15.14] was held in 1968 at Novosibirsk. The most remarkable topic in this conference was the report that Tokamak T-3 [15.14, Vol. 1, p. 157] had confined a plasma up to 30 times the Bohm time (several milliseconds) at an electron temperature of 1 keV. In Zeta experiments a quiescent period was found during a discharge and MHD stability of the magnetic field configuration of the quiescent period was discussed. This was the last report on Zeta, and HBTX succeeded this reversed field pinch experiment. Stellarator C [15.14, Vol. 1, pp. 479, 495] was still only confining plasmas to several times the Bohm time at electron temperatures of only several tens to a hundred eV. This was the last report on Stellarator C; this machine was converted into the ST tokamak before the next conference (Madison 1971). However, various aspects of stellarator research were still pursued. The magnetic coil systems of Clasp [15.14, Vol. 1, p. 465] were constructed accurately, and the confinement of high-energy electrons was examined using the $\beta$ decay of tritium. It was demonstrated experimentally that the electrons ran around the torus more than $10^7$ times and that the stellarator field had good charged-particle confinement properties. In WII the confinement of the barium plasma was tested, and resonant loss was observed when the magnetic surface was rational. Diffusion in a barium plasma in non-rational cases was classical. In 2X [15.14, Vol. 2, p. 225] a deuterium plasma was confined up to an ion temperature of 6–8 keV at a density of $n < 5 \times 10^{13}\,\mathrm{cm}^{-3}$ for up to $\tau = 0.2\,\mathrm{ms}$. Laser plasmas appeared at this conference.

At the Novosibirsk conference, toroidal confinement appeared to have the best overall prospects, and mainstream research shifted toward toroidal confinement. L.A. Artsimovich concluded this conference by saying: "We have rid ourselves of the gloomy spectre of the enormous losses embodied in Bohm's formula and have opened the way for further increases in plasma temperature leading to the physical thermonuclear level."

The Tokamak results were seen to be epoch-making if the estimates of the electron temperature were accurate. R.S. Pease, the director of the Culham Laboratory, and L.A. Artsimovich agreed to the visit of the British team of researchers to the Kurchatov Institute to measure the electron temperature of the T-3 plasma by laser scattering methods. The measurements supported the previous estimates by the tokamak group [15.20]. The experimental results of T-3 had a strong impact on the next phase of nuclear fusion research in various nations. At the Princeton Plasma Physics Laboratory, Stellarator C was converted to the ST tokamak device. Newly built were ORMAK at Oak Ridge National Laboratory, TFR at the Center for Nuclear Research, Fontaney aux Rose, Cleo at the Culham Laboratory, Pulsator at the Max Planck Institute for Plasma Physics, and JFT-2 at the Japan Atomic Energy Research Institute.

The Fourth International Conference was held in Madison, Wisconsin, in 1971 [15.15]. The main interest at Madison was naturally focused on the tokamak experiments. In T-4 [15.15, Vol. 1, p. 443], the electron temperature approached 3 keV at a confinement time around 10 ms. The ions were heated to around 600 eV by collision with electrons. ST [15.15, Vol. 1, pp. 451, 465] produced similar results.

## 15.3 The Trek to Large Tokamaks Since the Oil Crisis

Since then the IAEA conference has been held every two years: Tokyo in 1974 [15.16], Berchtesgarden in 1976, Innsburg in 1978, Brussels in 1980, Baltimore in 1982, London in 1984, Kyoto in 1986, Nice in 1988, Washington D.C. in 1990, Würzburg in 1992, Seville in 1994, Montreal in 1996, Yokohama in 1998, Sorrento in 2000, and Lyon in 2002. Tokamak research has made steady progress as the mainstream of magnetic confinement. Pease stated in his summary talk of the IAEA conference at Berchtesgarden in 1976 that: "one can see the surprisingly steady progress that has been maintained. Furthermore, looked at logarithmically, we have now covered the greater part of the total distance. What remains is difficult, but the difficulties are finite and can be summed up by saying that we do not yet have an adequate understanding or control of cross-field electron thermal conduction."

After the tokamaks of the first generation (T-4, T-6, ST, ORMAK, Alcator A, C. TFR, Pulsator, DITE, FT, JFT-2, JFT-2a, JIPP T-II, etc.), second generation tokamaks (T-10, PLT, PDX, ISX-B, Doublet III, ASDEX, etc.) began appearing around 1976. The energy confinement time of ohmically heated plasmas was approximately described by the Alcator scaling law ($\tau_{\rm E} \propto na^2$). The value of $n\tau_{\rm E}$ reached $2 \times 10^{13}\,{\rm cm}^{-3}{\rm s}$ in Alcator A in 1976. Heating experiments using neutral beam injection (NBI) in PLT achieved an ion temperature of 7 keV in 1978, and effective wave heating in an ion cyclotron range of frequency was demonstrated in TFR and PLT around 1980. The average $\beta$ value of 4.6% was realized in the Doublet III non-circular tokamak ($\kappa = 1.4$) in 1982 using 3.3 MW NBI. Non-inductive drives for plasma current have been pursued. Current drive by tangential injection of a neutral beam was proposed by Ohkawa in 1970 and demonstrated experimentally in DITE in 1980. Current drive by a lower hybrid wave was proposed by Fisch in 1978 and demonstrated in JFT-2 in 1980 and in Versator 2, PLT, Alcator C, JIPP T-II, Wega, T-7, and so on. Ramp-up experiments increasing the plasma current from 0 were succeeded by WT-2 and PLT in 1984. TRIAM-1M with superconducting toroidal coils sustained a plasma current of $I_{\rm p} = 22\,{\rm kA}$ ($n_{\rm e} \approx 2 \times 10^{18}\,{\rm m}^3$) during 70 minutes by LHW in 1990.

The suppression of impurity ions by a divertor was demonstrated in JFT-2a (DIVA) in 1978 and investigated in detail by ASDEX and Doublet III (1982). At that time the energy confinement time had deteriorated compared with the ohmic heating case as the heating power of NBI was increased (according to the Kaye–Goldston scaling law). However, the improved mode (named H mode) of the confinement time, increased by a factor of about 2 compared with the ordinary mode (L mode), was found in the divertor configuration of ASDEX in 1982. The H mode was also observed in Doublet III, PDX, JFT-2M, and DIII-D. Much progress had thus been made to solve many critical issues of tokamaks.

On the basis of these achievements, experiments started on the third generation of large tokamaks, with TFTR (United States) at the end of 1982, JET (European Community) in 1983 and JT-60 (Japan) in 1985. Originally these large tokamaks were planned in the early 1970s. TFTR achieved $n_{\rm DT}(0)\tau_{\rm E} \sim 1.2 \times 10^{19}\,{\rm m}^{-3}{\rm s}$, $\kappa T_{\rm i}(0) = 44\,{\rm keV}$ by supershot (H mode-like). JET achieved $n_{\rm D}(0)\tau_{\rm E} \sim 3.2 \times 10^{19}\,{\rm m}^{-3}{\rm s}$, $\kappa T_{\rm i}(0) = 18.6\,{\rm keV}$ by H mode with divertor configuration. JT-60 drove a plasma current of 1.7 MA ($\bar{n}_{\rm e} = 0.3 \times 10^{13}\,{\rm cm}^{-3}$) by lower hybrid wave ($P_{\rm RF} = 1.2\,{\rm MW}$) in 1986 and upgraded to JT60U in 1991 [15.21]. JT60U achieved $n_{\rm D}(0)\tau_{\rm E} \sim 3.4 \times 10^{19}\,{\rm m}^{-3}{\rm s}$, $\kappa T_{\rm i}(0) = 45\,{\rm keV}$ by high $\beta_{\rm p}$ H mode. A high performance confinement mode with negative magnetic shear was demonstrated in TFTR, DIII-D, JT60U, JET, Tore Supra [15.22] and T10.

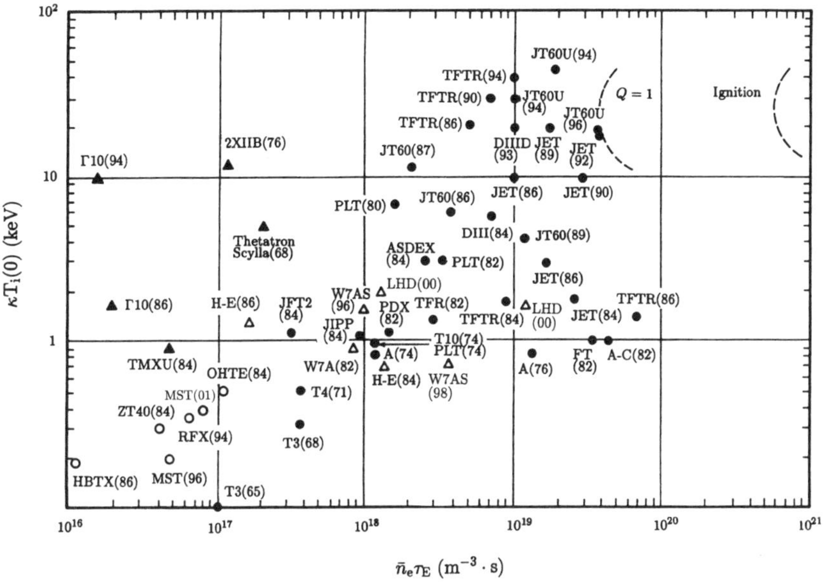


**Fig. 15.1.** Development of confinement experiments represented in a $\bar{n}_e\tau_E$–$\kappa T_i(0)$ diagram, where $\bar{n}_e$ is the line average electron density, $\tau_E$ is the energy confinement time $\tau_E \equiv W/(P_{tot} - dW/dt - L_{thr})$, and $T_i(0)$ is the ion temperature. Tokamak (*black circle*), stellarator (*white triangle*), RFP (*white circle*), tandem mirror, mirror, theta pinch (*black triangle*). $Q = 1$ is the critical condition. $W$ is the total energy of the plasma, $P_{tot}$ the total heating power, and $L_{thr}$ the shine-through of neutral beam heating

JET performed a preliminary tritium injection experiment in 1991 $[n_T/(n_D + n_T) \simeq 0.11]$ [15.23] and produced 1.7 MW ($Q \sim 0.11$) of fusion power using 15 MW of NBI. An extensive deuterium–tritium experiment was carried out on TFTR in 1994 [15.24]. Fusion power of 9.3 MW ($Q \sim 0.27$) was obtained using 34 MW of NBI in supershot ($I_p = 2.5$ MA). JET set records of DT fusion output of 16.1 MW ($Q \sim 0.62$) using 25.7 MW of input power (22.3 MW NBI + 3.1 MW ICRF) [15.25] in 1998. Pumped divertors were installed in JET, JT60U, DIIID, ASDEX-U and others in attempt to suppress impurity ions and the heat load on the divertor plate. These large tokamaks are now aiming at the scientific demonstration of required conditions with regard to several critical issues in fusion reactors, including plasma transport, steady operation, divertor and impurity control, and so on.

Based on the development of tokamak research, the design activities of tokamak reactors have been carried out. The International Tokamak Reactor (INTOR) [15.26] (1979–1987) and The International Thermonuclear Experimental Reactor (ITER) [15.27] (1988–2001) are collaborative efforts among

Euratom, Japan, The United States of America and the Russian Federation under the auspices of the IAEA. The status of ITER in 2000 [15.28] is described in Sect. 16.10.

## 15.4 Alternative Approaches

Potential theoretical advantages of the spherical tokamak have been outlined by Peng and Strickler [15.29], in which the aspect rato $A = R/a$ of the standard tokamak is substantially reduced toward unity. Predicted advantages include a naturally high elongation ($\kappa_s \sim 2$), high toroidal beta and tokamak-like confinement. These predictions have been confirmed experimentally, in particular by the START device [15.30] at Culham ($R/a \approx 0.3/0.28 = 1.31$, $I_p \approx 0.25\,\mathrm{MA}$, $B_t \approx 0.15\mathrm{T}$). The toroidal beta reached 40% and observed confinement times follow similar scaling to standard tokamaks. Spherical tokamak (ST) experiments were also conducted by Globus-M (Ioffe Physico-Technical Inst.), Pegasus (Univ. Wisconsin), TST, TS-3 (Univ. Tokyo). The next generation ST projects MAST (Culham) and NSTX (Princeton PPL) started experiments in 1999–2000 [15.31]. Potential merits of possible ST reactors are also discussed in [15.32].

Non-tokamak confinement systems have been investigated intensively to catch up with the achievements of tokamaks. The stellarator program proceeded from small-scale experiments (Wendelstein IIb, Clasp, Uragan-1, L-1, JIPP-I, Heliotron D) to middle-scale experiments (Wendelstein VIIA, Cleo, Uragan-2, L-2, JIPP T-II, Heliotron E). Plasmas with $T_e \sim T_i$ roughly several hundred eV to 1 keV, $n_e \sim 10^{13}\,\mathrm{cm}^{-3}$ were sustained by NBI heating without an ohmic heating current, and the possibility of steady-state operation of stellarators was demonstrated by WVIIA and Heliotron E. Confinement time scaling in a currentless plasma was studied in Heliotron E, CHS, ATF and WVII AS. The large helical device LHD started experiments in 1998 [15.33] and an advanced stellarator WVII-X is under construction.

The reversed field pinch (RFP) configuration was found in the stable quiescent period of Zeta discharge just before the shutdown in 1968. In 1974, J.B. Taylor pointed out that the RFP configuration is the minimum energy state under the constraint of the conservation of magnetic helicity in 1974 (see Sect. 17.1). RFP experiments have been conducted in HBTX-1B, ETA-BETA 2, TPE-1RM, TPE-1RM15, TPE-1RM20, ZT-40M, OHTE, REPUTE-1, STP-3M, MST. An average $\beta$ of 10–15% was realized. ZT-40M demonstrated that the RFP configuration can be sustained by relaxation phenomena (the so-called dynamo effect) as long as the plasma current is sustained (1982). The next-step projects RFX and TPE-RX are currently underway.

Spheromak configurations have been studied by S-1, CTX, and CTCC-1, and field reversed configurations have been studied by FRX, TRX, LSX, NUCTE and PIACE.

In mirror research, 2XIIB confined a plasma with an ion temperature of 13 keV and $n\tau_E \times 10^{11}$ cm$^{-3}$s in 1976 [15.34]. However, suppression of end loss is absolutely necessary. The concept of a tandem mirror, in which end losses are suppressed by electrostatic potential, was proposed in 1976–1977 [15.35, 15.36]. TMX, TMX-U and GAMMA 10 are typical tandem mirror projects. Mirror research is reviewed in [15.37]. The bumpy torus is the toroidal linkage of many mirrors to avoid end loss and this method was pursued in EBT and NBT.

Inertial confinement research has made great advances in the implosion experiment by using an Nd glass laser as energy driver. Gekko XII (30 kJ, 1 ns, 12 beams), Nova (100 kJ, 1 ns, 10 beams), Omega X (4 kJ, 1 ns, 24 beams), and Octal (2 kJ, 1 ns, 8 beams) investigated implosion using $\lambda = 1.06\,\mu$m laser light and its higher harmonics $\lambda = 0.53\,\mu$m and $0.35\,\mu$m. It was shown that a short wavelength is favorable because of the better absorption and less preheating of the core. A high-density plasma, 200–600 times as dense as the solid state, was produced by laser implosion (1990). Based on Nova results, the Lawrence Livermore National Laboratory is preparing the National Ignition Facility (NIF) [15.38, 15.39] (1.8 MJ, 20 ns, 0.35 μm, 192 beams, Nd glass laser system).

Controlled nuclear fusion research has been making steady progress through international collaboration and competition. A summary of the progress of magnetic confinement is given in Fig. 15.1 which shows the $\bar{n}_e\tau_E$–$T_i(0)$ diagram. TFTR demonstrated $Q \sim 0.27$ DT experiments and JET demonstrated $Q \sim 0.62$ DT experiments. JET and JT60U achieved the equivalent break-even condition with a D–D plasma, that is, the extrapolated D–T fusion power output would be the same as the heating input power ($Q_{\rm equiv} = 1$).

# 16 Tokamaks

The word 'tokamak' is said to be a contraction of the Russian words for current, vessel, magnet, and coil. Tokamaks are axisymmetric, with the plasma current itself giving rise to the poloidal field essential to the equilibrium of toroidal plasmas. In a tokamak, the toroidal field used to stabilize against MHD instabilities is strong enough to satisfy the Kruskal–Shafranov condition. This characteristic is quite different from that of the reversed field pinch, with its relatively weak toroidal field. There are excellent reviews and textbooks on tokamak experiments [16.1, 16.2], equilibrium [16.3], and diagnostics [16.4, 16.5].

## 16.1 Tokamak Devices

The constructions of the large tokamak devices JET, JT60U and TFTR are shown in Figs. 16.1, 16.2, and 16.3 as typical examples.

The toroidal field coils, equilibrium field coils (also called poloidal field coils, which produce the vertical field and shaping field), ohmic heating coils (primary windings of the current transformer), and vacuum vessel can be seen in the figures. Sometimes the term 'poloidal field coils' means both the equilibrium field coils and the ohmic heating coils. By raising the current of the primary windings of the current transformer (ohmic heating coils), a current is induced in the plasma, which acts as the secondary winding. In the JET device, the current transformer is of the iron-core type. The air-core type of current tranformer is utilized in JT60U and TFTR. The vacuum vessel is usually made of thin stainless steel or inconel so that it has enough electric resistance in the toroidal direction. Therefore the voltage induced by the primary windings can penetrate it.

The thin vacuum vessel is called the liner. Before starting an experiment, the liner is outgassed by baking at a temperature of 150–400°C for a long time under high vacuum. Furthermore, before running an experiment, a plasma is run with a weak toroidal field in order to discharge-clean the wall of the liner. Inside the liner there is a diaphragm made of tungsten, molybdenum, or graphite that limits the plasma size and minimizes the interaction of the plasma with the wall. This diaphragm is called a limiter. Recently a divertor configuration was introduced instead of the limiter. In this case

**Table 16.1.** Tokamak parameters. $R$, $a$, $b$ in [m], $B_t$ in [T], and $I_p$ in [MA]

| | $R$ | $a(\times b)$ | $R/a$ | $B_t$ | $I_p$ | Remarks |
|---|---|---|---|---|---|---|
| T-4 | 1.0 | 0.17 | 5.9 | 5.0 | 0.3 | |
| T-10 | 1.5 | 0.39 | 3.8 | 5.0 | 0.65 | |
| PLT | 1.32 | 0.4 | 3.3 | 3.2 | 0.5 | |
| TFTR | 2.48 | 0.85 | 2.9 | 5.2 | 2.5 | Compact |
| JET | 2.96 | 1.25(×2.1) | 2.4 | 3.45 | 7 | Non-circular |
| JT60U | 3.4 | 1.1(×1.4) | 3.1 | 4.2 | 6 | Non-circular, JT60 upgraded |

the magnetic surface, including the separatrix point, determines the plasma boundary (see Sect. 16.5). A conducting shell surrounds the plasma outside the liner and is used to maintain the positional equilibrium or to stabilize MHD instabilities during the skin time scale. The magnitude of the vertical field is feedback-controlled to keep the plasma at the center of the liner at all times. Many improvements have been made in tokamak devices over the years. The accuracy of the magnetic field is also important to improve the plasma performance in tokamak and other toroidal devices. The parameters of typical tokamak devices are listed in Table 16.1.

Measurements by magnetic probes surrounding the plasma are a simple and useful way to monitor plasma behavior. As the magnetic probes detect MHD fluctuations, they are indispensable in the study of MHD instabilities. These small magnetic coils are called Mirnov coils. The loop voltage $V_L$ and the plasma current $I_p$ can be measured by the magnetic loop and Rogowsky coil, respectively [16.4]. Then the electron temperature can be estimated by the Spitzer formula from the resistivity of the plasma, which can be evaluated using $V_L$ and $I_p$. From (6.11), the poloidal beta ratio $\beta_p$ is given by

$$\beta_p \approx 1 + \frac{2B_\varphi}{B_\omega^2}\langle B_{\varphi V} - B_\varphi \rangle \,, \tag{16.1}$$

where $|B_{\varphi V} - B_\varphi| \ll |B_\varphi|$ and $B_\omega = \mu_0 I_p / 2\pi a$. Since the diamagnetic flux $\delta\Phi$ is

$$\delta\Phi = \pi a^2 \langle B_{\varphi V} - B_\varphi \rangle \,, \tag{16.2}$$

we have

$$\beta_p = \frac{p}{B_\omega^2/2\mu_0} \approx 1 + \frac{8\pi B_\varphi}{\mu_0^2 I_p^2}\delta\Phi \,. \tag{16.3}$$

Therefore measurement of the diamagnetic flux $\delta\Phi$ yields $\beta_p$ and the plasma pressure. Magnetic probes $g_1$, $g_2$ located around the plasma, as shown in Fig. 16.4a, can be used to determine the plasma position. Since the necessary magnitude $B_\perp$ of the vertical field for equilibrium is related to the quantity $\Lambda = \beta_p + l_i/2$, the value of $\Lambda$ can be estimated from $B_\perp$ ($l_i$ is the normalized internal inductance). The fluctuations in the soft X-ray (bremsstrahlung)

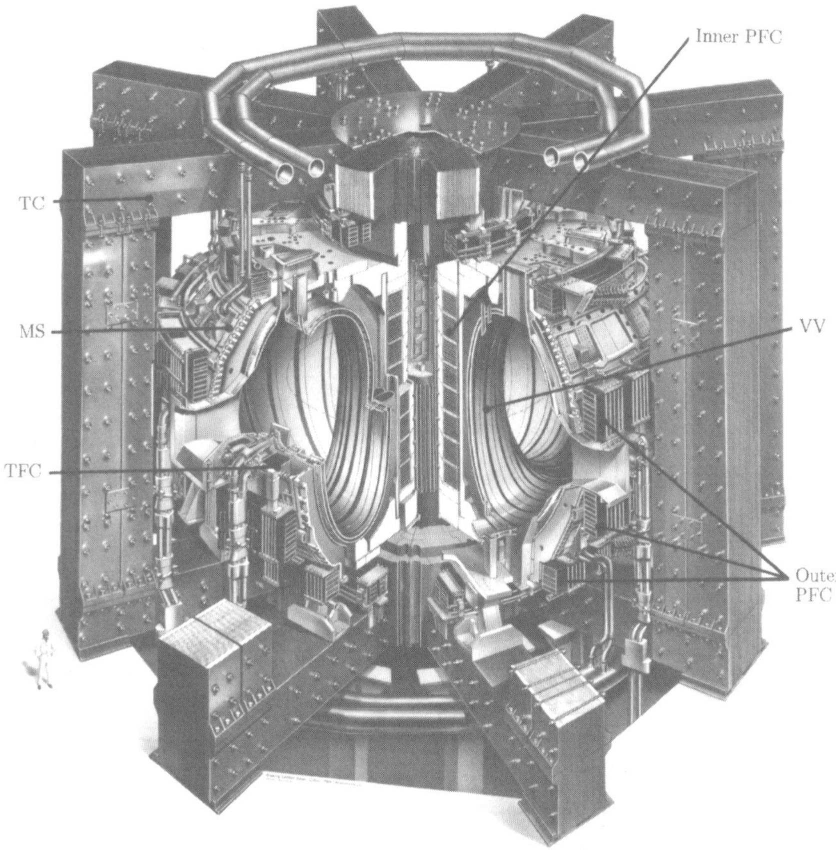


**Fig. 16.1.** Artist's drawing of JET (Joint European torus), JET Joint Undertaking, Abingdon, Oxfordshire, England. The toroidal field coils (TFC) are arranged around the vacuum vessel (VV). The outer poloidal field coils (Outer PFC, equilibrium field coils) and inner poloidal field coils (Inner PFC, ohmic heating coils) are wound in the toroidal direction outside the toroidal field coils (TFC). JET uses an iron-core current transformer (TC). The mechanical structures (MS) support the toroidal field coils against the large torque due to the equilibrium field

signal follow the fluctuations in electron temperature. The fluctuations occur at the rational surfaces $[q_s(r) = 1, 2, \ldots]$. The mode number and direction of propagation can be estimated by arrays of solid-state detectors, as shown in Fig. 16.4b. When the positions of the rational surfaces can be measured, the radial current profile can be estimated for use in studies of MHD stability.

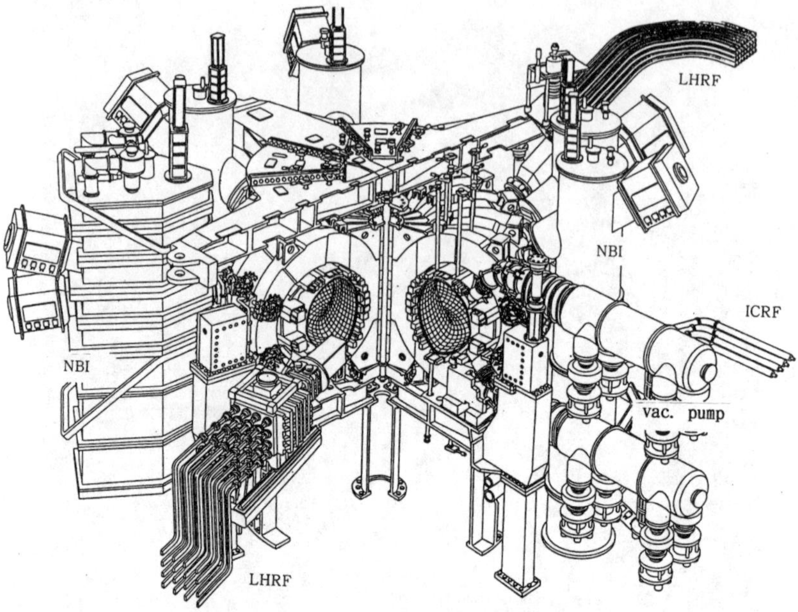


**Fig. 16.2.** Schematic view of JT60U, Japan Atomic Energy Research Institute

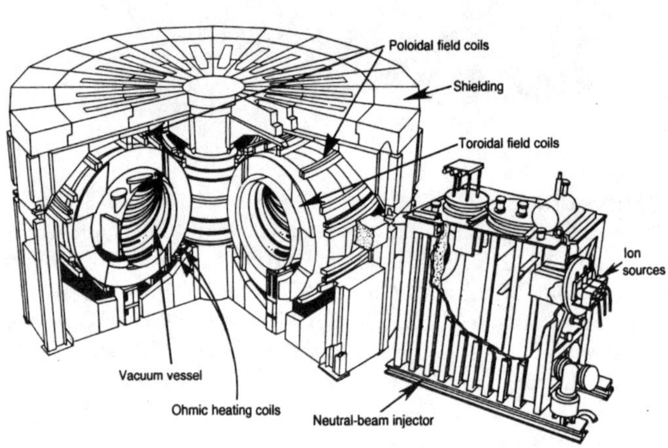


**Fig. 16.3.** Schematic view of TFTR (Tokamak Fusion Test Reactor), Plasma Physics Laboratory, Princeton University

## 16.2 Equilibrium

If the vertical field $B_\perp$ is uniform in space, the equilibrium is neutral with regard to changes in plasma position. When the lines of the vertical field are

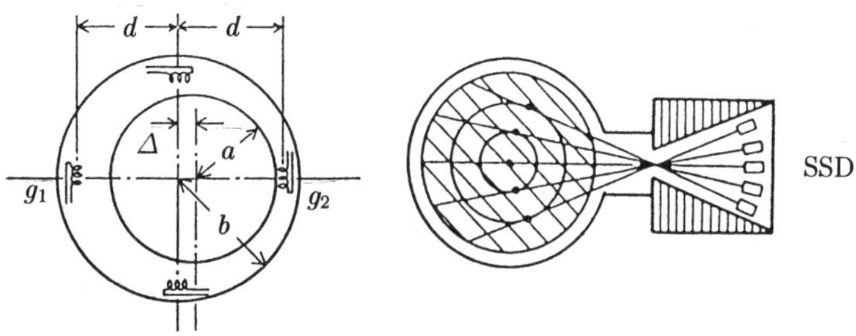


(a) (b)

**Fig. 16.4.** (**a**) Locations of magnetic probes around the plasma ($\Delta$ shown in the figure is negative). (**b**) Array of soft X-ray solid-state detectors. The main contribution of each detector to a signal comes from the emission at the peak temperature along the line of sight of the detector. The fluctuation in the electron temperature at this point can be detected

curved, as shown in Fig. 16.5, the plasma position is stable with regard to up and down motion. The $z$ component $F_z$ of the magnetic force applied to a plasma current ring with mass $M$ is

$$F_z = -2\pi R I_{\rm p} B_R \, .$$

From the relation $(\partial B_R/\partial z) - (\partial B_z/\partial R) = 0$,

$$M\frac{{\rm d}^2 z}{{\rm d}t^2} = -2\pi R I_{\rm p}\frac{\partial B_R}{\partial z} z = 2\pi I_{\rm p} B_z \left(-\frac{R}{B_z}\frac{\partial B_z}{\partial R}\right) z \, .$$

As $I_{\rm p} B_z < 0$, the stability condition for decay index $n$ is

$$n \equiv -\frac{R}{B_z}\frac{\partial B_z}{\partial R} > 0 \, . \tag{16.4}$$

The horizontal component $F_R$ of the magnetic force is

$$M\frac{{\rm d}^2(\Delta R)}{{\rm d}t^2} = F_R = 2\pi R I_{\rm p}(B_z - B_\perp)\Delta R \, .$$

The amount of $B_\perp$ necessary for plasma equilibrium [see (6.30)] is

$$B_\perp = \frac{-\mu_0 I_{\rm p}}{4\pi R}\left(\ln\frac{8R}{a} + \Lambda - \frac{1}{2}\right) , \qquad \Lambda = \frac{l_{\rm i}}{2} + \beta_{\rm p} - 1 \, .$$

When the plasma is ideally conductive, the magnetic flux inside the plasma ring is conserved and

$$\frac{\partial}{\partial R}(L_{\rm p} I_{\rm p}) + 2\pi R B_\perp = 0 \, .$$

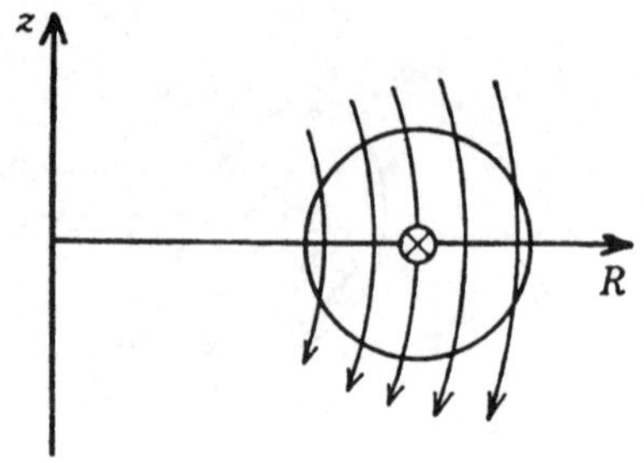


**Fig. 16.5.** Vertical field for plasma equilibrium

Here the self-inductance is $L_{\mathrm{p}} = \mu_0 R\big[\ln(8R/a) + l_{\mathrm{i}}/2 - 2\big]$. Therefore the equation of motion is

$$M\frac{\mathrm{d}^2(\Delta R)}{\mathrm{d}t^2} = 2\pi I_{\mathrm{p}} B_{\perp}\left(\frac{3}{2} - n\right)\Delta R\,,$$

under the assumption $\ln(8R/a) \gg 1$. Then the stability condition for horizontal movement is

$$\frac{3}{2} > n\,. \tag{16.5}$$

The poloidal beta limit of a circular tokamak is $\beta_{\mathrm{p}} = 0.5R/a$, as given by (6.37). The same poloidal beta limit is derived by similar considerations for the elongated tokamak with horizontal radius $a$ and vertical radius $b$. When the length of the circumference along the poloidal direction is denoted by $2\pi aK$ for the elongated plasma and the average poloidal field is $\bar{B}_{\mathrm{p}} = \mu_0 I_{\mathrm{p}}/(2\pi aK)$, the ratio of the poloidal and toroidal fields is $\bar{B}_{\mathrm{p}} = \mu_0 I_{\mathrm{p}}/(2\pi aK)$, where $K$ is given approximately by $K = \big\{\big[1 + (b/a)^2\big]/2\big\}^{1/2}$. The beta limit of an elongated tokamak is therefore

$$\beta \le 0.5K^2\frac{a}{Rq_{\mathrm{I}}^2}\,, \tag{16.6}$$

which is $K^2$ times as large as that of a circular one. In order to elongate the plasma cross-section, the decay index $n$ of the vertical field must be negative, and the elongated plasma is positionally unstable in the up–down motion. Therefore feedback control of the variable horizontal field is necessary to keep the plasma position vertically stable [16.6].

## 16.3 MHD Stability and Density Limit

A possible MHD instability in the low-beta tokamak is kink modes, which were treated in Sect. 8.3. Kink modes can be stabilized by tailoring the current profile and by appropriate choice of the safety factor $q_a$. When the

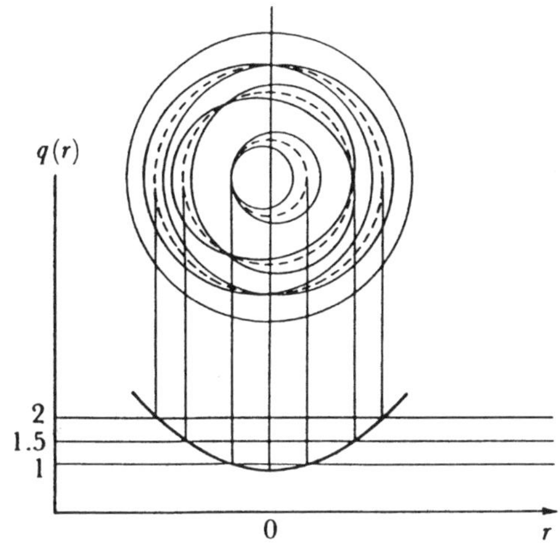


**Fig. 16.6.** Magnetic islands of $m = 1$, $m = 3/2$, $m = 2$ appear at $q(r) = 1$, $3/2$, $2$

plasma pressure is increased, the beta value is limited by the ballooning modes (Sect. 8.5). This instability is a mode localized in the bad curvature region driven by a pressure gradient. The beta limit of the ballooning mode is given by $\beta_{\max} \sim 0.28(a/Rq_a)$ of (8.138). The $\beta$ limit by kink and ballooning modes depends on the radial profile of the plasma current (shear) and the shape of the plasma cross-section. The limit of the average beta, $\beta_c = \langle p \rangle/(B^2/2\mu_0)$, of the optimized condition is derived by MHD simulation codes to be $\beta_c(\%) = \beta_N I_p\,(\mathrm{MA})/a(\mathrm{m})B_t(\mathrm{T})$ ($\beta_N = 2 \sim 3.5$) [16.7, 16.8]. $\beta_{\max}$ of (8.138) is consistent with the result of MHD simulation.

Even if a plasma is ideally MHD stable, tearing modes can be unstable for a finite resistive plasma. When $\Delta'$ is positive at the rational surfaces (see Sect. 9.1) in which the safety factor $q(r)$ is rational $q(r) = 1$, $3/2$, $2$, tearing modes grow and magnetic islands are formed, as shown in Fig. 16.6. When the profile of the plasma current is peaked, the safety factor at the center becomes $q(0) < 1$ and the tearing mode with $m = 1$, $n = 1$ grows at the rational surface $q(r) = 1$. The hot core of the plasma is then pushed out when the reconnection of magnetic surfaces occurs (Fig. 16.7) and the current profile is flattened. The thermal energy in the central hot core is lost in this way [16.2, 16.9]. Since the electron temperature in the central part is higher than in the outer region and the resistance in the central part is smaller, the current profile peaks again and the same process is repeated. This type of phenomenon is called *internal disruption* or *minor disruption*.

The stable operational region of a tokamak with plasma current $I_p$ and density $n_e$ is limited. With Greenward density defined by

$$n_G\ (10^{20}\mathrm{m}^{-3}) \equiv \frac{I_p\ (\mathrm{MA})}{\pi a\ (\mathrm{m})^2}\,, \tag{16.7}$$

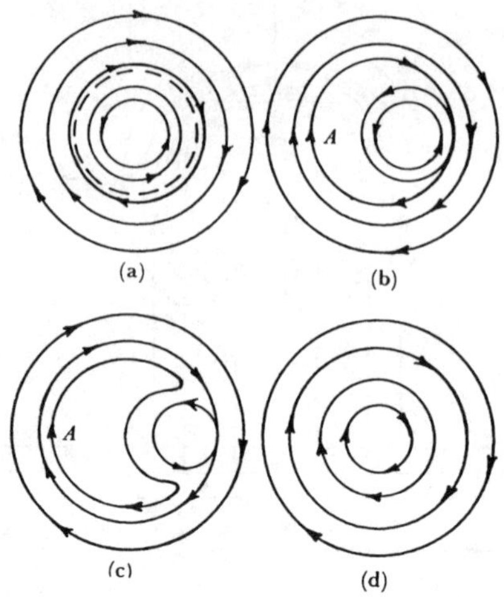


**Fig. 16.7.** The hot core in the center is expelled by the reconnection of magnetic surfaces

an empirical scaling of the normalized Greenward density or Greenward–Hugill–Murakami parameter $N_{\mathrm{GHM}}$

$$N_{\mathrm{GHM}} \equiv \frac{\langle n_{\mathrm{e}} \rangle}{n_{\mathrm{G}}} < 1 \tag{16.8}$$

holds for most tokamak experiments [16.10], where $n_{20}$ is the electron density in units of $10^{20}\mathrm{m}^{-3}$. $N_{\mathrm{GHM}}$ can be expressed in the alternative form [see (16.11)]

$$N_{\mathrm{GHM}} = \frac{0.628}{K^2} \frac{\langle n_{20} \rangle}{B_{\mathrm{t}}(\mathrm{T})/R(\mathrm{m})} q_{\mathrm{I}} \,. \tag{16.9}$$

The upper limit of the electron density depends critically on the plasma wall interaction and tends to increase as the heating power increases, although the scaling $N_{\mathrm{GHM}} < 1$ does not reflect the power dependence. When hydrogen ice pellets are injected into a plasma for fueling from the high field side of ASDEX-U with advanced divertor [16.11], $N_{\mathrm{GHM}}$ goes up to $\sim 1.5$. Therefore there is a possibility of further increasing $N_{\mathrm{GHM}}$. The safety factor $q_a$ at the plasma boundary is $q_a > 3$ in most cases.

Beyond the stable region ($N_{\mathrm{GHM}} < 1$, $1/q_a < 1/3$–$1/2$), a strong instability, called *disruptive instability*, occurs in typical operations. Negative spikes appear in the loop voltage due to the rapid expansion of the current channel (flattened current profile), that is, the rapid reduction of the internal inductance. The thermal energy of the plasma is suddenly lost. The electron temperature drops rapidly and the plasma resistance increases. A positive pulse appears in the loop voltage. Then the plasma discharge is terminated rapidly. In some cases, the time scale of disruption is much faster than the time scale (9.27) predicted by the resistive tearing mode. Possible mechanisms of disruptive instability under discussion are overlapping of magnetic islands of $m = 2/n = 1$ $[q(r) = 2]$ and $m = 3/n = 2$ $[q(r) = 1.5]$, or reconnection of $m = 2/n = 1$, $m = 1/n = 1$ magnetic islands. Reviews of MHD instabilities in tokamak plasmas and plasma transport are given in [16.12–16.15].

## 16.4 Beta Limit of Elongated Plasma

The output power density of nuclear fusion is proportional to $n^2\langle\sigma v\rangle$. Since $\langle\sigma v\rangle$ is proportional to $T_{\mathrm{i}}^2$ in the region near $T_{\mathrm{i}} \sim 10\,\mathrm{keV}$, the fusion output power is proportional to the square of the plasma pressure $p = n\kappa T$. Therefore, the higher the beta ratio $\beta = p/(B^2/2\mu_0)$, the more economical the possible fusion reactor. The average beta of $\langle\beta\rangle \sim 3\%$ was realized by NBI experiments in ISX-B, JFT-2, and PLT. All these tokamaks have a circular plasma cross-section. The theoretical upper limit of the average beta $\beta_{\mathrm{c}}$ of an elongated tokamak plasma due to kink and ballooning instability is [16.7, 16.8]

$$\beta_{\mathrm{c}}(\%) \simeq \beta_{\mathrm{N}} I_{\mathrm{p}}(\mathrm{MA})/a(\mathrm{m})B_{\mathrm{t}}(\mathrm{T}) \,. \tag{16.10}$$

$\beta_{\mathrm{N}}$ is called the Troyon factor or normalized beta ($\beta_{\mathrm{N}} = 2$–$3.5$). Using the definitions

$$\bar{B}_{\mathrm{p}} \equiv \frac{\mu_0 I_{\mathrm{p}}}{2\pi a K}\,, \qquad q_{\mathrm{I}} \equiv K\frac{a}{R}\frac{B_{\mathrm{t}}}{\bar{B}_{\mathrm{p}}}\,, \tag{16.11}$$

the critical beta reduces to

$$\beta_{\mathrm{c}}(\%) = 5\beta_{\mathrm{N}} K^2 \frac{a}{R q_{\mathrm{I}}}\,, \tag{16.12}$$

where $2\pi K a$ is the the length of the circumference of the plasma boundary, $K$ is given approximately by

$$K^2 \simeq \frac{1}{2}(1+\kappa_{\mathrm{s}}^2)\,,$$

and $\kappa_{\mathrm{s}}$ is the ratio of the vertical radius $b$ to the horizontal radius $a$. The safety factor $q_\psi$ at a magnetic surface $\psi$ is given by

$$q_\psi = \frac{1}{2\pi}\oint \frac{B_{\mathrm{t}}}{RB_{\mathrm{p}}}\mathrm{d}l = \frac{1}{2\pi\mathrm{d}\psi}\oint B_{\mathrm{t}}\frac{\mathrm{d}\psi}{RB_{\mathrm{p}}}\mathrm{d}l = \frac{1}{2\pi\mathrm{d}\psi}\oint B_{\mathrm{t}}\mathrm{d}s\mathrm{d}l = \frac{1}{2\pi}\frac{\mathrm{d}\Phi}{\mathrm{d}\psi}\,,$$

where $\Phi$ is the toroidal flux through the magnetic surface $\psi$. It should be noted that $q_{\mathrm{I}}$ differs from $q_\psi$ in the finite aspect ratio. The approximate formula used for the effective safety factor at the plasma boundary is

$$\begin{aligned} q_{\mathrm{eff}} = &\frac{a^2 B}{(\mu_0/2\pi)RI}\frac{1+\kappa_{\mathrm{s}}^2}{2}\left[1+\epsilon^2\left(1+\frac{\bar{\Lambda}^2}{2}\right)\right] \\ &\times\left[1.24 - 0.54\kappa_{\mathrm{s}} + 0.3(\kappa_{\mathrm{s}}^2+\delta^2) + 0.13\delta\right]\,, \end{aligned} \tag{16.13}$$

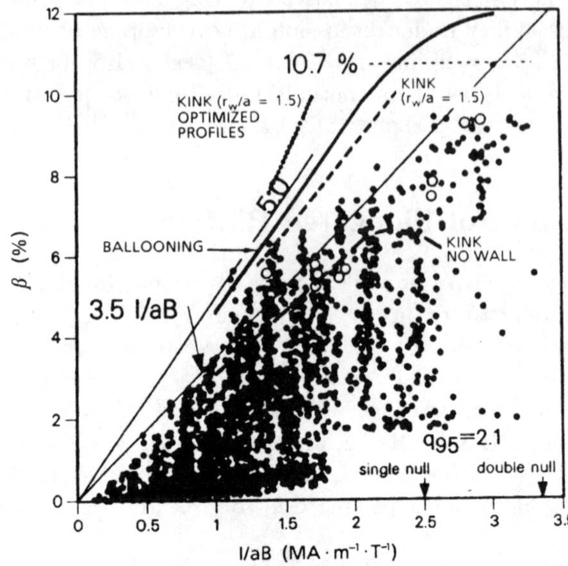


**Fig. 16.8.** Observed beta versus $I/aB$ for DIII-D. Various $\beta$ limit calculations are summarized in the curves with different assumptions on the location of the conducting wall $(r_w/a)$ [16.17]

including the divertor configuration (see Sect. 16.5) [16.16]. The notation is $\epsilon = a/R$, $\bar{\Lambda} \equiv \beta_p + l_i/2$ [see (6.21)] and $\delta = \Delta/a$ is the triangularity of the plasma shape (see Fig. 16.9 and Sect. 16.10).

In the non-circular tokamak DIII-D, $\langle\beta\rangle = 11\%$ was realized in 1990 [16.17], in which $a = 0.45\,\mathrm{m}$, $B_t = 0.75\,\mathrm{T}$, $I_p = 1.29\,\mathrm{MA}$, $I_p/aB_t = 3.1\,\mathrm{MA/Tm}$, $\beta_N \sim 3.6$, $\kappa_s = 2.35$, and $R = 1.43\,\mathrm{m}$. Figure 16.8 shows the experimental data for DIII-D on the observed beta versus $I_p/aB_t$.

## 16.5 Impurity Control, Scrape-Off Layer and Divertor

Radiation power loss $P_{\mathrm{brems}}$ by bremsstrahlung due to electron collisions with ions per unit volume is

$$P_{\mathrm{brems}} = 1.5 \times 10^{-38} Z_{\mathrm{eff}} n_e^2 \left(\frac{\kappa T_e}{e}\right)^{1/2} \quad (\mathrm{W/m^3}) .$$

The loss time due to bremsstrahlung defined by $\tau_{\mathrm{brems}} = (3/2) n_e \kappa T_e / P_{\mathrm{brems}}$ is

$$\tau_{\mathrm{brems}} = 0.16 \frac{1}{Z_{\mathrm{eff}} n_{20}} \left(\frac{\kappa T_e}{e}\right)^{1/2} \quad (\mathrm{s}) ,$$

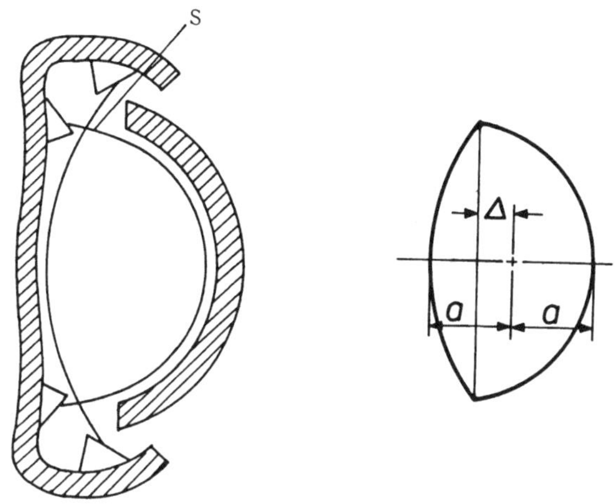


**Fig. 16.9.** Divertor configuration using the separatrix S of the magnetic surface (*left*). Definition of the triangularity $\delta = \Delta/a$ (*right*)

where $n_{20}$ is in units of $10^{20}\,\mathrm{m}^{-3}$ and $\kappa T_{\mathrm{e}}/e$ is in units of eV. When $n_{\mathrm{e}} \sim 10^{20}\,\mathrm{m}^{-3}$ and $\kappa T_{\mathrm{e}} \sim 10\,\mathrm{keV}$, then we have $\tau_{\mathrm{brems}} \sim 16/Z_{\mathrm{eff}}(\mathrm{s})$. Therefore if radiation losses such as bremsstrahlung, recombination radiation, and line spectrum emission are much enhanced by impurity ions, the fusion core plasma cannot be realized even due to radiation losses alone. When the temperature of the plasma increases, the ions from the plasma hit the walls of the vacuum vessel and impurity ions are sputtered. When the sputtered impurities penetrate the plasma, the impurities are highly ionized and yield a large amount of radiation loss, which causes radiation cooling of the plasma. Therefore impurity control is one of the most important subjects in fusion research.

Light impurities such as C and O can be removed by baking and discharge-cleaning the vacuum vessel. The sputtering of heavy atoms (Fe, etc.) of the wall material itself can be avoided by covering the metal wall with carbon tiles. Furthermore a divertor, as shown in Fig. 16.9, is very effective in reducing the plasma–wall interaction. Plasmas in the *scrape-off layer* (SOL) flow at the velocity of sound along the lines of magnetic force just outside the *separatrix* S into the neutralized plates, where the plasmas are neutralized. Even if the material of the neutralized plates is sputtered, the atoms are ionized within the divertor regions near the neutralized plates. Since the thermal velocity of the heavy ions is much smaller than the flow velocity of the plasma (which is the same as the thermal velocity of hydrogen ions), they are unlikely to flow back into the main plasma. In the divertor region the electron temperature of the plasma becomes low because of impurity radiation cooling. Because of pressure equilibrium along the lines of magnetic force,

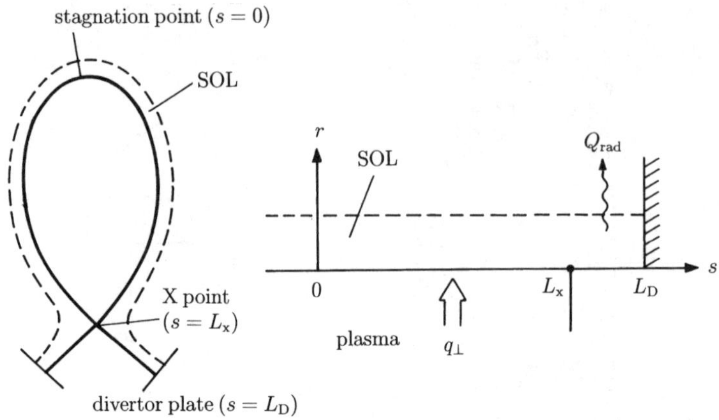


**Fig. 16.10.** *Left*: configuration of the scrape-off layer (SOL) and divertor. (*Right*): coordinates of the slab model

the density in the divertor region near the neutralized plates becomes high. Therefore the velocity of ions from the plasma into the neutralized plates is collisionally damped and sputtering is suppressed. A decrease in the impurity radiation in the main plasma can be observed using a divertor configuration.

However, the scrape-off layer of the divertor is not broad and most of the total energy loss is concentrated in the narrow region of the target divertor plate. The severe heat load to the divertor plate is one of the most critical issues of reactor design. Physical processes in the scrape-off layer and divertor region are being actively investigated both experimentally and theoretically [16.18].

Let us consider thermal transport in the scrape-off layer. It is assumed that thermal transport parallel to the magnetic line of force is dominated by classical electron thermal conduction, whilst thermal transport perpendicular to the magnetic field is anomalous thermal diffusion. We use a slab model, as shown in Fig. 16.10 and omit the Boltzmann constant in front of the temperature. Then we have

$$\nabla q_\parallel + \nabla q_\perp + Q_{\mathrm{rad}} = 0\,, \tag{16.14}$$

$$q_\parallel = -\kappa_{\mathrm{c}}\frac{\partial T_{\mathrm{e}}}{\partial s} = -\kappa_0 T_{\mathrm{e}}^{5/2}\frac{\partial T_{\mathrm{e}}}{\partial s} = -\frac{2}{7}\kappa_0\frac{\partial T_{\mathrm{e}}^{7/2}}{\partial s}\,, \tag{16.15}$$

$$q_\perp = -n\left(\chi_\perp^{\mathrm{e}}\frac{\partial T_{\mathrm{e}}}{\partial r} + \chi_\perp^{\mathrm{i}}\frac{\partial T_{\mathrm{i}}}{\partial r}\right) - \frac{3}{2}D(T_{\mathrm{e}}+T_{\mathrm{i}})\frac{\partial n}{\partial r}\,, \tag{16.16}$$

$$\kappa_{\mathrm{c}} \sim n\lambda_{\mathrm{ei}}^2\nu_{\mathrm{ei}} = 2.8\times 10^3\mathrm{m}^{-1}W(\mathrm{eV})^{-7/2}T_{\mathrm{e}}^{5/2}(\mathrm{eV})^{5/2}\,.$$

Here $q_\parallel$ and $q_\perp$ are heat fluxes in the directions parallel and perpendicular to the magnetic field and $Q_{\mathrm{rad}}$ is the radiation loss. $\kappa_{\mathrm{c}}$ is the heat conductivity,

$\chi_\perp^{\rm e}$, $\chi_\perp^{\rm i}$ are thermal diffusion coefficients, and $D$ is the particle diffusion coefficient. The stagnation point for heat flow is set at $s = 0$ and the X point of the separatrix and divertor plate are set at $s = L_{\rm x}$ and $s = L_{\rm D}$, respectively. Then the boundary conditions at $s = 0$ and $s = L_{\rm D}$ are

$$q_{\parallel 0} = 0 \,, \tag{16.17}$$

$$\begin{aligned} q_{\parallel \rm D} &= \gamma T_{\rm D} n_{\rm D} u_{\rm D} + \frac{1}{2} m_{\rm i} u_{\rm D}^2 n_{\rm D} u_{\rm D} + \xi n_{\rm D} u_{\rm D} \\ &= n_{\rm D} M_{\rm D} c_{\rm s} \left[ (\gamma + M_{\rm D}^2) T_{\rm D} + \xi \right] \,, \end{aligned} \tag{16.18}$$

where $u_{\rm D}$ is the flow velocity of the plasma at the divertor plate and $M_{\rm D}$ is the Mach number $M_{\rm D} = u_{\rm D}/c_{\rm s}$. The sheath energy transfer coefficient is $\gamma \approx 7$ and the ionization energy is $\xi \approx 20 \sim 27\,{\rm eV}$. The sound velocity is $c_{\rm s} = \tilde{c}_{\rm s} T_{\rm D}^{1/2}$, $\tilde{c}_{\rm s} = 0.98(2/A_{\rm i})^{1/2} 10^4\,{\rm ms}^{-1}({\rm eV})^{-1/2}$, where $A_{\rm i}$ is the ion atomic mass. The first and second terms of (16.18) are the power flux into the sheath and the third term is the power consumed within the recycling process. The equations of particles and momentum along the magnetic lines of force are

$$\frac{\partial (nu)}{\partial s} = S_{\rm i} - S_{\rm cx,r} - \nabla_\perp (n u_\perp) \approx S_{\rm i} - S_{\rm cx,r} \,, \tag{16.19}$$

$$mnu \frac{\partial u}{\partial s} = -\frac{\partial p}{\partial s} - mu S_{\rm m} \,, \tag{16.20}$$

where $S_{\rm m} = n n_0 \langle \sigma v \rangle_{\rm m}$ is the loss of momentum of the plasma flow by collision with neutrals, $S_{\rm i} = n n_0 \langle \sigma v \rangle_{\rm i}$ is the ionization term, and $S_{\rm cx,r} = n n_0 \langle \sigma v \rangle_{\rm cx,r}$ is the ion loss by charge exchange and radiation recombination. Equations (16.19) and (16.20) reduce to

$$\frac{\partial (nmu^2 + p)}{\partial s} = -mu(S_{\rm m} + S_{\rm cx,r}) + mu S_{\rm i} \,. \tag{16.21}$$

The flow velocities at $s = 0$ and $s = L_{\rm D}$ are $u_0 = 0$ and $u_{\rm D} = M_{\rm D} c_{\rm s}$, $M_{\rm D} \approx 1$, respectively. Equations (16.14) and (16.15) and the boundary conditions (16.17) and (16.18) reduce to

$$\frac{2\kappa_0}{7} \frac{\partial^2}{\partial s^2} T_{\rm e}^{7/2} = \nabla_\perp q_\perp + Q_{\rm rad} \,, \tag{16.22}$$

$$\frac{2\kappa_0}{7} \left[ T_{\rm e}^{7/2}(s) - T_{\rm eD}^{7/2} \right] = \int_{L_{\rm D}}^{s} {\rm d}s' \int_0^{s'} (\nabla_\perp q_\perp + Q_{\rm rad}) {\rm d}s'' \,. \tag{16.23}$$

When $\nabla_\perp q_\perp = {\rm const.}$, $Q_{\rm rad} = 0$ in $0 < s < L_{\rm x}$ and $\nabla_\perp q = 0$, $Q_{\rm rad} = {\rm const.}$ in $L_{\rm x} < s < L_{\rm D}$, we have

$$\frac{2\kappa_0}{7} \left[ T_{\rm e}^{7/2}(s) - T_{\rm eD}^{7/2} \right] = 0.5(-\nabla_\perp q_\perp)(2 L_{\rm x} L_{\rm D} - L_{\rm x}^2 - s^2) + 0.5 Q_{\rm rad} (L_{\rm D} - L_{\rm x})^2$$

for $(0 < s < L_{\rm x})$. When the radiation term is negligible, $T_{\rm e0} \equiv T_{\rm e}(0)$ becomes

$$T_{\rm e0}^{7/2} = T_{\rm eD}^{7/2} + \frac{7}{4\kappa_0}\left(\frac{2L_{\rm D}}{L_{\rm x}} - 1\right)(-\nabla_\perp q_\perp)L_{\rm x}^2 \,.$$

If $T_{\rm eD} < 0.5T_{\rm e0}$ and $L_{\rm D} - L_{\rm x} \ll L_{\rm x}$, we have

$$T_{\rm e0} \approx 1.17\left[\frac{(-\nabla_\perp q_\perp)L_{\rm x}^2}{\kappa_0}\right]^{2/7} = 1.17\left(\frac{q_\perp L_{\rm x}^2}{\kappa_0 \lambda_q}\right)^{2/7} , \qquad (16.24)$$

where $1/\lambda_q \equiv -\nabla_\perp q_\perp / q_\perp$. When the scale lengths of the temperature and density gradients are $\lambda_T$ and $\lambda_n$ respectively, where $T(r) = T\exp(-r/\lambda_T)$, $n(r) = n\exp(-r/\lambda_n)$, and assuming $\chi_\perp^{\rm i} \ll \chi_\perp^{\rm e}$ and $D \sim \chi_\perp^{\rm e}$, equation (16.16) becomes

$$q_\perp = n\chi_\perp^{\rm e}\frac{T_{\rm e}}{\lambda_T}\left[1 + \frac{3}{2}\left(1 + \frac{T_{\rm i}}{T_{\rm e}}\right)\frac{\lambda_T}{\lambda_n}\right] . \qquad (16.25)$$

Consequently, if $\chi^{\rm e}$ is known as a function $\chi^{\rm e}(T_{\rm e}, n, B)$, then $\lambda_T$ is given as $\lambda_T(T_{\rm e}, n, B, q_\perp)$.

Let us consider the relations between $n_{\rm s}$, $T_{\rm es}$, and $T_{\rm is}$ at the stagnation point $s = 0$ and $n_{\rm D}$, $T_{\rm D}$ at the divertor plate $s = L_{\rm D}$. The momentum flux in the divertor region decreases due to collisions with neutrals, charge exchange and ionization, and becomes smaller than at the stagnation point:

$$f_p \equiv \frac{2(1 + M_{\rm D}^2)n_{\rm D}T_{\rm D}}{n_{\rm s}(T_{\rm es} + T_{\rm is})} < 1 \,. \qquad (16.26)$$

The power flux to the divertor plate is reduced by radiation loss from the power flux $q_\perp L_{\rm x}$ into the scrape-off layer through the separatrix with length $L_{\rm x}$:

$$\int_0^\infty q_\parallel {\rm d}r = (1 - f_{\rm rad})q_\perp L_{\rm x} \,, \qquad (16.27)$$

where $f_{\rm rad}$ is the fraction of radiation loss. Equations (16.27) and (16.18) reduce to

$$M_{\rm D} n_{\rm D} \tilde{c}_{\rm s} T_{\rm D}^{1/2}\left[\frac{(\gamma + M_{\rm D}^2)T_{\rm D}}{3/2\lambda_T + 1/\lambda_n} + \frac{\xi}{1/(2\lambda_T) + 1/\lambda_n}\right] = (1 - f_{\rm rad})q_\perp L_{\rm x} \,,$$

that is,

$$(1 - f_{\rm rad})q_\perp L_{\rm x} = \frac{\tilde{c}_{\rm s} f_p \lambda_T}{1.5 + \lambda_T/\lambda_n} n_{\rm s} \frac{T_{\rm es} + T_{\rm is}}{2} G(T_{\rm D}) \,, \qquad (16.28)$$

$$G(T_{\rm D}) \equiv \frac{M_{\rm D}}{1 + M_{\rm D}^2}(\gamma + M_{\rm D})T_{\rm D}^{1/2}\left(1 + \frac{1}{\gamma + M_{\rm D}}\frac{\bar{\xi}}{T_{\rm D}}\right) . \qquad (16.29)$$

The curve of $G(T_{\rm D})$ as a function of $T_{\rm D}$ is shown in Fig. 16.11 and $G(T_{\rm D})$ has a minimum at $T_{\rm D} = \bar{\xi}/(\gamma + M_{\rm D}^2)$. When $M_{\rm D} \approx 1$, $\gamma \approx 7$, $\xi = 24\,{\rm eV}$, $G(T_{\rm D})$ is

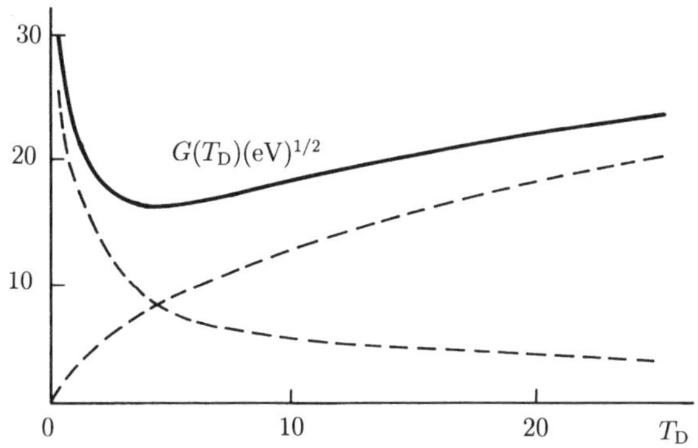


**Fig. 16.11.** Dependence of $G(T_{\rm D})\,({\rm eV})^{1/2}$ on $T_{\rm D}({\rm eV})$

$$G_{\rm D} = 4T_{\rm D}^{1/2}\left(1 + \frac{4.5}{T_{\rm D}}\right) .$$

$G(T_{\rm D})$ is roughly proportional to $T_{\rm D}^{1/2}$ when $T_{\rm D} > 15\,{\rm eV}$ in this case. Since $T_{\rm es}$ depends on $n_{\rm s}$ through $\lambda_q^{-2/7}$, as can be seen from (16.24), the dependence of $T_{\rm es}$ on $n_{\rm s}$ is very weak. From (16.28) and (16.26), we have roughly the following relations:

$$T_{\rm D} \propto n_{\rm s}^{-2} , \qquad n_{\rm D} \propto n_{\rm s}^{3} , \qquad (16.30)$$

and the density $n_{\rm D}$ at the divertor increases non-linearly with the density $n_{\rm s}$ of the upstream scrape-off layer.

When the upstream density $n_{\rm s}$ increases while keeping the left-hand side of (16.28) constant, the solution $T_{\rm D}$ of (16.28) cannot exist beyond a threshold density, since $G(T_{\rm D})$ has the minimum value (Fig. 16.11). This is related to the phenomenon of detached plasma above a threshold of upstream density [16.18].

The heat load $\phi_{\rm D}$ of the divertor normal to the magnetic flux surface is given by

$$\phi_{\rm D} \approx \frac{(1 - f_{\rm rad})P_{\rm sep}}{2\pi R 2\lambda_{\phi D}} = (1 - f_{\rm rad})\pi K \frac{a}{\lambda_T} q_\perp \left(1.5 + \frac{\lambda_T}{\lambda_n}\right) \frac{B_{\theta {\rm D}}}{B_\theta} , \qquad (16.31)$$

where $P_{\rm sep}$ is the total power flux across the separatrix surface and $\lambda_{\phi D}$ is the radial width of heat flux at the divertor plate, i.e.,

$$P_{\rm sep} = 2\pi a K 2\pi R q_\perp , \qquad \lambda_{\phi D} = \lambda_T \frac{1}{1.5 + \lambda_T/\lambda_n} \frac{B_\theta}{B_{\theta {\rm D}}} .$$

The term $B_\theta / B_{\theta {\rm D}} = 2 \sim 3$ is the ratio of separations of magnetic flux surfaces at the stagnation point and the divertor plate. If the divertor plate

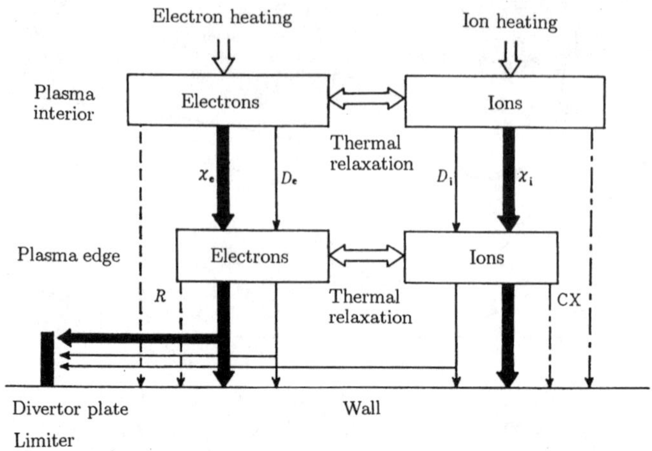


**Fig. 16.12.** Energy flow of ions and electrons in a plasma. *Bold arrows*: thermal conduction ($\chi$). *White arrows*: convective loss ($D$). *Dashed arrows*: radiation loss ($R$). *Dot-dashed arrows*: charge exchange loss (CX)

is inclined at an angle $\alpha$ relative to the magnetic flux surface, the heat load of the inclined divertor plate becomes $\sin\alpha$ times that of the divertor normal to the magnetic flux surface.

## 16.6 Confinement Scaling of L Mode

The energy flow of ions and electrons inside the plasma is shown in Fig. 16.12. If we denote the heating power into the electrons per unit volume by $P_{\mathrm{he}}$ and the radiation loss and the energy relaxation of electrons with ions by $R$ and $P_{\mathrm{ei}}$, respectively, then the time derivative of the electron thermal energy per unit volume is given by

$$\frac{\mathrm{d}}{\mathrm{d}t}\left(\frac{3}{2}n_{\mathrm{e}}\kappa T_{\mathrm{e}}\right) = P_{\mathrm{he}} - R - P_{\mathrm{ei}} + \frac{1}{r}\frac{\partial}{\partial r}r\left(\chi_{\mathrm{e}}\frac{\partial \kappa T_{\mathrm{e}}}{\partial r} + D_{\mathrm{e}}\frac{3}{2}\kappa T_{\mathrm{e}}\frac{\partial n_{\mathrm{e}}}{\partial r}\right),$$

where $\chi_{\mathrm{e}}$ is the electron thermal conductivity and $D_{\mathrm{e}}$ is the electron diffusion coefficient. Concerning the ions, the same relation is derived, but instead of the radiation loss, the charge exchange loss $L_{\mathrm{ex}}$ of ions with neutrals must be taken into account, whence

$$\frac{\mathrm{d}}{\mathrm{d}t}\left(\frac{3}{2}n_{\mathrm{i}}\kappa T_{\mathrm{i}}\right) = P_{\mathrm{hi}} - L_{\mathrm{cx}} + P_{\mathrm{ei}} + \frac{1}{r}\frac{\partial}{\partial r}r\left(\chi_{\mathrm{i}}\frac{\partial \kappa T_{\mathrm{i}}}{\partial r} + D_{\mathrm{i}}\frac{3}{2}\kappa T_{\mathrm{i}}\frac{\partial n_{\mathrm{i}}}{\partial r}\right).$$

The experimental results for ohmic heating and heating by neutral beam injection can be explained by classical processes. The efficiency of wave heating can be estimated fairly accurately by theoretical analysis. Radiation and

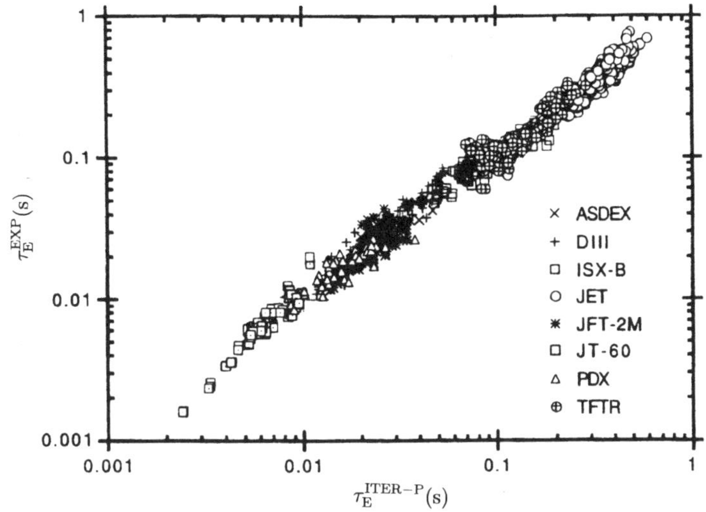


**Fig. 16.13.** Comparison of confinement scaling $\tau_E^{ITER-P}$ with experimental data on the energy confinement time $\tau_E^{EXP}$ for the L mode [16.20]

charge exchange loss are classical processes. In order to evaluate the energy balance of the plasma experimentally, the fundamental quantities $n_e(r,t)$, $T_i(r,t)$, $T_e(r,t)$, etc., must be be measured. According to the many experimental results, the energy relaxation between ions and electrons is classical, and the observed ion thermal conductivities in some cases are around 2–3 times the neoclassical thermal conductivity,

$$\chi_{i,nc} = n_i f(q,\varepsilon) q^2 (\rho_{\Omega i})^2 \nu_{ii} \ ,$$

where $f = 1$ in the Pfirsch–Schlüter region and $f = \epsilon_t^{-3/2}$ in the banana region, and the observed ion thermal conductivities in some other cases are anomalous. The electron thermal conduction estimated from the experimental results is always anomalous and is much larger than its neoclassical counterpart (by more than one order of magnitude). In most cases the energy confinement time of the plasma is determined mainly by electron thermal conduction loss. The energy confinement time $\tau_E$ is defined by

$$\tau_E \equiv \frac{\int (3/2)(n_e \kappa T_e + n_i \kappa T_i) dV}{P_{in}} \ .$$

The energy confinement time $\tau_{OH}$ of an ohmically heated plasma is well described by Alcator (neo-Alcator) scaling as follows (units are $10^{20}\,\mathrm{m}^{-3}$):

$$\tau_{OH}(\mathrm{s}) = 0.103 q^{0.5} \bar{n}_{e20} a^{1.04} R^{2.04} \ .$$

However, the linear dependence of $\tau_{OH}$ on the average electron density $\bar{n}_e$ deviates in the high-density region $n_e > 2.5 \times 10^{20}\,\mathrm{m}^{-3}$ and $\tau_{OH}$ tends to saturate. When the plasma is heated by high-power NBI or wave heating, the energy confinement time degrades as the heating power increases. Kaye and Goldston examined many experimental results for NBI heated plasmas and derived the so-called Kaye–Goldston scaling on the energy confinement time [16.19], i.e.,

$$\tau_{\mathrm{E}} = \left(\frac{1}{\tau_{\mathrm{OH}}^2} + \frac{1}{\tau_{\mathrm{AUX}}^2}\right)^{-1/2}, \quad \tau_{\mathrm{AUX}}(\mathrm{s}) = 0.037\kappa_{\mathrm{s}}^{0.5} I_{\mathrm{p}} P_{\mathrm{tot}}^{-0.5} a^{-0.37} R^{1.75}, \tag{16.32}$$

where the units are MA, MW, and m, and $\kappa_{\mathrm{s}}$ is the elongation ratio of non-circularity and $P_{\mathrm{tot}}$ is the total heating power in MW.

The ITER team assembled data from larger and more recent experiments. Analysis of the data base of L mode experiments (see next section) led to the proposal of the following ITER-P scaling [16.20],

$$\tau_{\mathrm{E}}^{\mathrm{ITER-P}}(\mathrm{s}) = 0.048 I_{\mathrm{p}}^{0.85} R^{1.2} a^{0.3} \bar{n}_{20}^{0.1} B^{0.2} \left(\frac{A_{\mathrm{i}}\kappa_{\mathrm{s}}}{P}\right)^{1/2}, \tag{16.33}$$

where units are MA, m, T, MW, and the units of $\bar{n}_{20}$ are $10^{20}\,\mathrm{m}^{-3}$. $P$ is the heating power corrected for radiation $P_{\mathrm{R}}$ ($P = P_{\mathrm{tot}} - P_{\mathrm{R}}$). A comparison of confinement scaling $\tau_{\mathrm{E}}^{\mathrm{ITER-P}}$ with experimental data for the L mode is presented in Fig. 16.13. For burning plasmas, the heating power is roughly equal to $\alpha$ particle fusion output power $P_\alpha \approx 0.04 n_{\mathrm{DT20}}^2 T^2 A a^3 \kappa_{\mathrm{s}}$ (MW, $10^{20}\mathrm{m}^{-3}$, keV, m) at around $T \sim 10\,\mathrm{keV}$ (see Sect. 16.8). It is interesting to note that $n_{\mathrm{DT}} T \tau_{\mathrm{E}}$ depends mainly only on the product of $A I_{\mathrm{p}}$ for the Goldston and L mode scalings ($A = R/a$ is the aspect ratio).

## 16.7 H Mode and Improved Confinement Modes

An improved confinement state, the so-called *H mode*, was found in the ASDEX [16.21, 16.22] experiments with divertor configuration. When the NBI heating power is larger than a threshold value in the divertor configuration, the $\mathrm{D}_\alpha$ line of deuterium (atom flux) in the edge region of the deuterium plasma decreases suddenly (time scale 100 μs) during discharge, and recycling of deuterium atoms near the boundary decreases. At the same time there is a marked change in the edge radial electric field $E_r$ (toward the negative). Furthermore, the electron density and the thermal energy density increase, and the energy confinement time of NBI heated plasma is improved by a factor of about 2. The H mode was observed in PDX, JFT-2, DIII-D, JET, JT60U, and so on. The confinement state following Kaye–Goldston scaling is called the *L mode*. In the H mode, the gradients of electron temperature and electron density become steep just inside the plasma boundary determined by the separatrix. In the spontaneous H mode, $E_r$ becomes more

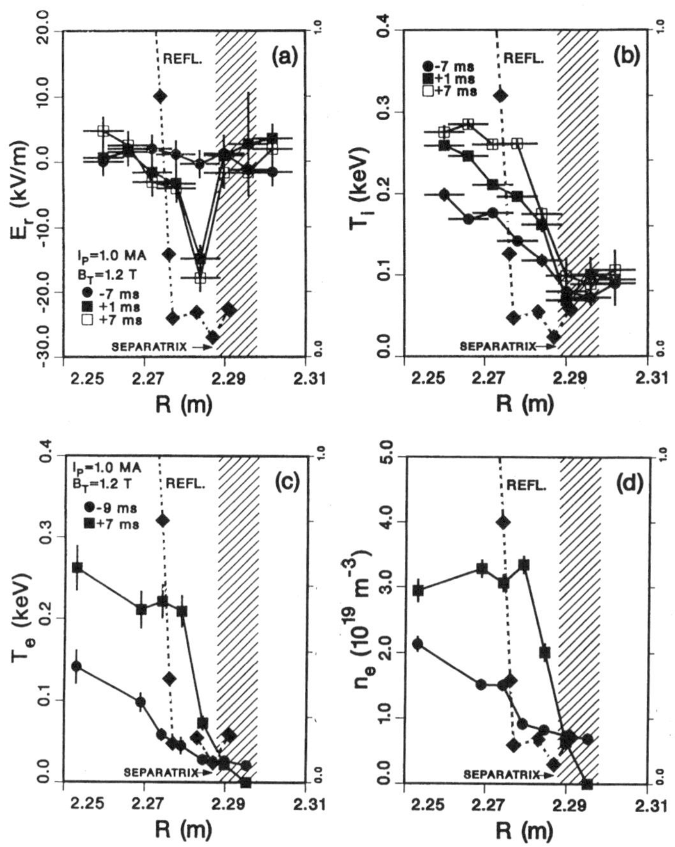


**Fig. 16.14.** Plots of various edge plasma profiles at times spanning the L–H transition in DIII-D. (**a**) $E_r$ profile. (**b**) Profiles of the ion temperature measured by CVII charge exchange recombination spectroscopy. (**c**) and (**d**) Profiles of electron temperature and electron density measured by Thomson scattering [16.24]

negative (inward) (see Fig. 16.14) [16.23, 16.24]. Theoretical and experimental studies on the L–H transition or bifurcation have been actively carried out [16.25, 16.26]. The radial electric field near the plasma boundary is driven by several mechanisms such as momentum injection due to NBI, or ion orbit loss near the plasma boundary, or non-ambipolar flux.

The radial electric field causes plasma rotation with velocity $v_\theta = -E_r/B$ in the poloidal direction and velocity $v_\phi = -(E_r/B)(B_\theta/B)$ in the toroidal direction. If a gradient of $E_r$ exists, sheared poloidal rotation and sheared toroidal rotation are generated. The importance of sheared flow for suppression of edge turbulence and for improved confinement was pointed out in [16.27].

Let us consider the following fluid model:

$$\left[\frac{\partial}{\partial t} + (\boldsymbol{v}_0 + \tilde{\boldsymbol{v}})\cdot\boldsymbol{\nabla} + L_{\rm d}\right]\tilde{\xi} = \tilde{s} ,$$

where $\tilde{\xi}$ is the fluctuating field, $\boldsymbol{v}_0$ is taken to be the equilibrium $\boldsymbol{E}\times\boldsymbol{B}$ flow, $\tilde{s}$ represents a driving source of the turbulence, and $L_{\rm d}$ is an operator responsible for dissipation of turbulence. The mutual correlation function $\langle\tilde{\xi}(1)\tilde{\xi}(2)\rangle$ of the fluctuating field $\tilde{\xi}(1)$ at a point 1 and $\tilde{\xi}(2)$ at a point 2 is given by [16.28]

$$\left[\frac{\partial}{\partial t} + (v'_\theta - v_\theta/r_+)r_+\frac{\partial}{\partial y_-} - \frac{\partial}{\partial r_+}D(r_+,y_-)\frac{\partial}{\partial r_+} + L_{\rm d}\right]\langle\tilde{\xi}(1)\tilde{\xi}(2)\rangle = T , \tag{16.34}$$

where $D$ is the radial diffusion coefficient of turbulence, $T$ is the driving term and $r_+ = (r_1+r_2)/2$, $\theta_- = \theta_1 - \theta_2$, $y_- = r_+\theta_-$. The decorrelation time $\tau_{\rm d}$ in the poloidal direction is the time in which the relative poloidal displacement between point 1 and point 2 due to sheared flow becomes the space correlation length of the turbulence $k_{0k}^{-1}$, that is,

$$k_{0k}\delta y \sim 1 , \qquad \delta y = v'_\theta(\Delta r)\tau_{\rm d} , \qquad \tau_{\rm d} = \frac{1}{v'_\theta \Delta r k_{0k}} .$$

The decorrelation rate $\omega_{\rm s}$ in the poloidal direction is

$$\omega_{\rm s} = \frac{1}{\tau_{\rm d}} = (\Delta r k_{0k})v'_\theta .$$

When $\Delta r$ is the radial correlation length of the turbulence, the radial decorrelation rate $\Delta\omega_{\rm t}$ is given by

$$\Delta\omega_{\rm t} = \frac{D}{(\Delta r)^2} .$$

Since there is strong mutual interaction between radial and poloidal decorrelation processes, the decorrelation rate $1/\tau_{\rm corr}$ becomes a hybrid of two decorrelation rates, that is,

$$\frac{1}{\tau_{\rm corr}} = (\omega_{\rm s}^2\Delta\omega_{\rm t})^{1/3} = \left(\frac{\omega_{\rm s}}{\Delta\omega_{\rm t}}\right)^{2/3}\Delta\omega_{\rm t} . \tag{16.35}$$

The decorrelation rate $1/\tau_{\rm corr}$ becomes $(\omega_{\rm s}/\Delta\omega_{\rm t})^{2/3}$ times as large as $\Delta\omega_{\rm t}$, where $\Delta\omega_{\rm t}$ is the decorrelation rate of the turbulence in the case of shearless flow. Since the saturation level of the fluctuating field $\tilde{\xi}$ is

$$|\tilde{\xi}|^2 \sim T \times \tau_{\rm corr} ,$$

the saturation level of the fluctuating field reduces to

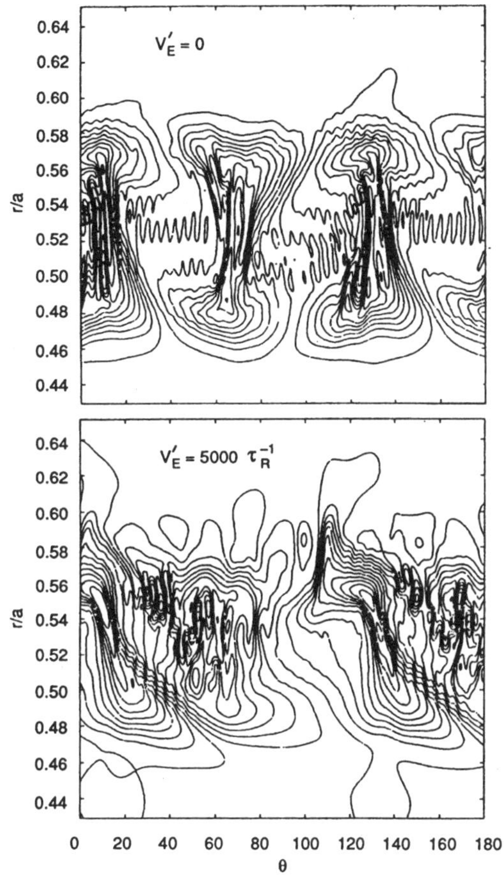


**Fig. 16.15.** Snapshot of the equidensity contours for shearless (*top*) and strongly sheared (*bottom*) flows [16.27]. The *vertical axis* represents the radial coordinate $r/a$ and the *horizontal axis* represents the poloidal angle $\theta$ (in degrees)

$$\frac{|\tilde{\xi}|^2}{|\tilde{\xi}_0|^2} \sim \left(\frac{\Delta\omega_{\rm t}}{\omega_{\rm s}}\right)^{2/3} \sim \left[\frac{1}{({\rm d}v_\theta/{\rm d}r)t_0}\right]^{2/3} \frac{1}{(k_{0y}\Delta r)^2} , \quad t_0^{-1} \equiv \langle k_{0y}^2\rangle D ,$$

where $|\tilde{\xi}_0|$ is the level in the case of shearless flow. The effect of sheared flow on the saturated resistive pressure-gradient-driven turbulence is shown in Fig. 16.15. The coupling between poloidal and radial decorrelation in shearing fluctuation is evident in this figure. Since the thermal diffusion coefficient is proportional to $|\tilde{\xi}|^2$, the thermal diffusion is reduced, i.e., a thermal barrier is formed near the plasma edge.

Theoretical studies on H mode physics are being actively pursued. In addition to the standard H mode as observed in ASDEX and elsewhere, other types of improved confinement modes have been observed. In the TFTR experiment [16.29], outgassing of deuterium from the wall and the carbon limiter located on the inner (high-field) side of the vacuum torus was extensively carried out before the experiments. Then balanced neutral beam injections of co-injection (beam direction parallel to the plasma current) and counter-injection (beam direction opposite to that of co-injection) were applied to the deuterium plasma, and an improved confinement 'supershot' was observed. In supershot, the electron density profile is strongly peaked $[n_{\rm e}(0)/\langle n_{\rm e}\rangle = 2.5\text{–}3]$.

In the DIII-D experiment, the VH mode [16.30] was observed, in which the region of strong radial electric field was expanded from the plasma edge to the plasma interior ($r/a \sim 0.6$), and $\tau_{\rm E}/\tau_{\rm E}^{\rm ITER-P}$ becomes 3.6.

In the JT60U experiment, the high beta-poloidal H mode [16.31] was observed, in which $\beta_{\rm p}$ was high (1.2–1.6) and the density profile was peaked $[n_{\rm e}(0)/\langle n_{\rm e}\rangle = 2.1\text{–}2.4]$. Furthermore the edge thermal barrier of the H mode was formed.

Hinton et al. [16.32] pointed out that the peaked pressure and density profiles induce a gradient in the radial electric field. From the radial component of the equation of motion (5.7) of the ion fluid or (5.28), we have

$$E_r \simeq B_{\rm p}u_{\rm t} - B_{\rm t}u_{\rm p} + \frac{1}{en_{\rm i}}\frac{{\rm d}p_{\rm i}}{{\rm d}r} \,. \tag{16.36}$$

The derivative of $E_r$ with respect to $r$ is

$$\frac{{\rm d}E_r}{{\rm d}r} \sim -\frac{1}{en_{\rm i}^2}\frac{{\rm d}n_{\rm i}}{{\rm d}r}\frac{{\rm d}p_{\rm i}}{{\rm d}r} \,,$$

since the contribution from the other terms is small in typical experimental conditions for the H mode.

Recently, a high performance mode of negative magnetic shear configuration has been demonstrated in DIII-D, TFTR, JT60U, JET and Tore Supra [16.33]. As described in Sect. 8.5, the ballooning mode is stable in the negative shear region:

$$S = \frac{r}{q}\frac{{\rm d}q}{{\rm d}r} < 0 \,. \tag{16.37}$$

An example of radial temperature and density profiles and $q$ profile of JT60U is shown in Fig. 16.16. Combining the central heating and magnetic negative shear, steep gradients in temperature and density appear at around the $q$ minimum point. This internal transport barrier is formed by the effects of the negative magnetic shear (see Sects. 7.3 and 14.6) and $\boldsymbol{E}\times\boldsymbol{B}$ flow shear. As a measure of the high performance of the improved confinement mode, the ratio of observed energy confinement time $\tau_{\rm E}^{\rm EXP}$ to ITER-P scaling $\tau_{\rm E}^{\rm ITER-P}$, known as the $H_{\rm L}$ factor, is widely used:

$$H_{\rm L} \equiv \frac{\tau_{\rm E}^{\rm EXP}}{\tau_{\rm E}^{\rm ITER-P}} \,. \tag{16.38}$$

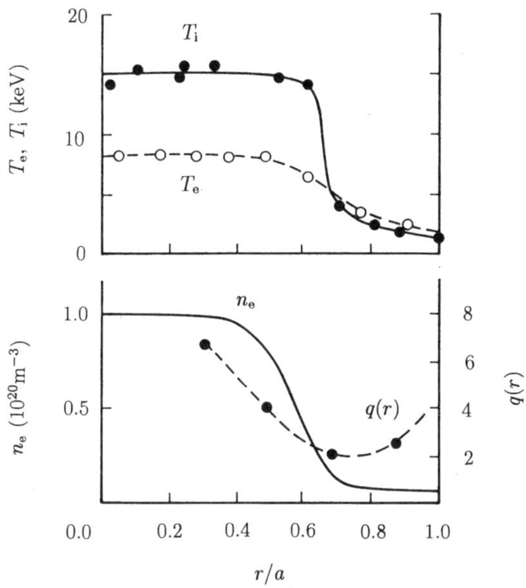


**Fig. 16.16.** Radial profiles of ion and electron temperatures, and density and $q$ profiles in the negative magnetic shear configuration of JT60U

Observed $H_L$ factors are in the range 2–3.

The ITER H mode database working group assembled standard experimental data for the H mode from ASDEX, ASDEX-U, DIII-D, JET, JFT-2M, PDX, PBX, Alcator C-Mod, and so on. Results of regression analysis of H mode experiments led to the following thermal energy confinement time [16.34]:

$$\tau_{\rm E,th}^{\rm IPB98y2} = 0.0562 I_{\rm p}^{0.98} B_{\rm t}^{0.15} P^{-0.69} M_{\rm i}^{0.19} R^{1.97} \bar{n}_{\rm e19}^{0.41} \epsilon^{0.58} \kappa^{0.78} \,, \qquad (16.39)$$

where units of s, MA, T, MW, amu, m, $10^{19}$m$^{-3}$ are used and the total heating power corrected for shine-through of NBI heating, orbit loss and charge exchange loss, less the time derivative of stored energy. This scaling is used when edge-localized-modes (ELM) exist. The heating power threshold scaling $P_{\rm LH}$, which defines the lowest boundary of the H mode operating window, is

$$P_{\rm LH} = 2.84 M_{\rm i}^{-1} B_{\rm t}^{0.82} \bar{n}_{\rm e20}^{0.58} R^{1.00} a^{0.81} \,. \qquad (16.40)$$

In most hot plasma experiments, neutral beam injections are used to heat the plasma. With improved confinement mode operations, such as H mode, supershot and high $\beta_{\rm p}$ mode in large tokamaks, fusion grade plasmas are produced by neutral beam injection. The plasma parameters of typical shots of JET [16.35], JT60U [16.31] and TFTR [16.29] are listed in Table 16.2.

**Table 16.2.** Plasma parameters of large tokamaks JET [16.35], JT60U [16.31], and TFTR [16.29]. $n_i(0)\tau_E^{tot}T_i(0)$ is the fusion triple product. $\kappa_s$ is the ratio of the vertical radius to the horizontal radius. $q$ is the effective safety factor near the plasma boundary with different definitions. $q_{95}$ is the safety factor at the 95% flux surface. $q_{eff}$ and $q^*$ are defined in (16.13) and [16.29], respectively. $q_I$ is the factor defined in (16.11). $E_{NB}$ is the particle energy of neutral beam injection

| | JET<br>ELM free<br>No. 26087 | JT60U<br>ELMy<br>No. E21140 | TFTR<br>supershot |
|---|---|---|---|
| $I_p$ (MA) | 3.1 | 2.2 | 2.5 |
| $B_t$ (T) | 2.8 | 4.4 | 5.1 |
| $R/a$ (m/m) | 3.15/1.05 | 3.05/0.72 | ~2.48/0.82 |
| $\kappa_s$ | 1.6 | 1.7 | 1 |
| $q$ | $q_{95} = 3.8$ | $q_{eff} = 4.6$ | $q^* = 3.2$ |
| $q_I$ | 2.8 | 3.0 | 2.8 |
| $n_e(0)$ ($10^{19}$m$^{-3}$) | 5.1 | 7.5 | 8.5 |
| $n_e(0)/\langle n_e \rangle$ | 1.45 | 2.4 | – |
| $n_i(0)$ ($10^{19}$m$^{-3}$) | 4.1 | 5.5 | 6.3 |
| $T_e(0)$ (keV) | 10.5 | 10 | 11.5 |
| $T_e(0)/\langle T_e \rangle$ | 1.87 | – | – |
| $T_i$ (keV) | 18.6 | 30 | 44 |
| $W_{dia}$ (MJ) | 11.6 | 7.5 | 6.5 |
| $\mathrm{d}W_{dia}/\mathrm{d}t$ (MJ/s) | 6.0 | – | 7.5 |
| $Z_{eff}$ | 1.8 | 2.2 | 2.2 |
| $\beta_p$ | 0.83 | 1.2 | ~1.1 |
| $\beta_t$ (%) | 2.2 | ~1.3 | ~1.2 |
| $g$ (Troyon factor) | 2.1 | ~1.9 | 2 |
| $P_{NB}$ (MW) | 14.9 | 24.8 | 33.7 |
| $E_{NB}$ (keV) | 135, 78 | 95 | 110 |
| $\tau_E^{tot} = W/P_{tot}$ (s) | 0.78 | 0.3 | 0.2 |
| $H = \tau_E^{tot}/\tau_E^{ITER-P}$ | ~3.0 | ~2.1 | ~2.0 |
| $n_i(0)\tau_E^{tot}T_i(0)$ ($10^{20}$keVm$^{-3}$s) | 5.9 | 5 | 5.5 |
| $n_T(0)/[n_T(0) + n_D(0)]$ | 0 | 0 | 0.5 |
| $P_{fusion}$ (MW) | – | – | 9.3 |

In the present neutral beam source, the positive hydrogen ions are accelerated and then passed through the cell filled with neutral hydrogen gas, where ions are converted to a fast neutral beam by charge exchange (electron attachment). However, the conversion ratio of positive hydrogen ions to neutrals becomes small when the ion energy is larger than 100 keV (2.5% at 200 keV of $H^+$). On the other hand, the conversion ratio of negative hydrogen ions ($H^-$) to neutrals (electron stripped) does not decrease in the high energy range (~60%). A neutral beam source with a *negative ion source* is being developed as a high-efficiency source.

Wave heating is another method of plasma heating, described in detail in Chap. 12. A similar heating efficiency was observed for wave heating in ICRF (ion cyclotron range of frequency) and for NBI in PLT. In the JET ICRF experiments, the parameters $\kappa T_{\rm i}(0) = 5.4\,{\rm keV}$, $\kappa T_{\rm e}(0) = 5.6\,{\rm keV}$, $n_{\rm e}(0) = 3.7 \times 10^{13}\,{\rm cm}^{-3}$, $\tau_{\rm E} \sim 0.3\,{\rm s}$ were obtained by $P_{\rm ICRF} = 7\,{\rm MW}$.

## 16.8 Non-Inductive Current Drive

As long as the plasma current is driven by electromagnetic induction of the current transformer in a tokamak device, the discharge is necessarily a pulsed operation with finite duration. If the plasma current can be driven in a non-inductive way, a steady-state tokamak reactor is possible in principle. Current drive by neutral beam injection has been proposed by Ohkawa [16.36] and current drive by traveling wave has been proposed by Wort [16.37]. The momenta of particles injected by NBI or of traveling waves is transferred to the charged particles of the plasma and the resulting charged particle flow produces the plasma current. Current drive by NBI was demonstrated by DITE, TFTR, etc. Current drive by a lower hybrid wave (LHW), proposed by Fisch, was demonstrated by JFT-2, JIPPT-II, WT-2, PLT, Alcator C, Versator 2, T-7, Wega, JT-60, and so on. Current drive by electron cyclotron waves was demonstrated by Cleo, T-10, WT-3, Compass-D, DIII-D, TCV, and others.

### 16.8.1 Lower Hybrid Current Drive

The theory of current drive by waves is described here according to Fisch and Karney [16.38]. When a wave is traveling along the line of magnetic force, the velocity distribution function near the phase velocity of the wave is flattened by the diffusion in velocity space. Denoting the diffusion coefficient in the velocity space of the wave by $D_{\rm rf}$, the Fokker–Planck equation is given by [16.39]

$$\frac{\partial f}{\partial t} + \boldsymbol{v} \cdot \nabla_r f + \left(\frac{\boldsymbol{F}}{m}\right) \cdot \nabla_v f = \frac{\partial}{\partial v_z}\left(D_{\rm rf}\frac{\partial f}{\partial v_z}\right) + \left(\frac{\delta f}{\delta t}\right)_{\rm F.P.}, \qquad (16.41)$$

where $(\delta f/\delta t)_{\rm F.P.}$ is the Fokker–Planck collision term

$$\left(\frac{\delta f}{\delta t}\right)_{\rm F.P.} = -\sum_{\rm i,e}\left[\frac{1}{v^2}\frac{\partial}{\partial v}(v^2 J_{\rm v}) + \frac{1}{v\sin\theta}\frac{\partial}{\partial\theta}(\sin\theta J_\theta)\right], \tag{16.42}$$

$$J_{\rm v} = -D_\parallel \frac{\partial f}{\partial v} + Af\,, \qquad J_\theta = -D_\perp \frac{1}{v}\frac{\partial f}{\partial\theta}\,. \tag{16.43}$$

When the velocity $v$ of a test particle is greater than the thermal velocity $v_T^*$ of field particles ($v > v_T^*$), the diffusion tensor in velocity space $D_\parallel, D_\perp$ and the coefficient of dynamic friction $A$ reduce to

$$D_\parallel = \frac{v_T^{*2}\nu_0}{2}\left(\frac{v_T^*}{v}\right)^3, \quad D_\perp = \frac{v_T^{*2}\nu_0}{2}\frac{v_T^*}{2v}, \quad A = -D_\parallel \frac{m}{m^*}\frac{v}{v_T^{*2}},$$

where $v_T^*$ and $\nu_0$ are given by

$$v_T^{*2} = \frac{T^*}{m^*}, \quad \nu_0 = \left(\frac{qq^*}{\epsilon_0}\right)^2 \frac{n^*\ln\Lambda}{2\pi v_T^{*3} m^2} = \Pi^{*4}\frac{\ln\Lambda}{2\pi v_T^{*3} n^*},$$

and $\Pi^{*2} \equiv qq^*n^*/(\epsilon_0 m)$. $(v, \theta, \psi)$ are spherical coordinates in velocity space. $v_T^*$, $q^*$, $n^*$ are the thermal velocity, charge, and density of field particles, respectively, and $v, q, n$ are the quantities relating to test particles. Let us consider the electron distribution function in a homogeneous case in space without external force ($\boldsymbol{F} = 0$). Electron–electron and electron–ion collision terms (ion charge number $Z$) are taken into account. When dimensionless quantities $\tau = \nu_{0\rm e}t$, $u = v/v_{T_{\rm e}}^*$, $w = v_z/v_{T_{\rm e}}^*$, $D(w) = D_{\rm rf}/v_{T_{\rm e}}^{*2}\nu_{0\rm e}$ are introduced, the Fokker–Planck equation reduces to

$$\frac{\partial f}{\partial\tau} = \frac{\partial}{\partial w}\left[D(w)\frac{\partial f}{\partial w}\right] + \frac{1}{2u^2}\frac{\partial}{\partial u}\left(\frac{1}{u}\frac{\partial f}{\partial u} + {}^{`}f\right) + \frac{1+Z}{4u^3}\frac{1}{\sin\theta}\frac{\partial}{\partial\theta}\left(\sin\theta\frac{\partial f}{\partial\theta}\right).$$

When Cartesian coordinates in velocity space $(v_x, v_y, v_z) = (v_1, v_2, v_3)$ are used instead of spherical coordinates in velocity space, the Fokker–Planck collision term in Cartesian coordinates is given as follows (assuming $v > v_T^*$):

$$A_i = -D_0 v_T^* \frac{m}{m^*}\frac{v_i}{v^3}, \tag{16.44}$$

$$D_{ij} = \frac{D_0}{2}\frac{v_T^*}{v^3}\left[(v^2\delta_{ij} - v_iv_j) + \frac{v_T^{*2}}{v^2}(3v_iv_j - v^2\delta_{ij})\right], \tag{16.45}$$

$$J_i = A_i f - \sum_j D_{ij}\frac{\partial f}{\partial v_j}, \tag{16.46}$$

$$D_0 \equiv \frac{(qq^*)^2 n^*\ln\Lambda}{4\pi\epsilon_0^2 m^2 v_T^*} = \frac{v_T^{*2}\nu_0}{2}, \tag{16.47}$$

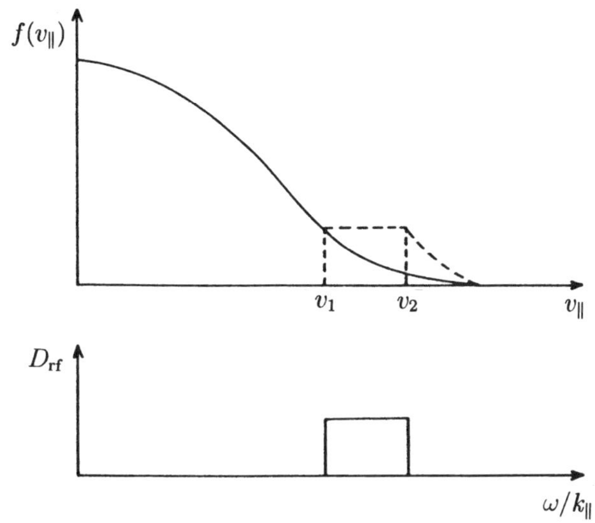


**Fig. 16.17.** The electron distribution function $f(v_\parallel)$ is flattened in the region from $v_1 = c/N_1$ to $v_2 = c/N_2$ due to the interaction with the lower hybrid wave, for which the spectrum of the parallel index $N_\parallel$ ranges from $N_1$ to $N_2$

$$\left(\frac{\delta f}{\delta t}\right)_{\mathrm{F.P.}} = -\boldsymbol{\nabla}_v \cdot \boldsymbol{J} \ .$$

$A_i$ is the *coefficient of dynamic friction* and $D_{ij}$ is the component of the *diffusion tensor*. Let us assume that the distribution function of the velocities $v_x, v_y$ perpendicular to the line of magnetic force is Maxwellian. Then the one-dimensional Fokker–Planck equation on the distribution function $F(w) = \int f \mathrm{d}v_x \mathrm{d}v_y$ of the parallel velocity $w = v_z/v_{T_e}^*$ can be deduced by integrating over $(v_x, v_y)$:

$$\int\int \left(\frac{\delta f}{\delta t}\right)_{\mathrm{F.P.}} \mathrm{d}v_x \mathrm{d}v_y = \int\int (-\boldsymbol{\nabla}_v \cdot \boldsymbol{J}) \mathrm{d}v_x \mathrm{d}v_y$$
$$= \int\int \frac{\partial}{\partial v_z}\left(-A_z f + \sum_j D_{zj} \frac{\partial f}{\partial v_j}\right) \mathrm{d}v_x \mathrm{d}v_y \ .$$

When $|v_z| \gg |v_x|, |v_y|$, the approximation $v \approx |v_z|$ can be used. The resulting one-dimensional Fokker–Planck equation for $F(w)$ is

$$\frac{\partial F}{\partial \tau} = \frac{\partial}{\partial w}\left[D(w)\frac{\partial F}{\partial w}\right] + \left(1 + \frac{Z}{2}\right)\frac{\partial}{\partial w}\left(\frac{1}{w^3}\frac{\partial}{\partial w} + \frac{1}{w^2}\right) F(w) \ .$$

The steady-state solution is

$$F(w) = C \exp \int^{w} \frac{-w\mathrm{d}w}{1 + w^3 D(w)/(1 + Z/2)} ,$$

which is shown schematically in Fig. 16.17. When $D(w) = 0$, this solution is Maxwellian. $F(w)$ is asymmetric with respect to $w = 0$, so that a current is induced. The induced current density $J$ is

$$J = env_{T_e}^{*} j ,$$

where $j = \int wF(w)\mathrm{d}w$, and

$$j \approx \frac{w_1 + w_2}{2} F(w_1)(w_2 - w_1) . \tag{16.48}$$

On the other hand, this current tends to dissipate by Coulomb collisions. Dissipated energy must be supplied by the input energy from the wave in order to sustain the current. The required input power $P_\mathrm{d}$ is

$$P_\mathrm{d} = -\int \frac{nmv^2}{2} \left(\frac{\delta f}{\delta t}\right)_{\mathrm{F.P.}} \mathrm{d}\boldsymbol{v} = \int \frac{nmv^2}{2} \frac{\partial}{\partial v_z} \left(D_\mathrm{rf} \frac{\partial f}{\partial v_z}\right) \mathrm{d}\boldsymbol{v}$$

$$= nmv_{T_e}^{*2} \nu_0 \int \frac{w^2}{2} \frac{\partial}{\partial w} \left[D(w) \frac{\partial F}{\partial w}\right] \mathrm{d}w = nmv_{T_e}^{*2} \nu_0 p_\mathrm{d} ,$$

where $p_\mathrm{d}$ is given by use of the steady-state solution of $F(w)$, under the assumption $w^3 D(w) \gg 1$, as follows:

$$p_\mathrm{d} = \left(1 + \frac{Z}{2}\right) F(w_1) \ln \left(\frac{w_2}{w_1}\right) \approx \left(1 + \frac{Z}{2}\right) F(w_1) \frac{w_2 - w_1}{w_1} ,$$

$$\frac{j}{p_\mathrm{d}} = \frac{1.5}{1 + 0.5 Z_\mathrm{i}} \frac{2}{3} w^2 . \tag{16.49}$$

More accurately, this ratio is [16.38]

$$\frac{j}{p_\mathrm{d}} = \frac{1.12}{1 + 0.12 Z_\mathrm{i}} 1.7 w^2 . \tag{16.50}$$

The ratio of the current density $J$ and the input power $P_\mathrm{d}$ per unit volume required to sustain the current is given by

$$\frac{J}{P_\mathrm{d}} = \frac{env_{\mathrm{Te}}^{*} j}{nT_\mathrm{e} \nu_0 p_\mathrm{d}} = 0.16 \frac{(\kappa T_\mathrm{e})_\mathrm{keV}}{n_{19}} \langle w^2 \rangle \frac{1.12}{1 + 0.12 Z_\mathrm{i}} \left(\frac{\mathrm{A/m^2}}{\mathrm{W/m^3}}\right) \tag{16.51}$$

where $(\kappa T)_\mathrm{keV}$ is the electron temperature in keV units and

$$\frac{I_\mathrm{CD}}{W_\mathrm{LH}} = \frac{1}{2\pi R} \frac{\int J 2\pi r \mathrm{d}r}{\int P_\mathrm{d} 2\pi r \mathrm{d}r} .$$

The current drive efficiency $\eta_\mathrm{LH}^\mathrm{T}$ of LHCD is

$$\eta_{\rm LH}^{\rm T} \equiv \frac{Rn_{19}I_{\rm CD}}{W_{\rm LH}} = \frac{\int \eta_{\rm LH}(r)P_{\rm d}(r)2\pi r{\rm d}r}{\int P_{\rm d}(r)2\pi r{\rm d}r} \left(10^{19}\frac{\rm A}{\rm Wm^2}\right) ,$$

where $\eta_{\rm LH}(r)$ is the local current drive efficiency given by

$$\eta_{\rm LH}(r) = \frac{Rn_{19}J(r)}{2\pi RP_{\rm d}(r)} = 0.026(\kappa T_{\rm e})_{\rm keV}\langle w^2\rangle\frac{1.12}{1+0.12Z_{\rm i}}\left(10^{19}\frac{\rm A}{\rm Wm^2}\right) , \tag{16.52}$$

and $R$ is the major radius in meters. The average square $\langle w^2\rangle$ of the ratio of the phase velocity (in the direction of the magnetic field) of traveling waves to the electron thermal velocity is of the order of 20–50. In the JT60U experiment (1994), a plasma current of $I_{\rm p} = 3\,{\rm MA}$ was driven by a lower hybrid wave with $W_{\rm LH} = 4.8\,{\rm MW}$ when $n = 1.2\times10^{19}\,{\rm m}^{-3}$, $\langle\kappa T_{\rm e}\rangle \sim 2\,{\rm keV}$, $R = 3.5\,{\rm m}$ and $B_{\rm t} = 4\,{\rm T}$ ($\eta_{\rm LH} \sim 2.6$). These results are consistent with the theoretical results.

The required current drive power is proportional to the density, and the current cannot be driven beyond a threshold density in the case of lower hybrid current drive because of the accessibility condition (see Sect. 12.5). Other possible methods, such as drive in the electron cyclotron range of frequencies (see Sect. 16.8.2), fast wave, and neutral beam injection (see Sect. 16.8.3) are also being studied.

A ramp-up experiment taking the plasma current from zero was first carried out by WT-2 and PLT and others by applying a lower hybrid wave to the target plasma produced by electron cyclotron heating and other types of heating. When the plasma current is ramped up in the low-density plasma and the density is increased after the plasma current reaches a specified value, all the available magnetic flux of the current transformer can be used solely to sustain the plasma current, so that the discharge duration can be increased several times.

### 16.8.2 Electron Cyclotron Current Drive

Electron cyclotron current drive (ECCD) relies on the generation of an asymmetric resistivity due to the selective heating of electrons moving in a particular toroidal direction. N.J. Fisch and A.H. Boozer [16.40] suggested that the collisionality of the plasma might be altered in such a way that, for example, electrons moving to the left collide less frequently with ions than electrons moving to the right. A net electric current would result, with electrons moving, on average, to the left and ions moving to the right.

Consider the displacement in velocity space of a small number, $\delta f$, of electrons from coordinates with subscript 1 to coordinates with subscript 2 as shown in Fig. 16.18. The energy expended to produce this displacement is given by

$$\Delta E = (E_2 - E_1)\delta f ,$$

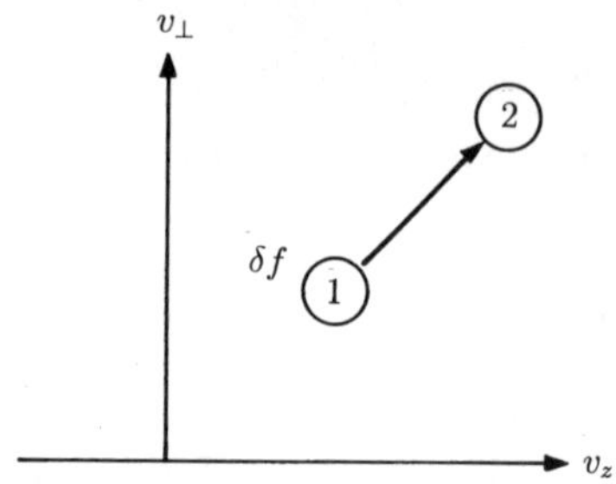


**Fig. 16.18.** Displacement in velocity space of a small number, $\delta f$, of electrons from coordinates with subscript 1 to coordinates with subscript 2

where $E_i$ is the kinetic energy associated with velocity-space location $i$. Electrons at different coordinates will lose their momentum parallel to the magnetic field, which is in the $z$ direction, at a rate $\nu_1$, but now lose it at a rate $\nu_2$. The $z$-directed current density is then given by

$$j(t) = -e\delta f\left[v_{z2}\exp(-\nu_2 t) - v_{z1}\exp(-\nu_1 t)\right] . \tag{16.53}$$

Consider the time-smoothed current $J$ over a time interval $\delta t$ which is large compared with both $1/\nu_1$ and $1/\nu_2$, so that

$$J = \frac{1}{\Delta t}\int_0^{\Delta t} j(t)\mathrm{d}t = -\frac{e\delta f}{\Delta t}\left(\frac{v_{z2}}{\nu_2} - \frac{v_{z1}}{\nu_1}\right) .$$

Therefore the input power density $P_\mathrm{d}$ required to induce the current density is

$$P_\mathrm{d} = \frac{\Delta E}{\Delta t} = \frac{E_2 - E_1}{\Delta t}\delta f .$$

The ratio $J/P_\mathrm{d}$ becomes

$$\frac{J}{P_\mathrm{d}} = -e\frac{v_{z2}/\nu_2 - v_{z1}/\nu_1}{E_2 - E_1} \Longrightarrow -e\frac{\boldsymbol{s}\cdot\boldsymbol{\nabla}(v_z/\nu)}{\boldsymbol{s}\cdot\boldsymbol{\nabla}E} , \tag{16.54}$$

where $\boldsymbol{s}$ is the unit vector in the direction of the displacement in velocity space. Let us estimate $\nu$ of (16.54). The deceleration rate of the momentum of a test electron by collision with electrons and ions is expressed by [see (2.14) and (2.17)]

$$\frac{\mathrm{d}p}{\mathrm{d}t} = -\frac{p}{\tau_{\mathrm{ee}\|}} - \frac{p}{\tau_{\mathrm{ei}\|}} = -\left(1 + \frac{Z_\mathrm{i}}{2}\right)\frac{\nu_0}{u^3}p ,$$

where

$$\nu_0 = \left(\frac{e^2 n_\mathrm{e}}{\epsilon_0 m_\mathrm{e}}\right)^2 \frac{\ln\Lambda}{2\pi n_\mathrm{e} v_{T_\mathrm{e}}^3} , \qquad u \equiv \frac{v}{v_{T_\mathrm{e}}} ,$$

and $v_{T_e} = (\kappa T_e/m_e)^{1/2}$ is the electron thermal velocity. Therefore we have

$$\frac{\mathrm{d}p}{\mathrm{d}t} = -\nu_M p\,, \qquad \nu_M \equiv (2+Z_i)\frac{\nu_0}{2u^3}\,.$$

In order to estimate $\mathrm{d}u/\mathrm{d}t$, we must use the energy relaxation time $\tau_{ee}^{\epsilon}$ [see (2.27)]

$$\frac{\mathrm{d}E}{\mathrm{d}t} = -\frac{E}{\tau_{ee}^{\epsilon}}\,, \qquad E = \frac{m_e}{2}u^2 v_{T_e}^2\,,$$

that is

$$\frac{\mathrm{d}u}{\mathrm{d}t} = -\frac{u}{2\tau_{ee}^{\epsilon}} = -\frac{\nu_0}{2u^3}u\,.$$

Each term in (16.53) for $j(t)$ must be modified as follows:

$$j(t) = j_0 \exp\left(-\int \nu_M \mathrm{d}t\right) = j_0\left[\frac{u(t)}{u_0}\right]^{2+Z_i}\,, \tag{16.55}$$

because

$$-\int \nu_M \mathrm{d}t = -\int \nu_M \frac{\mathrm{d}t}{\mathrm{d}u}\mathrm{d}u = (2+Z_i)\int \frac{\mathrm{d}u}{u} = (2+Z_i)\ln\frac{u(t)}{u_0}\,.$$

Then the integral of $j(t)$ in (16.55) reduces to

$$\int_0^\infty j(t)\mathrm{d}t = j_0\int_{u_0}^{0}\left[\frac{u(t)}{u_0}\right]^{2+Z_i}\frac{\mathrm{d}t}{\mathrm{d}u}\mathrm{d}u = \frac{j_0}{\nu_0}\frac{2u_0^3}{5+Z_i}\,.$$

Accordingly, $\nu$ in (16.54) is

$$\nu = \nu_0\frac{5+Z_i}{2u^3}\,, \tag{16.56}$$

and

$$\frac{J}{P_d} = \frac{en_e v_{T_e}}{n_e T_e \nu_0}\frac{j}{p_d}\,, \qquad \frac{j}{p_d} \equiv \frac{4}{5+Z_i}\frac{\boldsymbol{s}\cdot\boldsymbol{\nabla}(u^3 w)}{\boldsymbol{s}\cdot\boldsymbol{\nabla}u^2}\,,$$

where $w \equiv v_z/v_{T_e}$. In the case of ECCD, we have $j/p_d \approx 6wu/(5+Z_i)$ and

$$\frac{J}{P_d} = \frac{en_e v_{T_e}}{n_e T_e \nu_0}\frac{\langle 6wu\rangle}{5+Z_i} = 0.096\frac{(\kappa T_e)_{keV}}{n_{19}}\frac{\langle 6wu\rangle}{5+Z_i}\,. \tag{16.57}$$

The ratio of driven current $I_{CD}$ to ECCD power $W_{EC}$ is

$$\frac{I_{CD}}{W_{EC}} = \frac{1}{2\pi R}\frac{\int J 2\pi r\mathrm{d}r}{\int P_d 2\pi r \mathrm{d}r}\,,$$

and the current drive efficiency $\eta_{EC}^{T}$ of ECCD is

$$\eta_{EC}^{T} \equiv \frac{Rn_{19}I_{CD}}{W_{CD}} = \frac{\int \eta_{EC}(r)P_d(r)2\pi r\mathrm{d}r}{\int P_d 2\pi r\mathrm{d}r}\,,$$

where $\eta_{EC}(r)$ is the local current drive efficiency given by

$$\eta_{EC}(r) = \frac{Rn_{19}J(r)}{2\pi R P_d} = 0.015(\kappa T_e)_{keV}\frac{\langle 6wu\rangle}{5+Z_i}\left(10^{19}\frac{\mathrm{A}}{\mathrm{Wm}^2}\right)\,. \tag{16.58}$$

### 16.8.3 Neutral Beam Current Drive

When a fast neutral beam is injected into a plasma, it changes to a fast ion beam by charge exchange or ionization processes. When the fast ions have higher energy than $E_{\rm cr} = m_{\rm b}v_{\rm cr}^2/2$ given by (2.33), they are decelerated, mainly by electrons in the plasma, while the fast ions with $E < E_{\rm cr}$ are decelerated mainly by ions in the plasma. The distribution function of the ion beam can be obtained by solving the Fokker–Planck equations. The Fokker–Planck collision term (16.42) of the fast ions with $E \gg E_{\rm cr}$ is dominated by the dynamic friction term in (16.43) due to electrons. The dynamic friction term due to electrons on the fast ions in the case $v < v_T^*$ is given by [16.39]

$$A = -\frac{v}{2\tau_{\rm be}^{\epsilon}} .$$

Then the Fokker–Planck equation reduces to

$$\frac{\partial f_{\rm b}}{\partial t} + \frac{\partial}{\partial v}\left(\frac{-v f_{\rm b}}{2\tau_{\rm be}^{\epsilon}}\right) = \phi\delta(v - v_{\rm b}) , \tag{16.59}$$

where $v_{\rm b}$ is the initial injection velocity and $\tau_{\rm be}^{\epsilon}$ is the energy relaxation time of beam ions and electrons as described by (2.34). The right-hand side is the source term of beam ions. The steady-state solution of the Fokker–Planck equation is

$$f_{\rm b} \propto 1/v .$$

However, the dynamic friction term due to ions or the diffusion term dominates the collision term in the region $v < v_{\rm cr}$. Therefore the approximate distribution function of the ion beam is given by $f_{\rm b} \propto v^2/(v^3 + v_{\rm cr}^3)$, i.e.,

$$f_{\rm b}(v) = \frac{n_{\rm b}}{\ln\left[1 + (v_{\rm b}/v_{\rm cr})^3\right]^{1/3}} \frac{v^2}{v^3 + v_{\rm cr}^3} \qquad (v \le v_{\rm b}) , \tag{16.60a}$$

$$f_{\rm b}(v) = 0 \qquad (v > v_{\rm b}) . \tag{16.60b}$$

The ion injection rate $\phi$ per unit time per unit volume required to maintain the steady-state condition of the beam is derived by substituting the solution for $f_{\rm b}(v)$ into the Fokker–Planck equation:

$$\phi = \frac{n_{\rm b}}{2\tau_{\rm be}^{\epsilon}} \frac{\left[1 + (v_{\rm cr}/v_{\rm b})^3\right]^{-1}}{\left\{\ln\left[1 + (v_{\rm b}/v_{\rm cr})^3\right]\right\}^{1/3}} .$$

The required power is then

$$P_{\rm b} = \frac{m_{\rm b}v_{\rm b}^2}{2}\phi \approx \frac{m_{\rm b}v_{\rm b}^2 n_{\rm b}}{4\ln(v_{\rm b}/v_{\rm cr})\tau_{\rm be}^{\epsilon}} . \tag{16.61}$$

The average velocity of the decelerating ion beam is

$$\bar{v}_{\mathrm{b}} = v_{\mathrm{b}} \left[ \ln(v_{\mathrm{b}}/v_{\mathrm{cr}}) \right]^{-1} . \tag{16.62}$$

Then the current density $J$ driven by the fast ion beam consists of terms due to fast ions and bulk ions and electrons of the plasma:

$$J = Z_{\mathrm{i}} e n_{\mathrm{i}} \bar{v}_{\mathrm{i}} + Z_{\mathrm{b}} e n_{\mathrm{b}} \bar{v}_{\mathrm{b}} - e n_{\mathrm{e}} \bar{v}_{\mathrm{e}} , \qquad n_{\mathrm{e}} = Z_{\mathrm{i}} n_{\mathrm{i}} + Z_{\mathrm{b}} n_{\mathrm{b}} ,$$

where $\bar{v}_{\mathrm{i}}$ and $\bar{v}_{\mathrm{e}}$ are the average velocities of ions with density $n_{\mathrm{i}}$ and electrons with density $n_{\mathrm{e}}$, respectively. The electrons of the plasma receive momentum by collision with fast ions and lose it by collision with plasma ions, i.e.,

$$m_{\mathrm{e}} n_{\mathrm{e}} \frac{\mathrm{d}\bar{v}_{\mathrm{e}}}{\mathrm{d}t} = m_{\mathrm{e}} n_{\mathrm{e}} (\bar{v}_{\mathrm{b}} - \bar{v}_{\mathrm{e}}) \nu_{\mathrm{eb}\|} + m_{\mathrm{e}} n_{\mathrm{e}} (\bar{v}_{\mathrm{i}} - \bar{v}_{\mathrm{e}}) \nu_{\mathrm{ei}\|} = 0 ,$$

so that

$$(Z_{\mathrm{i}}^2 n_{\mathrm{i}} + Z_{\mathrm{b}}^2 n_{\mathrm{b}}) \bar{v}_{\mathrm{e}} = Z_{\mathrm{b}}^2 n_{\mathrm{b}} \bar{v}_{\mathrm{b}} + Z_{\mathrm{i}}^2 n_{\mathrm{i}} \bar{v}_{\mathrm{i}} .$$

Since $n_{\mathrm{b}} \ll n_{\mathrm{i}}$,

$$n_{\mathrm{e}} \bar{v}_{\mathrm{e}} = \frac{Z_{\mathrm{b}}^2}{Z_{\mathrm{i}}} n_{\mathrm{b}} \bar{v}_{\mathrm{b}} + Z_{\mathrm{i}} n_{\mathrm{i}} \bar{v}_{\mathrm{i}} ,$$

so that [16.36]

$$J = \left( 1 - \frac{Z_{\mathrm{b}}}{Z_{\mathrm{i}}} \right) Z_{\mathrm{b}} e n_{\mathrm{b}} \bar{v}_{\mathrm{b}} . \tag{16.63}$$

The driven current density consists of the fast ion beam term $Z_{\mathrm{b}} e n_{\mathrm{b}} \bar{v}_{\mathrm{b}}$ and the term corresponding to electrons dragged by the fast ion beam, viz., $-Z_{\mathrm{b}}^2 e n_{\mathrm{b}} \bar{v}_{\mathrm{b}} / Z_{\mathrm{i}}$. Then the ratio $J/P_{\mathrm{d}}$ becomes

$$\frac{J}{P_{\mathrm{d}}} = (1 - Z_{\mathrm{b}}/Z_{\mathrm{i}}) \frac{Z_{\mathrm{b}} e n_{\mathrm{b}} \bar{v}_{\mathrm{b}}}{m_{\mathrm{b}} n_{\mathrm{b}} v_{\mathrm{b}} \bar{v}_{\mathrm{b}} / 4\tau_{\mathrm{be}}^{\epsilon}} = \frac{2 e Z_{\mathrm{b}} (2\tau_{\mathrm{be}}^{\epsilon})}{m_{\mathrm{b}} v_{\mathrm{b}}} \left( 1 - \frac{Z_{\mathrm{b}}}{Z_{\mathrm{i}}} \right) . \tag{16.64}$$

When the charge number of the beam ions is equal to that of the plasma ions, that is, when $Z_{\mathrm{b}} = Z_{\mathrm{i}}$, the current density becomes zero for linear (cylindrical) plasmas. For toroidal plasmas, the motion of circulating electrons is disturbed by collision with the trapped electrons (banana electrons), and the dragged electron term is reduced. Thus $J/P_{\mathrm{d}}$ becomes [16.41]

$$\frac{J}{P_{\mathrm{d}}} = \frac{2 e Z_{\mathrm{b}} (2\tau_{\mathrm{be}}^{\epsilon})}{m_{\mathrm{b}} v_{\mathrm{b}}} \left\{ 1 - \frac{Z_{\mathrm{b}}}{Z_{\mathrm{i}}} \left[ 1 - G(Z_{\mathrm{eff}}, \epsilon) \right] \right\} , \tag{16.65}$$

$$G(Z_{\mathrm{eff}}, \epsilon) = \left( 1.55 + \frac{0.85}{Z_{\mathrm{eff}}} \right) \epsilon^{1/2} - \left( 0.2 + \frac{1.55}{Z_{\mathrm{eff}}} \right) \epsilon , \tag{16.66}$$

where $\epsilon$ is the inverse aspect ratio. When the effect of the pitch angle of the ionized beam is taken into account, (16.65) must be multiplied by the factor $\xi \equiv v_{\|}/v = R_{\mathrm{tang}}/R_{\mathrm{ion}}$, where $R_{\mathrm{tang}}$ is the minimum value of $R$ along the neutral beam path and $R_{\mathrm{ion}}$ is the $R$ value of the ionization position.

The driving efficiency calculated by the bounce average Fokker–Planck equation [16.41] becomes

$$\frac{J}{P_{\rm d}} = \frac{2eZ_{\rm b}(2\tau_{\rm be}^{\epsilon})}{m_{\rm b}v_{\rm b}} \left\{1 - \frac{Z_{\rm b}}{Z_{\rm i}}[1 - G(Z_{\rm eff}, \epsilon)]\right\} \xi_0 F_{\rm nc} x_{\rm b} J_0(x_{\rm b}, y) \,,$$

$$\frac{J}{P_{\rm d}} = \frac{2eZ_{\rm b}(2\tau_{\rm be}^{\epsilon})}{m_{\rm b}v_{\rm cr}} \left\{1 - \frac{Z_{\rm b}}{Z_{\rm i}}[1 - G(Z_{\rm eff}, \epsilon)]\right\} \xi_0 F_{\rm nc} J_0(x_{\rm b}, y) \,, \qquad (16.67)$$

where

$$x_{\rm b} \equiv \frac{v_{\rm b}}{v_{\rm cr}} \,, \quad y = 0.8\frac{Z_{\rm eff}}{A_{\rm b}} \,, \quad J_0(x, y) = \frac{x^2}{x^3 + (1.39 + 0.61y^{0.7})x^2 + (4 + 3y)}$$

and $F_{\rm nc} = 1 - b\epsilon^{\sigma}$ is the correction factor [16.42]. Finally, we have

$$\frac{J}{P_{\rm d}}\left(\frac{\rm Am}{\rm W}\right) = \frac{15.8(\kappa T_{\rm e})_{\rm keV}\xi_0}{Z_{\rm b} n_{\rm e19}} \left[1 - \frac{Z_{\rm b}}{Z_{\rm i}}(1 - G)\right] (1 - b\epsilon^{\sigma}) J_0(x_{\rm b}, y) \,. \quad (16.68)$$

The local current drive efficiency $\eta_{\rm NB}$ of the neutral beam current drive (NBCD) is

$$\eta_{\rm NB} \equiv \frac{R n_{\rm e19} J}{2\pi R P_{\rm d}} \left(10^{19} \frac{\rm A}{\rm Wm^2}\right)$$

$$= 2.52(\kappa T_{\rm e})_{\rm keV}\xi_0 \left[1 - \frac{Z_{\rm b}}{Z_{\rm i}}(1 - G)\right] (1 - b\epsilon^{\sigma}) J_0(x_{\rm b}, y) \,. \quad (16.69)$$

When $Z_{\rm b} = 1$, $Z_{\rm eff} = 1.5$, $A_{\rm b} = 2$, $x_{\rm b}^2 = 4$, then $(1 - b\epsilon^{\sigma})J_0 \sim 0.2$. When $\langle\epsilon\rangle \sim 0.15$, then $\eta_{\rm NB} \sim 0.29(\kappa T_{\rm e})_{\rm keV}(10^{19}{\rm A/Wm^2})$. Current drive by NBI has been demonstrated by the DITE, TFTR JT60U and JET experiments.

When the application of a current drive to a fusion grade plasma with $n_{\rm e} \sim 10^{20}\,{\rm m}^{-3}$ is considered, the necessary input power for any current drive of the full plasma current occupies a considerable amount of the fusion output. Therefore a substantial part of the plasma current must be driven by the bootstrap current, as described in the next section.

### 16.8.4 Bootstrap Current

It was predicted theoretically that radial diffusion induces a current in the toroidal direction and that the current can be large in the banana region [16.43–16.46]. This current, known as the *bootstrap current*, was later well confirmed experimentally. This is an important process which can provide the means to sustain the plasma current in the tokamak in a steady state.

As described in Sect. 7.2, electrons in the collisionless region $\nu_{\rm ei} < \nu_{\rm b}$ make a complete circuit of the banana orbit. When a density gradient exists, there is a difference in particle number on neighboring orbits passing through a point A, as shown in Fig. 16.19.

The difference is $({\rm d}n_{\rm t}/{\rm d}r)\Delta_{\rm b}$, where $\Delta_{\rm b}$ is the width of the banana orbit. As the component of velocity parallel to the magnetic field is $v_{\|} = \epsilon^{1/2} v_T$, the current density due to trapped electrons with density $n_{\rm t}$ is

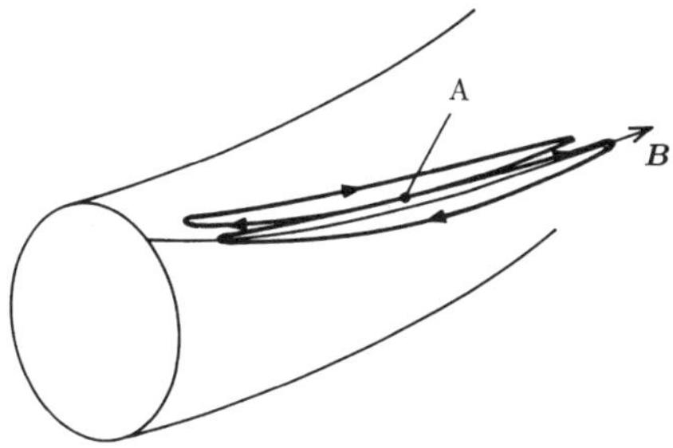


**Fig. 16.19.** Banana orbits of trapped electrons which induce the bootstrap current

$$j_{\mathrm{banana}} = -(ev_{\parallel})\left(\frac{\mathrm{d}n_{\mathrm{t}}}{\mathrm{d}r}\Delta_{\mathrm{b}}\right) = -\epsilon^{3/2}\frac{1}{B_{\mathrm{p}}}\frac{\mathrm{d}p}{\mathrm{d}r} .$$

The untrapped electrons start to drift in the same direction as the trapped electrons due to the collisions between them and the drift becomes steady state due to the collisions with ions. The drift velocity $V_{\mathrm{untrap}}$ of untrapped electrons in the steady state is given by

$$m_{\mathrm{e}}V_{\mathrm{untrap}}\nu_{\mathrm{ei}} = \frac{\nu_{\mathrm{ee}}}{\epsilon}m_{\mathrm{e}}\left(\frac{j_{\mathrm{banana}}}{-en_{\mathrm{e}}}\right) ,$$

where $\nu_{\mathrm{ee}}/\epsilon$ is the effective collision frequency between trapped and untrapped electrons. The current density due to the drift velocity $V_{\mathrm{untrap}}$ is

$$j_{\mathrm{boot}} \approx -\epsilon^{1/2}\frac{1}{B_{\mathrm{p}}}\frac{\mathrm{d}p}{\mathrm{d}r} . \tag{16.70}$$

This current is called the *bootstrap current*. When the average poloidal beta $\beta_{\mathrm{p}} = \langle p\rangle/(B_{\mathrm{p}}^2/2\mu_0)$ is used, the ratio of the total bootstrap current $I_{\mathrm{b}}$ to the plasma current $I_{\mathrm{p}}$ to form $B_{\mathrm{p}}$ is given by

$$\frac{I_{\mathrm{b}}}{I_{\mathrm{p}}} \sim c\left(\frac{a}{R}\right)^{1/2}\beta_{\mathrm{p}} , \tag{16.71}$$

where $c \sim 0.3$ is constant. This value can be near 1 if $\beta_{\mathrm{p}}$ is high ($\beta_{\mathrm{p}} \sim R/a$) and the pressure profile is peaked. Experiments on the bootstrap current have been carried out in TFTR, JT60U and JET. 70–80% of $I_{\mathrm{p}} = 1\,\mathrm{MA}$ was bootstrap driven in high $\beta_{\mathrm{p}}$ operation.

As the bootstrap current profile is hollow, it can produce a negative magnetic shear $q$ profile, which is stable against ballooning. The MHD stability of the hollow current profile is analyzed in detail in [16.47].

## 16.9 Neoclassical Tearing Mode

Much attention has been focused on tokamak operational pressure limits imposed by non-ideal MHD instabilities, such as the effects of bootstrap-current-driven magnetic islands. At high $\beta_{\rm p}$ (poloidal beta) and low collisionality, the pressure gradient in the plasma gives rise to a bootstrap current (see Sect. 16.8.4). If an island develops, the pressure within the island tends to flatten out, thereby removing the drive for the bootstrap current. This gives rise to a helical 'hole' in the bootstrap current, which increases the size of the island (see Fig. 16.22).

Tearing instability was treated in the slab model in Sect. 9.1. The zeroth-order magnetic field $\boldsymbol{B}_0$ depends only on $x$ and is given by

$$\boldsymbol{B}_0 = B_{0y}(x)\boldsymbol{e}_y + B_{0z}\boldsymbol{e}_z\,, \quad |B_{0y}(x)| \ll |B_{0z}|\,, \quad B_{0z} = \text{const.}$$

The basic equations are

$$\rho\left[\frac{\partial \boldsymbol{v}}{\partial t} + (\boldsymbol{v}\cdot\boldsymbol{\nabla})\boldsymbol{v}\right] = -\boldsymbol{\nabla}p + \boldsymbol{j}\times\boldsymbol{B}\,, \tag{16.72}$$

$$-\boldsymbol{E} = \boldsymbol{v}\times\boldsymbol{B} - \eta\boldsymbol{j} = \frac{\partial \boldsymbol{A}}{\partial t}\,, \qquad \boldsymbol{A} = (0, 0, -\psi)\,, \tag{16.73}$$

$$B_x = -\frac{\partial\psi}{\partial y}\,, \qquad B_y = \frac{\partial\psi}{\partial x}\,,$$

$$-\frac{\partial\psi}{\partial t} = (v_x B_y - v_y B_x) - \eta j_z = (\boldsymbol{v}\cdot\boldsymbol{\nabla})\psi - \eta j_z\,, \tag{16.74}$$

$$\nabla^2\psi = \mu_0 j_z\,. \tag{16.75}$$

Since

$$\boldsymbol{v} = \frac{\boldsymbol{E}\times\boldsymbol{B}}{B^2} = \left(\frac{E_y}{B_{0z}}, -\frac{E_x}{B_{0z}}, 0\right) = \left(-\frac{1}{B_{0z}}\frac{\partial\phi}{\partial y}, \frac{1}{B_{0z}}\frac{\partial\phi}{\partial x}, 0\right)\,,$$

a stream function $\varphi$ can be introduced such that

$$v_x = -\frac{\partial\varphi}{\partial y}\,, \qquad v_y = \frac{\partial\varphi}{\partial x}\,.$$

Furthermore, if the $z$ component $w_z = (\nabla\times\boldsymbol{v})_z$ of the vorticity is introduced, then $w_z = \nabla^2\varphi$. The rotation of (16.72) yields

$$\rho\frac{\partial w_z}{\partial t} + (\boldsymbol{v}\cdot\boldsymbol{\nabla})w_z = \left[\boldsymbol{\nabla}\times(\boldsymbol{j}\times\boldsymbol{B})\right]_z = (\boldsymbol{B}\cdot\boldsymbol{\nabla})j_z - (\boldsymbol{j}\cdot\boldsymbol{\nabla})B_z = (\boldsymbol{B}\cdot\boldsymbol{\nabla})j_z\,. \tag{16.76}$$

The relations $\boldsymbol{\nabla}\cdot\boldsymbol{B} = 0$, $\boldsymbol{\nabla}\cdot\boldsymbol{j} = 0$ were used here. The zeroth-order flux function $\psi_0$ and the first-order perturbation $\tilde{\psi}$ are

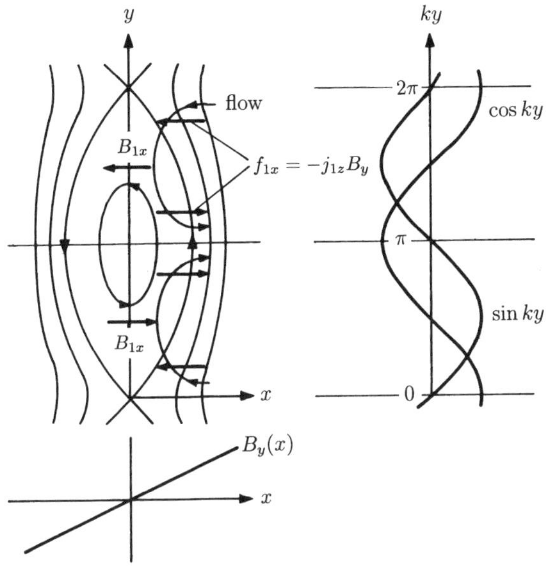


**Fig. 16.20.** Tearing mode structure in the singular layer

$$\psi_0(x) = B'_{0y}\frac{x^2}{2}\,, \quad \boldsymbol{B}_0 = (0, B'_{0y}x, B_{0z})\,,$$

$$\tilde{\psi}(y,t) = \frac{B_{1x}(t)}{k}\cos ky\,, \quad \boldsymbol{B}_1 = \left(B_{1x}(t)\sin ky, 0, 0\right)\,, \quad \tilde{\psi}_{\mathrm{A}}(t) \equiv \frac{B_{1x}(t)}{k}\,,$$

$$\psi = \psi_0(x) + \tilde{\psi}(y,t) = B'_{0y}\frac{x^2}{2} + \frac{B_{1x}(t)}{k}\cos ky\,. \tag{16.77}$$

$x = 0$ is the location of the singular layer. The separatrix of islands is given by

$$B'_{0y}\frac{x^2}{2} + \frac{B_{1x}(t)}{k}\cos ky = \frac{B_{1x}(t)}{k}\,, \quad x_{\mathrm{s}} = 2\left(\frac{B_{1x}}{kB'_{0y}}\right)^{1/2}\,,$$

and the full width $w$ of the island is

$$w = 4\left(\frac{B_{1x}}{kB'_{0y}}\right)^{1/2} = 4\left[\frac{\tilde{\psi}_{\mathrm{A}}(t)}{B'_{0y}}\right]^{1/2}\,. \tag{16.78}$$

The perturbation $B_{1x}(t)\sin ky$ growing with the growth rate $\gamma$ induces a current $j_{1z} = E_{1z}/\eta = \gamma B_{1x}/\eta k$, which provides the $x$ direction linear force $f_{1x} = -j_{1z}B'_{0y}x$ indicated in Fig. 16.20. This drives the flow pattern of narrow vortices. Moving away from the resistive singular layer, the induced electric field produces a flow $v_x = -E_z/B_y = -\gamma B_{1x}\cos ky/(kB'_{0y}x)$. For incompressible flow (in strong equilibrium field $B_{0z}$), this requires a strongly

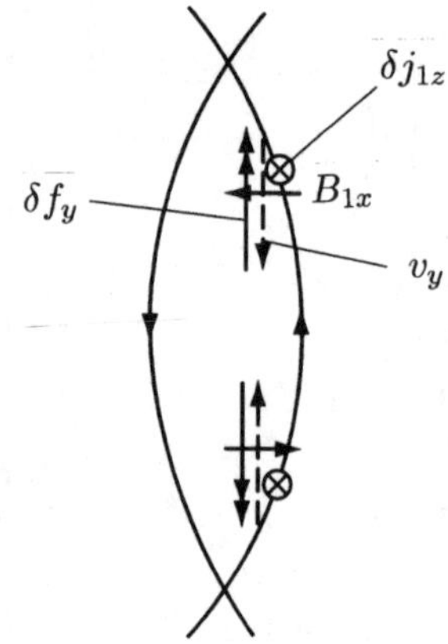


**Fig. 16.21.** Nonlinear forces decelerating $v_y$ flow in tearing mode

sheared flow $v_y(x)$ over the layer $x \sim x_{\mathrm{T}}$, that is the narrow vortex pattern shown in Fig. 16.20, and we have

$$v_y x_{\mathrm{T}} \sim \frac{v_x}{k}, \qquad v_y \sim \frac{v_x}{k x_{\mathrm{T}}} \sim \frac{\gamma B_{1x}}{k^2 B'_{0y} x_{\mathrm{T}}}.$$

If this shear flow is to be driven against inertia by the torque produced by the linear forces, we require

$$x_{\mathrm{T}} j_{1z} B_{0y} = \frac{\gamma \rho v_y}{k}, \quad B_{0y} = B'_{0y} x_{\mathrm{T}} \longrightarrow x_{\mathrm{T}}^4 = \frac{\gamma \rho}{j_{1z} k B'_{0y}} \frac{\gamma B_{1x}}{k^2 B'_{0y}} = \frac{\gamma \rho \eta}{(k B'_{0y})^2},$$

since $j_{1z} = E_z/\eta = \gamma B_{1x}/\eta k$. Thus the width of the perturbation is [16.48]

$$x_{\mathrm{T}} = \frac{(\gamma \rho \eta)^{1/4}}{(k B'_{0y})^{1/2}}. \tag{16.79}$$

This is consistent with the results (9.26) and (9.27) obtained by the linear theory of the tearing instability, described in Sect. 9.1. (Note that $\varepsilon$ was used in Sect. 9.1 instead of $x_{\mathrm{T}}$.)

Rutherford showed that the growth of the mode is drastically slowed down and perturbation grows only linearly in time when non-linear effects are taken into account [16.48]. The vortex flow will induce the second-order $y$-independent eddy current $\delta j_{1z} = -v_y B_{1x}/\eta \sim \gamma B_{1x}^2/(\eta k^2 B'_{0y} x_{\mathrm{p}}^2)$. The $y$-direction third-order non-linear forces $\delta f_y \sim \delta j_z B_{1x}$ indicated in Fig. 16.21 provide a torque opposing vortex flow and decelerate the $v_y$ flow.

We restrict ourselves to the case where the inertia may be neglected in (16.76):

$$(\boldsymbol{B}\cdot\boldsymbol{\nabla})j_z = -\frac{\partial\psi}{\partial y}\frac{\partial j_z}{\partial x} + \frac{\partial\psi}{\partial x}\frac{\partial j_z}{\partial y} = 0 \quad \longrightarrow \quad j_z = j_z(\psi) \, .$$

Equation (16.74) yields

$$\frac{\partial\psi_0}{\partial t} + \frac{\partial\tilde{\psi}}{\partial t} = -v_x B'_{0y} x + \eta j_{1z} \, , \qquad \frac{\partial\psi_0}{\partial t} = \eta j_{0z} \, ,$$

$$\longrightarrow \quad \frac{\partial\tilde{\psi}}{\partial t} = -\frac{\partial\varphi}{\partial y} B'_{0y} x + \eta j_{1z} - \eta j_{0z} \, . \tag{16.80}$$

We may eliminate $\varphi$ from (16.80) by dividing by $x$ and averaging over $y$ at constant $\psi$. From (16.77), $x$ is given by

$$x = \left(\frac{2}{B'_{0y}}(\psi - \tilde{\psi})\right)^{1/2} = \left(\frac{2}{B'_{0y}}\right)^{1/2} \tilde{\psi}_{\mathrm{A}}^{1/2} (W - \cos ky)^{1/2} \, , \quad W \equiv \frac{\psi}{\tilde{\psi}_{\mathrm{A}}} \, , \tag{16.81}$$

and

$$\left\langle \frac{1}{(\psi - \tilde{\psi})^{1/2}} \right\rangle [\eta j_{1z}(\psi) - \eta j_{0z}(\psi)] = \left\langle \frac{\partial\tilde{\psi}(y,t)/\partial t}{\left[\psi - \tilde{\psi}(y,t)\right]^{1/2}} \right\rangle \, ,$$

$$j_{1z}(\psi) = j_{0z}(\psi) + \frac{1}{\eta} \left\langle \frac{\partial\tilde{\psi}/\partial t}{(\psi - \tilde{\psi})^{1/2}} \right\rangle \left\langle (\psi - \tilde{\psi})^{-1/2} \right\rangle^{-1} \, , \tag{16.82}$$

where

$$\langle f \rangle \equiv \frac{k}{2\pi} \int_0^{2\pi/k} f \mathrm{d}y \, .$$

For the outer solution we require the discontinuity in the logarithmic derivatives across the singularity. We must match the logarithmic discontinuity from the solution within the singular layer to that of the outer solution:

$$\Delta' \equiv \left( \left.\frac{\partial\tilde{\psi}_{\mathrm{A}}}{\partial x}\right|_{+0} - \left.\frac{\partial\tilde{\psi}_{\mathrm{A}}}{\partial x}\right|_{-0} \right) \frac{1}{\tilde{\psi}_{\mathrm{A}}} = \frac{\partial}{\partial x} \ln \tilde{\psi}_{\mathrm{A}} \Big|_{-0}^{+0} \, .$$

We utilize $\nabla^2\tilde{\psi} = \mu_0 j_{1z}$, together with $\partial^2\tilde{\psi}/\partial x^2 \approx \mu_0 j_{1z}$ and

$$\Delta'\tilde{\psi}_{\mathrm{A}} = \frac{\partial\tilde{\psi}_{\mathrm{A}}}{\partial x} = 2\mu_0 \left\langle \cos ky \int_{-\infty}^{\infty} j_{1z} \mathrm{d}x \right\rangle \, , \tag{16.83}$$

$$\mathrm{d}x = \left(\frac{1}{2B'_{0y}}\right)^{1/2} \frac{\mathrm{d}\psi}{(\psi - \tilde{\psi})^{1/2}} \, .$$

Inserting (16.82) into (16.83) yields

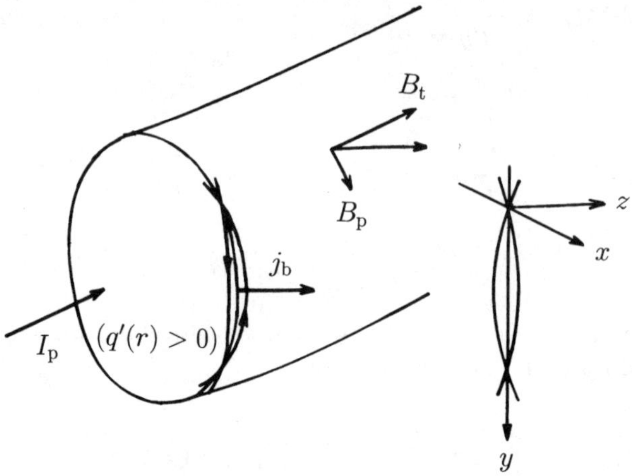


**Fig. 16.22.** Coordinates in the slab model and coordinates in the toroidal plasma. The coordinates $(x, y, z)$ correspond to the radial direction $(r - r_s)$, the poloidal direction $(\sim r\theta)$, and the direction of the magnetic field at the rational surface in the toroidal plasma, respectively. *Arrows* in the island indicate the direction of the magnetic field $B_p - (nr/mR)B_t$ [see (16.86)]

$$\Delta'\tilde{\psi}_A = 2\frac{\mu_0}{\eta(2B'_{0y})^{1/2}}\int_{x=-\infty}^{x=\infty}\left\langle\frac{\partial\tilde{\psi}/\partial t}{(\psi-\tilde{\psi})^{1/2}}\right\rangle\left\langle(\psi-\tilde{\psi})^{-1/2}\right\rangle^{-1}\left\langle\frac{\cos ky}{(\psi-\tilde{\psi})^{1/2}}\right\rangle \mathrm{d}\psi$$

$$= \frac{4\mu_0}{\eta(2B'_{0y})^{1/2}}\int_{\psi_{\min}}^{\infty}\mathrm{d}\psi\left\langle\frac{\partial\tilde{\psi}_A\cos ky/\partial t}{(\psi-\tilde{\psi})^{1/2}}\right\rangle\left\langle(\psi-\tilde{\psi})^{-1/2}\right\rangle^{-1}\left\langle\frac{\cos ky}{(\psi-\tilde{\psi})^{1/2}}\right\rangle .$$

Since

$$\int \mathrm{d}\psi\left\langle\frac{\cos ky}{(\psi-\tilde{\psi})^{1/2}}\right\rangle^2\frac{1}{\left\langle(\psi-\tilde{\psi})^{-1/2}\right\rangle}$$

$$= \int\left\langle\frac{\cos ky}{(W-\cos ky)^{1/2}}\right\rangle^2\frac{\mathrm{d}W\tilde{\psi}_A^{1/2}}{\left\langle(W-\cos ky)^{-1/2}\right\rangle} \equiv A\tilde{\psi}_A^{1/2} ,$$

we obtain

$$\Delta'\tilde{\psi}_A = \frac{4\mu_0 A}{\eta(2B'_{0y})^{1/2}}\frac{\partial\tilde{\psi}_A}{\partial t}\tilde{\psi}_A^{1/2}$$

and

$$\frac{\partial}{\partial t}\tilde{\psi}_A^{1/2} = \frac{\eta(2B'_{0y})^{1/2}}{8\mu_0 A}\Delta' .$$

Taking note of (16.78), the time variation of the island width reduces to

$$\frac{\mathrm{d}w}{\mathrm{d}t} = \frac{1}{2^{1/2}A}\frac{\eta}{\mu_0}\Delta' \approx \frac{\eta}{\mu_0}\Delta' \,, \quad \tau_{\mathrm{R}}\frac{\mathrm{d}}{\mathrm{d}t}\frac{w}{r_{\mathrm{s}}} = \Delta' r_{\mathrm{s}} \,, \quad \tau_{\mathrm{R}} \equiv \frac{\mu_0 r_{\mathrm{s}}^2}{\eta} \,. \tag{16.84}$$

Let us consider a toroidal plasma as shown in Fig. 16.22. The magnetic field

$$B_{\mathrm{p}} - \frac{nr}{nR}B_{\mathrm{t}} = \left[\frac{1}{q(r)} - \frac{1}{q_{\mathrm{s}}}\right]\frac{r}{R}B_{\mathrm{t}} \qquad \left(q_{\mathrm{s}} = \frac{m}{n}\right)$$

corresponds to $B_{0y}$ in the slab model near the singular radius. The coordinates $(x, y, z)$ in the slab model correspond to the radial direction $(r - r_{\mathrm{s}})$, the poloidal direction $(\sim r\theta)$, and the direction of the magnetic field at the rational surface in the toroidal plasma, respectively. The flux function is

$$\psi(x, y) = \int_0^{r-r_{\mathrm{s}}} \left[\frac{1}{q(r)} - \frac{1}{q_{\mathrm{s}}}\right]\frac{r}{R}B_{\mathrm{t}}\mathrm{d}x + \frac{B_{1x}}{k}\cos ky \tag{16.85}$$

and the magnetic field is given by

$$B_{1x} = -\frac{\partial \psi}{\partial y} = B_{1x}\sin ky \,,$$

$$B_{0y} = \frac{\partial \psi}{\partial x} = \left[\frac{1}{q(r)} - \frac{1}{q_{\mathrm{s}}}\right]\frac{r}{R}B_{\mathrm{t}} = -\frac{q'}{q}B_{\mathrm{p}}x = B'_{0y}x \,. \tag{16.86}$$

Equation (16.85) reduces to

$$\psi(x, y) = B'_{0y}\frac{x^2}{2} + \frac{B_{1x}}{k}\cos ky \,. \tag{16.87}$$

The change $\delta j_{1z}^{\mathrm{b}}$ in the bootstrap current induces the change $\delta\psi_{\mathrm{b}}$ in the flux function and the electric field $E_z$ given by

$$E_z = \frac{\partial \psi_{\mathrm{b}}}{\partial t} = \eta\delta j_{1z}^{\mathrm{b}} \,.$$

The discontinuity of the logarithmic derivative due to $\delta j_{1z}^{\mathrm{b}}$ is

$$\Delta'_{\mathrm{b}} = \frac{1}{\tilde{\psi}_{\mathrm{A}}}\left(\left.\frac{\partial\tilde{\psi}_{\mathrm{A}}^{\mathrm{b}}}{\partial r}\right|_{r_{\mathrm{s}}+} - \left.\frac{\partial\tilde{\psi}_{\mathrm{A}}^{\mathrm{b}}}{\partial r}\right|_{r_{\mathrm{s}}-}\right) = \frac{1}{\tilde{\psi}_{\mathrm{A}}}\int_{r_{\mathrm{s}}-}^{r_{\mathrm{s}}+} \mu_0\delta j_{1z}^{\mathrm{b}}\mathrm{d}r \,,$$

where

$$\tilde{\psi}_{\mathrm{A}} = \frac{B_{1x}}{k} = \frac{w^2 B'_{0y}}{16} \,,$$

so that

$$\Delta'_{\mathrm{b}} = \frac{16}{w^2 B'_{0y}}\int_{r_{\mathrm{s}}-}^{r_{\mathrm{s}}+} \mu_0\delta j_{1z}^{\mathrm{b}}\mathrm{d}r \,.$$

Due to the flattening of the pressure profile caused by formation of the island, $\delta j_{1z}^{\mathrm{b}}$ is given by [see (16.70)]

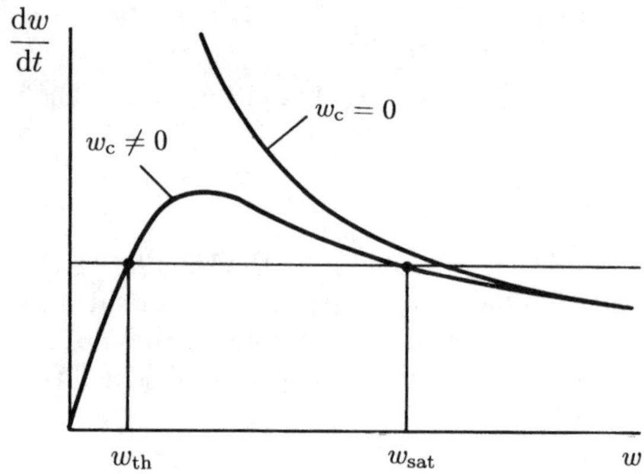


**Fig. 16.23.** The curve of (16.90). $w_{th}$ is the threshold width of the island for the onset of the neoclassical tearing mode, and $w_{sat}$ is the saturated width

$$\delta j_{1z}^{b} = 0 - \left( -\frac{\epsilon_s^{1/2}}{B_p} \frac{dp}{dr} \right) = \frac{\epsilon_s^{1/2}}{B_p} \frac{dp}{dr} . \qquad (16.88)$$

This is called the helical hole of the bootstrap current. Thus the discontinuity of the logarithmic derivative due to $\delta j_{1z}^{b}$ reduces to

$$\Delta_b' r_s = \frac{16\mu_0}{w^2 B_{0y}'} \left( \frac{\epsilon_s^{1/2}}{B_p} \frac{dp}{dr} \right)_{r_s} w r_s = \frac{8 r_s}{w} \frac{p}{B_p^2/2\mu_0} \epsilon_s^{1/2} \frac{L_q}{L_p} ,$$

$$B_{0y}' = -\frac{q'}{q} B_p \equiv -\frac{B_p}{L_q} , \quad \frac{dp}{dr} \equiv -\frac{p}{L_p} .$$

Then the time variation of the island width is given by

$$\tau_R \frac{d}{dt} \frac{w}{r_s} = \Delta' r_s + a \epsilon_s^{1/2} \beta_p \frac{L_q}{L_p} \frac{r_s}{w} , \quad a \sim 8 . \qquad (16.89)$$

The first term of right-hand side of (16.89) is the Rutherford term and the second is the destabilizing term of the bootstrap current. This is the equation for the neoclassical tearing mode. When transport and the effect of the ion polarization current across the island are taken into account, a reduction in the bootstrap current takes place. Then the term due to the bootstrap current is modified to

$$\tau_R \frac{d}{dt} \frac{w}{r_s} = \Delta' r_s + a_1 \beta_p \epsilon_s^{1/2} \frac{L_q}{L_p} \frac{r_s w}{w^2 + w_c^2} + a_2 \beta_p \left( \frac{L_q}{L_p} \right)^2 \frac{r_s \rho^2}{w^3} , \qquad (16.90)$$

where $w_c$ is the effect of transport across the island, parametrizing the magnitude of the contribution of the $\chi_\perp/\chi_\parallel$ model [16.49] and given by the relation

$$w_c = 1.8 r_s \left(\frac{8RL_q}{r_s^2 n}\right)^{1/2} \left(\frac{\chi_\perp}{\chi_\parallel}\right)^{1/4} .$$

The third term of the ion polarization current on the left-hand side of (16.90) is given in [16.50]. Figure 16.23 shows the curve of (16.90). When the effect of $w_c$ is included, there is a threshold $w_{th}$ for the onset of the neoclassical tearing mode. When $w$ becomes large, the destabilizing term in the bootstrap current becomes weak and the island width is saturated. The neoclassical tearing mode can be controlled by local current drive in the rational (singular) surface [16.51].

## 16.10 Tokamak Reactors

Although many parameters are needed to specify a tokamak device, there are also many relations and constraints between them [16.52]. If the plasma radius $a$, toroidal field $B_t$ and ratio $Q$ of fusion output power to auxiliary heating power are specified, the other parameters of the tokamak are determined by means of scaling laws for the electron density, beta, energy confinement time and burning condition, when the cylindrical safety factor $q_I$ (or effective safety factor at the plasma boundary $q_{eff}$, defined later), the elongation ratio $\kappa_s$, and triangularity $\delta$ of the plasma cross-section are given. From the definition of $q_I$, we have

$$q_I \equiv \frac{Ka}{R}\frac{B_t}{B_p} = \frac{5K^2 a B_t}{A I_p} , \qquad B_p = \frac{\mu_0 I_p}{2\pi K a} = \frac{I_p}{5Ka} ,$$

and the plasma current is

$$I_p = \frac{5K^2 a B_t}{A q_I} ,$$

where $K^2 = (1+\kappa_s^2)/2$ ($I_p$ in MA, $B_t$ in T, and $a$ in m). The aspect ratio $A = R/a$ will be given as a function of $a$ and $B_t$ in (16.97). The effective safety factor $q_{eff}$ at the plasma boundary is given approximately by [16.16]

$$q_{eff} = q_I f_{As}(A) ,$$

$$f_{As}(A) \approx \left(1 + \frac{1+\bar{\Lambda}^2/2}{A^2}\right)\left[1.24 - 0.54\kappa_s + 0.3(\kappa_s^2 + \delta^2) + 0.13\delta\right] ,$$

where $\bar{\Lambda} = \beta_p + l_i/2$. The volume average electron density $n_{20}$ in units of $10^{20}\,\mathrm{m}^{-3}$ is

$$n_{20} = N_G \frac{I_p}{\pi a^2} , \tag{16.91}$$

where $N_G$ is the Greenward normalized density. The beta ratio of the thermal plasma, viz.,

$$\beta_{\mathrm{th}} \equiv \frac{\langle p \rangle}{B_{\mathrm{t}}^2/2\mu_0} = 0.0403(1 + f_{\mathrm{DT}} + f_{\mathrm{He}} + f_{\mathrm{I}}) \frac{\langle n_{20} T \rangle}{B_{\mathrm{t}}^2} ,$$

is expressed by

$$\beta_{\mathrm{th}} = 0.01 \beta_{\mathrm{N}} \frac{I_{\mathrm{p}}}{a B_{\mathrm{t}}} , \tag{16.92}$$

where $\beta_{\mathrm{N}}$ is the normalized beta. The symbols $f_{\mathrm{DT}}$, $f_{\mathrm{He}}$ and $f_{\mathrm{I}}$ denote the ratios of fuel DT, He and impurity density to electron density, respectively, and the unit of $T$ is keV. $\langle X \rangle$ indicates the volume average of $X$. The thermal energy of the plasma $W$ is

$$W = \frac{3}{2} \beta_{\mathrm{th}} \frac{B_{\mathrm{t}}^2}{2\mu_0} V = 0.5968 \beta_{\mathrm{th}} B_{\mathrm{t}}^2 V ,$$

where $W$ is in units of MJ and the plasma volume $V$ is in units of $\mathrm{m}^{-3}$. The plasma shape with elongation ratio $\kappa_{\mathrm{s}}$ and triangularity $\delta$ is given by

$$R = R_0 + a \cos(\theta + \delta \sin\theta) , \qquad z = a \kappa_{\mathrm{s}} \sin\theta .$$

The plasma volume $V$ is given by

$$V \approx 2\pi^2 a^2 R \kappa_{\mathrm{s}} f_{\mathrm{shape}} ,$$

where $f_{\mathrm{shape}}$ is a correction factor due to the triangularity, viz.,

$$f_{\mathrm{shape}} = 1 - \frac{\delta}{8} + \frac{\delta^4}{192} - \frac{a}{4R} \left( \delta - \frac{\delta^3}{3} \right) .$$

We utilize the thermal energy confinement scaling of IPB98y2 [16.34]:

$$\tau_{\mathrm{E}} = 0.0562 \times 10^{0.41} H_{\mathrm{y2}} I^{0.93} B_{\mathrm{t}}^{0.15} M^{0.19} n_{20}^{0.41} a^{1.97} A^{1.39} \kappa_{\mathrm{s}}^{0.78} P^{-0.69} , \tag{16.93}$$

where $M(= 2.5)$ is the average ion mass unit and $P$ is the power loss in MW due to transport, equal to the necessary absorbed heating power subtracted from the radiation loss power $P_{\mathrm{rad}}$. The total $\alpha$ particle fusion output power $P_\alpha$ is

$$P_\alpha = \frac{Q_\alpha}{4} \langle n_{\mathrm{DT}}^2 \langle \sigma v \rangle_v \rangle V ,$$

where $Q_\alpha = 3.515\,\mathrm{MeV}$. $\langle \sigma v \rangle_v$ is a function of $T$ and a fitting equation for $\langle \sigma v \rangle_v$ is given in (1.5). Since the fusion rate $\sigma v$ near $T = 10\,\mathrm{keV}$ is approximated by

$$\langle \sigma v \rangle_v \approx 1.1 \times 10^{-24} T_{\mathrm{keV}}^2 \, (\mathrm{m}^3/s) ,$$

the following $\Theta$ ratio is introduced:

$$\Theta(\langle T \rangle) \equiv \frac{\langle n_{\mathrm{DT}}^2 \langle \sigma v \rangle_v \rangle}{1.1 \times 10^{-24} \langle n_{\mathrm{DT}}^2 T^2 \rangle} .$$

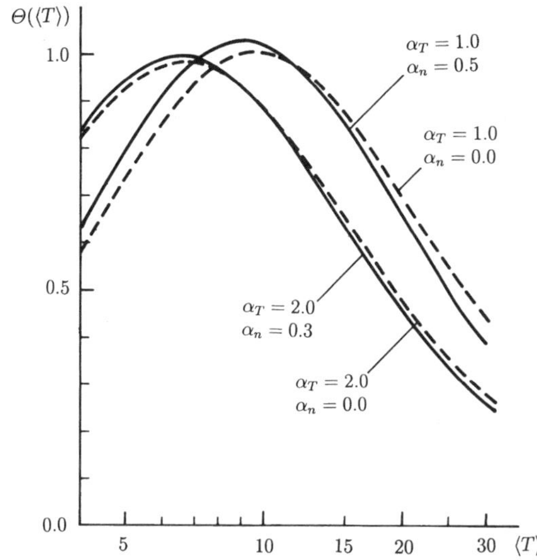


**Fig. 16.24.** $\Theta$ is function of the average temperature $\langle T\rangle$ (keV) in cases with profile parameters $(\alpha_T = 1.0, \alpha_n = 0.0)$, $(\alpha_T = 2.0, \alpha_n = 0.0)$, $(\alpha_T = 1.0, \alpha_n = 0.5)$ and $(\alpha_T = 2.0, \alpha_n = 0.3)$

$\Theta$ is a function of the average temperature $\langle T\rangle$ in keV and the density and temperature profiles have a peak of around 1 near $\langle T\rangle \approx$ 8–10 keV. The curves of $\Theta$ versus $\langle T\rangle$ for

$$n(\rho) = \frac{\langle n\rangle(1-\rho^2)^{\alpha_n}}{1+\alpha_n}\,, \qquad T(\rho) = \frac{\langle T\rangle(1-\rho^2)^{\alpha_T}}{1+\alpha_T}$$

are shown in Fig. 16.24 [16.53]. Then $P_\alpha$ reduces to

$$P_\alpha = 0.9551\frac{f_{\mathrm{prof}}f_{\mathrm{DT}}^2}{(1+f_{\mathrm{DT}}+f_{\mathrm{He}}+f_{\mathrm{I}})^2}\beta_{\mathrm{th}}^2 B_{\mathrm{t}}^4 \Theta V\,, \tag{16.94}$$

where $f_{\mathrm{prof}} \equiv \langle n^2T^2\rangle/\langle nT\rangle^2 \approx (\alpha_n+\alpha_T+1)^2/(2\alpha_n+2\alpha_T+1)$ is the profile effect of temperature and density. When the absorbed auxiliary heating power is denoted by $P_{\mathrm{aux}}$ and the heating efficiency of $\alpha$ heating is $f_\alpha$, the total heating power is $f_\alpha P_\alpha + P_{\mathrm{aux}}$. When the ratio of radiation loss power to total heating power is $f_{\mathrm{R}}$, the heating power to sustain the burning plasma is given by

$$P = (1-f_{\mathrm{R}})(f_\alpha P_\alpha + P_{\mathrm{aux}})\,.$$

When the $Q$ ratio is defined as the ratio of total fusion output power $P_{\mathrm{n}}+P_\alpha = 5P_\alpha$ ($P_{\mathrm{n}}$ is the neutron output power) to absorbed auxiliary heating power $P_{\mathrm{aux}}$, $Q$ is

$$Q = \frac{5P_\alpha}{P_{\mathrm{aux}}} .$$

Then $P$ reduces to

$$P = (1 - f_{\mathrm{R}}) \left( f_\alpha + \frac{5}{Q} \right) P_\alpha .$$

Therefore the burning condition is

$$\frac{W}{\tau_{\mathrm{E}}} = (1 - f_{\mathrm{R}}) \left( f_\alpha + \frac{5}{Q} \right) P_\alpha . \tag{16.95}$$

and (16.93)–(16.95) reduce to [16.53]

$$a = 3.22 f_{\mathrm{shape}}^{1.64} \left[ \frac{(1 + f_{\mathrm{DT}} + f_{\mathrm{He}} + f_{\mathrm{I}})^2}{(1 - f_{\mathrm{R}})(f_\alpha + 5/Q) f_{\mathrm{prof}} f_{\mathrm{DT}}^2 \Theta} \right]^{0.738} \tag{16.96}$$

$$\times \frac{\beta_{\mathrm{N}}^{0.905} \left[ q_{\mathrm{eff}} / f_{\mathrm{As}}(A) \right]^{2.29}}{H_{\mathrm{y2}}^{2.36} N_{\mathrm{G}}^{0.976} \kappa_{\mathrm{s}}^{0.214} K^{4.57}} \frac{A^{0.619}}{B_{\mathrm{t}}^{1.74}} .$$

Therefore the aspect ratio $A$ is given as a function of

$$A = C^{1.616} a^{1.616} B_{\mathrm{t}}^{2.81} , \tag{16.97}$$

where $C$ is the coefficient of $A^{0.619}/B_{\mathrm{t}}^{1.74}$ in (16.96). When the distance between the plasma separatrix and the conductor of the toroidal field coil is $\Delta$ and the maximum field of the toroidal field coil is $B_{\mathrm{max}}$ (see Fig. 16.25), there is a constraint

$$\frac{B_{\mathrm{t}}}{B_{\mathrm{max}}} = \frac{R - a - \Delta}{R} = 1 - \left( 1 + \frac{\Delta}{a} \right) \frac{1}{A}$$

and

$$1 - \frac{2}{A} < \frac{B_{\mathrm{t}}}{B_{\mathrm{max}}} < 1 - \frac{1}{A} ,$$

under the assumption $a > \Delta > 0$. By specification of $\Delta$ and $B_{\mathrm{max}}$, $B_{\mathrm{t}}$ is a function of $a$.

The ratio $\xi$ of the flux swing $\Delta\Phi$ of the ohmic heating coil and the flux of the plasma ring $L_{\mathrm{p}} I_{\mathrm{p}}$ is given by

$$\xi \equiv \frac{\Delta\Phi}{L_{\mathrm{p}} I_{\mathrm{p}}} = \frac{5 B_{\mathrm{max}}(\mathrm{T}) \left[ (R_{\mathrm{OH}} + d_{\mathrm{OH}})^2 + 0.5 d_{\mathrm{OH}}^2 \right]}{\left[ \ln(8A/\kappa_{\mathrm{s}}^{1/2}) + l_{\mathrm{i}} - 2 \right] R I_{\mathrm{p}}(\mathrm{MA})} ,$$

where $R_{\mathrm{OH}} = R - (a + \Delta + d_{\mathrm{TF}} + d_{\mathrm{s}} + d_{\mathrm{OH}})$, $d_{\mathrm{TF}}$ and $d_{\mathrm{OH}}$ being the thickness of the TF and OH coil conductors, respectively, and $d_{\mathrm{s}}$ the separation of the TF and OH coil conductors in meters (see Fig. 16.25). The average current densities $j_{\mathrm{TF}}$ and $j_{\mathrm{OH}}$ of the TF and OH coil conductors in $\mathrm{MA/m^2 = A/(mm)^2}$ are

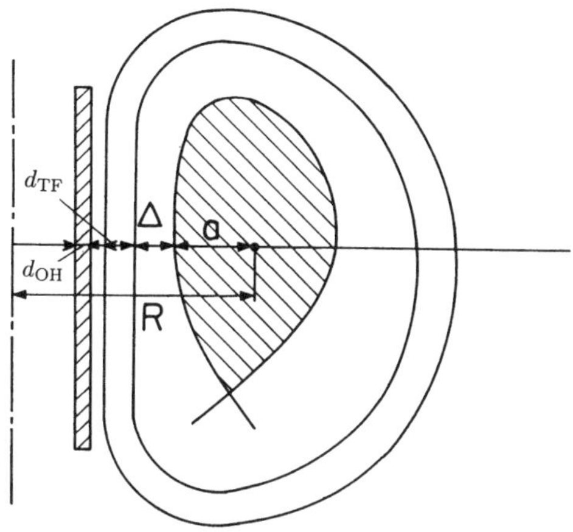


**Fig. 16.25.** Geometry of plasma, toroidal field coil, and central solenoid of the current transformer in a tokamak

$$j_{TF}(\mathrm{MA/m^2}) = \frac{2.5}{\pi}\frac{B_{max}(\mathrm{T})}{d_{TF}}\frac{1}{1-0.5d_{TF}/(R-a-\Delta)},$$

$$j_{OH}(\mathrm{MA/m^2}) = \frac{2.5}{\pi}\frac{B_{max}(\mathrm{T})}{d_{OH}}.$$

When parameters $a$, $B_t$, $A$ and the other dimensionless parameters are specified instead of $a$, $B_t$, $Q$ as is shown in Table 16.3, then the $Q$ value and the other parameters can be evaluated and are shown in Table 16.4.

The conceptual design of tokamak reactors has been actively pursued in the wake of tokamak experimental research. INTOR (International Tokamak

**Table 16.3.** Specified design parameters. The specified value of $\beta_N$ is the normalized beta of a thermal plasma which does not include the contribution of energetic ion components. ($H_{y2} = 1.0$, $\alpha_T = 1.0$, and $\alpha_n = 0.1$ are assumed)

| $a$(m) | $B_t$(T) | $A$ | $q_{eff}$ | $\kappa_s$ | $\delta$ | $f_R$ | $f_\alpha$ | $\beta_N$ | $N_G$ | $f_{DT}$ | $f_{He}$ | $f_I$ |
|---|---|---|---|---|---|---|---|---|---|---|---|---|
| 2.0 | 5.3 | 3.1 | 3.38 | 1.7 | 0.35 | 0.3 | 0.95 | 1.63 | 0.85 | 0.82 | 0.04 | 0.02 |

**Table 16.4.** Reduced parameters. Unit of $\langle T\rangle$ is keV and units of $P_n$, $P_\alpha$, $P_{aux}$ and $P_{rad}$ are MW.

| $Q$ | $R$(m) | $I_p$(MA) | $\tau_E(s)$ | $n_{20}$ | $\langle T\rangle$ | $P_n$ | $P_\alpha$ | $P_{aux}$ | $P_{rad}$ | $q_I$ | $\Theta$ |
|---|---|---|---|---|---|---|---|---|---|---|---|
| 10.2 | 6.2 | 15.0 | 3.8 | 1.01 | 8.1 | 324 | 81 | 40 | 35 | 2.22 | 0.99 |

**Table 16.5.** Parameters of ITER in the outline design of 2000

| | | | |
|---|---|---|---|
| $I_\mathrm{p}$ | 15 MA | $Z_\mathrm{eff}$ | 1.65 |
| $B_\mathrm{t}$ | 5.3 T | $f_\mathrm{DT}$ | $\sim 82\%$ |
| $R$ | 6.2 m | $f_\mathrm{He}$ | 4.1% |
| $a$ | 2.0 m | $f_\mathrm{Be}$ | 2% |
| $R/a$ | 3.1 | $f_\mathrm{Ar}$ | 0.12% |
| $\kappa_\mathrm{s}$ | 1.7 | $f_\mathrm{R}$ | 0.39 |
| $\langle n_\mathrm{e}\rangle$ | $1.01 \times 10^{20}\mathrm{m}^{-3}$ | $\beta_\mathrm{t}$ | 2.5% |
| $\langle 0.5 \times (T_\mathrm{e} + T_\mathrm{i})\rangle$ | 8.5 keV | $\beta_\mathrm{p}$ | 0.67 |
| $W_\mathrm{thermal}$ | 325 MJ | $\beta_\mathrm{N}$ | 1.77 |
| $W_\mathrm{fast}$ | 25 MJ | $N_\mathrm{G}$ | 0.85 |
| $\tau_\mathrm{E}^\mathrm{tr}$ | 3.7 s | $H_\mathrm{y2} = \tau_\mathrm{E}^\mathrm{tr}/\tau_\mathrm{E}^\mathrm{IPB98y2}$ | 1.0 |
| $P_\mathrm{fus}(P_\alpha)$ | 410 MW (82 MW) | $q_{95}$ | 3.0 |
| $P_\mathrm{aux}$ | 41MW | $q_\mathrm{I}$ | 2.22 |
| $P_\mathrm{rad}$ | 48MW | $l_\mathrm{i}$ | 0.86 |

Reactor) [16.54] and ITER (International Thermonuclear Experimental Reactor) [16.55, 16.56] are representative of international activity in this field. ITER aims [16.56] to achieve extended burn in inductively driven plasmas with $Q \sim 10$ and to demonstrate steady-state operation using non-inductive drive with $Q \sim 5$.

The main parameters of ITER in 2000 are given in Table 16.5. $\tau_\mathrm{E}^\mathrm{tr}$ is the energy confinement time corrected for radiation loss, $Q = 10$, $\kappa_\mathrm{s}$ is the ratio of the vertical radius to the horizontal radius, and $q_{95}$ is the safety factor at the 95% flux surface. The maximum field of the toroidal field coils is $B_\mathrm{max} = 11.8\,\mathrm{T}$. The number of toroidal field coils is 18 and the configuration involves a single null divertor. The one-turn loop voltage is $V_\mathrm{loop} = 89\,\mathrm{mV}$. The inductive pulse flat-top under $Q = 10$ condition is several hundred seconds. $P_\mathrm{fus}$ is the total fusion output power. $N_\mathrm{G}$ is defined in (16.7). $f_\mathrm{R}$ is the fraction of radiation loss and $f_\mathrm{DT}$, $f_\mathrm{Be}$, $f_\mathrm{He}$, and $f_\mathrm{He}$ are the ratios of the DT, Be, He, and Ar densities to the electron density.

A cross-section of the ITER outline design in 2000 is shown in Fig. 16.26.

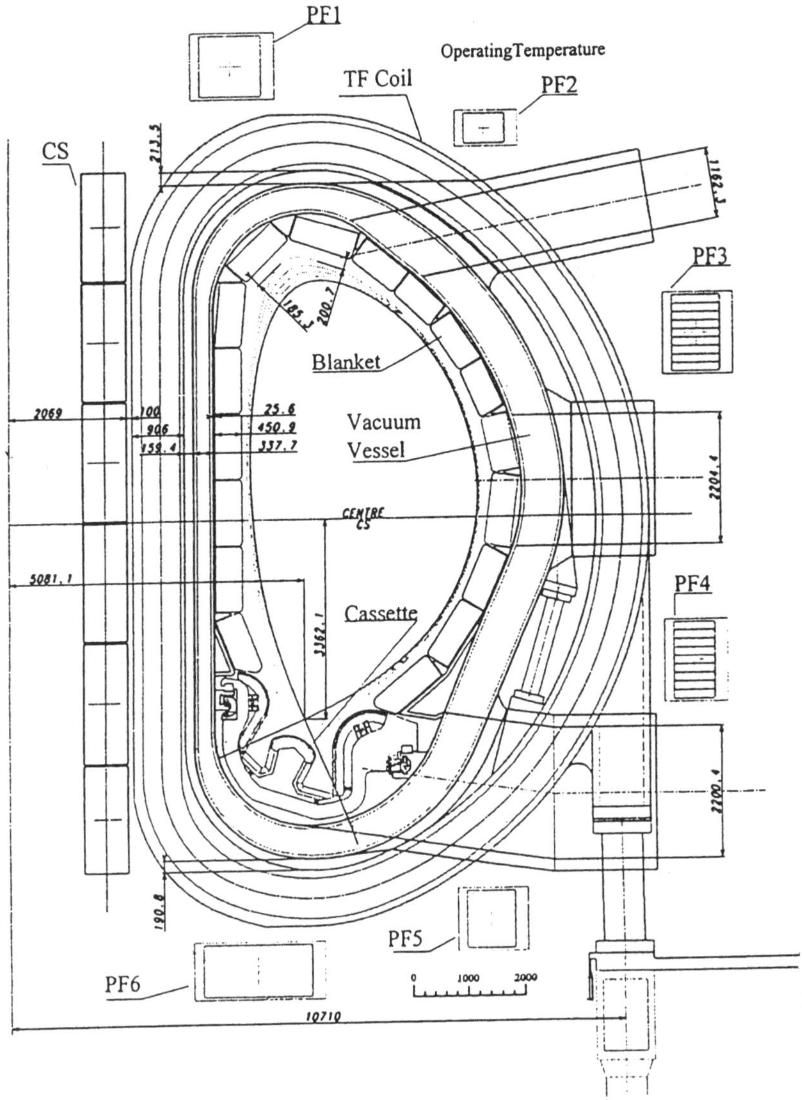


Fig. 16.26. Poloidal cross-section of ITER outline design in 2000 [16.56]

# 17 RFP Stellarator

## 17.1 Reversed Field Pinch

### 17.1.1 Reversed Field Pinch Configuration

Reversed field pinch (RFP) is an axisymmetric toroidal field used as a tokamak. The magnetic field configuration is composed of the poloidal field $B_p$ produced by the toroidal component of the plasma current and the toroidal field $B_t$ produced by the external toroidal field coil and the poloidal component of the plasma current. The particle orbit loss is as small as in a tokamak. However, RFP and tokamaks have quite different characteristics. In RFP, the magnitudes of the poloidal field $B_p$ and the toroidal field $B_t$ are comparable and the safety factor

$$q_s(r) = \frac{r}{R}\frac{B_z(r)}{B_\theta(r)}$$

is much less than 1 [$q_s(0) \sim a/(R\Theta)$, $\Theta \sim 1.6$]. The radial profile of the toroidal field is shown in Fig. 17.1. The direction of the boundary toroidal field is reversed with respect to the direction of the on-axis field, and the magnetic shear is strong. Therefore high-beta ($\langle\beta\rangle = 10$–$20\%$) plasmas can be confined in an MHD stable way. Since the plasma current can be larger than the Kruskal–Shafranov limit ($q < 1$), there is a possibility of reaching the ignition condition by ohmic heating alone (although it depends on the confinement scaling).

RFP started in an early phase of nuclear fusion research. A stable quiescent phase of discharge was found in Zeta in Harwell in 1968 [17.1]. The configuration of the magnetic field in the quiescent phase was the reversed field pinch configuration, as shown in Fig. 17.1. The electron temperature, the energy confinement time, and the average beta of Zeta were $\kappa T_e = 100$–$150\,\text{eV}$, $\tau_E = 2\,\text{ms}$, $\langle\beta\rangle \sim 10\%$ at the time of the IAEA conference at Novosibirsk. However, the epoch-making result of tokamak T-3 with high electron temperature ($\kappa T_e = 1\,\text{keV}$, $\tau_E =$ several ms, $\beta \sim 0.2\%$) was also presented in the same conference, and Zeta was shut down because of the better confinement characteristics in tokamaks. On the other hand, RFP can confine higher beta plasma and has been actively investigated to improve the confinement characteristics (ZT-40 M, OHTE, HBTX1-B, TPE-1RM 20, MST and RFX,

TPE-RX) [17.2–17.5]. The important issues of RFP are confinement scaling and impurity control in the high-temperature region.

### 17.1.2 MHD Relaxation

Even if the plasma is initially MHD unstable in the formation phase, it has been observed in RFP experiments that the plasma turns out to be a stable RFP configuration irrespective of the initial condition. J.B. Taylor pointed out in 1974 that the RFP configuration is a minimum energy state by relaxation processes under certain constraints [17.6].

We introduce a physical quantity referred to as *magnetic helicity* to study this subject. Using the scalar and vector potentials $\phi$, $\boldsymbol{A}$ of the electric and magnetic fields $\boldsymbol{E}$, $\boldsymbol{B}$, the magnetic helicity $K$ is defined by the integral of the scalar product $\boldsymbol{A}\cdot\boldsymbol{B}$ over the volume $V$ surrounded by a magnetic surface:

$$K = \int_V \boldsymbol{A}\cdot\boldsymbol{B}\,\mathrm{d}\boldsymbol{r}\,, \tag{17.1}$$

where $\mathrm{d}\boldsymbol{r} \equiv \mathrm{d}x\,\mathrm{d}y\,\mathrm{d}z$. Since

$$\boldsymbol{E} = -\nabla\phi - \frac{\partial \boldsymbol{A}}{\partial t}\,, \qquad \boldsymbol{B} = \nabla\times\boldsymbol{A}\,,$$

we find from Maxwell's equations [17.7]

$$\begin{aligned}\frac{\partial}{\partial t}(\boldsymbol{A}\cdot\boldsymbol{B}) &= \frac{\partial \boldsymbol{A}}{\partial t}\cdot\boldsymbol{B} + \boldsymbol{A}\cdot\frac{\partial \boldsymbol{B}}{\partial t} = (-\boldsymbol{E} - \nabla\phi)\cdot\boldsymbol{B} - \boldsymbol{A}\cdot(\nabla\times\boldsymbol{E}) \\ &= -\boldsymbol{E}\cdot\boldsymbol{B} - \nabla\cdot(\phi\boldsymbol{B}) + \nabla\cdot(\boldsymbol{A}\times\boldsymbol{E}) - \boldsymbol{E}\cdot(\nabla\times\boldsymbol{A}) \\ &= -\nabla\cdot(\phi\boldsymbol{B} + \boldsymbol{E}\times\boldsymbol{A}) - 2(\boldsymbol{E}\cdot\boldsymbol{B})\,.\end{aligned}$$

When the plasma is surrounded by a perfect conductive wall, then the conditions $(\boldsymbol{B}\cdot\boldsymbol{n}) = 0$, $\boldsymbol{E}\times\boldsymbol{n} = 0$ hold, where $\boldsymbol{n}$ is the unit outward vector normal to the wall, and we find

$$\frac{\partial K}{\partial t} = \frac{\partial}{\partial t}\int_V \boldsymbol{A}\cdot\boldsymbol{B}\mathrm{d}\boldsymbol{r} = -2\int_V \boldsymbol{E}\cdot\boldsymbol{B}\mathrm{d}\boldsymbol{r}\,. \tag{17.2}$$

The right-hand side of (17.2) is the loss term of the magnetic helicity. When Ohm's law

$$\boldsymbol{E} + \boldsymbol{v}\times\boldsymbol{B} = \eta\boldsymbol{j}$$

is applicable, the loss term reduces to

$$\frac{\partial K}{\partial t} = -2\int_V \eta\boldsymbol{j}\cdot\boldsymbol{B}\mathrm{d}\boldsymbol{r}\,. \tag{17.3}$$

When $\eta = 0$, the magnetic helicity is conserved. In other words, if a plasma is perfectly conductive, the integral of $K$ over the volume surrounded by

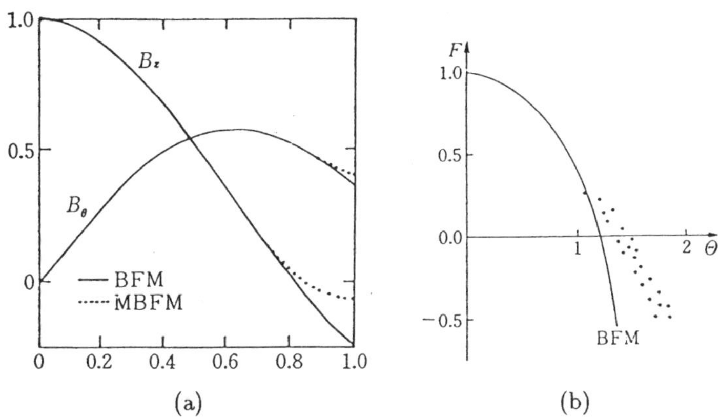


**Fig. 17.1.** (**a**) Toroidal field $B_z(r)$ and poloidal field $B_\theta(r)$ of RFP. The radial profiles of the Bessel function model (BFM) and the modified Bessel function model (MBFM) are shown. (**b**) $F$–$\Theta$ curve

arbitrary closed magnetic surfaces is constant. However, if there is small resistivity in the plasma, local reconnections of the lines of magnetic force are possible and the plasma can relax to a more stable state, whereupon the magnetic helicity may change locally. However, J.B. Taylor postulates that the global magnetic helicity $K_{\mathrm{T}}$ integrated over the whole region of the plasma changes much more slowly. It is assumed that $K_{\mathrm{T}}$ is constant within the time scale of relaxation processes. Under the constraint of invariant $K_{\mathrm{T}}$,

$$\delta K_{\mathrm{T}} = \int \boldsymbol{B}\cdot\delta\boldsymbol{A}\mathrm{d}\boldsymbol{r} + \int \delta\boldsymbol{B}\cdot\boldsymbol{A}\mathrm{d}\boldsymbol{r} = 2\int \boldsymbol{B}\cdot\delta\boldsymbol{A}\mathrm{d}\boldsymbol{r} = 0 ,$$

the condition of minimum energy of the magnetic field, viz.,

$$(2\mu_0)^{-1}\delta\int (\boldsymbol{B}\cdot\boldsymbol{B})\,\mathrm{d}\boldsymbol{r} = \mu_0^{-1}\int \boldsymbol{B}\cdot\nabla\times\delta\boldsymbol{A}\mathrm{d}\boldsymbol{r} = \mu_0^{-1}\int (\nabla\times\boldsymbol{B})\cdot\delta\boldsymbol{A}\mathrm{d}\boldsymbol{r} ,$$

can be obtained by the method of undetermined multipliers, and we have

$$\nabla\times\boldsymbol{B} - \lambda\boldsymbol{B} = 0 . \tag{17.4}$$

This solution is the minimum energy state in the force-free or pressureless plasma ($\boldsymbol{j}\times\boldsymbol{B} = \nabla p = 0$, $\boldsymbol{j}\|\boldsymbol{B}$). The axisymmetric solution in cylindrical coordinates is

$$B_r = 0 , \quad B_\theta = B_0 J_1(\lambda r) , \quad B_z = B_0 J_0(\lambda r) , \tag{17.5}$$

and is called a *Bessel function model* (BFM). The profiles of $B_\theta(r)$ and $B_z(r)$ are shown in Fig. 17.1a. In the region $\lambda r > 2.405$, the toroidal field $B_z$ is reversed. The pinch parameter $\Theta$ and the field reversal ratio $F$ are commonly used to characterize the RFP magnetic field as follows:

$$\Theta = \frac{B_\theta(a)}{\langle B_z \rangle} = \frac{(\mu_0/2)I_{\rm p}a}{\int B_z 2\pi r\,{\rm d}r}\,, \qquad F = \frac{B_z(a)}{\langle B_z \rangle}\,, \tag{17.6}$$

where $\langle B_z \rangle$ is the volume average of the toroidal field. The values of $F$ and $\Theta$ for the Bessel function model are

$$\Theta = \frac{\lambda a}{2}\,, \qquad F = \frac{\Theta J_0(2\Theta)}{J_1(2\Theta)}\,, \tag{17.7}$$

and the $F$–$\Theta$ curve is plotted in Fig. 17.1b. The quantity

$$\lambda = \frac{\mu_0 \boldsymbol{j}\cdot\boldsymbol{B}}{B^2} = \frac{(\boldsymbol{\nabla}\times\boldsymbol{B})\cdot\boldsymbol{B}}{B^2} = \text{const.}$$

is constant in the Taylor model. The RFP fields observed in experiments deviate from the Bessel function model due to the finite beta effect and the imperfect relaxation state. The $\lambda$ value is no longer constant in the outer region of the plasma and tends to 0 on the boundary. The solution of the relation $\boldsymbol{\nabla}\times\boldsymbol{B} - \lambda\boldsymbol{B} = 0$ with $\lambda(r)$ is called the modified Bessel function model (MBFM).

The stability condition of the local MHD mode [17.8] is

$$\frac{1}{4}\left(\frac{q_{\rm s}'}{q_{\rm s}}\right)^2 + \frac{2\mu_0 p'}{rB_z^2}(1-q_{\rm s}^2) > 0\,. \tag{17.8}$$

This formula indicates that the strong shear can stabilize the RFP plasma in the $p'(r) < 0$ region, but that the flat pressure profile, $p'(r) \sim 0$, is preferable in the central region of weak shear. When $q_s^2 < 1$, the local MHD mode is unstable near $q_s' = 0$ (*pitch minimum*).

When the effect of finite resistivity of a plasma is taken into account, according to the classical process of magnetic dissipation, the RFP configuration is expected to be sustainable only during the period $\tau_{\rm cl} = \mu_0\sigma a^2$, where $\sigma$ is the specific conductivity. However, ZT-40M experiments [17.9] demonstrated that RFP discharge was sustained more than three times ($\sim 20\,$ms) as long as $\tau_{\rm cl}$. This is clear evidence that a regeneration process of the toroidal flux exists during the relaxation process, which is consumed by classical magnetic dissipation, so that the RFP configuration can be sustained as long as the plasma current is sustained.

When there are fluctuations in plasmas, the magnetic field $\boldsymbol{B}$ in the plasma, for example, is expressed by the sum $\boldsymbol{B} = \langle\boldsymbol{B}\rangle_{\rm t} + \tilde{\boldsymbol{B}}$ of the time average $\langle\boldsymbol{B}\rangle_{\rm t}$ and the fluctuation term $\tilde{\boldsymbol{B}}$. The time average of Ohm's law $\eta\boldsymbol{j} = \boldsymbol{E} + \boldsymbol{v}\times\boldsymbol{B}$ reduces to

$$\langle\eta\boldsymbol{j}\rangle_{\rm t} = \langle\boldsymbol{E}\rangle_{\rm t} + \langle\boldsymbol{v}\rangle_{\rm t}\times\langle\boldsymbol{B}\rangle_{\rm t} + \langle\tilde{\boldsymbol{v}}\times\tilde{\boldsymbol{B}}\rangle_{\rm t}\,, \tag{17.9}$$

where $\langle\ \rangle_t$ denotes the time average. A new term $\langle \tilde{\boldsymbol{v}} \times \tilde{\boldsymbol{B}} \rangle_t$ appears due to fluctuations. Since the time average of the toroidal flux $\Phi_z = \int B_z \mathrm{d}S$ within the plasma cross-section is constant during the quasi-stationary state, the time average of the electric field in the $\theta$ direction is 0 ($\oint E_\theta \mathrm{d}l = -\mathrm{d}\Phi_z/\mathrm{d}t = 0$) and $\langle \boldsymbol{v}_r \rangle_t = 0$. Steady-state RFP plasmas require the condition

$$\langle \eta j_\theta \rangle_t = \langle (\tilde{\boldsymbol{v}} \times \tilde{\boldsymbol{B}})_\theta \rangle_t \,. \tag{17.10}$$

In other words, resistive dissipation is compensated by the effective electric field due to the fluctuations. This process is called the MHD dynamo mechanism. Active research has been carried out on the relaxation process [17.10–17.12].

When the electron mean free path is very long, local relations such as Ohm's law may not be applicable. To replace the MHD dynamo theory, the kinetic dynamo theory has been proposed [17.13], in which anomalous transport of electron momentum across magnetic surfaces plays an essential role in sustaining the RFP configuration.

Magnetic fluctuations of the dynamo sustaining the poloidal plasma current, on the other hand, enhance electron diffusion, since the electron diffusion coefficient is given by $D_\mathrm{e} \sim v_{T_\mathrm{e}} a \langle (\delta B_r/B)^2 \rangle$ (see Sect. 7.4) and the energy confinement of RFP deteriorates. In MST devices, a pulsed poloidal current drive (PPCD) is applied [17.14] and the poloidal plasma current (between the magnetic axis and the plasma edge) $I_\theta = 2\pi R a \langle j_\theta \rangle$ is increased transiently. The spatial distribution of the plasma current density is flattened and MHD fluctuations decrease. As a result, the energy confinement time is greatly improved (by a factor of $\sim 5$).

### 17.1.3 Confinement

The energy confinement time $\tau_\mathrm{E}$ in an ohmically heated plasma can be obtained from the energy balance equation:

$$\frac{(3/2)\langle n\kappa(T_\mathrm{e} + T_\mathrm{i})\rangle_\mathrm{v} 2\pi R \pi a^2}{\tau_\mathrm{E}} = V_z I_\mathrm{p} \,,$$

where $V_z$ is the loop voltage and $I_\mathrm{p}$ is the plasma current. The notation $\langle\ \rangle_\mathrm{v}$ indicates the volume average. Using the definition of the poloidal beta,

$$\beta_\theta \equiv \frac{\langle n\kappa(T_\mathrm{e} + T_\mathrm{i})\rangle_\mathrm{v}}{B_\theta^2/2\mu_0} = \frac{8\pi^2 a^2 \langle n\kappa(T_\mathrm{e} + T_\mathrm{i})\rangle_\mathrm{v}}{\mu_0 I_\mathrm{p}^2} \,,$$

the energy confinement time is given by

$$\tau_\mathrm{E} = \frac{3\mu_0}{8} R \beta_\theta \frac{I_\mathrm{p}}{V_z} \,. \tag{17.11}$$

Therefore the scalings of $\beta_\theta$ and $V_z$ are necessary for the scaling of $\tau_\mathrm{E}$. In order to apply a loop voltage to the RFP plasma, a cut in the toroidal direction is necessary in the shell conductor surrounding the plasma. In this case, the contribution of the surface integral must be added to the equation (17.1) giving the magnetic helicity:

$$\frac{\partial K}{\partial t} = -2\int \boldsymbol{E}\cdot\boldsymbol{B}\,\mathrm{d}\boldsymbol{r} - \int (\phi\boldsymbol{B} + \boldsymbol{E}\times\boldsymbol{A})\cdot\boldsymbol{n}\,\mathrm{d}S\,.$$

The induced electric field in the (conductive) shell surface is zero and is concentrated between the two edges of the shell cut. The surface integral consists of the contribution $2V_z\Phi_z$ from the shell cut and the contribution from the other part of the surface $S_-$, i.e.,

$$\frac{\partial K}{\partial t} = -2\int \eta\boldsymbol{j}\cdot\boldsymbol{B}\,\mathrm{d}\boldsymbol{r} + 2V_z\Phi_z - \int_{\mathrm{S}_-} (\phi\boldsymbol{B} + \boldsymbol{E}\times\boldsymbol{A})\cdot\boldsymbol{n}\,\mathrm{d}S\,, \qquad (17.12)$$

where $\Phi_z$ is the volume average of the toroidal magnetic flux $\Phi_z = \pi a^2\langle B_z\rangle_{\mathrm{v}}$. In the quasi-steady state, the time average $\langle\partial K/\partial t\rangle_{\mathrm{t}}$ is zero. Then the time average of (17.2) yields

$$V_z = \frac{\int\langle\eta\boldsymbol{j}\cdot\boldsymbol{B}\rangle_{\mathrm{t}}\mathrm{d}\boldsymbol{r} + (1/2)\int_{\mathrm{S}_-}\langle\phi\boldsymbol{B} + \boldsymbol{E}\times\boldsymbol{A}\rangle_{\mathrm{t}}\cdot\boldsymbol{n}\,\mathrm{d}S}{\langle\Phi_z\rangle_{\mathrm{t}}} = \frac{2\pi R}{\pi a^2}\eta_0 I_{\mathrm{p}}\zeta + V_{\mathrm{B}}\,,$$

$$V_{\mathrm{B}} = \frac{2\pi R}{a}\frac{\langle\langle(\phi\boldsymbol{B} + \boldsymbol{E}\times\boldsymbol{A})\cdot\boldsymbol{n}\rangle_{\mathrm{t}}\rangle_{\mathrm{S}_-}}{\langle\langle B_z\rangle_{\mathrm{t}}\rangle_{\mathrm{v}}}\,,$$

where $\langle\ \ \rangle_{\mathrm{S}_-}$ is the average in the surface region $\mathrm{S}_-$. The notation $\zeta$ is a dimensionless factor determined by the radial profiles of the specific resistivity and magnetic field as follows:

$$\zeta \equiv \frac{\langle\langle\eta\boldsymbol{j}\cdot\boldsymbol{B}\rangle_{\mathrm{t}}\rangle_{\mathrm{v}}}{\eta_0\langle\langle j_z\rangle_{\mathrm{t}}\rangle_{\mathrm{v}}\langle\langle B_z\rangle_{\mathrm{t}}\rangle_{\mathrm{v}}} = \frac{\langle\langle\eta\boldsymbol{j}\rangle_{\mathrm{t}}\cdot\langle\boldsymbol{B}\rangle_{\mathrm{t}}\rangle_{\mathrm{v}} + \langle\langle\widetilde{(\eta\boldsymbol{j})}\cdot\widetilde{\boldsymbol{B}}\rangle_{\mathrm{t}}\rangle_{\mathrm{v}}}{\eta_0\langle\langle j_z\rangle_{\mathrm{t}}\rangle_{\mathrm{v}}\langle\langle B_z\rangle_{\mathrm{t}}\rangle_{\mathrm{v}}}\,.$$

Here $\eta_0$ is the specific resistivity at the plasma center. When the fluctuation term is negligible, the value $\zeta$ of the modified Bessel function model is $\zeta \sim 10$, but the value is generally $\zeta > 10$ due to fluctuations. The value of $V_{\mathrm{B}}$ is 0 when the whole plasma boundary is a conductive shell. In reality, the plasma boundary is a liner or protecting material for the liner. Lines of magnetic force can cross the wall by the magnetic fluctuation or shift of plasma position ($\boldsymbol{B}\cdot\boldsymbol{n} \neq 0$, $\boldsymbol{E} \neq 0$). Then the term $V_{\mathrm{B}}$ has a finite value. Substituting $V_z$ into the equation for the energy confinement time $\tau_{\mathrm{E}}$ gives

$$\frac{1}{\tau_{\mathrm{E}}} = \frac{8}{3\beta_\theta}\left[\left(\frac{\eta_0}{\mu_0 a^2}\right)2\zeta + \frac{V_{\mathrm{B}}/2\pi R}{aB_\theta(a)}\right]\,.$$

When plasmas become hot, the resistive term becomes small and the fluctuation term and contribution from $V_{\mathrm{B}}$ are no longer negligible. The experimental scaling in the region $I_{\mathrm{p}} < 0.5\,\mathrm{MA}$ is $I_{\mathrm{p}}/\pi a^2\langle n\rangle_{\mathrm{v}} = (1\text{–}5)\times 10^{-14}\,\mathrm{A\,m}$, $\beta_\theta \sim 0.1$, $[\kappa T_{\mathrm{e}}(0)]_{\mathrm{keV}} \sim I_{\mathrm{p}}(\mathrm{MA})$.

### 17.1.4 Oscillating Field Current Drive

RFP plasmas tend to be modelled by the modified Bessel function model due to the non-linear phenomena of MHD relaxation. Oscillating field current drive (OFCD) was proposed [17.15] to sustain the plasma current, and preliminary experiments have been done [17.16]. If the terms $V_z$ and $\Phi_z$ of the second term on the right-hand side of the magnetic helicity balance equation (17.12) are modulated according to $V_z(t) = \tilde{V}_z \cos\omega t$, $\Phi_z(t) = \Phi_{z0} + \tilde{\Phi}_z \cos\omega t$, a direct current component $\tilde{V}_z \tilde{\Phi}_z$ appears in the product $2V_z\Phi_z$ and compensates the resistive loss of the magnetic helicity. The period of the oscillating field must be longer than the characteristic time of relaxation and shorter than the magnetic diffusion time. Furthermore, the disturbing effect of the oscillating field on the RFP plasma must be evaluated.

## 17.2 Stellarator

A stellarator field can provide a steady-state magnetohydrodynamic equilibrium configuration of the plasma only by the external field produced by the coils outside the plasma. The rotational transform, which is needed to confine the toroidal plasma, is formed by the external coils so that the stellarator has the merit of steady-state confinement. Although Stellarator C [17.17] was rebuilt as the ST tokamak in 1969 at the Princeton Plasma Physics Laboratory, confinement experiments by Wendelstein 7A, 7AS, Heliotron-E, and ATF are being carried out, for the benefits of steady-state confinement without current-driven instabilities. The large helical device LHD started experiments in 1998 and the advanced stellarator WVII-X is under construction.

### 17.2.1 Helical Field

Let us consider a magnetic field with helical symmetry. Using cylindrical coordinates $(r,\ \theta,\ z)$, we can express the field in terms of $(r,\ \varphi \equiv \theta - \delta\alpha z)$, where $\alpha > 0$, $\delta = \pm 1$. A magnetic field in a current-free region ($\boldsymbol{j} = 0$) can be expressed in terms of a scalar potential $\phi_B$, satisfying $\Delta\phi_B = 0$, and we can write

$$\phi_B = B_0 z + \frac{1}{\alpha}\sum_{l=1}^{\infty} b_l I_l(l\alpha r)\sin(l\varphi)\ , \tag{17.13}$$

$$\varphi \equiv \theta - \delta\alpha z\ .$$

The field components $(B_r,\ B_\theta,\ B_z)$ of $\boldsymbol{B} = \boldsymbol{\nabla}\phi_B$ are given by

$$B_r = \sum_{l=1}^{\infty} l b_l I_l'(l\alpha r)\sin(l\varphi)\ , \tag{17.14}$$

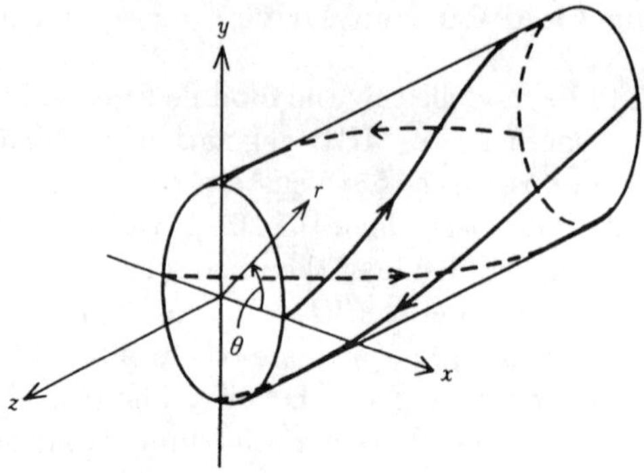


**Fig. 17.2.** Current of helical coils

$$B_\theta = \sum_{l=1}^{\infty} \left(\frac{1}{\alpha r}\right) l b_l I_l(l\alpha r)\cos(l\varphi) \ , \tag{17.15}$$

$$B_z = B_0 - \delta \sum_{l=1}^{\infty} l b_l I_l(l\alpha r)\cos(l\varphi) \ . \tag{17.16}$$

The vector potential corresponding to this field has components

$$A_r = -\frac{\delta}{\alpha^2 r}\sum_{l=1}^{\infty} b_l I_l(l\alpha r)\sin(l\varphi) \ , \quad A_\theta = \frac{B_0}{2}r - \frac{\delta}{\alpha}\sum_{l=1}^{\infty} b_l I_l'(l\alpha r)\cos(l\varphi) \ ,$$

$$A_z = 0 \ .$$

Using these, we can write

$$B_r = -\frac{\partial A_\theta}{\partial z} \ , \quad B_\theta = \frac{\partial A_r}{\partial z} \ , \quad B_z = \frac{1}{r}\frac{\partial (rA_\theta)}{\partial r} - \frac{1}{r}\frac{\partial A_r}{\partial \theta} \ .$$

The magnetic surface $\psi = A_z + \delta\alpha r A_\theta = \delta\alpha r A_\theta = \text{const.}$ is given by

$$\psi(r,\varphi) = B_0\frac{\delta\alpha r^2}{2} - r\sum_{l=1}^{\infty} b_l I_l'(l\alpha r)\cos(l\varphi) = \text{const.} \tag{17.17}$$

Such a helically symmetric field can be produced by a helical current distribution, as shown in Fig. 17.2. Let the magnetic fluxes in the $z$ and $\theta$ directions inside the magnetic surface be denoted by $\Phi$ and $X$, where $X$ is the integral over the pitch along $z$, i.e., over $2\pi/\alpha$. Then these may be expressed by

$$\Phi = \int_0^{2\pi}\int_0^{r(\varphi)} B_z(r,\,\varphi) r\,\mathrm{d}r\,\mathrm{d}\theta \ ,$$

$$X = \int_0^{2\pi/\alpha}\int_0^{r(\varphi)} B_\theta(r,\,\varphi)\,\mathrm{d}r\,\mathrm{d}z = \frac{1}{\alpha}\int_0^{2\pi}\int_0^{r(\varphi)} B_\theta(r,\,\varphi)\,\mathrm{d}r\,\mathrm{d}\theta \ .$$

Since $\alpha r B_z - \delta B_\theta = \alpha \partial(r A_\theta)/\partial r = \delta \partial\psi/\partial r$, we find that

$$\Phi - \delta X = \frac{2\pi\psi}{\delta\alpha} .$$

Let us consider only one harmonic component of the field. The scalar potential and the magnetic surface are expressed by

$$\phi_B = B_0 z + \frac{b}{\alpha} I_l(l\alpha r)\sin(l\theta - \delta l\alpha z) ,$$

$$\psi = \frac{B_0}{2\delta\alpha}\left[(\alpha r)^2 - \frac{2\delta(\alpha r)b}{B_0} I_l'(l\alpha r)\cos(l\theta - \delta l\alpha z)\right] = \frac{B_0}{2\delta\alpha}(\alpha r_0)^2 .$$

The singular points $(r_\mathrm{s},\, \theta_\mathrm{s})$ in the $z = 0$ plane are given by

$$\frac{\partial\psi}{\partial r} = 0 , \qquad \frac{\partial\psi}{\partial\theta} = 0 .$$

Since the modified Bessel function $I_l(x)$ satisfies

$$I_l''(x) + \frac{I_l'(x)}{x} - (1 + l^2/x^2) I_l = 0 ,$$

the singular points are given by

$$\sin(l\theta_\mathrm{s}) = 0 , \qquad \alpha r\left\{1 - \frac{\delta b l}{B_0}\left[1 + \frac{1}{(\alpha r_\mathrm{s})^2}\right] l_l(l\alpha r_\mathrm{s})\cos(l\theta_\mathrm{s})\right\} = 0 ,$$

or

$$\theta_\mathrm{s} = \begin{cases} \dfrac{2\pi}{l}(j-1) , & \delta b/B_0 > 0 , \\ \dfrac{2\pi}{l}\left(j - \dfrac{1}{2}\right) , & \delta b/B_0 < 0 , \end{cases} \qquad j = 1, \ldots, l ,$$

$$\left|\frac{\delta b l}{B_0}\right| = \frac{1}{[1 + (\alpha r_\mathrm{s})^{-2}] I_l(l\alpha r_\mathrm{s})} .$$

The magnetic surfaces for $l = 1, 2, 3$ are shown in Fig. 17.3. The magnetic surface which passes through the hyperbolic singular point or X point is called the *separatrix*. When $x \ll 1$, the modified Bessel function is

$$I_l(x) \approx \frac{1}{l!}\left(\frac{x}{2}\right)^l .$$

The magnetic surfaces in the region $\alpha r \ll 1$ are expressed by

$$(\alpha r)^2 - \frac{\delta b (l/2)^{l-1}}{B_0 (l-1)!}(\alpha r)^l \sin l(\theta - \delta\alpha z) = \text{const.}$$

The magnitude $B$ is given by

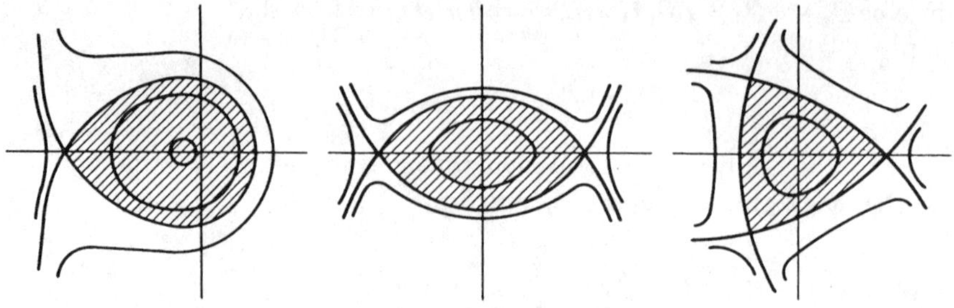


**Fig. 17.3.** Magnetic surfaces of the helical field, showing X points and separatrices

$$\left(\frac{B}{B_0}\right)^2 = 1 - 2\frac{\delta lb}{B_0} I_l \cos(l\varphi) + \left(\frac{lb}{B_0}\right)^2 \left\{ I_l^2 \left[1 + \frac{1}{(\alpha r)^2}\right] \cos^2(l\varphi) + (I_l')^2 \sin^2(l\varphi) \right\} .$$

The magnitude $B$ at the X point $(r_\mathrm{s},\ \theta_\mathrm{s})$ is given by

$$\left(\frac{B}{B_0}\right)^2 = 1 - \frac{(\alpha r)^2}{1 + (\alpha r)^2} ,$$

and at the point $(r_\mathrm{s},\ \theta_\mathrm{s} + \pi/l)$ by

$$\left(\frac{B}{B_0}\right)^2 = 1 + \frac{(\alpha r)^2}{1 + (\alpha r)^2} .$$

Therefore the magnitude $B$ is small at X points.

Let us estimate the rotational transform angle $\iota$. As the line of magnetic force is expressed by

$$\frac{\mathrm{d}r}{B_r} = \frac{r\mathrm{d}\theta}{B_\theta} = \frac{\mathrm{d}z}{B_z} ,$$

the rotational transform angle is given by

$$\frac{r\iota}{2\pi R} = \left\langle \frac{r\mathrm{d}\theta}{\mathrm{d}z} \right\rangle = \left\langle \frac{B_\theta}{B_z} \right\rangle = \left\langle \frac{(1/\alpha r) lb I_l(l\alpha r) \cos l(\theta - \delta z)}{B_0 - lb I_l(l\alpha r) \cos l(\theta - \delta z)} \right\rangle .$$

Here $r$ and $\theta$ are the values on the line of magnetic force and are functions of $z$, and $\langle\ \rangle$ denotes the average over $z$. In a vacuum field,

$$\oint B_\theta \mathrm{d}l = \int (\nabla \times \boldsymbol{B}) \cdot \mathrm{d}\boldsymbol{S} = 0 ,$$

so that the rotational transform angle is 0 to first order in $b/B_0$. However the first-order components of $B_\theta$ and $B_z$ resonate to yield the second-order rotational transform angle. The average method gives the formula for the *rotational transform angle* [17.18, 17.19]:

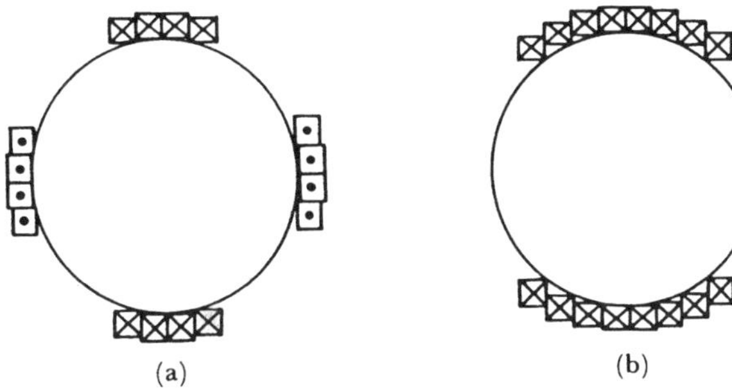


**Fig. 17.4.** Cross-sectional views of helical coils in the $l = 2$ case. (**a**) standard stellarator. (**b**) heliotron/torsatron

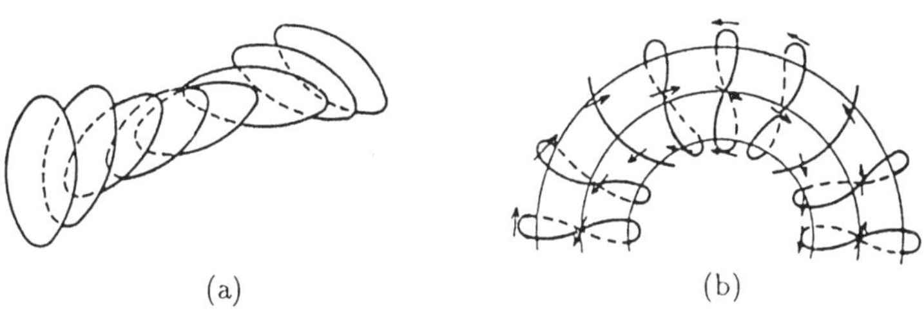


**Fig. 17.5.** (**a**) Arrangement of elliptical coils used to produce an $l = 2$ linear helical field. (**b**) twisted toroidal coils that produce the $l = 2$ toroidal helical field

$$\frac{\iota}{2\pi} = \delta \left(\frac{b}{B}\right)^2 \frac{l^3}{2} \left[\frac{\mathrm{d}}{\mathrm{d}x}\left(\frac{I_l I_l'}{x}\right)\right]_{x=l\alpha r} \frac{R}{r} . \tag{17.18}$$

Using the expansion

$$I_l(x) = \left(\frac{x}{2}\right)^l \left[\frac{1}{l!} + \frac{1}{(l+1)!}x^2 + \frac{1}{2!(l+2)!}x^4 + \cdots\right] ,$$

we find

$$\frac{\iota}{2\pi} = \delta \left(\frac{b}{B}\right)^2 \left(\frac{1}{2^l l!}\right)^2 l^5 (l-1)\alpha R \left[(l\alpha r)^{2(l-2)} + \cdots\right] \quad (l \geq 2) . \tag{17.19}$$

An example of the analysis of the toroidal helical field is given in [17.20].

### 17.2.2 Stellarator Devices

Familiar helical fields are of pole number $l = 2$ or $l = 3$. The three-dimensional magnetic axis system of Heliac has the $l = 1$ component. When the ratio of

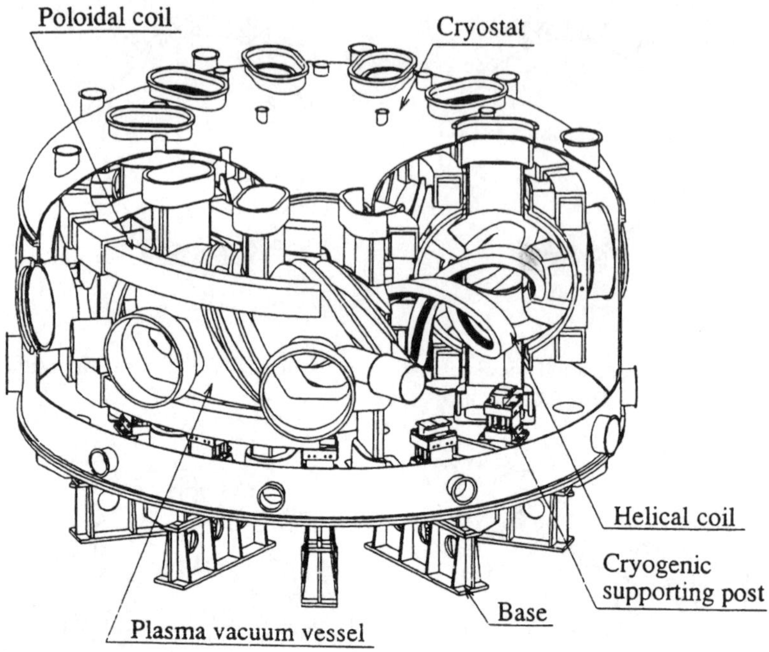


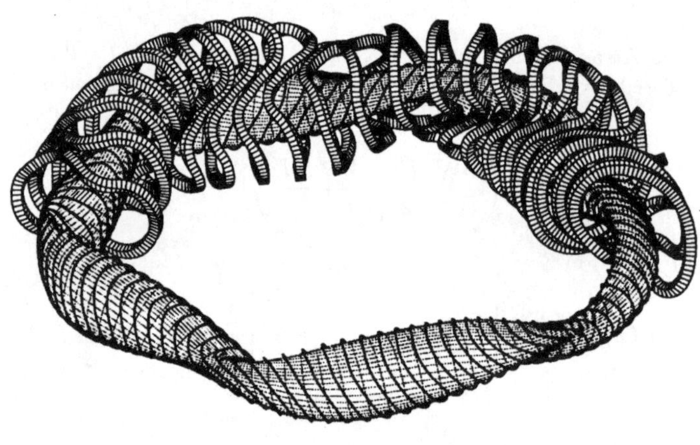

**Fig. 17.6.** *Upper*: schematic view of the LHD device in Toki ($R = 3.9\,\mathrm{m}$, $a \sim 0.6\,\mathrm{m}$, $B = 3\,\mathrm{T}$) [17.44]. *Lower*: modular coil system and a magnetic surface of the optimized stellerator Wendelstein 7-X under construction in Greifswald ($R = 5.5\,\mathrm{m}$, $a = 0.55\,\mathrm{m}$, $B = 3\,\mathrm{T}$) [17.43]

the minor radius $a_h$ of a helical coil to the helical pitch length $R/m$ ($R$ is the major radius and $m$ the number of field periods) is much less than 1, i.e., $ma_h/R \ll 1$, the rotational transform angle is $\iota_2(r) = \text{const.}$ for $l = 2$ and $\iota_3(r) = \iota(r/a)^2$ for $l = 3$. In this case the shear is small for the $l = 2$ configuration, and $\iota_3(r)$ is very small in the central region for the $l = 3$ configuration. However, if $ma_h/R \sim 1$, then $\iota_2(r) = \iota_0 + \iota_2(r/z)^2 + \cdots$, so that the shear can be large even when $l = 2$.

The arrangement of coils in the $l = 2$ case is shown in Fig. 17.4. Figure 17.4a is the standard type of stellarator [17.21, 17.22] and Fig. 17.4b is a heliotron/torsatron type [17.23, 17.24]. Helical fields are usually produced by the toroidal field coils and the helical coils. In the heliotron/torsatron configuration the current directions of the helical coils are the same, so that the toroidal field and the helical field can be produced by the helical coils alone [17.27,17.28]. Therefore, if the pitch is properly chosen, closed magnetic surfaces can be formed even without toroidal field coils [17.29, 17.30]. The typical devices of this type are Heliotron E, ATF and LHD. The device LHD is shown in the upper part of Fig. 17.6.

When elliptical coils are arranged as shown in Fig. 17.5a, an $l = 2$ helical field can be obtained [17.25]. The currents produced by the twisted toroidal coil system shown in Fig. 17.5b can simulate the currents of the toroidal field coils and the helical coils taken together [17.26]. Typical devices of this modular coil type are Wendelstein 7AS and 7X. The modular coil system of Wendelstein 7X is shown in the lower part of Fig. 17.6.

For linear helical fields, the magnetic surface $\Psi = rA_\theta$ exists due to its helical symmetry. However, the existence of magnetic surfaces in toroidal helical fields has not yet been proven in the strict mathematical sense. According to numerical calculations, the magnetic surfaces exist in the central region near the magnetic axis, but in the outer region the lines of magnetic force behave ergodically and the magnetic surfaces are destroyed. Although the helical coils have a relatively complicated structure, the lines of magnetic force can be traced by computer, and the design of helical field devices becomes less elaborate. The effect of the geometrical error on the helical field can be estimated and accurate coil windings are possible with numerically controlled devices ($\Delta l/R < 0.05$–$0.1\%$).

### 17.2.3 Neoclassical Diffusion in Helical Field

To analyze classical diffusion due to Coulomb collision, one must study the orbits of charged particles. In a helical field, or even in a tokamak, the toroidal field is produced by a finite number of coils, and there is an asymmetric inhomogeneous term in the magnitude $B$ of the magnetic field,

$$\frac{B}{B_0} \approx 1 - \epsilon_h \cos(l\theta - m\varphi) - \epsilon_t \cos\theta \;, \tag{17.20}$$

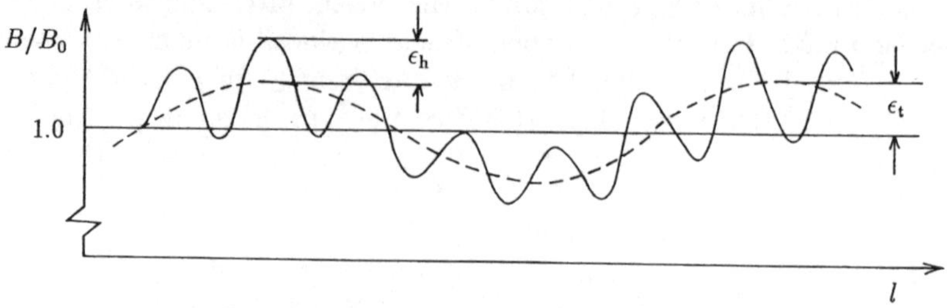


**Fig. 17.7.** Variation of the magnitude $B$ along the length $l$ of the line of magnetic force

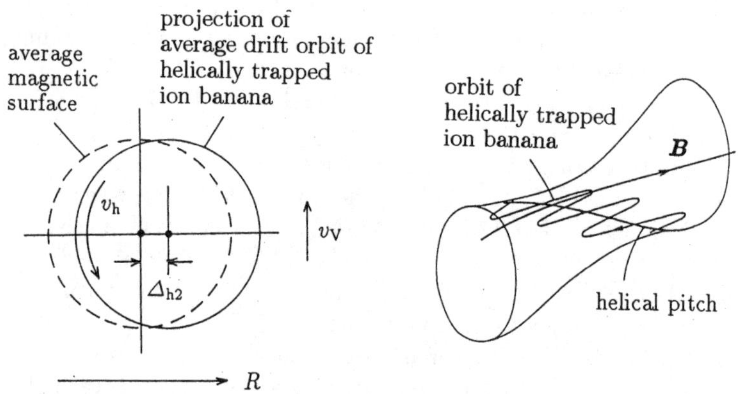


**Fig. 17.8.** Orbit of a helical banana ion trapped in a helical ripple

in addition to the toroidal term $-\epsilon_t \cos\theta$. The variation of $B$ along lines of magnetic force is shown in Fig. 17.7. Particles trapped by the inhomogeneous field of helical ripples drift across the magnetic surfaces and contribute to the particle diffusion in addition to the banana particles, as was discussed for tokamaks. The curvature of the line of magnetic force near the helically trapped region is convex outward and is denoted by $R_h$. Helically trapped particles drift in the poloidal direction ($\theta$ direction) due to $\nabla B$ drift with the velocity $v_h \approx mv_\perp^2/(qBR_h)$ (see Fig. 17.8). The angular velocity of poloidal rotation is

$$w_h = \frac{v_h}{r} \approx \frac{r}{R_h}\frac{kT}{qBr^2} \,. \tag{17.21}$$

In the case of a linear helical field ($\epsilon_t = 0$), helically trapped particles rotate along the magnetic surface. However, in the case of a toroidal helical field, the toroidal drift is superposed and the toroidal drift velocity is $v_v = kT/(qBR)$ in the vertical direction (see Sect. 3.5). When the effective collision time $(\nu_{\rm eff})^{-1} = (\nu/\epsilon_h)^{-1}$ is shorter than one period $(\omega_h)^{-1}$ of poloidal rotation, the deviation of the orbit of the helical banana ion from the magnetic surface is

$$\Delta_{\rm h1} = v_{\rm v}\frac{\epsilon_{\rm h}}{\nu} = \epsilon_{\rm h}\frac{kT}{qBR}\frac{1}{\nu} \, .$$

Then the coefficient of particle diffusion becomes [17.31]

$$D_{\rm h1} \sim \epsilon_{\rm h}^{1/2}\Delta_{\rm h1}^2\nu_{\rm eff} = \epsilon_{\rm h}^{3/2}\left(\frac{kT}{qBR}\right)^2\frac{1}{\nu} = \epsilon_t^2\epsilon_{\rm h}^{3/2}\left(\frac{kT}{qBr^2}\frac{1}{\nu}\right)\left(\frac{kT}{qB}\right) \, .$$

Since $R_{\rm h} \sim r/\epsilon_{\rm h}$, the other expression is

$$D_{\rm h1} \sim \gamma_{\rm h}\epsilon_{\rm h}^{1/2}\epsilon_{\rm t}^2\left(\frac{\omega_{\rm h}}{\nu}\right)\left(\frac{kT}{qB}\right) \qquad (\nu/\epsilon > \omega_{\rm h}) \, , \qquad (17.22)$$

where $\gamma_{\rm h}$ is a coefficient of order $O(1)$ (Fig. 17.9).

When the effective collision time $(\nu_{\rm eff})^{-1}$ is longer than $(\omega_{\rm h})^{-1}$, the deviation $\Delta_{\rm h2}$ of the orbit from the magnetic surface is

$$\Delta_{\rm h2} \approx \frac{v_{\rm v}}{\omega_{\rm h}} \approx \frac{R_{\rm h}}{R}r \sim \frac{\epsilon_{\rm t}}{\epsilon_{\rm h}}r \, ,$$

and the particle diffusion coefficient $D_{\rm h2}$ in this region becomes (Fig. 17.9)

$$D_{\rm h2} \approx \epsilon_{\rm h}^{1/2}\Delta_{\rm h2}^2\nu_{\rm eff} = \left(\frac{\epsilon_{\rm t}}{\epsilon_{\rm h}}\right)^2\frac{1}{\epsilon_{\rm h}^{1/2}}r^2\nu \qquad (\nu/\epsilon_{\rm h} < \omega_{\rm h}) \, .$$

When a particle is barely trapped in a local helical mirror, the particle moves very slowly near the reflection point where the magnetic field is locally maximum and the field line is concave outward. The effective curvature, which the particle feels in time average, becomes negative (concave). The orbit of the trapped particle in this case becomes the so-called *superbanana* [17.31]. However, this theoretical treatment is based on the assumption concerning the longitudinal adiabatic invariant $J_{\|} =$ const. along the orbit of the helically trapped particle. The adiabatic invariance is applicable when the poloidal rotation angle, during one period of back and forth motion in the helical local mirror, is small. As the single period of back and forth motion of the barely trapped particles becomes long, the adiabatic invariance may not be applicable in this case. The orbit plotted from numerical calculations shows that the superbanana does not appear [17.32] in the realistic case when $\epsilon_{\rm h} \sim \epsilon_{\rm t}$. If a particle orbit crosses the wall, the particle is lost. This is called orbit loss. A loss region in velocity space appears due to orbit loss in some cases [17.33]. When a radial electric field appears, the angular frequency of the poloidal drift rotation becomes $\omega_{\rm h} + \omega_{\rm E}$ $(\omega_{\rm E} = E_r/B_0)$ and the orbit is affected by the radial electric field.

The thermal diffusion coefficient $\chi_{\rm h1}$ due to helically trapped particles in the region $\nu/\epsilon_{\rm h} > \omega_{\rm h}$ is given by

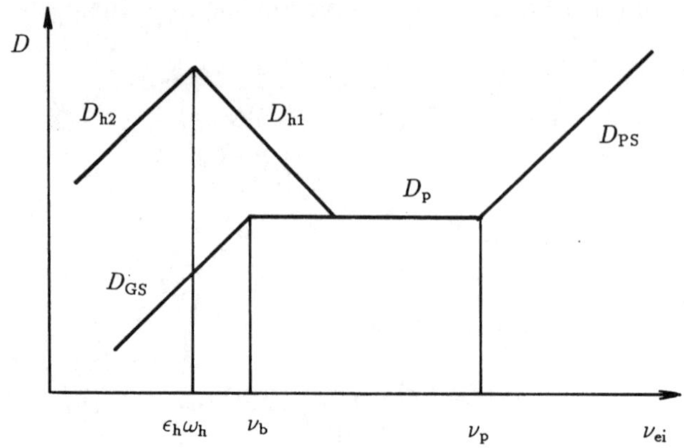


**Fig. 17.9.** Dependence of the neoclassical diffusion coefficient of the helical field on the collision frequency. $\nu_{\mathrm{p}} = (\iota/2\pi)v_{T_{\mathrm{e}}}/R$, $\nu_{\mathrm{b}} = \epsilon_{\mathrm{t}}^{3/2}\nu_{\mathrm{p}}$, $\omega_{\mathrm{h}} = \epsilon_{\mathrm{h}}\kappa T_{\mathrm{e}}/(qBr^2)$

$$\chi_{\mathrm{h1}} \sim \gamma_{\mathrm{T}}\epsilon_{\mathrm{t}}^2\epsilon_{\mathrm{h}}^{3/2}\left(\frac{kT}{qBr}\right)^2\frac{1}{\nu} \qquad (\gamma_{\mathrm{T}} \sim 50)\ . \tag{17.23}$$

Since $\nu \propto T^{-1.5}$, this means $\chi_{\mathrm{h1}} \propto T^{3.5}$. This may suggest that the thermal conduction loss becomes large in hot plasmas and the suppression of helical ripple loss is very important [17.31–17.36].

Since toroidal helical systems lose helical symmetry as well as axisymmetry, the generalized momentum correponding to the cyclic coordinate is no longer conserved [angular momentum $mr^2\dot{\theta} + qrA_\theta = \mathrm{const.}$ for the axisymmetric system $\boldsymbol{A}(r,z)$ and $m(\dot{z} + \alpha r^2\dot{\theta}) + q(A_z + \alpha rA_\theta) = \mathrm{const.}$ for the helically symmetric system $\boldsymbol{A}(r, \theta - \alpha z)$]. Therefore the orbit loss of energetic ions produced by heating or fusion-produced alpha particles with 3.5 MeV becomes large and heating efficiencies may deteriorate.

There are active efforts to design quasi-axisymmetric [17.37], quasi-helically symmetric [17.38] and quasi-omnigenous [17.39] stellerators by use of Boozer magnetic coordinates [17.40].

### 17.2.4 Confinement of Stellarator System

After Stellarator C [17.41], the basic experiments were carried out in small but accurate stellarator devices (Clasp, Proto Cleo, Wendelstein IIb, JIPP I, Heliotron D, L1, Uragan 1). Alkali plasmas, or afterglow plasmas produced by wave heating or gun injection, were confined quiescently. The effect of shear on the stability and confinement scaling were investigated.

The $l = 2$ stellarators with long helical pitch, such as Wendelstein IIa or JIPP I-b, have nearly constant rotational transform angles and the shears are small. When the transform angle is rational, $\iota/2\pi = n/m$, a line of magnetic force comes back to the initial position after $m$ turns of the torus and is closed. If electric charges are localized in some place, they cannot be dispersed uniformly within the magnetic surface in the case of rational surfaces. A resistive drift wave or resistive MHD instabilities are likely to be excited, and convective loss is also possible [17.42]. An enhanced loss is observed in the rational case. This is called resonant loss. Resonant loss can be reduced by the introduction of shear.

Medium-scale stellarator devices (Wendelstein VIIA, Cleo, JIPP T-II, Heliotron-E, L2, Uragan 2, Uragan 3) have been constructed. The confinement time of the ohmically heated plasmas ($T_e < 1\,\text{keV}$) is similar to that of tokamaks with the same scale. When the rotational transform angle is larger than $\iota_h/2\pi > 0.14$, the major disruption observed in tokamaks is suppressed (W VIIA, JIPP T-II). NBI heating or wave heatings, which were developed in tokamaks, have been applied to plasma production in helical devices. In Wendelstein VIIA, a target plasma was produced by ohmic heating; then the target plasma was sustained by NBI heating while the plasma current was gradually decreased, and finally a high-temperature plasma with $\kappa T_i \sim$ several hundred eV, $n_e \sim$ several $10^{13}\,\text{cm}^{-3}$ was confined without plasma current (1982). In Heliotron-E, a target plasma was produced by electron cyclotron resonance heating with $\kappa T_e \sim 800\,\text{eV}$, $n_e \sim 0.5 \times 10^{13}\,\text{cm}^{-3}$, and the target plasma was heated by NBI heating with 1.8 MW to the plasma with $\kappa T_i \sim 1\,\text{keV}$, $n_e = 2 \times 10^{13}\,\text{cm}^{-3}$ (1984). The average beta, $\langle\beta\rangle \sim 2\%$, was obtained for $B = 0.94\,\text{T}$ and NBI power $P_{NB} \sim 1\,\text{MW}$. These experimental results demonstrate the possibility of steady-state confinement by stellarator configurations [17.43]. Experimental scaling laws of energy confinement time are presented from Heliotron-E group [17.44] as follows;

$$\tau_E^{LHD} = 0.17a^{2.0}R^{0.75}n_{20}^{0.69}B^{0.84}P^{-0.58}\,, \tag{17.24}$$

where the units of $n_{20}$ are $10^{20}\,\text{m}^{-3}$. W7AS group presented W7AS confinement scaling [17.45] of

$$\tau_E^{W7AS} = 0.115A^{0.74}a^{2.95}n_{19}^{0.5}B^{0.73}P^{-0.54}(\iota/2\pi)^{0.43}\,. \tag{17.25}$$

The scaling law of the international stellarator database is [17.45]

$$\tau_E^{ISS95} = 0.079a^{2.21}R^{0.65}n_{19}^{0.51}B^{0.83}P^{-0.59}(\iota/2\pi)^{0.4}\,, \tag{17.26}$$

where the units of $n_{19}$ are $10^{19}\,\text{m}^{-3}$ and $\iota/2\pi$ is the value at $r = (2/3)a$. Units are s, m, T, MW.

High confinement operation by NBI was observed in WV7-AS and an improvement factor of about 2 is obtained in H mode and HDH mode (Hig Density H mode) [17.46]. The advanced stellarator Wendelstein 7-X [17.47] with superconductor modular coils is under construction (refer the lower figure of fig.17.6). Large helical device (LHD) [17.48] with superconductor helical coils started experiments in 1998 (refer the upper figure of fig.17.6). The relations of observed radial electric field and the transition or bifurcation of improved confinement mode are actively investigated [17.49].

### Open End Systems

Open end magnetic field systems [17.50–17.52] have a simpler configuration than toroidal systems. The attainment of absolute minimum-$B$ configurations is possible with mirror systems, whereas only average minimum-$B$ configurations can be realized in toroidal systems. Although absolute minimum-$B$ configurations are MHD stable, the velocity distribution of the plasma becomes non-Maxwellian due to end losses, and the plasma confined in mirrors will be prone to velocity-space instabilities.

The *particle confinement time* $\tau_{\mathrm{p}}$ of a mirror field is essentially determined by the diffusion time to the loss cone in the velocity space, i.e., the ion–ion Coulomb collision time $\tau_{\mathrm{ii}}$, and is given by [17.53]

$$\tau_{\mathrm{p}} = \tau_{\mathrm{ii}} \ln R_{\mathrm{M}} \ , \tag{17.27}$$

where $R_{\mathrm{M}}$ is the mirror ratio. Therefore the most critical issue for open-end systems is the suppression of end loss. The end plug of the mirror due to electrostatic potential has been studied by tandem mirrors [17.54–17.56].

# 18 Inertial Confinement

The characteristic of inertial confinement is that the extremely high-density plasma is produced within a short period by means of an intense energy driver, such as a laser or particle beam, so that fusion reactions can occur before the plasma starts to expand [18.1]. Magnetic confinement plays no part in this process, which has come to be called inertial confinement. For fusion conditions to be reached by inertial confinement, a small solid deuterium–tritium pellet must be compressed to a particle density $10^3$–$10^4$ times that of the solid pellet particle density $n_s = 5 \times 10^{22}\,\mathrm{cm}^{-3}$. One cannot expect the laser light pressure or the momentum carried by the particle beam to compress the solid pellet: they are too small. A more feasible method of compression involves irradiating the pellet from all sides, as shown in Fig. 18.1. The plasma is produced on the surface of the pellet and heated instantaneously. The plasma expands immediately. The reaction of the outward plasma jet accelerates and compresses the inner pellet inward like a spherical rocket. This process is called implosion. The study of implosion processes is one of the most important current issues, and theoretical and experimental research is being carried out intensively.

## 18.1 Pellet Gain

The pellet gain $G_{\mathrm{pellet}}$ is the ratio of the output nuclear fusion energy $E_{\mathrm{NF}}$ to the input driver energy $E_{\mathrm{L}}$ delivered to the pellet. The heating efficiency $\eta_{\mathrm{h}}$ of the incident driver is defined as the conversion ratio of the driver energy $E_{\mathrm{L}}$ to the internal energy $E_{\mathrm{fuel}}$ of the compressed pellet core. Denote the density and volume of the compressed core plasma by $n$ and $V$, respectively, and assume that $\langle \varepsilon_{\mathrm{fuel}} \rangle$ is the average energy of a fuel particle. Then we have the following relation:

$$E_{\mathrm{fuel}} = \langle \epsilon_{\mathrm{fuel}} \rangle n V = \eta_{\mathrm{h}} E_{\mathrm{L}} \,. \tag{18.1}$$

The densities $n_{\mathrm{D}}$ and $n_{\mathrm{T}}$ of the deuterium and tritium are decreased by the D–T fusion reaction ($n_{\mathrm{D}} = n_{\mathrm{T}} = n/2$) and

$$\frac{1}{n_{\mathrm{D}}} \frac{\mathrm{d}n_{\mathrm{D}}}{\mathrm{d}t} = -n_{\mathrm{T}} \langle \sigma v \rangle \,, \qquad n(t) = n_0 \frac{1}{1 + n_0 \langle \sigma v \rangle t/2} \,.$$

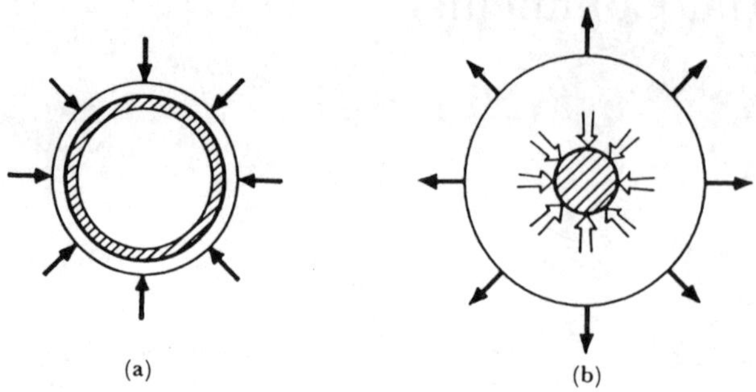


**Fig. 18.1.** Conceptual drawing of implosion. (**a**) Irradiation from all sides by laser or particle beam. (**b**) Expansion of plasma from the pellet surface and implosion due to the reaction of the outward plasma jet

When the plasma is confined during the time $\tau$, the fuel-burn ratio $f_{\rm B}$ is given by

$$f_{\rm B} \equiv \frac{n_0 - n(\tau)}{n_0} = \frac{n_0\langle\sigma v\rangle\tau/2}{1 + n_0\langle\sigma v\rangle\tau/2} = \frac{n_0\tau}{2/\langle\sigma v\rangle + n_0\tau}\,, \tag{18.2}$$

and the fusion output energy $E_{\rm NF}$ is

$$E_{\rm NF} = f_{\rm B} n V \frac{Q_{\rm NF}}{2}\,. \tag{18.3}$$

When the core gain $G_{\rm core}$ is defined by

$$G_{\rm core} \equiv \frac{E_{\rm NF}}{E_{\rm fuel}}\,, \tag{18.4}$$

the pellet gain $G_{\rm pellet}$ is

$$G_{\rm pellet} \equiv \frac{E_{\rm NF}}{E_{\rm L}} = \eta_{\rm h} G_{\rm core}\,, \tag{18.5}$$

which reduces to

$$G_{\rm pellet} = \eta_{\rm h}\left(\frac{Q_{\rm NF}}{2\langle\epsilon_{\rm fuel}\rangle}\right)\frac{n_0\tau}{2/\langle\sigma v\rangle + n_0\tau}\,. \tag{18.6}$$

Let us consider the energy balance of a possible inertial fusion reactor. The conversion efficiency of the thermal-to-electric energy is $\eta_{\rm el}$, and the conversion efficiency of the electric energy to output energy of the driver is denoted by $\eta_{\rm L}$. Then at least

$$\eta_{\rm el}\eta_{\rm L} G_{\rm pellet} > 1$$

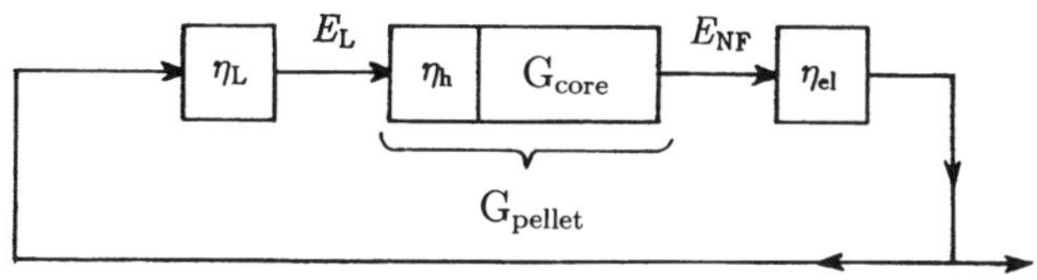


**Fig. 18.2.** Energy flow diagram of an inertial confinement reactor

is necessary to obtain usable net energy from the reactor (see Fig. 18.2). (When $\eta_L \sim 0.1$, $\eta_{el} \sim 0.4$ are assumed, $G_{pellet} > 25$ is necessary.) Therefore we find from (18.6)

$$n\tau > \frac{4\langle\epsilon_{fuel}\rangle}{\eta_{el}\eta_L\eta_h Q_{NF}\langle\sigma v\rangle} \frac{1}{1 - 2\langle\epsilon_{fuel}\rangle/(\eta_{el}\eta_L\eta_h Q_{NF})} \, . \tag{18.7}$$

The confinement time $\tau$ is the characteristic expansion time, expressed by [18.2]

$$\tau \approx \frac{r}{3c_s} \, , \qquad c_s^2 = \frac{5}{3}\frac{p}{\rho_m} = \frac{10}{3}\frac{\kappa T}{m_i} \, , \tag{18.8}$$

where $c_s$ is the sound velocity. Since the volume $V$ of the core is

$$V = \frac{4\pi r^3}{3} \, ,$$

equation (18.1) reduces to

$$E_L = \frac{4\pi}{\eta_h} n r^3 \langle\epsilon_{fuel}\rangle \, . \tag{18.9}$$

Equations (18.2) and (18.8) yield

$$\tau = \frac{f_B}{(1 - f_B)} \frac{2}{n\langle\sigma v\rangle} \, , \tag{18.10}$$

$$r = 5.5 \left(\frac{\kappa T}{m_i}\right)^{1/2} \tau \, . \tag{18.11}$$

The fuel-burn ratio $f_B$ now reduces to

$$f_B = \frac{\rho_m r}{6c_s m_i/\langle\sigma v\rangle + \rho_m r} \, , \qquad \beta(T) \equiv \frac{6c_s m_i}{\langle\sigma v\rangle} \, , \tag{18.12}$$

where $m_i$ is the average mass of D and T particles ($m_i = 2.5m_p$, where $m_p$ is the proton mass). When $\kappa T = 10\,\text{keV}$, then $\langle\sigma v\rangle \simeq 1.1 \times 10^{-16}\,\text{cm}^3\text{s}^{-1}$ and $\beta(T) \approx 26\,\text{g/cm}^2$. The plasma density is expressed by the ratio with respect

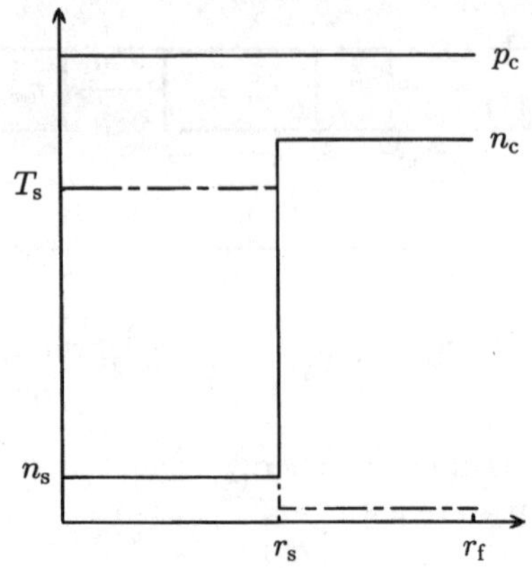


**Fig. 18.3.** Pressure, temperature and density profiles for the isobar model ignition configuration

to the solid density $n_{\rm solid} = 5 \times 10^{22}\,{\rm cm}^{-3}$ or the mass density $\rho_{\rm m\,solid} = m_{\rm i} n_{\rm solid} = 0.21\,{\rm g/cm}^3$. Equations (18.10), (18.11) and (18.9) yield

$$\tau = 0.36 \times 10^{-6} f'_{\rm B} \left(\frac{n_{\rm solid}}{n}\right) \quad ({\rm s})\,, \tag{18.13}$$

$$r = 1.2 f'_{\rm B} \left(\frac{n_{\rm solid}}{n}\right) \times 10^2 \quad ({\rm cm})\,, \tag{18.14}$$

$$E_{\rm L} = \frac{1}{\eta_{\rm h}} \langle \epsilon_{\rm fuel} \rangle \frac{4\pi}{3} n V \propto \frac{\langle \epsilon_{\rm fuel} \rangle}{\eta_{\rm h}} \left(\frac{n_{\rm solid}}{n}\right)^2 {f'_{\rm B}}^3\,, \tag{18.15}$$

$$G_{\rm pellet} \sim \eta_{\rm h} f_{\rm B} \frac{Q_{\rm NF}/2}{\langle \epsilon_{\rm fuel} \rangle} \sim \left(\frac{\eta_{\rm h}}{0.1}\right) \left(\frac{f_{\rm B}}{0.1}\right) \frac{88}{\langle \epsilon_{\rm fuel} \rangle_{\rm eV}} \times 10^3\,, \tag{18.16}$$

where

$$f'_{\rm B} \equiv \frac{f_{\rm B}}{1 - f_{\rm B}} = \frac{\rho_m r}{\beta({\rm T})}\,.$$

Equation (18.14) is equivalent to

$$r \rho_{\rm m} \sim 26 f'_{\rm B} \quad {\rm g/cm}^2\,.$$

Let us estimate the internal energy of the fuel $E_{\rm fuel}$. At the time of compression, almost all the inward-going energy has been converted into internal energy and the pressure is nearly uniform over the total (hot and cold) fuel region at ignition (isobar model) [18.3]. But the central spark region becomes hot and is surrounded by the cold compressed fuel (see Fig. 18.3). The internal energy of solid cold DT fuel per unit volume is given by the product of the Fermi energy $\varepsilon_{\rm F} = (\hbar^2/2m_{\rm e}) \times (3\pi^2 n)^{2/3}$, where $\hbar = h/2\pi$ is Planck's constant and $m_{\rm e}$ is the electron mass, and the density with a factor of 3/5. The

Fermi energy of electrons with solid density $n = 5\times10^{22}\,\mathrm{cm}^{-3}$ is $\varepsilon_{\mathrm{F,s}} = 4.9\,\mathrm{eV}$. If preheating occurs before compression starts, the energy density of the cold fuel is

$$n\langle\varepsilon_{\mathrm{fuel}}\rangle = \frac{3}{5}\alpha n\varepsilon_{\mathrm{F}} = \frac{3}{5}\alpha n\left(\frac{\hbar^2}{2m_{\mathrm{e}}}\right)(3\pi^2 n)^{2/3}\,. \tag{18.17}$$

$\alpha = 2$–$3$ is called the preheating factor. Then the fuel energy is

$$E_{\mathrm{fuel}} = 3n_{\mathrm{s}}\kappa T_{\mathrm{s}}V_{\mathrm{s}} + \frac{3}{5}\alpha\varepsilon_{\mathrm{F}}n_{\mathrm{c}}V_{\mathrm{c}} = \frac{3}{5}\alpha\varepsilon_{\mathrm{F}}n_{\mathrm{c}}V_{\mathrm{f}}\,, \tag{18.18}$$

because of the pressure balance $2n_{\mathrm{s}}\kappa T_{\mathrm{s}} = (2/5)\alpha\varepsilon_{\mathrm{F}}$, where $V_{\mathrm{s}}$, $r_{\mathrm{s}}$ and $V_{\mathrm{f}}$, $r_{\mathrm{f}}$ are the volumes and radii of the spark region and the total region of the fuel, respectively. $V_{\mathrm{c}}$ is the volume of the cold compressed region $V_{\mathrm{c}} = V_{\mathrm{f}} - V_{\mathrm{s}}$. $T_{\mathrm{s}}$ and $n_{\mathrm{s}}$ are the temperature and density of the spark region and $n_{\mathrm{c}}$ is the density of the cold compressed region. Then the fusion output is

$$E_{\mathrm{NF}} \sim \frac{f_{\mathrm{B}}}{2}n_{\mathrm{c}}V_{\mathrm{c}}Q_{\mathrm{NF}}\,, \quad f_{\mathrm{B}} \sim \frac{\rho_{\mathrm{c}}(r_{\mathrm{f}} - r_{\mathrm{s}})}{\beta(T) + \rho_{\mathrm{c}}(r_{\mathrm{f}} - r_{\mathrm{s}})}\,. \tag{18.19}$$

Here the contribution from the spark region is neglected ($n_{\mathrm{s}}V_{\mathrm{s}} \ll n_{\mathrm{c}}V_{\mathrm{c}}$). The core gain $G_{\mathrm{core}}$ is

$$G_{\mathrm{core}} \sim \left[\frac{\rho_{\mathrm{c}}(r_{\mathrm{f}} - r_{\mathrm{s}})}{\beta(T) + \rho_{\mathrm{c}}(r_{\mathrm{f}} - r_{\mathrm{s}})}\right]\frac{Q_{\mathrm{NF}}/2}{(3/5)\alpha\varepsilon_{\mathrm{F}}}\frac{V_{\mathrm{c}}}{V_{\mathrm{f}}}\,. \tag{18.20}$$

The ignition condition is given by [18.4, 18.5]

$$\rho_{\mathrm{s}}r_{\mathrm{s}} > 0.3\text{–}0.4\,\mathrm{g/cm^2} \quad \text{when } \kappa T \approx 5\,\mathrm{keV}\,, \tag{18.21}$$

and the slowing-down length $\lambda_\alpha$ of alpha particles is given by [18.6, 18.7] $\rho\lambda_\alpha = 0.015T_{\mathrm{keV}}^{5/4}\,\mathrm{g/cm^2}$ and $\rho_{\mathrm{s}}r_{\mathrm{s}} > \rho\lambda_\alpha$ is required.

Let us take the example of $n_{\mathrm{c}} = 2000n_{\mathrm{solid}}$. In this case $\rho_{\mathrm{c}} = 420\,\mathrm{g/cm^3}$ and $\varepsilon_{\mathrm{F}} = 786\,\mathrm{eV}$. The mass density of the spark region is $\rho_{\mathrm{s}} = \alpha\varepsilon_{\mathrm{F}}\rho_{\mathrm{c}}/5\kappa T_{\mathrm{s}} = 26.4\,\mathrm{g/cm^3}$ ($\kappa T_{\mathrm{s}} = 5\,\mathrm{keV}$, $\alpha = 2$). From the ignition condition we choose $r_{\mathrm{s}} = 0.015\,\mathrm{cm}$. The value of the fuel radius is chosen as $r_{\mathrm{c}} = 0.03\,\mathrm{cm}$. Then we have $\langle\varepsilon_{\mathrm{fuel}}\rangle = 0.6\alpha\varepsilon_{\mathrm{F}} = 943\,\mathrm{eV}$, assuming $\alpha = 2$ and $E_{\mathrm{fuel}} = 1.7\,\mathrm{MJ}$. The driver energy becomes $E_{\mathrm{L}} = 17\,\mathrm{MJ}$ under the assumption that $\eta_{\mathrm{h}} = 0.1$ and the pellet gain is $G_{\mathrm{pellet}} = 3.2\times10^2$ for $f_{\mathrm{B}} \sim 0.34$ ($\beta(\mathrm{T}) \sim 12\mathrm{g/cm^3}$) for $T \sim 17$ keV.

The critical issue for an inertial fusion reactor is how to produce extremely high-density plasmas by implosion. Hence, the optimum design of fuel pellet structures and materials is important. Technological issues for energy drivers involve increasing the efficiency of laser drivers and improving the focusing of light ions or heavy ion beams.

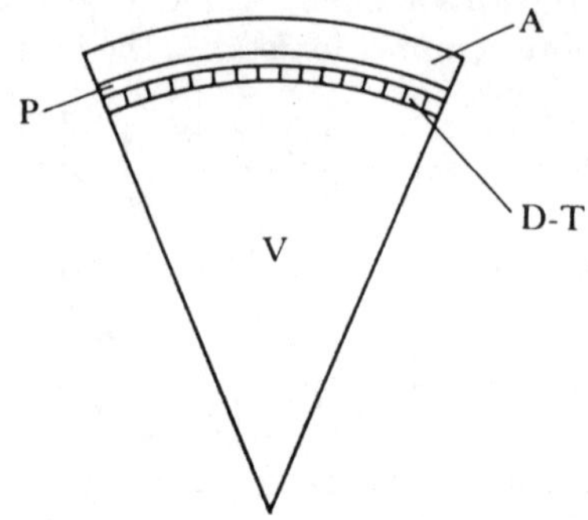


**Fig. 18.4.** Pellet structure. A: ablator, P: pusher, D-T: solid D–T fuel, V: vacuum

## 18.2 Implosion

A typical pellet structure is shown in Fig. 18.4. Outside the spherical shell of deuterium–tritium fuel, there is a pusher cell, which plays the role of a piston during compression; an ablator cell with low-$Z$ material surrounds the pusher cell and the fuel. The heating efficiency $\eta_{\mathrm{h}}$ is the conversion ratio of the driver energy to the thermal energy of the compressed core fuel. The heating efficiency depends on the interaction of the driver energy with the ablator, the transport process of the particles, and the energy and motion of the plasma fluid. The driver energy is absorbed on the surface of the ablator and the plasma is produced and heated.

The plasma then expands and the inner deuterium–tritium fuel shell is accelerated inward by the reaction of the outward plasma jet. The implosion takes place at the center. Therefore the heating efficiency $\eta_{\mathrm{h}}$ is the product of three terms, i.e., the absorption ratio $\eta_{\mathrm{ab}}$ of the driver energy by the ablator, the conversion ratio $\eta_{\mathrm{hydro}}$ of the absorbed driver energy to the kinetic energy of hydrodynamic fluid motion, and the conversion ratio $\eta_{\mathrm{T}}$ of the kinetic energy of the hydrodynamic motion to the internal energy of the compressed core:

$$\eta_{\mathrm{h}} = \eta_{\mathrm{ab}}\eta_{\mathrm{hydro}}\eta_{\mathrm{T}}.$$

The internal energy of the solid deuterium–tritium fuel per unit volume is given by $(3/5)n\varepsilon_{\mathrm{F}}$. The internal energy of solid deuterium–tritium per unit mass $w_0$ can be estimated to be $w_0 = 1.1 \times 10^8$ J/kg. If the preheating occurs before the compression starts, the initial internal energy is increased to $\alpha_{\mathrm{p}} w_0$, and then the solid deuterium–tritium fuel is compressed adiabatically. Using the equation of state for an ideal gas, the internal energy $w$ after compression is

$$w = \alpha_{\mathrm{p}} w_0 \left(\frac{\rho}{\rho_0}\right)^{2/3},$$

where $\rho_0$ and $\rho$ are the mass densities before and after compression. If the preheating is well suppressed and $\alpha_\mathrm{p}$ is of the order of 2–3, the internal energy per unit mass after 2000 × compression is $w \sim (4.5) \times 10^{10}$ J/kg. This value $w$ corresponds to the kinetic energy of unit mass with velocity $v \sim 3 \times 10^5$m/s ($w = v^2/2$). Therefore, if the spherical fuel shell is accelerated to this velocity and if the kinetic energy is converted with good efficiency $\eta_\mathrm{T}$ into the internal energy of the fuel core at the center, then compression with 2000 times mass density of the solid deuterium–tritium is possible.

When the pellet is irradiated from all sides by the energy driver, the plasma expands with velocity $u$ from the surface of the ablator. Then the spherical shell with mass $M$ is accelerated inward by the reaction with the ablation pressure $P_\mathrm{a}$. The inward velocity $v$ of the spherical shell can be analyzed by the rocket model with an outward plasma jet [18.8, 18.9]:

$$\frac{\mathrm{d}(Mv)}{\mathrm{d}t} = -\frac{\mathrm{d}M}{\mathrm{d}t}u = SP_\mathrm{a} \,, \tag{18.22}$$

where $S$ is the surface area of the shell. When the average mass density and the thickness of the spherical shell are denoted by $\rho$ and $\Delta$, respectively, the mass $M$ is $M = \rho S \Delta$. The outward velocity $u$ of the expanding plasma is usually much larger than the inward velocity $v$ of the spherical shell, and $u$ is almost constant. The change in the sum of the kinetic energies of the plasma jet and the spherical shell is equal to the absorbed power of the energy driver:

$$\eta_\mathrm{ab} I_\mathrm{L} S = \frac{\mathrm{d}}{\mathrm{d}t}\left(\frac{1}{2}Mv^2\right) + \frac{1}{2}\left(-\frac{\mathrm{d}M}{\mathrm{d}t}\right)u^2 \,, \tag{18.23}$$

where $I_\mathrm{L}$ is the input power per unit area of the energy driver. From (18.22) and (18.23), the absorbed energy $E_\mathrm{a}$ reduces to

$$E_\mathrm{a} = \int \eta_\mathrm{ab} I_\mathrm{L} S \,\mathrm{d}t \approx \frac{1}{2}(\Delta M)u^2 \,, \tag{18.24}$$

where the approximations $u \gg v$, and $u =$ const. are used. The quantity $\Delta M$ is the absolute value of the change in the mass of the spherical shell. The pressure $P_\mathrm{a}$ is estimated from (18.22) and (18.23) as follows:

$$P_\mathrm{a} = \frac{u}{S}\left(-\frac{\mathrm{d}M}{\mathrm{d}t}\right) \approx 2\eta_\mathrm{ab} I_\mathrm{L} \frac{1}{u} \,. \tag{18.25}$$

Then the conversion ratio $\eta_\mathrm{hydro}$ of the absorbed energy to the kinetic energy of the spherical shell is

$$\eta_\mathrm{hydro} = \frac{1}{2E_\mathrm{a}}(M_0 - \Delta M)v^2 = \frac{M_0 - \Delta M}{\Delta M}\left(\frac{v}{u}\right)^2 \,.$$

Since the rocket equation (18.22) implies $v/u = -\ln\left[(M_0 - \Delta M\right] / M_0)$, the conversion ratio $\eta_\mathrm{hydro}$ is

$$\eta_{\text{hydro}} = \left(\frac{M_0}{\Delta M} - 1\right)\left[\ln\left(1 - \frac{\Delta M}{M_0}\right)\right]^2 \approx \frac{\Delta M}{M_0} , \tag{18.26}$$

assuming that $\Delta M/M_0 \ll 1$.

The final inward velocity of the accelerated spherical shell must still be larger than $v \sim 3 \times 10^5\,\text{m/s}$. The necessary ablation pressure $P_\text{a}$ can be obtained from (18.22) with the relation $S = 4\pi r^2$, and the approximation $M \approx M_0 P_\text{a} \approx$ const. as follows:

$$\frac{\text{d}v}{\text{d}t} = \frac{4\pi P_\text{a}}{M_0} r^2 = \frac{P_\text{a}}{\rho_0 r_0^2 \Delta_0} r^2 , \qquad v = -\frac{\text{d}r}{dt} .$$

Integration of $v\text{d}v/\text{d}t$ gives

$$P_\text{a} = \frac{3}{2}\rho_0 v^2 \frac{\Delta_0}{r_0} , \tag{18.27}$$

where $\rho_0$, $r_0$, and $\Delta_0$ are the mass density, the radius, and the thickness of the spherical shell at the initial conditions, respectively. When $r_0/\Delta_0 = 20$ and $\rho_0 = 0.21\,\text{g/cm}^3$, the necessary ablation pressure is $P_\text{a} = 1.4 \times 10^{12}\,\text{Newton/m}^2 = 14\,\text{Mbar}$ ($1\,\text{atm} = 1.013\,\text{bar}$) in order to achieve the velocity $v = 3 \times 10^5\,\text{m/s}$. Therefore the energy flux intensity $I_\text{L}$ required from the driver is

$$\eta_\text{ab} I_\text{L} = \frac{P_\text{a} u}{2} . \tag{18.28}$$

For the evaluation of the velocity $u$ of the expanding plasma, the interaction of the driver energy and the ablator cell must be taken into account. In this section the case of the laser driver is described. Let the sound velocity of the plasma at the ablator surface be $c_\text{s}$ and the mass density be $\rho_\text{c}$. The energy extracted by the plasma jet from the ablator surface per unit time is $4\rho_\text{c} c_\text{s}^3$ and this must be equal to the absorbed power $\eta_\text{ab} I_\text{L}$. The plasma density is around the cutoff density corresponding to the laser light frequency (wavelength), i.e.,

$$u \sim 4c_\text{s} , \qquad \eta_\text{ab} I_\text{L} \sim 4 m_\text{DT} n_\text{c} c_\text{s}^3 ,$$

where $m_\text{DT} = 2.5 \times 1.67 \times 10^{-27}\,\text{kg}$ is the average mass of deuterium and tritium, and the cutoff density is $n_\text{c} = 1.1 \times 10^{27}/\lambda^2\,(\mu\text{m})\,\text{m}^{-3}$, with $\lambda$ the wavelength of the laser in units of μm. From (18.28), we have

$$P_\text{a} = 13 \left[\frac{(\eta_\text{ab} I_\text{L})_{14}}{\lambda(\mu\text{m})}\right]^{2/3} \quad (\text{Mbar}) , \tag{18.29}$$

where $(\eta_\text{ab} I_\text{L})_{14}$ is the value in $10^{14}\,\text{W/cm}^2$. This scaling is consistent with the experimental results in the range $1 < (\eta_\text{ab} I_\text{L})_{14} < 10$.

Most implosion research is carried out by the laser driver. The observed absorption rate $\eta_\text{ab}$ tends to decrease according to the increase in laser light intensity $I_\text{L}$.

The absorption rate is measured for an Nd glass laser with wavelength 1.06 μm (red), second harmonic 0.53 μm (green), and third harmonic 0.35 μm (blue). The absorption is better for shorter wavelengths, and it is $\eta_{\mathrm{ab}} \approx 0.9$–$0.8$ for $\lambda = 0.35\,\mu\mathrm{m}$ in the range $I_{\mathrm{L}} = 10^{14}$–$10^{15}\,\mathrm{W/cm^2}$. The conversion ratio $\eta_{\mathrm{hydro}}$ determined by experiment is 0.1–0.15. The conversion ratio $\eta_{\mathrm{T}}$ is expected to be $\eta_{\mathrm{T}} > 0.5$. In order to compress the fuel to extremely high density, it is necessary to avoid the preheating the inner pellet during the implosion process, since the pressure of the inner part of the pellet must be kept as low as possible before compression. When laser light with a long wavelength ($CO_2$ laser $\lambda = 10.6\,\mu\mathrm{m}$) is used, high-energy electrons are produced by the laser–plasma interaction and these penetrate into the inner part of the pellet and preheat it. However, the production of high-energy electrons is much lower in short-wavelength experiments.

## 18.3 MHD Instabilities

In the accelerating phase of the implosion process, the low density plasma ablating from the surface of the ablator accelerates the high density fuel, so that Rayleigh–Taylor (RT) instability is likely to occur at the ablation front. Furthermore, the boundary of the central spark region (low density) and main pusher fuel (high density) can be unstable in the decelerating phase near the stagnation time of the implosion process. The resulting RT instability may cause mixing of the fuel and ablator material in the accelerating phase, and mixing of the pusher fuel into the central spark region in the decelerating phase, thereby severely degrading the pellet performance.

Let us consider the case where a fluid with mass density $\rho_{\mathrm{h}}$ is supported against acceleration $g$ by a fluid with lower density of $\rho_{\mathrm{l}}$. In this case the growth rate of Rayleigh–Taylor instability is given by [18.10]

$$\gamma = (\alpha_{\mathrm{A}} g k)^{1/2}\,, \qquad \alpha_{\mathrm{A}} = \frac{\rho_{\mathrm{h}} - \rho_{\mathrm{l}}}{\rho_{\mathrm{h}} + \rho_{\mathrm{l}}}\,, \tag{18.30}$$

where $k$ is the wave number of the perturbation in the direction perpendicular to the acceleration. When $\rho_{\mathrm{h}} \gg \rho_{\mathrm{l}}$, the growth rate is $\gamma = (gk)^{1/2}$. When the density gradient is finite with scale length $L$, the growth rate becomes $\gamma \sim (\alpha_{\mathrm{A}} g/L)^{1/2}$ when $kL \gg 1$.

The dispersion relations of the RT perturbation near the ablation front that are widely used constitute an analytical fit to numerical simulation and are given by [18.11, 18.12]

$$\gamma = \left(\frac{kg}{1 + kL}\right)^{1/2} - \hat{\beta} k V_{\mathrm{a}}\,, \tag{18.31}$$

where $\hat{\beta}$ is a constant with $\hat{\beta} = 1$–$3$ and $V_{\mathrm{a}}$ is the flow velocity across the ablation front in the frame moving with the ablation front. The first term in

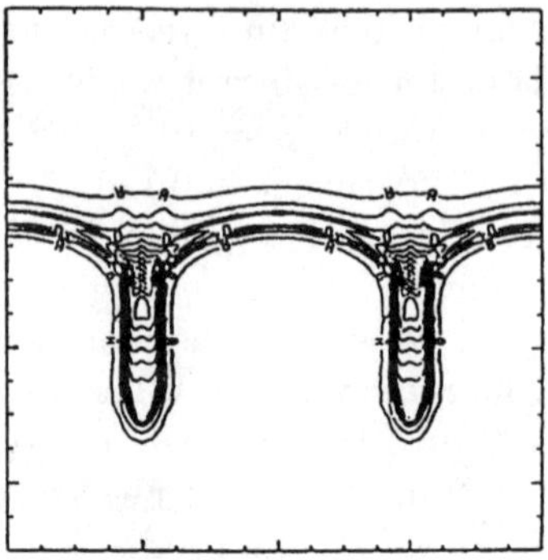

**Fig. 18.5.** Numerical simulation of the side-on image of the spatial distribution of a perturbed sample of planar target [18.2]. The ablated low density plasma is in the upper region and the acceleration is upward

(18.31) is the usual one with the correction for the finite density gradient, and the second one is a stabilizing term due to the convective effect, as will be seen in the following.

Let us consider the case where the region of high density is $x < 0$ and the acceleration $g$ is in the positive $x$ direction. The wave number in the $y$ direction is $k$. The irrotational incompressible flow velocity $(v_x, v_y)$ is given by

$$v_x = \frac{\partial \phi}{\partial y}\,, \quad v_y = -\frac{\partial \phi}{\partial x}\,, \quad \Delta\phi = 0\,,$$

so that the stream function $\phi$ is

$$\phi = \phi_0 \exp(-k|x| + iky) \exp \gamma_0 t\,,$$

in the frame moving with the ablation front. When the fluid flows with velocity $V_{\rm a}$ (positive $x$ direction), the coordinate of the fluid element is $x = x_0 + V_{\rm a}t$ and $\phi$ is expressed by $(x > 0)$

$$\begin{aligned}\phi &= \phi_0 \exp\left[-k(x_0 + V_{\rm a}t) + iky\right] \exp \gamma_0 t \\ &= \phi_0 \exp(-kx_0 + iky) \exp(\gamma_0 - kV_{\rm a})t\,. \end{aligned} \qquad (18.32)$$

This equation demonstrates the stabilizing effect of the convective fluid. Figure 18.5 shows the results of numerical simulation [18.13], in good agreement with experiment.

When a shock encounters a fluid discontinuity, transmitted and reflected shocks are generated. These are then refracted by any perturbations at the fluid interface. The modulated shocks produce pressure variations in the upstream and downstream fluids that reinforce the initial interfacial perturbations and cause them to grow. This type of instability is called a *Richtmyer–Meshkov instability* [18.14, 18.15].

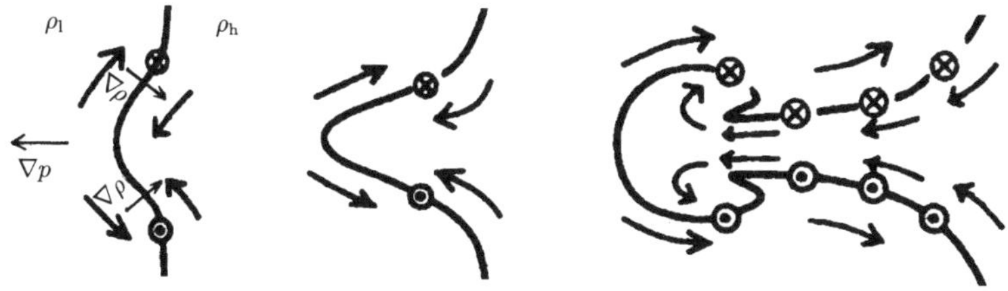


**Fig. 18.6.** Growth of perturbation due to Rayleigh–Taylor instability and Richtmyer–Meshkov instability [18.17]

When an impulse of acceleration $g(t) = \Delta U \delta(t)$ is applied, a single sinusoidal perturbation $\xi$ obeys $\mathrm{d}\xi/\mathrm{d}t = \alpha_{\mathrm{A}}^{*} k \Delta U \xi_0^{*}$, where $\alpha_{\mathrm{A}}^{*}$ and $\xi_0^{*}$ are the post-shock values at $t = 0_+$. With a constant $k$, the amplitude grows linearly in time and eventually saturates when the amplitude becomes large, $k\xi \sim 1$. Thus, short wavelength modes grow quickly, but they saturate and will be overtaken by longer wavelength, slower-growing modes [18.16].

When a shock encounters the perturbed fluid discontinuity, the pressure gradient $\boldsymbol{\nabla} p$ and density gradient $\boldsymbol{\nabla} \rho$ are not necessarily parallel and a flow vortex can be induced. The equation for an ideal fluid is

$$\frac{\mathrm{d}\boldsymbol{u}}{\mathrm{d}t} = -\frac{1}{\rho} \boldsymbol{\nabla} p \,. \tag{18.33}$$

Noting the vortex $\boldsymbol{\omega} = \boldsymbol{\nabla} \times \boldsymbol{u}$, the rotation of (18.33) reduces to

$$\frac{\mathrm{d}\boldsymbol{\omega}}{\mathrm{d}t} = (\boldsymbol{\omega} \cdot \boldsymbol{\nabla})\boldsymbol{u} - \boldsymbol{\omega}(\boldsymbol{\nabla} \cdot \boldsymbol{u}) + \frac{1}{\rho^2} \boldsymbol{\nabla} \rho \times \boldsymbol{\nabla} p \,. \tag{18.34}$$

The third term on the left-hand side of (18.34) induces the vortex and helps the growth of the perturbation of RT instability [18.17], as shown in Fig. 18.6.

In conclusion, effective implosion requires high quality spherical symmetry of the irradiating laser light intensity and target structures. Moreover, the limit of the radius, the density of the compressed fuel core, and the required laser input energy are determined by the extent to which RT mixing can be minimized.

## 18.4 Fast Ignition

Ultra-high intensity lasers with petawatt output ($10^{15}$ W) have been developed by the technology of chirp pulse amplification [18.18] and a new approach called fast ignition [18.19] is being actively studied. One scenario for hot ignition has three phases. First a fuel capsule is imploded as in the usual approach to assemble a high density fuel configuration. Second, a hole is

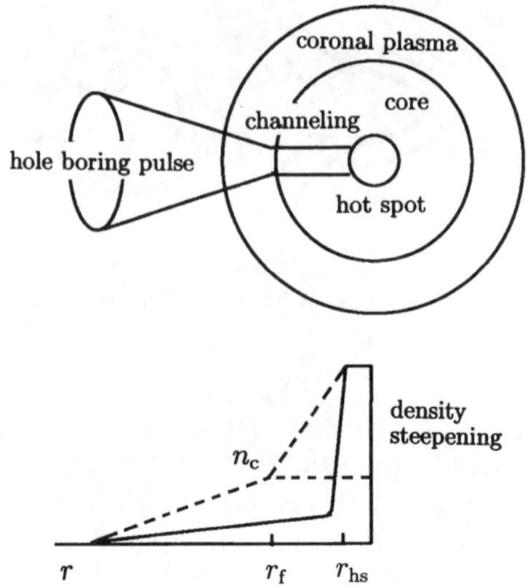


**Fig. 18.7.** A configuration for efficient heating of the imploded core plasma using an ultra-intense laser as fast ignitor

bored through the capsule corona composed of ablated material, as the critical density is pushed close to the high density core of the capsule by the ponderomotive force associated with the high intensity laser light. Finally, the fuel is ignited by suprathermal electrons, produced in the high-intensity laser–plasma interactions, which then propagate from critical density to this high density core (see Fig. 18.7). This new scheme, if realized, could separate the process of implosion and self-spark of the central spot and also drastically reduce the difficulties involved with implosion, allowing lower quality fabrication and less stringent beam quality and symmetry requirements from the implosion driver.

Fuel ignition requires the hot spot to reach an average temperature of 5 keV within a fuel areal density of $(\rho r)_{hs} = 0.3$–$0.4\,\mathrm{g/cm^2}$ (see Sect. 18.1). The mass of the hot spot region is

$$M_{hs}(\mathrm{g}) = \frac{4\pi\rho_{hs}r_{hs}^3}{3} = 4\pi\frac{(\rho r)_{hs}^3}{3\rho_{hs}^2} \sim 4.2\frac{(0.4)^3}{\rho_{hs}^2} = \frac{0.27}{\rho_{hs}^2(\mathrm{g})}\,,$$

where $\rho_{hs}$, $r_{hs}$ are the mass density and radius of the hot spot region, respectively. The thermal energy of the heated fuel is

$$E_{hs} = \frac{M_{hs}}{m_i}3\kappa T = 31\frac{T_{keV}}{\rho_{hs}^2} \qquad (\mathrm{MJ})\,.$$

The energy $E_f$ of the cold imploded fuel before fast ignition is

$$E_{\mathrm{f}} = \frac{3}{5}\alpha\varepsilon_{\mathrm{F}}\frac{M_{\mathrm{f}}}{m_{\mathrm{i}}} = 0.33\alpha\rho_{\mathrm{f}}^{2/3}M_{\mathrm{f}} \qquad (\mathrm{MJ})\ ,$$

where $M_{\mathrm{f}}$ is the mass of the main fuel in units of g and the mass density $\rho_{\mathrm{f}}$ is in units of $\mathrm{g/cm^3}$.

When the ignition energy is injected sufficiently rapidly, the hot spot and main fuel are not in pressure equilibrium and the uniform density model can be used. Then the total energy $E_{\mathrm{fuel}}$ of the fuel is given by

$$E_{\mathrm{fuel}} = 31\frac{T_{\mathrm{hs}}}{\rho^2} + 0.33\alpha\rho^{2/3}M_{\mathrm{f}} \qquad (\mathrm{MJ})\ , \tag{18.35}$$

and the fusion output energy $E_{\mathrm{NF}}$ is

$$E_{\mathrm{NF}} = f_{\mathrm{B}}\frac{M_{\mathrm{f}}}{m_{\mathrm{i}}}\frac{Q_{\mathrm{NF}}}{2} = 334\times10^3 f_{\mathrm{B}}M_{\mathrm{f}} \qquad (\mathrm{MJ})\ . \tag{18.36}$$

Therefore the core gain $G_{\mathrm{core}}$ reduces to

$$G_{\mathrm{core}} = \frac{334M_{\mathrm{f}}\rho r_{\mathrm{f}}/\left[\beta(T)+\rho r_{\mathrm{f}}\right]}{31T_{\mathrm{hs}}/\rho^2 + 0.33\alpha\rho^{2/3}M_{\mathrm{f}}}\ . \tag{18.37}$$

Let us consider the example of $\rho_{\mathrm{f}} = 1000\rho_{\mathrm{solid}} = 210\,\mathrm{g/cm^3}$. We choose $r_{\mathrm{hs}} = 0.002\,\mathrm{cm}$ and $r_{\mathrm{f}} = 0.01\,\mathrm{cm}$. Then we have $M_{\mathrm{f}} = 0.88\times10^{-3}\,\mathrm{g}$, $E_{\mathrm{fuel}} = (3.5+20.5) = 24\,\mathrm{kJ}$, $f_{\mathrm{B}} = 0.22$, $E_{\mathrm{NF}} = 65.3\,\mathrm{MJ}$, $G_{\mathrm{core}} = 2.7\times10^3$, assuming $\alpha = 2$, $\kappa T_{\mathrm{hs}} = 5\,\mathrm{keV}$, $\beta(T) \sim 7\,\mathrm{g/cm^2}$, where $\beta(T)$ is about $7\,\mathrm{g/cm^2}$ for $\kappa T = 30$–$40\,\mathrm{keV}$, and $E_{\mathrm{L}} = 240\,\mathrm{kJ}$, $G_{\mathrm{pellet}} = 2.7\times10^2$, under the assumption that $\eta_{\mathrm{h}} = 0.1$.

Let us estimate the laser power required for fast ignition. The energy needed to ignite the hot spot region ($\rho_{\mathrm{m}}r > 0.4\,\mathrm{g/cm^2}$, $\rho_{\mathrm{m}} \sim 10^3\rho_{\mathrm{solid}}$, $\kappa T \sim 5\,\mathrm{keV}$) is at least 7–8 kJ, taking heating efficiency into account. The confinement time $\tau = r/(3c_{\mathrm{s}})$ is of the order of 8 ps. Therefore a power of $10^{15}\,\mathrm{W} = 1$ petawatt is necessary. When the radius of the hot spot is 0.02 mm, the intensity of the laser beam becomes $10^{20}\,\mathrm{W/cm^2}$.

The ponderomotive force $\boldsymbol{F}_{\mathrm{p}}$ of the laser beam is given by [see (18.40) at the end of this chapter]

$$\boldsymbol{F}_{\mathrm{p}} = -\frac{\omega_{\mathrm{p}}^2}{\omega^2}\boldsymbol{\nabla}\frac{\epsilon_0\langle\boldsymbol{E}^2\rangle}{2}\ . \tag{18.38}$$

A laser beam of finite diameter causes a radially directed ponderomotive force in a plasma. This force moves plasma out of the beam, so that the plasma frequency $\omega_{\mathrm{p}}$ is lower and the dielectric constant $\epsilon$ is higher inside the beam than outside. The plasma acts as a convex lens resulting in the self-focus of the laser beam. The ponderomotive force will also push the critical surface forward, resulting in the laser channel forming an over-dense plasma [18.20]. The pressure of the cold imploded fuel with density of $n = 10^3 n_{\mathrm{solid}}$ is $2/5\alpha\varepsilon_{\mathrm{F}}n \sim 3.2\times10^{15}\,\mathrm{Pa}$ ($1\,\mathrm{bar} = 10^5\,\mathrm{Pa} \sim 1\,\mathrm{atm}$) and the pressure due to

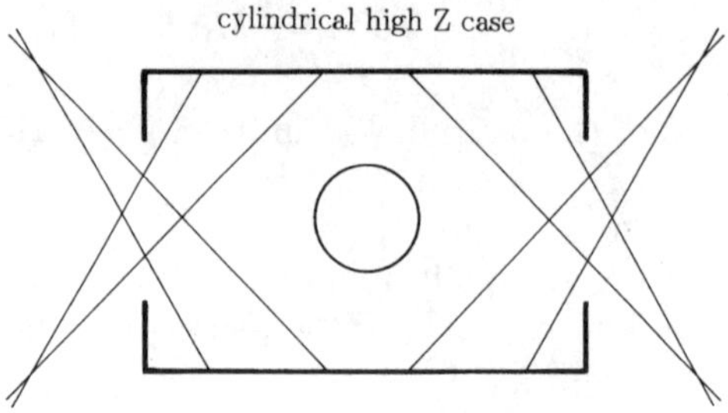


**Fig. 18.8.** Configuration of Hohlraum target

the ponderomotive force of the laser beam with intensity $I_{\rm L} = 10^{24}\,{\rm W/m^2}$ is $\epsilon_0 \langle \boldsymbol{E}^2 \rangle /2 = I_{\rm L}/c \sim 3 \times 10^{15}\,{\rm Pa}$. The critical surface can therefore be pushed into the core plasma. Once the channel is created in the plasma with a critical (cutoff) density, the laser light heats the plasma by $\boldsymbol{J} \times \boldsymbol{B}$ heating [18.21], in which the oscillating component of the ponderomotive force [see (18.39)] can lead to heating, and also by 'not-so-resonant' resonance absorption [18.22]. The interaction of dense plasma and ultra-high intensity laser light is being actively studied in experiments [18.23] and computer simulations.

## Hohlraum Target

The implosion process just described is for directly irradiated pellets. The other case is indirectly irradiated pellets. The outer cylindrical case surrounds the fuel pellet, as shown in Fig. 18.8. The inner surface of the outer cylindrical case is irradiated by the laser light, and the laser energy is converted into X-ray energy and plasma energy. The converted X-rays and the plasma particles irradiate the inner fuel pellet and implosion occurs. The X-ray and plasma energy is confined between the outer cylindrical case and the inner fuel pellet and is used for effective implosion. This type of pellet is called a Hohlraum target [18.2]. In this configuration, the possibility of using a heavy ion beam as energy driver is being examined [18.24].

Recent activities in inertial confinement fusion including the NIF (National Ignition Facility) are well described in [18.25].

## Ponderomotive Force

The electron equation of motion in the electromagnetic wave $\boldsymbol{E}(\boldsymbol{r},t) = \widehat{\boldsymbol{E}}(\boldsymbol{r}) \cos(\boldsymbol{k}\cdot\boldsymbol{r} - \omega t)$ is

$$m \frac{{\rm d}\boldsymbol{v}}{{\rm d}t} = -e(\boldsymbol{E} + \boldsymbol{v} \times \boldsymbol{B}) \, .$$

Here we assume that $\widehat{\boldsymbol{E}}(\boldsymbol{r})$ varies slowly on the scale of the wavelength. Using the notation $\alpha \equiv \boldsymbol{k}\cdot\boldsymbol{r} - \omega t$, the magnetic field $\boldsymbol{B}$ is given by

$$\frac{\partial \boldsymbol{B}}{\partial t} = -\boldsymbol{\nabla} \times \boldsymbol{E} = \boldsymbol{\nabla} \times \widehat{\boldsymbol{E}} \cos\alpha + \boldsymbol{k} \times \widehat{\boldsymbol{E}} \sin\alpha \ ,$$

$$\boldsymbol{B} = \frac{\boldsymbol{\nabla} \times \widehat{\boldsymbol{E}}}{\omega} \sin\alpha + \frac{\boldsymbol{k} \times \widehat{\boldsymbol{E}}}{\omega} \cos\alpha \ .$$

To first order, we neglect the second-order term $\boldsymbol{v} \times \boldsymbol{B}$, whence

$$m\frac{\mathrm{d}\boldsymbol{v}_1}{\mathrm{d}t} = -e\boldsymbol{E}(\boldsymbol{r}_0, t) = -e\widehat{\boldsymbol{E}}(\boldsymbol{r}_0)\cos(\boldsymbol{k}\cdot\boldsymbol{r}_0 - \omega t) \ ,$$

$$\boldsymbol{v}_1 = \frac{e\widehat{\boldsymbol{E}}(\boldsymbol{r}_0)}{m\omega}\sin(\boldsymbol{k}\cdot\boldsymbol{r}_0 - \omega t) \ , \qquad \boldsymbol{r}_1 = \frac{e\widehat{\boldsymbol{E}}(\boldsymbol{r}_0)}{m\omega^2}\cos(\boldsymbol{k}\cdot\boldsymbol{r}_0 - \omega t) \ ,$$

where $\boldsymbol{r}_0$ is the initial position. We expand $\boldsymbol{E}(\boldsymbol{r}, t)$ about $\boldsymbol{r}_0$:

$$\begin{aligned}\boldsymbol{E}(\boldsymbol{r}, t) &= \boldsymbol{E}(\boldsymbol{r}_0, t) + (\boldsymbol{r}_1\cdot\boldsymbol{\nabla})\boldsymbol{E}(\boldsymbol{r}, t)|_{\boldsymbol{r}_0} \\ &= \boldsymbol{E}(\boldsymbol{r}_0, t) + (\boldsymbol{r}_1\cdot\boldsymbol{\nabla})\widehat{\boldsymbol{E}}\cos\alpha_0 - \widehat{\boldsymbol{E}}(\boldsymbol{r}_1\cdot\boldsymbol{k})\sin\alpha_0 \ ,\end{aligned}$$

where $\alpha_0 \equiv \boldsymbol{k}\cdot\boldsymbol{r}_0 - \omega t$. To second order, we must add the term $\boldsymbol{v}_1 \times \boldsymbol{B}$:

$$\begin{aligned} m\frac{\mathrm{d}\boldsymbol{v}_2}{\mathrm{d}t} &= -e\Big[(\boldsymbol{r}_1\cdot\boldsymbol{\nabla})\widehat{\boldsymbol{E}}\cos\alpha_0 - \widehat{\boldsymbol{E}}(\boldsymbol{r}_1\cdot\boldsymbol{k})\sin\alpha_0\Big] \\ &\quad -e\boldsymbol{v}_1 \times \left(\boldsymbol{\nabla} \times \widehat{\boldsymbol{E}}\sin\alpha_0 + \frac{\boldsymbol{k} \times \widehat{\boldsymbol{E}}}{\omega}\cos\alpha_0\right) \\ &= -\frac{e^2}{m\omega^2}\Big[(\widehat{\boldsymbol{E}}\cdot\boldsymbol{\nabla})\widehat{\boldsymbol{E}}\cos^2\alpha_0 + \widehat{\boldsymbol{E}} \times \boldsymbol{\nabla} \times \widehat{\boldsymbol{E}}\sin^2\alpha_0\Big] \\ &\quad -\frac{e^2}{m\omega^2}\Big[-(\widehat{\boldsymbol{E}}\cdot\boldsymbol{k})\widehat{\boldsymbol{E}} + \widehat{\boldsymbol{E}} \times \boldsymbol{k} \times \widehat{\boldsymbol{E}}\Big]\sin\alpha_0\cos\alpha_0 \\ &= -\frac{e^2}{2m\omega^2}\Bigg\{\boldsymbol{\nabla}\frac{\widehat{\boldsymbol{E}}^2}{2}\big[1 - \cos 2(\boldsymbol{k}\cdot\boldsymbol{r}_0 - \omega t)\big] + 2(\widehat{\boldsymbol{E}}\cdot\boldsymbol{\nabla})\widehat{\boldsymbol{E}}\cos 2(\boldsymbol{k}\cdot\boldsymbol{r}_0 - \omega t) \\ &\quad +\boldsymbol{k}\widehat{\boldsymbol{E}}^2\sin 2(\boldsymbol{k}\cdot\boldsymbol{r}_0 - \omega t) - 2(\boldsymbol{k}\cdot\widehat{\boldsymbol{E}})\widehat{\boldsymbol{E}}\sin 2(\boldsymbol{k}\cdot\boldsymbol{r}_0 - \omega t)\Bigg\} \ . \end{aligned} \qquad (18.39)$$

Here we have used the formulas

$$\widehat{\boldsymbol{E}} \times (\boldsymbol{\nabla} \times \widehat{\boldsymbol{E}}) = \frac{1}{2}\boldsymbol{\nabla}(\widehat{\boldsymbol{E}}\cdot\widehat{\boldsymbol{E}}) - (\widehat{\boldsymbol{E}}\cdot\boldsymbol{\nabla})\widehat{\boldsymbol{E}} \ ,$$

$$\widehat{\boldsymbol{E}} \times (\boldsymbol{k} \times \widehat{\boldsymbol{E}}) = \boldsymbol{k}\widehat{\boldsymbol{E}}^2 - (\boldsymbol{k}\cdot\widehat{\boldsymbol{E}})\widehat{\boldsymbol{E}} \ .$$

In the case of a transverse electromagnetic wave, the terms $(\widehat{\boldsymbol{E}}\cdot\nabla)\widehat{\boldsymbol{E}}$ and $(\boldsymbol{k}\cdot\widehat{\boldsymbol{E}})\widehat{\boldsymbol{E}}$ are negligible and the terms due to the Lorentz force dominate. The time average of $m\mathrm{d}\boldsymbol{v}_2/\mathrm{d}t$ becomes

$$m\left\langle\frac{\mathrm{d}\boldsymbol{v}_2}{\mathrm{d}t}\right\rangle = -\frac{e^2}{4m\omega^2}\nabla\widehat{\boldsymbol{E}}^2 .$$

This is the effective nonlinear force on a single electron. The nonlinear force on the plasma per unit volume is

$$nm\left\langle\frac{\mathrm{d}\boldsymbol{v}_2}{\mathrm{d}t}\right\rangle = -\frac{\omega_\mathrm{p}^2}{\omega^2}\nabla\frac{\epsilon_0\widehat{\boldsymbol{E}}^2}{4} = -\frac{\omega_\mathrm{p}^2}{\omega^2}\nabla\frac{\epsilon_0\langle\boldsymbol{E}^2\rangle}{2} , \tag{18.40}$$

where $\omega_\mathrm{p}$ is the electron plasma frequency. This force is called the ponderomotive force.

# References

## Chapter 1


1.1 W.R. Arnold, J.A. Phillips, G.A. Sawyer, E.J. Stovall, Jr. and J.C. Tuck: Phys. Rev. **93**, 483 (1954)
1.2 C.F. Wandel, T.H. Jensen and O. Kofoed-Hansen: Nucl. Instr. and Methods **4**, 249 (1959)
1.3 J.L. Tuck: Nucl. Fusion **1**, 201 (1961)
1.4 T. Takizuka and M. Yamagiwa: JAERI-M 87-066 (1987) Japan Atomic Energy Research Institute


## Chapter 2


2.1 D.V. Sivukhin: Reviews of Plasma Physics **4**, 93 (ed. by M.A. Leontovich) Consultant Bureau, New York (1966)
2.2 K. Miyamoto: *Plasma Physics for Nuclear Fusion* (revised edn.) Chap. 4, MIT Press, Cambridge, Mass. (1989)
2.3 L. Spitzer, Jr.: *Physics of Fully Ionized Gases*, Interscience, New York (1962)
2.4 T.H. Stix: Plasma Phys. **14**, 367 (1972)


## Chapter 3


3.1 A. I. Morozov and L. S. Solovev: *Rev. of Plasma Physics* **2**, 201 (ed. by M. A. Leontovich) Consultant Bureau, New York 1966
3.2 R. G. Littlejohn: Phys. Fluids **24**, 1730 (1981); R. G. Littlejohn: J. Plasma Phys. **29**, 111 (1983)


## Chapter 5


5.1 K. Miyamoto: *Plasma Physics for Nuclear Fusion* (revised edn.) Chap. 6, MIT Press, Cambridge, Mass. (1989)


## Chapter 6


6.1 L.S. Solovev: Sov. Physics JETP **26**, 400 (1968); N.M. Zueva and L.S. Solovev: Atomnya Energia **24**, 453 (1968)

6.2 R.H. Weening: Phys. Plasmas **7**, 3654 (2000)
6.3 V.S. Mukhovatov and V.D. Shafranov: Nucl. Fusion **11**, 605 (1971)
6.4 V.D. Shafranov: Plasma Physics, J. of Nucl. Energy **C5**, 251 (1963)
6.5 D. Pfirsch and A. Schlüter: MPI/PA/7/62, Max-Planck Institut für Physik und Astrophysik, München (1962)
6.6 V.D. Shafranov: Plasma Physics **13**, 757 (1971)

## Chapter 7

7.1 K. Miyamoto: *Plasma Physics for Nuclear Fusion* (revised edn.) Chap. 6, MIT Press, Cambridge, Mass. (1989)
7.2 D. Pfirsh and A. Schlüter: MPI/PA/7/62, Max-Planck Institute für Physik und Astrophysik München (1962)
7.3 A.A. Galeev and R.Z. Sagdeev: Sov. Phys. JETP **26**, 233 (1968)
7.4 B.B. Kadomtsev and O.P. Pogutse: Nucl. Fusion **11**, 67 (1971)
7.5 W. Horton: Phys. Rev. Lett. **37**, 1269 (1976)
7.6 S. Hamaguchi and W. Horton: Phys. Fluids **B4**, 319 (1992)
7.7 W. Horton, Jr., R. Esres, H. Kwak and Duk-In Choi: Phys. Fluids **21**, 1366 (1978)
7.8 J.Y. Kim and M. Wakatani: Phys. Rev. Lett. **73**, 2200 (1994)
7.9 Y. Kishimoto, J.Y. Kim, T. Fukuda, S. Ishida, T. Fujita, T. Tajima, W. Horton, G. Furnish, M.J. LeBrun: *6th IAEA Fusion Energy Conference* (Conf. Proceedings, Montreal, 1996) **2**, 581 (1997) IAEA Vienna; Y. Kishimoto: J. Plasma Fusion Res. **76**, 1280 (2000) (in Japanese)
7.10 A.B. Rechester and M.N. Rosenbluth: Phys. Rev. Lett. **40**, 38 (1978)

## Chapter 8

8.1 G. Bateman: *MHD Instabilities*, MIT Press, Cambridge Mass. (1978)
8.2 M. Kruskal and M. Schwarzschild: Proc. Roy. Soc. **A223**, 348 (1954)
8.3 M.N. Rosenbluth, N.A. Krall and N. Rostoker: Nucl. Fusion Suppl. **1**, 143 (1962)
8.4 M.N. Rosenbluth and C.L. Longmire: Ann. Phys. **1**, 120 (1957)
8.5 I.B. Berstein, E.A. Frieman, M.D. Kruskal and R.M. Kulsrud: Proc. Roy. Soc. **A244**, 17 (1958)
8.6 B.B. Kadmotsev: Reviews of Plasma Physics **2**, 153 (ed. by M.A. Loentovich) Consultant Bureau, New York (1966)
8.7 K. Miyamoto: *Plasma Physics for Nuclear Fusion* (revised edn.) Chap. 9, MIT Press, Cambridge, Mass. (1989)
8.8 M.D. Kruskal, J.L. Johnson, M.B. Gottlieb and L.M. Goldman: Phys. Fluids **1**, 421 (1958)
8.9 V.D. Shafranov: Sov. Phys. JETP **6**, 545 (1958)
8.10 B.R. Suydam: *Proc. 2nd U.N. International Conf. on Peaceful Uses of Atomic Energy*, Geneva, **31**, 157 (1958)
8.11 W.A. Newcomb: Ann. Phys. **10**, 232 (1960)
8.12 D.C. Robinson: Plasma Phys. **13**, 439 (1971)
8.13 K. Hain and R. Lüst: Z. Naturforsh. **13a**, 936 (1958)
8.14 K. Matsuoka and K. Miyamoto: Jpn. J. Appl. Phys. **18**, 817 (1979)

8.15 J.M. Greene and J.L. Johnson: Plasma Phys. **10**, 729 (1968)
8.16 G. Bateman: *MHD Instabilities*, MIT Press, Cambridge, Mass. (1978)
8.17 J.W. Connor, R.J. Hastie and J.B. Taylor: Proc. Roy. Soc. **A365** 1, (1979)
8.18 J.W. Connor, R.J. Hastie and J.B. Taylor: Phys. Rev. Lett. **40**, 396 (1978)
8.19 R.M. Kulsrud: Plasma Phys. Contr. Nucl. Fusion Research, (Conf. Proceedings, Culham 1965) **1**, 127 (1966) IAEA Vienna; H.P. Furth, J. Killeen and M.N. Rosenbluth: ibid. 103
8.20 J.A. Wesson, A. Sykes: Nucl. Fusion **25**, 85 (1985)
8.21 J.M. Green, M.S. Chance: Nucl. Fusion **21**, 453 (1981)
8.22 T. Ozeki, M. Azumi, S. Tokuda, S. Ishida: Nucl. Fusion **33**, 1025 (1993)
8.23 J. Wesson: *Tokamaks*, 2nd. edn., Sect. 3.7, Clarendon Press, Oxford (1997)
8.24 B.B. Kadomtsev and O.P. Pogutse: Reviews of Plasma Physics, **5**, 304 (ed. by M.A. Leontovich) Consultant Bureau, New York (1970)

## Chapter 9

9.1 H.P. Furth and J. Killeen: Phys. Fluids **6**, 459 (1963)
9.2 H.P. Furth, P.H. Rutherford and H. Selberg: Phys. Fluids **16**, 1054 (1973); A. Pletzer and R.L. Dewar: J. Plasma Phys. **45**, 427 (1991)
9.3 S.S. Moiseev and R.Z. Sagdeev: Sov. Phys. JETP **17**, 515 (1963), Sov. Phys. Tech. Phys. **9**, 196 (1964)
9.4 F.F. Chen: Phys. Fluids **8**, 912 and 1323 (1965)

## Chapter 10

10.1 T.H. Stix: *The Theory of Plasma Waves*, McGraw-Hill, New York (1962)
10.2 T.H. Stix: *Waves in Plasmas*, American Institute of Physics, New York, (1992)
10.3 W.P. Allis, S.J. Buchsbanm and A. Bers: *Waves in Anisotropic Plasmas*, MIT Press, Cambridge, Mass. (1963)
10.4 G. Bekefi: *Radiation Processes in Plasmas*, John Wiley and Son, Gordon and Breach Science Publishers, New York (1961)

## Chapter 11

11.1 L.D. Landau: J. Phys. (USSR) **10**, 45 (1946)
11.2 J.H. Malmberg, C.B. Wharton and W.E. Drummond: Plasma Phys. Contr. Nucl. Fusion Research (Conf. Proceedings, Culham 1965) **1**, 485 (1966) IAEA Vienna
11.3 T.H. Stix: *The Theory of Plasma Waves*, McGraw Hill, New York (1962); T.H. Stix: *Waves in Plasmas*, American Institute of Physics, New York, (1992)

## Chapter 12


12.1 T.H. Stix: *The Theory of Plasma Waves*, McGraw-Hill, New York (1962); T.H. Stix: *Waves in Plasmas*, American Institute of Physics, New York (1992)
12.2 B.D. Fried and S.D. Conte: *The Plasma Dispersion Function*, Academic Press, New York (1961)
12.3 K. Miyamoto: *Plasma Physics for Nuclear Fusion* (revised edn.), Chaps. 12, 13, MIT Press, Cambridge, Mass. (1989)
12.4 M. Porkolab: *Fusion* (ed. by E. Teller) **1**, Part B, 151, Academic Press, New York (1981)
12.5 J.E. Scharer, B.D. McVey and T.K. Mau: Nucl. Fusion **17**, 297 (1977)
12.6 T.H. Stix: Nucl. Fusion **15**, 737 (1975)
12.7 A. Fukuyama, S. Nishiyama, K. Itoh and S.I. Itoh: Nucl. Fusion **23**, 1005 (1983)
12.8 M. Ono, T. Watari, R. Ando, J.Fujita et al.: Phys. Rev. Lett. **54**, 2339 (1985)
12.9 T.H. Stix: Phys. Rev. Lett. **15**, 878 (1965)
12.10 V.M. Glagolev: Plasma Phys. **14**, 301 and 315 (1972)
12.11 M. Brambilla: Plasma Phys. **18**, 669 (1976)
12.12 S. Bernabei, M.A. Heald, W.M. Hooke, R.W. Motley, F.J. Paoloni, M. Brambilla and W.D. Getty: Nucl. Fusion **17**, 929 (1977)
12.13 S. Takamura: *Fundamentals of Plasma Heatings* (in Japanese), Nagoya Univ. Press (1986)
12.14 I. Fidone, G. Granata and G. Ramponi: Phys. Fluids **21**, 645 (1978)
12.15 R. Prator: Phys. Plasmas **11**, 2349 (2004)
12.16 R.J. Briggs: *Electron-Stream Interaction with Plasma*, MIT Press, Cambridge, Mass. (1964)
12.17 M.N. Rosenbluth and R.F. Post: Phys. Fluids **8**, 547 (1965)
12.18 E.G. Harris: Phys. Rev. Lett. **2**, 34 (1959)
12.19 E.G. Harris: *Physics of Hot Plasma*, 145 (ed. by B.J. Rye and J.B. Taylor) Oliver & Boyd, Edinburgh (1970)
12.20 N.A. Krall and M.N. Rosenbluth: Phys. Fluids **8**, 1488 (1965)
12.21 B.B. Kadomtsev and O.P. Pogutse: Nucl. Fusion **11**, 67 (1971)


## Chapter 13


13.1 L. Chen, R.B. White and M.N. Rosenbluth: Phys. Rev. Lett. **52**, 1122 (1984); Y.Z. Zhang, H.L. Berk and S.M. Mahajan: Nucl. Fusion **29**, 848 (1989)
13.2 V.D. Shafranov: Sov. Phys. Tech. Phys. **15**, 175 (1970)
13.3 M.N. Bussac, R. Pella, D. Edery and J.L. Soule: Phys. Rev. Lett. **35**, 1638 (1975)
13.4 G. Ara, B. Basu, B. Coppi, G. Laval, M.N. Rosenbluth and B.V. Waddell: Ann. Phys. **112**, 443 (1978)
13.5 B.N. Kuvshinov, A.B. Mikhailovskii and E.G. Tatarinov: Sov. J. Plasma Phys. **14**, 239 (1988)
13.6 P.J. Catto, W.M. Tang and D.E. Baldwin: Plasma Phys. **23**, 639 (1981)
13.7 A. Hasegawa and L. Chen: Phys. Fluids **19**, 1924 (1976)

13.8 H.L. Berk, J.W. Van Dam, Z. Guo and D.M. Lindberg: Phys. Fluids **B4** 1806 (1992)
13.9 R. Betti and J.P. Freidberg: Phys. Fluids **B4**, 1465 (1992)
13.10 M.N. Rosenbruth, H.L. Berek, J.W. Van Dam, D.M. Lindberg: Phys. Fluids **B4**, 2189 (1992)
13.11 J. Candy and N.M. Rosenbluth: Plasma Phys. Contr. Fusion **35**, 957 (1993); J. Candy and N.M. Rosenbluth: Phys. Plasmas **1**, 356 (1994); R.R. Mett and S.M. Mahajan: Phys. Fluids **B4**, 2885 (1992)
13.12 G.Y. Fu and C.Z. Cheng: Phys. Fluids **B2**, 985 (1990); H.L. Berk, J.W. Van Dam, D. Borba, J. Candy, G.T.A. Huysmans, S. Sharapov: Phys. Plasmas **2**, 3401 (1995)
13.13 F. Zonca and L. Chen: Phys. Fluids **B5**, 3668 (1993); F. Zonca and L. Chen: Phys. Plasmas **3**, 323 (1996)
13.14 M.S. Chu, J.M. Greene, L.L. Lao, A.D. Turnbull and M.S. Chance: Phys. Fluids **B4**, 3713 (1992); A.D. Turnbull, E.J. Strait, W.W. Heidbrink, M.S. Chu, H.H. Duong, J.M. Greene, L.L. Lao, T.S. Taylor, S.J. Thompson: Phys. Fluids **B5**, 2546 (1993)
13.15 ITER Physics Basis: Nucl. Fusion **39**, No.12, 2471 (1999)
13.16 E.J. Strait, W.W. Heidbrik, A.D. Turnbull, M.S. Chu and H.H. Duong: Nucl. Fusion **33**, 1849 (1993); King-Lap Wong: Plasma Phys. Contr. Fusion **41**, R1 (1999); A. Fukuyama and T. Ozeki: J. Plasma Fusion Res. **75**, 537 (1999) (in Japanese)

## Chapter 14

14.1 R.F. Post: Plasma Physics in the 20th Century, Chap. 22. In: *Twentieth Century Physics*, Vol. III, ed. by L.M. Brown, A. Pais and B. Pippard: IOP Publishing, Bristol and American Inst. of Physics Press, New York (1995)
14.2 R.W. Hockney and J.W. Eastwood: *Computer Simulation Using Particles*, McGraw-Hill, NewYork (1981)
14.3 T. Tajima: *Computational Plasma Physics with Application to Fusion and Astrophysics*, Addison-Wesley Publishing Company, Redwood City, Calif. (1989)
14.4 C.K. Birdsall and A.B. Langdon: *Plasma Physics via Computer Simulation*, IOP Pubishing, Bristol (1991)
14.5 D. Berger, R. Gruber and F. Troyon: Computer Phys. Commun. **11**, 313 (1976); D. Berger, L.C. Bernard, R. Gruber and F. Troyon: Plasma Phys. Contr. Nucl. Fusion Research (Conf. Proceedings, Berchtesgaden 1976) **2**, 411 (1977) IAEA Vienna
14.6 F. Troyon, R. Gruber, H. Saurenmann, S. Semenzato and S. Succi: Plasma Phys. Contr. Fusion **26**, 209 (1984)
14.7 M.S. Chance, R.L. Dewar, A.H. Glasser, J.M. Green, R.C. Grimm et al.: Plasma Phys. Contr. Nucl. Fusion Research (Conf. Proceedings, Tokyo 1974) **1**, 463 (1975) IAEA Vienna; W. Kerner: Nucl. Fusion **16**, 643 (1976)
14.8 D.D. Schnack, D.C. Barnes, Z. Mikic: J. Comput. Physics **70**, 330 (1987); D.D. Schnack, E.J. Caramana, R.A. Nebel: Phys. Fluids **28**, 321 (1985)

14.9 K. Kusano, T. Sato: Nucl. Fusion **26**, 1051 (1986)
14.10 W. Park, E.D. Fredrickson, A. Janos, J. Manickam, W.M. Tang: Phys. Rev. Lett. **75**, 1763 (1995)
14.11 T. Hayashi, N. Mizuguchi, T.H. Watanabe, Y. Todo, T. Sato: Nucl. Fusion **40**, 721 (2000); N. Mizuguchi, T. Hayashi, T. Sato: Phys. Plasma **7**, 940 (2000)
14.12 L. Villard, K. Appert, R. Gruber, J. Vaclavik: Computer Phys. Reports **4**, 95 (1986)
14.13 A. Jaun, K. Appert, J. Vaclavik, L. Villard: Computer Phys. Commun. **92**, 153 (1995)
14.14 L.-G. Eriksson, T. Hellsren, U. Willen: Nucl. Fusion **33**, 1037 (1993)
14.15 A. Fukuyama, K. Itoh, S.-I. Itoh: Computer Phys. Reports **4**, 137 (1986)
14.16 A. Jaun, A. Fosoli, J. Vaclavik, L. Villard: Nucl. Fusion **39**, 2095 (1999); A. Jaun, A. Fosoli, J. Vaclavik, L. Villard: Nucl. Fusion **40**, 1343 (2000)
14.17 A. Fukuyama, T. Akutsu: *19th IAEA Fusion Energy Conference* (Lyon 2000) TH/P3-14
14.18 C.Z. Cheng: Physics Reports **211**, 1 (1992); C.Z. Cheng, N.N. Gorelenkov, C.T. Hsu: Nucl Fusion **35**, 1639 (1995)
14.19 Y. Todo, T.-H. Watanabe, Hyoung-Bin Park, T. Sato: Nucl. Fusion **41**, 1153 (2001); Y. Todo, T. Sato, K. Watanabe, T.H. Watanabe, R. Horiuchi: Phys. Plasma **2**, 2711 (1995)
14.20 M.A. Beer and G.W. Hammett: Phys. Plasma **3**, 4046 (1996); W. Dorland and G.W. Hammett: Phys. Fluids **B5**, 812 (1993); A. Brizard: Phys. Fluids **B4**, 1213 (1992); G.W. Hammett and F.W. Perkins: Phys. Rev. Lett. **64**, 3019 (1990)
14.21 R.E. Waltz, G.M. Staebler, W. Dorland, G.W. Hammett, M. Kotschenreuther, J.A. Konings: Phys. Plasma **4**, 2482 (1997); R.E. Waltz, G.D. Kerbel, J. Milovich, G.W. Hammett: Phys. Plasma **2**, 2408 (1995); R.E. Waltz, R.R. Domingues, G.W. Hammett: Phys. Fluids **B4**, 3138 (1992)
14.22 E.A. Frieman, L. Chen: Phys. Fluids **25**, 502 (1982)
14.23 P.J. Catto, W.M. Tang, D.E. Baldwin: Plasma Phys. **23**, 639 (1981)
14.24 A.M. Dimits, G. Bateman, M.A. Beer, B.I. Cohen, W. Dorland et al.: Phys. Plasma **7**, 969 (2000)
14.25 W.W. Lee: Phys. Fluids **26**, 556 (1983); W.W. Lee: J. Comput. Physics **72**, 243 (1987)
14.26 R.D. Sydora: Phys. Fluids **B2**, 1455 (1990)
14.27 S.E. Parker, W.W. Lee, R.A. Santoro: Phys. Rev. Lett. **71**, 2042 (1993)
14.28 S.E. Parker, W.W. Lee: Phys. Fluids **B5**, 77 (1993)
14.29 A.M. Dimits, B.I. Cohen, N. Mattor, W.M. Nevins, D.E. Shumaker: Nucl. Fusion **40**, 661 (2000)
14.30 R. G. Littlejohn: Phys. Fluids **24**, 1730 (1981); R. G. Littlejohn: J. Plasma Phys. **29**, 111 (1983); A. Brizard: J. Plasma Phys. **41**, 541 (1989); H. Sugama: Phys. Plasmas **7**, 466 (2000)
14.31 A.B. Langdon and C.K. Birdsall: Phys. Fluids **13**, 2115 (1970); H. Okuda and C.K. Birdsall: Phys. Fluids **13**, 2123 (1970); H. Okuda: Phys. Fluids **15**, 1268 (1972)
14.32 H. Naitou: J. Plasma Fusion Res. **74**, 470 (1998) (in Japanese)
14.33 K. Nishimura, R. Horiuchi T. Sato: Phys. Plasmas **4**, 4035 (1997); R. Horiuchi, T. Sato: Phys. Fluids **B2**, 2652 (1990)

14.34 E.V. Belova, S.C. Jardin, H. Ji, M. Yamada, R. Kulsrud: Phys. Plasma **7**, 4996 (2000)
14.35 M.J. LeBrun, T. Tajima, M.G. Gray, G. Furnish, W. Horton: Phys. Fluids **B5**, 752 (1993)
14.36 Y. Kishimoto, J.-Y. Kim, W. Horton, T. Tajima, M.J. LeBrun, H. Shirai: Plasma Phys. Contr. Fusion **40**, A663 (1998)

## Chapter 15

15.1 A.S. Bishop: *Project Sherwood*, Addison Wesley, Reading Mass. (1958)
15.2 R.S. Pease: Plasma Phys. and Contr. Fusion **28**, 397 (1986)
15.3 L.A. Artsimovich: Sov. Phys. Uspekhi **91**, 117 (1967)
15.4 J.D. Lawson: Proc. Phys. Soc. B**70**, 6 (1957)
15.5 M.D. Kruskal and M. Schwarzschild: Proc. Roy. Soc. **A223**, 348 (1954)
15.6 M.N. Rosenbluth and C.L. Longmire: Ann. Phys. **1**, 120 (1957); I.B. Berstein, E.A. Frieman, M.D. Kruskal and R.M. Kulsrud: Proc. Roy. Soc. **A244**, 17 (1958)
15.7 Nature **181**, No. 4604, p. 217, 25 January (1958)
15.8 L. Spitzer, Jr.: Phys. Fluids **1**, 253 (1958)
15.9 *Proceedings of the Second United Nations International Conference on the Peaceful Uses of Atomic Energy* (Geneva 1958), **31**, Theoretical and Experimental Aspects of Controlled Nuclear Fusion; **32**, Controlled Fusion Devices, United Nations Publication, Geneva (1958)
15.10 *The Second Geneva Series on the Peaceful Uses of Atomic Energy* (series Ed. J.G. Beckerley, Nuclear Fusion) D. Van Nostrand, New York (1960)
15.11 S. Hayakawa, K. Kimura: J. Plasma Fusion Res. **57**, 201, 271 (1987) (in Japanese)
15.12 Plasma Phys. Contr. Nucl. Fusion Research (Conf. Proceedings, Salzburg 1961), Nucl. Fusion Suppl. (1962) [Translation of Russian Papers: U.S. Atomic Energy Commission, Division of Technical Information Office of Tech. Service, Depart. of Commerce, Washington D.C. (1963)]
15.13 Ibid. (Conf. Proceedings, Culham 1965) International Atomic Energy Agency (IAEA), Vienna (1966) [Translation of Russian Paper: U.S. Atomic Energy Commission, Division of Technical Information, Oak Ridge, Tenn. (1966)]
15.14 Ibid. (Conf. Proceedings, Novosibirsk 1968) (1969) IAEA Vienna [Translation of Russian Paper: Nucl. Fusion Suppl. (1969)]
15.15 Ibid. (Conf. Proceedings, Madison 1971) (1971) IAEA Vienna [Translation of Russian Paper: Nucl. Fusion Suppl. (1972)]
15.16 Ibid. (Conf. Proceedings, Tokyo 1974) (1975) IAEA Vienna
15.17 M.N. Rosenbluth and R.F. Post: Phys. Fluids **8**, 547 (1965)
15.18 E.G. Harris: Phys. Rev. Lett. **2**, 34 (1959)
15.19 A.B. Mikhailovskii and L.I. Rudakov: Sov. Phys. JETP **17**, 621 (1963); N.A. Krall and M.N. Rosenbluth: Phys. Fluids **8**, 1488 (1965)
15.20 M.J. Forrest, N.J. Peacock, D.C. Robinson, V.V. Sannikov and P.D. Wilcock: Culham Report CLM-R 107, July (1970)
15.21 JT60U Team: Plasma Phys. Contr. Nucl. Fusion Research (Conf. Proceedings, Seville 1994) **1**, 31 (1995) IAEA Vienna

15.22 O1-2, O1-6, O1-3, A5-5, O2-2: In: *16th IAEA Fusion Energy Conference* (Conf. Proceedings, Montreal 1996) **1**, 19, 95, 37, 487, 141 (1997) IAEA Vienna
15.23 JET Team: Nucl. Fusion **32**, 187 (1992)
15.24 TFTR Team: Plasma Phys. Contr. Nucl. Fusion Research (Conf. Proceedings, Seville 1994) **1**, 11 (1995) IAEA Vienna
15.25 JET Team: *17th IAEA Fusion Energy Conference* (Conf. Proceedings, Yokohama 1998) **1**, 29 (1999) IAEA Vienna
15.26 INTOR Team: Nucl. Fusion **23**, 1513, (1983)
15.27 ITER Physics Basis: Nucl. Fusion **39**, No.12, 2137–2638 (1999)
15.28 ITER-FEAT: Technical Basis for the ITER–FEAT Outline Design (December 1999) IAEA Vienna
15.29 Y-K.M. Peng, D.J. Strickler: Nucl.Fusion **26**, 769 (1986)
15.30 A. Sykes: *17th IAEA Fusion Energy Conference* (Conf. Proceedings, Yokohama 1998) **1**, 129 (1999) IAEA Vienna
15.31 MAST(OV4/1), NSTX(OV4/2): *18th IAEA Fusion Energy Conference* (Sorrento 2000)
15.32 R.J. Akers et al.: Nucl. Fusion **40**, 1223 (2000)
15.33 A. Iiyoshi et al.: Nucl. Fusion **39**, 1245 (1999)
15.34 F.H. Coensgen, W.F. Cummins, V.A. Finlayson, W.E. Nexsen, Jr. and T.C. Simonen: Plasma Phys. Contr. Nucl. Fusion Research (Conf. Proceedings, Madison 1971) **2**, 721 (1971) IAEA Vienna
15.35 G.I. Dimov, V.V. Zakaidakov and M.E. Kishinevskii: Sov. J. Plasma Phys. **2**, 326 (1976)
15.36 T.K. Fowler and B.G. Logan: Comments Plasma Phys. Contr. Fusion Res. **2**, 167 (1977)
15.37 T.K. Fowler (Ed.): Nucl. Fusion **9**, 3 (1969); M.S. Ioffe and B.B. Kadomotsev: Sov. Phys. Uspekhi **13**, 225 (1970); R.F. Post: Nucl. Fusion **27**, 1579 (1987)
15.38 J.D. Lindl, M.M. Marinak: *16th IAEA Fusion Energy Conference* (Conf. Proceedings, Montreal 1996) **3**, 43 (1997) IAEA Vienna
15.39 J.D. Lindl: *Inertial Confinement Fusion*, Springer/AIP Press New York (1998)

## Chapter 16

16.1 L.A. Artsimovich: Nucl. Fusion **12**, 215 (1972); H.P. Furth: Nucl. Fusion **15**, 487 (1975)
16.2 J. Wesson: *Tokamaks*, 2nd edn., Clarendon Press, Oxford (1997); ITER Physics Basis: Nucl. Fusion **39**, No.12, 2138–2638 (1999)
16.3 V.S. Mukhovatov and V.D. Shafranov: Nucl. Fusion **11**, 605 (1971)
16.4 Equip TFR: Nucl. Fusion **18**, 647 (1978)
16.5 J. Sheffield: *Plasma Scattering of Electromagnetic Radiation*, Academic Press, New York (1975)
16.6 For example: Y. Nagayama, Y. Ohki and K. Miyamoto: Nucl. Fusion **23**, 1447 (1983)
16.7 F. Troyon, R. Gruber, H. Saurenmann, S. Semenzato and S. Succi: Plasma Phys. Contr. Fusion **26**, 209 (1984)

16.8 A. Sykes, M.F. Turner and S. Patel: *Proc. 11th European Conference on Controlled Fusion Plasma Physics*, Part II, 363 (1983) Aachen; T. Tuda, M. Azumi, K. Itoh, G. Kurita, T. Takeda et al.: Plasma Phys. Contr. Nucl. Fusion Research (Conf. Proceedings, London 1984) **2**, 173 (1985) IAEA Vienna
16.9 B.B. Kadomtsev: Sov. J. Plasma Phys. **1**, 389 (1975)
16.10 M. Greenwald, J.L. Terry, S.M. Wolfe, S. Ejima, M.G. Bell, S.M. Kaye, G.H. Nelson: Nucl. Fusion **28**, 2199 (1988)
16.11 ASDEX-U Team: *16th IAEA Fusion Energy Conference* (Conf. Proceedings, Montreal 1996) **1**, 79 (1996) IAEA Vienna
16.12 J.A. Wesson: Nucl. Fusion **18**, 87 (1978)
16.13 B.B. Kadomtsev and O.P. Pogutse: Nucl. Fusion **11**, 67 (1971)
16.14 J.W. Connor and H.R. Wilson: Plasma Phys. Contr. Fusion **36**, 719 (1994)
16.15 F. Wagner and U. Stroth: Plasma Phys. Contr. Fusion **35**, 1321 (1993)
16.16 T.N. Todd: *Proceedings of 4th International Symposium on Heating in Toroidal Plasmas*, Rome **1**, 21,(1984) International School of Plasma Physics, Varenna; Y. Kamada, K. Ushigusa, O. Naito, Y. Neyatani, T. Ozeki et al.: Nucl. Fusion **34**, 1605 (1994)
16.17 DIII-D team: Plasma Phys. Contr. Nucl. Fusion Research (Conf. Proceedings, Washington D.C. 1990) **1**, 69 (1991) IAEA Vienna
16.18 K. Borrass: Nucl. Fusion: **31**, 1035 (1991); K. Borrass, R. Farengo, G.C. Vlases: Nucl. Fusion **36**, 1389 (1996); B. LaBombard, J.A. Goetz, I. Hutchinson, D. Jablonski, J. Kesner et al.: Nucl. Materials **241–243**, 149 (1997)
16.19 R.J. Goldston: Plasma Phys. Contr. Fusion **26**, 87 (1984); S.M. Kaye: Phys. Fluids **28**, 2327 (1985)
16.20 P.N. Yushimanov, T. Takizuka, K.S. Riedel, D.J.W.F. Kardaun, J.G. Cordey, S.M. Kaye, D.E. Post: Nucl. Fusion **30**, 1999 (1990); N.A. Uckan, P.N. Yushimanov, T. Takizuka, K. Borras, J.D. Callen et al.: Plasma Phys. Contr. Nucl. Fusion Research (Conf. Proceedings, Washington D.C. 1990) **3**, 307 (1991) IAEA Vienna
16.21 F. Wagner, G. Becker, K. Behringer et al.: Phys. Rev. Lett. **49**, 1408 (1982); F. Wagner, G. Becker, K. Behringer, D. Campbell, M. Keilhacker et al.: Plasma Phys. Contr. Nucl. Fusion Research (Conf. Proceedings, Baltimore 1982) **1**, 43 (1983) IAEA Vienna
16.22 ASDEX Team: Nucl. Fusion **29**, 1959 (1989)
16.23 R.J. Groebner: Phys. Fluids **B5**, 2343 (1993)
16.24 E.J. Doyle, C.L. Rettig, K.H. Burrell, P. Gohil, R.J. Groebner et al.: Plasma Phys. Contr. Nucl. Fusion Research (Conf. Proceedings, Würzburg 1992) **1**, 235 (1992) IAEA Vienna
16.25 S.I. Itoh and K. Itoh: Phys. Rev. Lett. **63**, 2369 (1988); K.C. Shaing and E.C. Crume: Phy. Rev. Lett. **63**, 2369 (1989)
16.26 K. Ida: Plasma Phys. Contr. Fusion **40**, 1429 (1998)
16.27 H. Bigrali, D.H. Diamond, Y.-B. Kim, B.A. Carreras, V.E. Lynch, F.L. Hinton et al.: Plasma Phys. Contr. Nucl. Fusion Research (Conf. Proceedings, Washington D.C. 1990) **2**, 191 (1991) IAEA Vienna
16.28 T.H. Dupree: Phys. Fluids **15**, 334 (1972); T. Boutros-Ghali and T.H. Dupree: Phys. Fluids: **24**, 1839 (1981)

16.29 TFTR Team: Plasma Phys. Contr. Nucl. Fusion Research (Conf. Proceedings, Seville 1994) **1**, 11 (1995) IAEA Vienna; ibid. (Conf. Proceedings, Washington D.C. 1990) **1**, 9 (1991) IAEA Vienna
16.30 T.S. Taylor, T.H. Osborne, K.H. Burrel et al.: Plasma Phys. and Contr. Nucl. Fusion Research (Conf. Proceedings, Würzburg 1992) **1**, 167 (1992) IAEA Vienna
16.31 JT60U Team: ibid. (Conf. Proceedings, Seville 1994) **1**, 31 (1995) IAEA Vienna
16.32 F.L. Hinton and G.M. Staebner: Phys. Fluids **B5**, 1281 (1993)
16.33 O1-6, O1-2, O1-3, A5-5, O2-2: In: *16th IAEA Fusion Energy Conference* (Conf. Proceedings, Montreal 1996) IAEA Vienna
16.34 ITER Physics Basis, Chap. 2 in Nucl. Fusion **39**, No.12 (1999)
16.35 JET Team: Nucl. Fusion **32**, 187 (1992)
16.36 T. Ohkawa: Nucl. Fusion **10**, 185 (1970)
16.37 D.J.H. Wort: Plasma Phys. **13**, 258 (1971)
16.38 N.J. Fisch: Phys. Rev. Lett. **41**, 873 (1978); C.F.F. Karney and N.J. Fisch: Phys. Fluids **22**, 1817 (1979)
16.39 D.V. Sivukhin: Reviews of Plasma Physics **4**, 93 (ed. by M.A. Leontobich) Consultant Bureau, New York (1966)
16.40 N.J. Fisch and A.H. Boozer: Phys. Rev. Lett. **45**, 720 (1980)
16.41 D.F.H. Start, J.G. Cordey and E.M. Jones: Plasma Phys. **22**, 303 (1980)
16.42 K. Okano: Nucl. Fusion **30**, 423 (1990)
16.43 R.J. Bickerton, J.W. Connor and J.B. Taylor: Nature Physical Science **229**, 110 (1971)
16.44 A.A. Galeev: Sov. Phys. JETP **32**, 752 (1971)
16.45 M.N. Rosenbluth, R.D. Hazeltine and F.L. Hinton: Phys. Fluids **15**, 116 (1972)
16.46 D.J. Sigmar: Nucl. Fusion **13**, 17 (1973)
16.47 T. Ozeki, M. Azumi, S. Tokuda, S. Ishida: Nucl. Fusion **33**, 1025 (1993)
16.48 P.H. Rutherford: Phys. Fluids **16**, 1903 (1973)
16.49 R. Fitzpatrick: Phys. Plasmas **2**, 825 (1995)
16.50 H.R. Wilson, J.W. Connor, R.H. Hastie, C.C. Hegna: Phys. Plasma **3**, 249 (1996); A.I. Smolyakov, A. Hirose, E. Lazzaro, G.B. Re, J.D. Callen: Phys. Plasma **2**, 1581 (1995); F.L. Hinton and J.R. Robertson: Phys. Fluids **27**, 1243 (1984)
16.51 D.A. Gates, B. Lloyd, A.W. Morris, G. McArdle, M.R. O'Brien et al.: Nucl. Fusion **37**, 1593 (1997)
16.52 ITER Team: Plasma Phys. Contr. Nucl. Fusion Research (Conf. Proceedings, Washington D.C. 1990) **3**, 413 (1991) IAEA Vienna
16.53 K. Miyamoto: Jour. Plasma Fusion Research **76**, 166 (2000)
16.54 INTOR Team: Nucl Fusion **23**, 1513 (1983)
16.55 ITER Team: *16th IAEA Fusion Energy Conference* (Conf. Proceedings, Montreal 1996) O1-1, F1–F5, IAEA Vienna
16.56 ITER Team: *18th IAEA Fusion Energy Conference* (Sorrento 2000) OV/1, ITER/1-6

## Chapter 17

17.1 D.C. Robinson and R.E. King: Plasma Phys. Contr. Nucl. Fusion Research (Conf. Proceedings, Novosibirsk 1968) **1**, 263 (1969) IAEA Vienna

17.2 H.A.B. Bodin and A.A. Newton: Nucl. Fusion **20**, 1255 (1980)
17.3 H.A.B. Bodin: Plasma Phys. Contr. Fusion **29**, 1297 (1987)
17.4 MST Team: Plasma Phys. Contr. Nucl. Fusion Research (Conf. Proceedings, Washington D.C. 1990) **2**, 519 (1991) IAEA Vienna; TPE-1RM20 Team: *19th IAEA Fusion Energy Conference* (Conf. Proceedings, Montreal 1996) **2**, 95 (1997) IAEA Vienna
17.5 EX4/3(RFX), EX4/4(TPE-RX): *17th IAEA Fusion Energy Conference* (Conf. Proceedings, Yokohama 1998) **1**, 367, 375 (1998) IAEA Vienna
17.6 J.B. Taylor: Phys. Rev. Lett. **33**, 1139 (1974)
17.7 T.H. Jensen and M.S. Chu: Phys. Fluids **27**, 2881 (1984)
17.8 V.D. Shafranov and E.I. Yurchenko: Sov. Phys. JETP **26**, 682 (1968)
17.9 D.A. Backer, M.D. Bausman, C.J. Buchenauer, L.C. Burkhardt, G. Chandler, J.N. Dimarco et al.: Plasma Phys. Contr. Nucl. Fusion Research (Conf. Proceeding, Baltimore 1982) **1**, 587 (1983) IAEA Vienna
17.10 D.D. Schnack, E.J. Caramana and R.A. Nebel: Phys. Fluids **28**, 321 (1985)
17.11 K. Kusano and T. Sato: Nucl. Fusion **26**, 1051 (1986)
17.12 K. Miyamoto: Plasma Phys. Contr. Fusion **30**, 1493 (1988)
17.13 A.R. Jacobson and R.W. Moses: Phys. Rev. **A29**, 3335 (1984); R.W. Moses, K.F. Schoenberg and D.A. Baker: Phys. Fluids **31**, 3152 (1988)
17.14 J.S. Sarff, S.A. Hokin, H. Ji, S.C. Prager, C.R. Sovinec: Phys. Rev. Lett. **72**, 3670 (1994); B.E. Chapman, J.K. Anderson, T.M. Biewer, D.L. Brower, S. Castillo et al.: Phys. Rev. Lett. **87**, 205001-1 (2001)
17.15 M.K. Bevir and J.W. Gray: *Proc. of Reversed Field Pinch Theory Workshop* (LANL Los Alamos 1981) Report No.8944-C, 176; M.K. Bevir and C.G. Gimblett: Phys. Fluids **28**, 1826 (1985)
17.16 K.F. Schoenberg, J.C. Ingraham, C.P. Munson, P.G. Weber et al.: Phys. Fluids **31**, 2285 (1988)
17.17 L. Spitzer, Jr.: Phys. Fluids **1**, 253 (1958)
17.18 A.I. Morozov and L.S. Solovév: Reviews of Plasma Physics **2**, 1 (ed. by M.A. Leontovich) Consultant Bureau, New York (1966)
17.19 K. Miyamoto: *Plasma Physics for Nuclear Fusion* (revised edn.) Chap. 2, MIT Press, Cambridge, Mass. (1989)
17.20 K. Nagasaki, K. Itoh, M. Wakatani and A. Iiyoshi: J. Phys. Soc. Japan **57**, 2000 (1988)
17.21 W VIIA Team: Plasma Phys. Contr. Nucl. Fusion Research (Conf. Proceedings, Baltimore 1982) **2**, 241 (1983) IAEA Vienna
17.22 E.D. Andryukhina, G.M. Batanov, M.S. Berezhetshij, M.A. Blokh, G.S. Vorosov et al.: Plasma Phys. Contr. Nucl. Fusion Reseach (Conf. Proceedings, London 1984) **2**, 409 (1985) IAEA Vienna
17.23 K. Uo, A. Iiyoshi, T. Obiki, O. Motojima, S. Morimoto, M. Wakatani et al.: Plasma Phys. Contr. Nucl. Fusion Research (Conf. Proceedings, London 1984) **2**, 383 (1985) IAEA Vienna
17.24 L. Garcia, B.A. Carreras and J.H. Harris: Nucl. Fusion **24**, 115 (1984)
17.25 Yu. N. Petrenlco and A.P. Popyyadukhin: *The 3rd International Symposium on Toroidal Plasma Confinements*, D 8, (1973) Garching
17.26 H. Wobig and S. Rehker: *Proceedings of the 7th Symposium on Fusion Technology* 345, (1972) Grenoble; S. Rehker and H. Wobig: *Proceedings of the 6th European Conference on Contr. Fusion Plasma Physics* 117 (1973) Moscow. IPP 2/215 Max Planck Inst. of Plasma Phys. (1973)

17.27 C. Gourdon, D. Marty, E.K. Maschke and J.P. Dumont: Plasma Phys. Contr. Nucl. Fusion Research (Conf. Proceedings, Novosibirsk 1968) **1**, 847 (1969) IAEA Vienna

17.28 K. Uo: Plasma Phys. **13**, 243 (1971)

17.29 A. Mohri: J. Phys. Soc. Japan **28**, 1549 (1970)

17.30 C. Gourdon, D. Marty, E.K. Maschke and J. Touche: Nucl. Fusion **11**, 161 (1971)

17.31 B.B. Kadomtsev and O.P. Pogutse: Nucl. Fusion **11**, 67 (1971)

17.32 J.A. Derr and J.L. Shohet: Phys. Rev. Lett. **44**, 1730 (1979)

17.33 M. Wakatani, S. Kodama, M. Nakasuga and K. Hanatani: Nucl. Fusion **21**, 175 (1981)

17.34 J.W. Connor and R.J. Hastie: Phys. Fluids **17**, 114 (1974)

17.35 L.M. Kovrizhnykh: Nucl. Fusion **24**, 851 (1984)

17.36 D.E. Hastings, W.A. Houlberg and K.C. Shaing: Nucl. Fusion **25**, 445 (1985)

17.37 M. C. Zarnstorff, L. A. Berry, A. Brooks, F. Fredrickson, G-Y. Fu et al: Plasma Phys. Contr. Fusion **43**, A237 (2001);
S. Okamura, K. Matsuoka, S. Nishimura, M. Isobe, C. Suzuki et al: Nucl. Fusion **44**, 575 (2004);
N. Nakajima, M. Yokoyama, M. Okamoto and J. Nührenberg: Plasma Phys. Reports **23**, 460 (1997)

17.38 D. A. Garren and A. H. Boozer: Phys. Fluids **B3**, 2822 (1991);
F. S. B. Anderson, A. F. Almagri, D. T. Anderson, P. G. Mathews, J. N. Talmadge and J. L. Shohet: Fusion Technol. **27**, 273 (1995)

17.39 D. A. Spong, S. P. Hirshman, J. C. Whitson, D. B. Batchelor, R. Sanchez, B. A. Carreras et al: Nucl. Fusion **40**, 563 (2000)

17.40 A. H. Boozer: Phys. Fluids **23**, 904 (1980), ibid **27**, 2441 (1984);
R. B. White, A. H. Boozer and R. Hay: Phys. Fluids **25**, 575 (1982);
R. B. White and M. S. Chance: Phys. Fluids **27**, 2455 (1984);
N. Nakajima, J. Todoroki and M. Okamoto: J. Plasma Fusion Res. **68**, 395 (1992)

17.41 K. M. Young: Plasma Phys. **16**, 119 (1974)

17.42 K. Miyamoto: Nucl. Fusion **18**, 243 (1978)

17.43 B. A. Carreras, G. Grieger, J. H. Harris, J. L. Johnson, J. F. Lyon, O. Motojima, F. Rau, H. Renner, J. A. Rome, K. Uo, M. Wakatani and H. Wobig: Nucl. Fusion **28**, 1613 (1988)

17.44 S. Sudo, Y. Takeiri, Z. Zushi, F. Sano, K. Itoh, K. Kondo and A. Iiyoshi: Nucl. Fusion **30**, 11 (1990)

17.45 U. Stroth, M. Murakami, R. A. Dory, H. Yamada, S. Okamura, F. Sano, T. Obiki: Nucl. Fusion **36**, 1063 (1996)

17.46 U. Stroth et al: Plasma Phys. Contr. Fusion **40**, 1551 (1998);
Wagner et al: *19th IAEA Fusion Energy Conference* (Lyon in 2003), OV/2-4.

17.47 M. Wanner and the W7-X Team: *19th IAEA Fusion Energy Conference* (Lyon in 2003), FT/2-3;
H. Wobig: Plasma Phys. Contr. Fusion **41**, A159 (1999)

17.48 A. Iiyoshi, A. Komori, A. Ejiri, M. Emoto, Funaba et al: Nucl Fusion **39**, 1245 (1999);
M. Fujiwara, O. Motojima, Y. Yamada, T. Watari, M. Okamoto et al: Plasma Phys. Contr. Fusion **39**, A261 (1997)

17.49 A. Fujisawa: Plasma Phys. Contr. Fusion **45**, R1 (2003)
17.50 ed. by T. K. Fowler: Nucl. Fusion **9**, 3 (1969)
17.51 M. S. Ioffe and B. B. Kadomotsev: Sov. Phys. Uspekhi **13**, 225 (1970)
17.52 R. F. Post: Nucl. Fusion **27**, 1579 (1987)
17.53 D. V. Sivukhim: *Reviews of Plasma Physics* **4**, 93 (ed. by M. A. Leontovich) Consultants Bureau, New York (1966)
17.54 G. I. Dimov, V. V. Zakaidakov and M. E. Kishinevskii: Sov. J. Plasma Phys. **2**, 326 (1976)
17.55 T. K. Fowler and B. G. Logan: *Comments Plasma Phys. Contr. Fusion Res.* **2**, 167 (1977)
17.56 V. P. Pastukhov: Nucl. Fusion **14**, 3 (1974)

## Chapter 18

18.1 J. Nuckolls, L. Wood, A. Thiessen and G. Zimmerman: Nature **239**, 139 September (1972)
18.2 J. Lindl: Phys. Plasmas **2**, 3933 (1995)
18.3 J. Meyer-Ter-Vehn: Nucl. Fusion **22**, 561 (1982)
18.4 R.E. Kidder: Nucl. Fusion **19**, 223 (1979)
18.5 S. Atzeni: Jpn J. Appl. Phys. **34**, 1980 (1995)
18.6 G.S. Fraley, E.J. Linnebur, R.J. Mason and R.L. Morse: Phys. Fluids **17**, 474 (1974)
18.7 S.Yu. Guskov, O.N. Krokhin, V.B. Rozanov: Nucl. Fusion **16**, 957 (1976)
18.8 R. Decoste, S.E. Bodner, B.H. Ripin, E.A. McLean, S.P. Obenshain and C.M. Armstrong: Phys. Rev. Lett. **42**, 1673 (1979)
18.9 K. Mima: J. Plasma Fusion Res. (Kakuyugo Kenkyu) **51**, 400 (1984) (in Japanese)
18.10 G. Bateman: *MHD Instabilities*, MIT Press, Cambridge, Mass. (1978)
18.11 H. Takabe, K. Mima, L. Montierth, R.L. Morse: Phys. Fluids **28**, 3676 (1985)
18.12 K.S. Budil, B.A. Remington, T.A. Peyser, K.O. Mikaelian, P.L. Miller et al.: Phys. Rev. Lett. **76**, 4536 (1996)
18.13 B.A. Remington, S.V. Weber, S.W. Haan, J.D. Kilkenny, S.G. Glendinning, R.J. Wallace, W.H. Goldstein, B.G. Willson, J.K. Nash: Phys. Fluids **B5**, 2589 (1993)
18.14 R.D. Richtmyer: Commun. Pure Appl. Math. **13**, 297 (1960)
18.15 E.E. Meshkov: Izv. Akad. Sci. USSR Fluid Dynamics: **4**, 101 (1969)
18.16 G. Dimonte, C.E. Frerking and M. Schnider: Phys. Rev. Lett. **74**, 4855 (1995)
18.17 H. Takabe: J. Plasma Fusion Res. **69**, 1285 (1993) (in Japanese)
18.18 D. Strickland and G. Mourou: Optics Comm. **56**, 219 (1985); G.A. Mourou, C.P.J. Barty and M.D. Perry: Physics Today, 22 January (1998)
18.19 M. Tabak, J. Hammer, M.E. Glinsky, W.L. Kruer, S.C. Wilks, J. Woodworth, E.M. Campbell and M.D. Perry: Phys. Plasmas **1**, 1626 (1994)
18.20 S.C. Wilks, W.L. Kruer, M. Tabak and A.B. Landon: Phys. Rev. Lett. **69**, 1383 (1992)
18.21 W.L. Kruer and K. Estabrook: Phys. Fluids 28, 430 (1985)
18.22 F. Brunel: Phys. Rev. Lett. **59**, 52 (1987)

18.23 R. Kodama, P.A. Norreys, K. Mima, A.E. Dangor, R.G. Evans et al.: Nature **412**, 798 August (2001); R. Kodama and the Fast Ignitor Consortium: Nature **418**, 933 August (2002)
18.24 R.G. Logan et al.: *19th IAEA Fusion Energy Conference* (Lyon 2002) OV/3-4
18.25 J. Lindl: *Inertial Confinement Fusion*, Springer/AIP Press, New York (1998)

# Index

accessibility of lower hybrid wave 193
adiabatic heating 19
adiabatic invariant 21
Alfvén velocity 60
Alfvén frequency gap 226
Alfvén resonance 225
Alfvén wave 158
antenna–plasma coupling 177
aspect ratio 39, 65
average minimum-$B$ 98
axial symmetry 34

ballooning mode 128
ballooning representation 127
banana
  particle 41
  region 82
  width 41
Bessel function model 321
beta ratio 62
Bohm diffusion coefficient 86
Boltzmann's equation 50
bootstrap current 303
  drive 302
break-even condition 11
bremsstrahlung 9
burning condition 10, 314

canonical variables 47
charge exchange 26, 293
charge separation 38, 78
circular polarization 152
circulating particle 39
CMA diagram 157
cold plasma 147
collision frequency 23
collision time 23
collisional region 82
collisionless drift instability 146
compressional Alfvén wave 60, 159, 160
conductive loss 76
connection length 77
continuum damping 229
convective loss 76, 88, 335
core-localized mode (CLM) 237
corona 4
Coulomb collision 21
cross-section of D–T, D–D, D–He$^3$ nuclear fusion 8
current drive
  bootstrap current *see* bootstrap current drive
  ECCD *see* electron cyclotron current drive
  LHCD *see* lower hybrid current drive
  NBCD *see* neutral beam current drive
  oscillating field CD *see* oscillating field current drive
curvature drift 17
cutoff 153
cyclotron
  damping 173
  frequency 15, 149
  velocity 173

Debye length 5
decay index 273
degenerate electron plasma 6
detached plasma 283
diamagnetism 62
dielectric constant 32
dielectric tensor
  of hot plasma 208

of bi-Maxwellian plasma 184, 210
of cold plasma 149
diffusion coefficient 75
due to fluctuation loss 84
diffusion tensor 295
dispersion equation
of cold plasma 151
of drift wave 212
of electrostatic wave 200, 210
dissipative drift instability *see* resistive drift instability
distribution function *see* velocity space distribution function
divertor 279
Dreicer field 27
drift frequency 85, 134, 144, 213
drift instability 202, 212
drift velocity of guiding center 16, 42
dynamic friction 295

ECCD *see* electron cyclotron current drive
ECH *see* electron cyclotron heating
effective collision frequency 81
electric displacement 31
electric intensity 31
electron beam instability 201
electron cyclotron current drive 297
electron cyclotron heating 196
electron cyclotron resonance 153
electron cyclotron wave 164
electron drift frequency 134
electron plasma frequency 14
electron plasma wave 14
electrostatic wave 164
ellipticity-induced Alfvén eigenmodes (EAE) 237
elongated plasma 277
energetic particle mode (EPM) 238
energy confinement time 77
of H mode tokamak 291
of Kaye–Goldston scaling 286
of L mode tokamak 286
of RFP 323
of stellarator 335
energy integral 102
of axisymmetric toroidal system 126
energy principle 102
equilibrium 61
ERATO code 240
Eta-i mode 135
excitation of wave 177
extraordinary wave 152

fast ignition 347
fast wave 153
Fermi acceleration 21
field reversed configuration 254
fishbone instability 215
flute instability *see* interchange instability
Fokker–Planck collision term 50, 294, 300
Fokker–Planck equation 50, 293, 300
full orbit particle model 251

Galeev–Sagdeev diffusion coefficient 81
Grad–Shafranov equation 64
Solovev solution 65
Solovev–Weening solution 66
gradient $B$ drift 17
gravitational interchange mode 142
Greenward density 275
Greenward–Hugill–Murakami parameter 276
group velocity 182
guiding center 16
gyro-Bohm diffusion coefficient 88
gyro-Landau-fluid model 245
gyrofluid model 245
gyrokinetic particle model 249

H mode 286
Hain–Lüst MHD equation 118
Hamiltonian equation of motion 36
Harris instability 202
helical hole 304
helical symmetry 34, 325
Hermite matrix 180
high beta-poloidal H mode 290
Hohlraum target 350
hollow current profiles 132
hoop force 68
horizontal positional stability 274
hybrid resonance 153

ICRF heating 189
ignition condition 11
implosion 342
inertial confinement 337
interchange instability 93, 345
intermediate region 82
internal disruption 275
INTOR 315
ion Bernstein wave 192
ion cyclotron resonance 153, 160
ion cyclotron wave 160
ion drift frequency 134
ion temperature gradient mode *see* ITG mode
ion–ion hybrid heating 191
ion–ion hybrid resonance 190
isobar model 340
ITER 315
ITG mode 135

kinetic Alfvén eigenmodes (KTAE) 237
kinetic Alfvén wave 225
kink instability 107
Kruskal–Shafranov condition 107

L mode 286
L wave 152
Lagrange equation of motion 35
Landau damping 169, 187
Langmuir wave 14
Larmor motion 15
Larmor radius 15
laser plasma 337
LHCD *see* lower hybrid current drive
LHH *see* lower hybrid heating
line of magnetic force 33
linearized equation of MHD 99
linearized Vlasov equation 203
Liouville's theorem 47
longitudinal adiabatic invariant 21
Lorentz condition 32
loss cone 20
loss-cone instability 202
lower hybrid current drive 293
lower hybrid heating 196
lower hybrid resonance 161, 192

macroscopic instabilities 91
magnetic
  axis 39
  fluctuation 89, 99, 204
  helicity 320
  induction 31
  intensity 31
  moment 19
  probe 270
  Reynolds number 57
  surface 33
  viscosity 57
  well depth 99
magnetoacoustic slow wave 60
magnetoacoustic wave 59
magnetohydrodynamic equation 51
magnetohydrodynamic instability 91
major axis 38
major radius 39
Maxwell distribution function 13
Maxwell's equations 31
mean free path 23
MHD equation *see* magnetohydrodynamic equation
MHD instability *see* magnetohydrodynamic instability
MHD model 240
MHD region 82
microscopic instability 199
minimum-$B$ condition 94
minor axis 39
minor disruption *see* internal disruption
minor radius 39
minority heating 192
Mirnov coil 270
mirror 19, 336
mirror ratio 19
mode conversion 178, 192
modified Bessel function model 322

NBCD *see* neutral beam current drive
NBI *see* neutral beam injection
negative dielectric constant 201
negative energy wave 202
negative shear 132, 290
neoclassical diffusion
  of stellarator 333
  of tokamak 82
neoclassical tearing mode 310

neutral beam current drive 301
neutral beam injection 26
  of negative ion source 293
normalized beta *see* Troyon factor
nuclear fusion reactions 7

Ohm's law 54
ohmic heating 28
open end system 336
orbit surface 43
ordinary wave 152
oscillating field current drive 325

paramagnetism 63
particle confinement time 75
  of mirror 336
pellet gain 337
permeability 32
PEST code 241
Pfirsch–Schlüter factor 79
Pfirsch–Schlüter current 71
pitch minimum 322
plasma dispersion function 185, 209
plasma frequency 149
plasma parameter 5
plateau region 82
Poisson's equation 248, 254
polarization 152
  current 45
  drift 45
poloidal beta 62, 66, 70
poloidal magnetic field 39
ponderomotive force 352
Poynting vector 178
preheating 342
pulsed poloidal current drive (PPCD) 323

quasi-linear theory of evolution in the distribution function 174

R wave 152
radiation loss 9, 284
rare collisional region 82
ray tracing 182
Rayleigh–Taylor instability *see* interchange instability
resistive drift instability 145
resistive instability 137
resonance 153, 160, 163
reversed field pinch 319
RFP *see* reversed field pinch
Richtmyer–Meshkov instability 346
rippling mode 142
Rogowsky coil 270
rotational transform angle 39, 328
runaway electron 27
Rutherford term 310

safety factor 108, 142, 277
sausage instability 106
scalar potential 31
scrape-off layer 279
separatrix 279, 327
Shafranov shift 133
shear Alfvén wave *see* torsional Alfvén wave
shear parameter 114
sheared flow 288
slow wave 153
slowing down time of ion beam 27
small solution 114
SOL *see* scrape-off layer
solid-state X-ray detector 271
specific electric resistivity 28
specific volume 97
spherical tokamak 266
stationary convective loss 88
stellarator 325
strongly coupled plasma 6
superbanana 333
superparticle 252
supershot 290
Suydam's criterion 114

TAE *see* toroidal Alfvén eigenmode
tandem mirror 336
tearing instability 142
thermal conductivity 76
thermal diffusion coefficient 76
thermal flux 76
tokamak device 269
tokamak reactor 315
toroidal Alfvén eigenmode 226
toroidal drift 38
toroidal precession velocity 224
toroidicity-induced Alfvén eigenmode *see* toroidal Alfvén eigenmode

torsional Alfvén wave 59, 159, 225
transit time damping 171, 187
translational symmetry 34
transversal adiabatic invariant 21
trapped particle *see* banana particle
trapped particle instability 202
triangularity-induced Alfvén eigenmodes (NAE) 237
Troyon factor 277

untrapped particle 41
upper hybrid resonance 162

vector potential 31
velocity space distribution function 13
velocity space instability *see* microscopic instability
vertical positional stability 273
VH mode 290
virial theorem 72
Vlasov's equation 50

Ware's pinch 45
wave heating 178
wave propagation 178
weakly coupled plasma 6
whistler wave 164

X point 327

Springer Series on

# ATOMIC, OPTICAL, AND PLASMA PHYSICS